T0236334

Lecture Notes in Computer Science 9790

Commenced Publication in 1973
Founding and Former Series Editors:
Gerhard Goos, Juris Hartmanis, and Jan van Leeuwen

More information about this series at http://www.springer.com/series/7407

Osvaldo Gervasi · Beniamino Murgante
Sanjay Misra · Ana Maria A.C. Rocha
Carmelo M. Torre · David Taniar
Bernady O. Apduhan · Elena Stankova
Shangguang Wang (Eds.)

Computational Science and Its Applications – ICCSA 2016

16th International Conference
Beijing, China, July 4–7, 2016
Proceedings, Part V

 Springer

Editors
Osvaldo Gervasi
University of Perugia
Perugia
Italy

Beniamino Murgante
University of Basilicata
Potenza
Italy

Sanjay Misra
Covenant University
Ota
Nigeria

Ana Maria A.C. Rocha
University of Minho
Braga
Portugal

Carmelo M. Torre
Polytechnic University
Bari
Italy

David Taniar
Monash University
Clayton, VIC
Australia

Bernady O. Apduhan
Kyushu Sangyo University
Fukuoka
Japan

Elena Stankova
Saint Petersburg State University
Saint Petersburg
Russia

Shangguang Wang
Beijing University of Posts
    and Telecommunications
Beijing
China

ISSN 0302-9743   ISSN 1611-3349 (electronic)
Lecture Notes in Computer Science
ISBN 978-3-319-42091-2   ISBN 978-3-319-42092-9 (eBook)
DOI 10.1007/978-3-319-42092-9

Library of Congress Control Number: 2016944355

LNCS Sublibrary: SL1 – Theoretical Computer Science and General Issues

Printed on acid-free paper

This Springer imprint is published by Springer Nature
The registered company is Springer International Publishing AG Switzerland

Preface

These multi-volume proceedings (LNCS volumes 9786, 9787, 9788, 9789, and 9790) consist of the peer-reviewed papers from the 2016 International Conference on Computational Science and Its Applications (ICCSA 2016) held in Beijing, China, during July 4–7, 2016.

ICCSA 2016 was a successful event in the series of conferences, previously held in Banff, Canada (2015), Guimares, Portugal (2014), Ho Chi Minh City, Vietnam (2013), Salvador, Brazil (2012), Santander, Spain (2011), Fukuoka, Japan (2010), Suwon, South Korea (2009), Perugia, Italy (2008), Kuala Lumpur, Malaysia (2007), Glasgow, UK (2006), Singapore (2005), Assisi, Italy (2004), Montreal, Canada (2003), (as ICCS) Amsterdam, The Netherlands (2002), and San Francisco, USA (2001).

Computational science is a main pillar of most present research as well as industrial and commercial activities and it plays a unique role in exploiting ICT innovative technologies. The ICCSA conference series has been providing a venue to researchers and industry practitioners to discuss new ideas, to share complex problems and their solutions, and to shape new trends in computational science.

Apart from the general tracks, ICCSA 2016 also included 33 international workshops, in various areas of computational sciences, ranging from computational science technologies to specific areas of computational sciences, such as computer graphics and virtual reality. The program also featured three keynote speeches and two tutorials.

The success of the ICCSA conference series, in general, and ICCSA 2016, in particular, is due to the support of many people: authors, presenters, participants, keynote speakers, session chairs, Organizing Committee members, student volunteers, Program Committee members, Steering Committee members, and many people in other various roles. We would like to thank them all.

We would also like to thank our sponsors, in particular NVidia and Springer for their very important support and for making the Best Paper Award ceremony so impressive.

We would also like to thank Springer for their continuous support in publishing the ICCSA conference proceedings.

July 2016

Shangguang Wang
Osvaldo Gervasi
Bernady O. Apduhan

Organization

ICCSA 2016 was organized by Beijing University of Post and Telecommunication (China), University of Perugia (Italy), Monash University (Australia), Kyushu Sangyo University (Japan), University of Basilicata (Italy), University of Minho, (Portugal), and the State Key Laboratory of Networking and Switching Technology (China).

Honorary General Chairs

Junliang Chen	Beijing University of Posts and Telecommunications, China
Antonio Laganà	University of Perugia, Italy
Norio Shiratori	Tohoku University, Japan
Kenneth C.J. Tan	Sardina Systems, Estonia

General Chairs

Shangguang Wang	Beijing University of Posts and Telecommunications, China
Osvaldo Gervasi	University of Perugia, Italy
Bernady O. Apduhan	Kyushu Sangyo University, Japan

Program Committee Chairs

Sen Su	Beijing University of Posts and Telecommunications, China
Beniamino Murgante	University of Basilicata, Italy
Ana Maria A.C. Rocha	University of Minho, Portugal
David Taniar	Monash University, Australia

International Advisory Committee

Jemal Abawajy	Deakin University, Australia
Dharma P. Agarwal	University of Cincinnati, USA
Marina L. Gavrilova	University of Calgary, Canada
Claudia Bauzer Medeiros	University of Campinas, Brazil
Manfred M. Fisher	Vienna University of Economics and Business, Austria
Yee Leung	Chinese University of Hong Kong, SAR China

International Liaison Chairs

Ana Carla P. Bitencourt	Universidade Federal do Reconcavo da Bahia, Brazil
Alfredo Cuzzocrea	ICAR-CNR and University of Calabria, Italy
Maria Irene Falcão	University of Minho, Portugal

Robert C.H. Hsu	Chung Hua University, Taiwan
Tai-Hoon Kim	Hannam University, Korea
Sanjay Misra	University of Minna, Nigeria
Takashi Naka	Kyushu Sangyo University, Japan
Rafael D.C. Santos	National Institute for Space Research, Brazil
Maribel Yasmina Santos	University of Minho, Portugal

Workshop and Session Organizing Chairs

Beniamino Murgante	University of Basilicata, Italy
Sanjay Misra	Covenant University, Nigeria
Jorge Gustavo Rocha	University of Minho, Portugal

Award Chair

Wenny Rahayu	La Trobe University, Australia

Publicity Committee Chair

Zibing Zheng	Sun Yat-Sen University, China
Mingdong Tang	Hunan University of Science and Technology, China
Yutao Ma	Wuhan University, China
Ao Zhou	Beijing University of Posts and Telecommunications, China
Ruisheng Shi	Beijing University of Posts and Telecommunications, China

Workshop Organizers

Agricultural and Environment Information and Decision Support Systems (AEIDSS 2016)

Sandro Bimonte	IRSTEA, France
André Miralles	IRSTEA, France
Thérèse Libourel	LIRMM, France
François Pinet	IRSTEA, France

Advances in Information Systems and Technologies for Emergency Preparedness and Risk Assessment (ASTER 2016)

Maurizio Pollino	ENEA, Italy
Marco Vona	University of Basilicata, Italy
Beniamino Murgante	University of Basilicata, Italy

Advances in Web-Based Learning (AWBL 2016)

Mustafa Murat Inceoglu	Ege University, Turkey

Bio- and Neuro-Inspired Computing and Applications (BIOCA 2016)

Nadia Nedjah   State University of Rio de Janeiro, Brazil
Luiza de Macedo Mourell   State University of Rio de Janeiro, Brazil

Computer-Aided Modeling, Simulation, and Analysis (CAMSA 2016)

Jie Shen   University of Michigan, USA and Jilin University,
                    China
Hao Chenina   Shanghai University of Engineering Science, China
Xiaoqiang Liun   Donghua University, China
Weichun Shi   Shanghai Maritime University, China
Yujie Liu   Southeast Jiaotong University, China

Computational and Applied Statistics (CAS 2016)

Ana Cristina Braga   University of Minho, Portugal
Ana Paula Costa Conceicao   University of Minho, Portugal
  Amorim

Computational Geometry and Security Applications (CGSA 2016)

Marina L. Gavrilova   University of Calgary, Canada

Computational Algorithms and Sustainable Assessment (CLASS 2016)

Antonino Marvuglia   Public Research Centre Henri Tudor, Luxembourg
Mikhail Kanevski   Université de Lausanne, Switzerland
Beniamino Murgante   University of Basilicata, Italy

Chemistry and Materials Sciences and Technologies (CMST 2016)

Antonio Laganà   University of Perugia, Italy
Noelia Faginas Lago   University of Perugia, Italy
Leonardo Pacifici   University of Perugia, Italy

Computational Optimization and Applications (COA 2016)

Ana Maria Rocha   University of Minho, Portugal
Humberto Rocha   University of Coimbra, Portugal

Cities, Technologies, and Planning (CTP 2016)

Giuseppe Borruso   University of Trieste, Italy
Beniamino Murgante   University of Basilicata, Italy

Databases and Computerized Information Retrieval Systems (DCIRS 2016)

Sultan Alamri   College of Computing and Informatics, SEU,
                    Saudi Arabia
Adil Fahad   Albaha University, Saudi Arabia
Abdullah Alamri   Jeddah University, Saudi Arabia

Data Science for Intelligent Decision Support (DS4IDS 2016)

Filipe Portela   University of Minho, Portugal
Manuel Filipe Santos   University of Minho, Portugal

Econometrics and Multidimensional Evaluation in the Urban Environment (EMEUE 2016)

Carmelo M. Torre   Polytechnic of Bari, Italy
Maria Cerreta   University of Naples Federico II, Italy
Paola Perchinunno   University of Bari, Italy
Simona Panaro   University of Naples Federico II, Italy
Raffaele Attardi   University of Naples Federico II, Italy

Future Computing Systems, Technologies, and Applications (FISTA 2016)

Bernady O. Apduhan   Kyushu Sangyo University, Japan
Rafael Santos   National Institute for Space Research, Brazil
Jianhua Ma   Hosei University, Japan
Qun Jin   Waseda University, Japan

Geographical Analysis, Urban Modeling, Spatial Statistics (GEO-AND-MOD 2016)

Giuseppe Borruso   University of Trieste, Italy
Beniamino Murgante   University of Basilicata, Italy
Hartmut Asche   University of Potsdam, Germany

GPU Technologies (GPUTech 2016)

Gervasi Osvaldo   University of Perugia, Italy
Sergio Tasso   University of Perugia, Italy
Flavio Vella   University of Rome La Sapienza, Italy

ICT and Remote Sensing for Environmental and Risk Monitoring (RS-Env 2016)

Rosa Lasaponara   Institute of Methodologies for Environmental Analysis,
                           National Research Council, Italy
Weigu Song   University of Science and Technology of China, China
Eufemia Tarantino   Polytechnic of Bari, Italy
Bernd Fichtelmann   DLR, Germany

7th International Symposium on Software Quality (ISSQ 2016)

Sanjay Misra   Covenant University, Nigeria

International Workshop on Biomathematics, Bioinformatics, and Biostatisticss (IBBB 2016)

Unal Ufuktepe   American University of the Middle East, Kuwait

Land Use Monitoring for Soil Consumption Reduction (LUMS 2016)

Carmelo M. Torre	Polytechnic of Bari, Italy
Alessandro Bonifazi	Polytechnic of Bari, Italy
Valentina Sannicandro	University of Naples Federico II, Italy
Massimiliano Bencardino	University of Salerno, Italy
Gianluca di Cugno	Polytechnic of Bari, Italy
Beniamino Murgante	University of Basilicata, Italy

Mobile Communications (MC 2016)

Hyunseung Choo	Sungkyunkwan University, Korea

Mobile Computing, Sensing, and Actuation for Cyber Physical Systems (MSA4IoT 2016)

Saad Qaisar	NUST School of Electrical Engineering and Computer Science, Pakistan
Moonseong Kim	Korean Intellectual Property Office, Korea

Quantum Mechanics: Computational Strategies and Applications (QM-CSA 2016)

Mirco Ragni	Universidad Federal de Bahia, Brazil
Ana Carla Peixoto Bitencourt	Universidade Estadual de Feira de Santana, Brazil
Vincenzo Aquilanti	University of Perugia, Italy
Andrea Lombardi	University of Perugia, Italy
Federico Palazzetti	University of Perugia, Italy

Remote Sensing for Cultural Heritage: Documentation, Management, and Monitoring (RSCH 2016)

Rosa Lasaponara	IRMMA, CNR, Italy
Nicola Masini	IBAM, CNR, Italy Zhengzhou Base, International Center on Space Technologies for Natural and Cultural Heritage, China
Chen Fulong	Institute of Remote Sensing and Digital Earth, Chinese Academy of Sciences, China

Scientific Computing Infrastructure (SCI 2016)

Elena Stankova	Saint Petersburg State University, Russia
Vladimir Korkhov	Saint Petersburg State University, Russia
Alexander Bogdanov	Saint Petersburg State University, Russia

Software Engineering Processes and Applications (SEPA 2016)

Sanjay Misra	Covenant University, Nigeria

Social Networks Research and Applications (SNRA 2016)

Eric Pardede   La Trobe University, Australia
Wenny Rahayu   La Trobe University, Australia
David Taniar   Monash University, Australia

Sustainability Performance Assessment: Models, Approaches, and Applications Toward Interdisciplinarity and Integrated Solutions (SPA 2016)

Francesco Scorza   University of Basilicata, Italy
Valentin Grecu   Lucia Blaga University on Sibiu, Romania

Tools and Techniques in Software Development Processes (TTSDP 2016)

Sanjay Misra   Covenant University, Nigeria

Volunteered Geographic Information: From Open Street Map to Participation (VGI 2016)

Claudia Ceppi   University of Basilicata, Italy
Beniamino Murgante   University of Basilicata, Italy
Francesco Mancini   University of Modena and Reggio Emilia, Italy
Giuseppe Borruso   University of Trieste, Italy

Virtual Reality and Its Applications (VRA 2016)

Osvaldo Gervasi   University of Perugia, Italy
Lucio Depaolis   University of Salento, Italy

Web-Based Collective Evolutionary Systems: Models, Measures, Applications (WCES 2016)

Alfredo Milani   University of Perugia, Italy
Valentina Franzoni   University of Rome La Sapienza, Italy
Yuanxi Li   Hong Kong Baptist University, Hong Kong,
                             SAR China
Clement Leung   United International College, Zhuhai, China
Rajdeep Niyogi   Indian Institute of Technology, Roorkee, India

Program Committee

Jemal Abawajy   Deakin University, Australia
Kenny Adamson   University of Ulster, UK
Hartmut Asche   University of Potsdam, Germany
Michela Bertolotto   University College Dublin, Ireland
Sandro Bimonte   CEMAGREF, TSCF, France
Rod Blais   University of Calgary, Canada
Ivan Blečić   University of Sassari, Italy
Giuseppe Borruso   University of Trieste, Italy
Yves Caniou   Lyon University, France

José A. Cardoso e Cunha	Universidade Nova de Lisboa, Portugal
Carlo Cattani	University of Salerno, Italy
Mete Celik	Erciyes University, Turkey
Alexander Chemeris	National Technical University of Ukraine KPI, Ukraine
Min Young Chung	Sungkyunkwan University, Korea
Elisete Correia	University of Trás os Montes e Alto Douro, Portugal
Gilberto Corso Pereira	Federal University of Bahia, Brazil
M. Fernanda Costa	University of Minho, Portugal
Alfredo Cuzzocrea	ICAR-CNR and University of Calabria, Italy
Florbela Maria da Cruz Domingues Correia	Intituto Politécnico de Viana do Castelo, Portugal
Vanda Marisa da Rosa Milheiro Lourenço	FCT from University Nova de Lisboa, Portugal
Carla Dal Sasso Freitas	Universidade Federal do Rio Grande do Sul, Brazil
Pradesh Debba	The Council for Scientific and Industrial Research (CSIR), South Africa
Hendrik Decker	Instituto Tecnológico de Informática, Spain
Adelaide de Fátima Baptista Valente Freitas	University of Aveiro, Portugal
Carina Soares da Silva Fortes	Escola Superior de Tecnologias da Saúde de Lisboa, Portugal
Frank Devai	London South Bank University, UK
Rodolphe Devillers	Memorial University of Newfoundland, Canada
Joana Dias	University of Coimbra, Portugal
Prabu Dorairaj	NetApp, India/USA
M. Irene Falcao	University of Minho, Portugal
Cherry Liu Fang	U.S. DOE Ames Laboratory, USA
Florbela Fernandes	Polytechnic Institute of Bragança, Portugal
Jose-Jesús Fernandez	National Centre for Biotechnology, CSIS, Spain
Mara Celia Furtado Rocha	PRODEB-Pós Cultura/UFBA, Brazil
Akemi Galvez	University of Cantabria, Spain
Paulino Jose Garcia Nieto	University of Oviedo, Spain
Marina Gavrilova	University of Calgary, Canada
Jerome Gensel	LSR-IMAG, France
Mara Giaoutzi	National Technical University, Athens, Greece
Andrzej M. Goscinski	Deakin University, Australia
Alex Hagen-Zanker	University of Cambridge, UK
Malgorzata Hanzl	Technical University of Lodz, Poland
Shanmugasundaram Hariharan	B.S. Abdur Rahman University, India
Tutut Herawan	Universitas Teknologi Yogyakarta, Indonesia
Hisamoto Hiyoshi	Gunma University, Japan
Fermin Huarte	University of Barcelona, Spain
Andrés Iglesias	University of Cantabria, Spain
Mustafa Inceoglu	Ege University, Turkey
Peter Jimack	University of Leeds, UK

Maurizio Pollino	Italian National Agency for New Technologies, Energy and Sustainable Economic Development, Italy
Alenka Poplin	University of Hamburg, Germany
Vidyasagar Potdar	Curtin University of Technology, Australia
David C. Prosperi	Florida Atlantic University, USA
Maria Emilia F. Queiroz Athayde	University of Minho, Portugal
Wenny Rahayu	La Trobe University, Australia
Jerzy Respondek	Silesian University of Technology, Poland
Ana Maria A.C. Rocha	University of Minho, Portugal
Maria Clara Rocha	ESTES Coimbra, Portugal
Humberto Rocha	INESC-Coimbra, Portugal
Alexey Rodionov	Institute of Computational Mathematics and Mathematical Geophysics, Russia
Jon Rokne	University of Calgary, Canada
Octavio Roncero	CSIC, Spain
Maytham Safar	Kuwait University, Kuwait
Chiara Saracino	A.O. Ospedale Niguarda Ca' Granda - Milano, Italy
Haiduke Sarafian	The Pennsylvania State University, USA
Jie Shen	University of Michigan, USA
Qi Shi	Liverpool John Moores University, UK
Dale Shires	U.S. Army Research Laboratory, USA
Takuo Suganuma	Tohoku University, Japan
Sergio Tasso	University of Perugia, Italy
Parimala Thulasiraman	University of Manitoba, Canada
Carmelo M. Torre	Polytechnic of Bari, Italy
Giuseppe A. Trunfio	University of Sassari, Italy
Unal Ufuktepe	American University of the Middle East, Kuwait
Toshihiro Uchibayashi	Kyushu Sangyo University, Japan
Mario Valle	Swiss National Supercomputing Centre, Switzerland
Pablo Vanegas	University of Cuenca, Equador
Piero Giorgio Verdini	INFN Pisa and CERN, Italy
Marco Vizzari	University of Perugia, Italy
Koichi Wada	University of Tsukuba, Japan
Krzysztof Walkowiak	Wroclaw University of Technology, Poland
Robert Weibel	University of Zurich, Switzerland
Roland Wismüller	Universität Siegen, Germany
Mudasser Wyne	SOET National University, USA
Chung-Huang Yang	National Kaohsiung Normal University, Taiwan
Xin-She Yang	National Physical Laboratory, UK
Salim Zabir	France Telecom Japan Co., Japan
Haifeng Zhao	University of California, Davis, USA
Kewen Zhao	University of Qiongzhou, China
Albert Y. Zomaya	University of Sydney, Australia

Reviewers

Abawajy, Jemal	Deakin University, Australia
Abuhelaleh, Mohammed	Univeristy of Bridgeport, USA
Acharjee, Shukla	Dibrugarh University, India
Andrianov, Sergei Nikolaevich	Universitetskii prospekt, Russia
Aguilar, José Alfonso	Universidad Autónoma de Sinaloa, Mexico
Ahmed, Faisal	University of Calgary, Canada
Alberti, Margarita	University of Barcelona, Spain
Amato, Alba	Seconda Universit degli Studi di Napoli, Italy
Amorim, Ana Paula	University of Minho, Portugal
Apduhan, Bernady	Kyushu Sangyo University, Japan
Aquilanti, Vincenzo	University of Perugia, Italy
Asche, Hartmut	Posdam University, Germany
Athayde Maria, Emlia Feijão Queiroz	University of Minho, Portugal
Attardi, Raffaele	University of Napoli Federico II, Italy
Azam, Samiul	United International University, Bangladesh
Azevedo, Ana	Athabasca University, USA
Badard, Thierry	Laval University, Canada
Baioletti, Marco	University of Perugia, Italy
Bartoli, Daniele	University of Perugia, Italy
Bentayeb, Fadila	Université Lyon, France
Bilan, Zhu	Tokyo University of Agriculture and Technology, Japan
Bimonte, Sandro	IRSTEA, France
Blecic, Ivan	Università di Cagliari, Italy
Bogdanov, Alexander	Saint Petersburg State University, Russia
Borruso, Giuseppe	University of Trieste, Italy
Bostenaru, Maria	"Ion Mincu" University of Architecture and Urbanism, Romania
Braga Ana, Cristina	University of Minho, Portugal
Canora, Filomena	University of Basilicata, Italy
Cardoso, Rui	Institute of Telecommunications, Portugal
Ceppi, Claudia	Polytechnic of Bari, Italy
Cerreta, Maria	University Federico II of Naples, Italy
Choo, Hyunseung	Sungkyunkwan University, South Korea
Coletti, Cecilia	University of Chieti, Italy
Correia, Elisete	University of Trás-Os-Montes e Alto Douro, Portugal
Correia Florbela Maria, da Cruz Domingues	Instituto Politécnico de Viana do Castelo, Portugal
Costa, Fernanda	University of Minho, Portugal
Crasso, Marco	National Scientific and Technical Research Council, Argentina
Crawford, Broderick	Universidad Catolica de Valparaiso, Chile

Cuzzocrea, Alfredo	University of Trieste, Italy
Cutini, Valerio	University of Pisa, Italy
Danese, Maria	IBAM, CNR, Italy
Decker, Hendrik	Instituto Tecnológico de Informática, Spain
Degtyarev, Alexander	Saint Petersburg State University, Russia
Demartini, Gianluca	University of Sheffield, UK
Di Leo, Margherita	JRC, European Commission, Belgium
Dias, Joana	University of Coimbra, Portugal
Dilo, Arta	University of Twente, The Netherlands
Dorazio, Laurent	ISIMA, France
Duarte, Júlio	University of Minho, Portugal
El-Zawawy, Mohamed A.	Cairo University, Egypt
Escalona, Maria-Jose	University of Seville, Spain
Falcinelli, Stefano	University of Perugia, Italy
Fernandes, Florbela	Escola Superior de Tecnologia e Gest ão de Bragança, Portugal
Florence, Le Ber	ENGEES, France
Freitas Adelaide, de Fátima Baptista Valente	University of Aveiro, Portugal
Frunzete, Madalin	Polytechnic University of Bucharest, Romania
Gankevich, Ivan	Saint Petersburg State University, Russia
Garau, Chiara	University of Cagliari, Italy
Garcia, Ernesto	University of the Basque Country, Spain
Gavrilova, Marina	University of Calgary, Canada
Gensel, Jerome	IMAG, France
Gervasi, Osvaldo	University of Perugia, Italy
Gizzi, Fabrizio	National Research Council, Italy
Gorbachev, Yuriy	Geolink Technologies, Russia
Grilli, Luca	University of Perugia, Italy
Guerra, Eduardo	National Institute for Space Research, Brazil
Hanzl, Malgorzata	University of Lodz, Poland
Hegedus, Peter	University of Szeged, Hungary
Herawan, Tutut	University of Malaya, Malaysia
Hu, Ya-Han	National Chung Cheng University, Taiwan
Ibrahim, Michael	Cairo University, Egipt
Ifrim, Georgiana	Insight, Ireland
Irrazábal, Emanuel	Universidad Nacional del Nordeste, Argentina
Janana, Loureio	University of Mato Grosso do Sul, Brazil
Jaiswal, Shruti	Delhi Technological University, India
Johnson, Franklin	Universidad de Playa Ancha, Chile
Karimipour, Farid	Vienna University of Technology, Austria
Kapcak, Sinan	American University of the Middle East in Kuwait, Kuwait
Kiki Maulana, Adhinugraha	Telkom University, Indonesia
Kim, Moonseong	KIPO, South Korea
Kobusińska, Anna	Poznan University of Technology, Poland

Korkhov, Vladimir	Saint Petersburg State University, Russia
Koutsomitropoulos, Dimitrios A.	University of Patras, Greece
Krishna Kumar, Chaturvedi	Indian Agricultural Statistics Research Institute (IASRI), India
Kulabukhova, Nataliia	Saint Petersburg State University, Russia
Kumar, Dileep	SR Engineering College, India
Laganà, Antonio	University of Perugia, Italy
Lai, Sen-Tarng	Shih Chien University, Taiwan
Lanza, Viviana	Lombardy Regional Institute for Research, Italy
Lasaponara, Rosa	National Research Council, Italy
Lazzari, Maurizio	National Research Council, Italy
Le Duc, Tai	Sungkyunkwan University, South Korea
Le Duc, Thang	Sungkyunkwan University, South Korea
Lee, KangWoo	Sungkyunkwan University, South Korea
Leung, Clement	United International College, Zhuhai, China
Libourel, Thérèse	LIRMM, France
Lourenço, Vanda Marisa	University Nova de Lisboa, Portugal
Machado, Jose	University of Minho, Portugal
Magni, Riccardo	Pragma Engineering srl, Italy
Mancini Francesco	University of Modena and Reggio Emilia, Italy
Manfreda, Salvatore	University of Basilicata, Italy
Manganelli, Benedetto	Università degli studi della Basilicata, Italy
Marghany, Maged	Universiti Teknologi Malaysia, Malaysia
Marinho, Euler	Federal University of Minas Gerais, Brazil
Martellozzo, Federico	University of Rome "La Sapienza", Italy
Marvuglia, Antonino	Public Research Centre Henri Tudor, Luxembourg
Mateos, Cristian	Universidad Nacional del Centro, Argentina
Matsatsinis, Nikolaos	Technical University of Crete, Greece
Messina, Fabrizio	University of Catania, Italy
Millham, Richard	Durban University of Technoloy, South Africa
Milani, Alfredo	University of Perugia, Italy
Misra, Sanjay	Covenant University, Nigeria
Modica, Giuseppe	Università Mediterranea di Reggio Calabria, Italy
Mohd Helmy, Abd Wahab	Universiti Tun Hussein Onn Malaysia, Malaysia
Murgante, Beniamino	University of Basilicata, Italy
Nagy, Csaba	University of Szeged, Hungary
Napolitano, Maurizio	Center for Information and Communication Technology, Italy
Natário, Isabel Cristina Maciel	University Nova de Lisboa, Portugal
Navarrete Gutierrez, Tomas	Luxembourg Institute of Science and Technology, Luxembourg
Nedjah, Nadia	State University of Rio de Janeiro, Brazil
Nguyen, Tien Dzung	Sungkyunkwan University, South Korea
Niyogi, Rajdeep	Indian Institute of Technology Roorkee, India

Oliveira, Irene	University of Trás-Os-Montes e Alto Douro, Portugal
Panetta, J.B.	Tecnologia Geofísica Petróleo Brasileiro SA, PETROBRAS, Brazil
Papa, Enrica	University of Amsterdam, The Netherlands
Papathanasiou, Jason	University of Macedonia, Greece
Pardede, Eric	La Trobe University, Australia
Pascale, Stefania	University of Basilicata, Italy
Paul, Padma Polash	University of Calgary, Canada
Perchinunno, Paola	University of Bari, Italy
Pereira, Oscar	Universidade de Aveiro, Portugal
Pham, Quoc Trung	HCMC University of Technology, Vietnam
Pinet, Francois	IRSTEA, France
Pirani, Fernando	University of Perugia, Italy
Pollino, Maurizio	ENEA, Italy
Pusatli, Tolga	Cankaya University, Turkey
Qaisar, Saad	NURST, Pakistan
Qian, Junyan	Guilin University of Electronic Technology, China
Raffaeta, Alessandra	University of Venice, Italy
Ragni, Mirco	Universidade Estadual de Feira de Santana, Brazil
Rahman, Wasiur	Technical University Darmstadt, Germany
Rampino, Sergio	Scuola Normale di Pisa, Italy
Rahayu, Wenny	La Trobe University, Australia
Ravat, Franck	IRIT, France
Raza, Syed Muhammad	Sungkyunkwan University, South Korea
Roccatello, Eduard	3DGIS, Italy
Rocha, Ana Maria	University of Minho, Portugal
Rocha, Humberto	University of Coimbra, Portugal
Rocha, Jorge	University of Minho, Portugal
Rocha, Maria Clara	ESTES Coimbra, Portugal
Romano, Bernardino	University of l'Aquila, Italy
Sannicandro, Valentina	Polytechnic of Bari, Italy
Santiago Júnior, Valdivino	Instituto Nacional de Pesquisas Espaciais, Brazil
Sarafian, Haiduke	Pennsylvania State University, USA
Schneider, Michel	ISIMA, France
Selmaoui, Nazha	University of New Caledonia, New Caledonia
Scerri, Simon	University of Bonn, Germany
Shakhov, Vladimir	Institute of Computational Mathematics and Mathematical Geophysics, Russia
Shen, Jie	University of Michigan, USA
Silva-Fortes, Carina	ESTeSL-IPL, Portugal
Singh, Upasana	University of Kwa Zulu-Natal, South Africa
Skarga-Bandurova, Inna	Technological Institute of East Ukrainian National University, Ukraine
Soares, Michel	Federal University of Sergipe, Brazil
Souza, Eric	Universidade Nova de Lisboa, Portugal
Stankova, Elena	Saint Petersburg State University, Russia

Stalidis, George	TEI of Thessaloniki, Greece
Taniar, David	Monash University, Australia
Tasso, Sergio	University of Perugia, Italy
Telmo, Pinto	University of Minho, Portugal
Tengku, Adil	La Trobe University, Australia
Thorat, Pankaj	Sungkyunkwan University, South Korea
Tiago Garcia, de Senna Carneiro	Federal University of Ouro Preto, Brazil
Tilio, Lucia	University of Basilicata, Italy
Torre, Carmelo Maria	Polytechnic of Bari, Italy
Tripathi, Ashish	MNNIT Allahabad, India
Tripp, Barba	Carolina, Universidad Autnoma de Sinaloa, Mexico
Trunfio, Giuseppe A.	University of Sassari, Italy
Upadhyay, Ashish	Indian Institute of Public Health-Gandhinagar, India
Valuev, Ilya	Russian Academy of Sciences, Russia
Varella, Evangelia	Aristotle University of Thessaloniki, Greece
Vasyunin, Dmitry	University of Amsterdam, The Netherlans
Vijaykumar, Nandamudi	INPE, Brazil
Villalba, Maite	Universidad Europea de Madrid, Spain
Walkowiak, Krzysztof	Wroclav University of Technology, Poland
Wanderley, Fernando	FCT/UNL, Portugal
Wei Hoo, Chong	Motorola, USA
Xia, Feng	Dalian University of Technology (DUT), China
Yamauchi, Toshihiro	Okayama University, Japan
Yeoum, Sanggil	Sungkyunkwan University, South Korea
Yirsaw, Ayalew	University of Botswana, Bostwana
Yujie, Liu	Southeast Jiaotong University, China
Zafer, Agacik	American University of the Middle East in Kuwait, Kuwait
Zalyubovskiy, Vyacheslav	Russian Academy of Sciences, Russia
Zeile, Peter	Technische Universitat Kaiserslautern, Germany
Žemlička, Michal	Charles University, Czech Republic
Zivkovic, Ljiljana	Republic Agency for Spatial Planning, Belgrade
Zunino, Alejandro	Universidad Nacional del Centro, Argentina

Sponsoring Organizations

ICCSA 2016 would not have been possible without the tremendous support of many organizations and institutions, for which all organizers and participants of ICCSA 2016 express their sincere gratitude:

Springer International Publishing AG, Switzerland
(http://www.springer.com)

NVidia Co., USA
(http://www.nvidia.com)

Beijing University of Post and Telecommunication, China
(http://english.bupt.edu.cn/)

State Key Laboratory of Networking and Switching Technology, China

University of Perugia, Italy
(http://www.unipg.it)

University of Basilicata, Italy
(http://www.unibas.it)

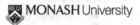

Monash University, Australia
(http://monash.edu)

Kyushu Sangyo University, Japan
(www.kyusan-u.ac.jp)

Universidade do Minho, Portugal
(http://www.uminho.pt)

Contents – Part V

Discovering Popular Events on Twitter

Sartaj Kanwar[1], Rajdeep Niyogi[1(✉)], and Alfredo Milani[2]

[1] Department of Computer Science and Engineering,
Indian Institute of Technology Roorkee, Roorkee 247667, India
kanwarsartaj@gmail.com, rajdpfec@iitr.ac.in
[2] Department of Mathematics and Computer Science,
University of Perugia, Perugia 06123, Italy
milani@unipg.it

Abstract. Event detection in twitter is the process of discovering popular events using messages generated by the twitter users. Event detection is an interesting research topic. Tweets are focused and may contain short forms. The tweets are noisy because there may be personal messages by the user also. In this paper we propose an algorithm to find top k popular events using keywords contained in the tweets. This paper classifies the popular events into different categories and the timeline is provided for every event. The timeline is useful to check when the event was popular. Geotagging is also done to find where the event was popular. We have implemented the algorithm using 14,558 users and 5,27,548 tweets over a period of 10 months (22 June, 2015 to 25 April, 2016). The results are quite promising.

Keywords: Twitter · Event detection · Social media

1 Introduction

Twitter is an electronic medium that has gained popularity in recent time. Twitter allows users to post their messages (known as Tweets) on various events. Length of tweets are restricted to be less than 140 characters. Twitter gained popularity as it has more than 100 million users in 2012 who has posted 340 million tweets per day. In 2013 Twitter was among the ten top most visited website and given the title of "SMS of the internet". To post tweets on twitter, user has to register himself on the twitter. Users who have not registered themselves on twitter can only read the messages. The relationship between twitter users are of follower and followee type.

Twitter users have unilateral relationship, i.e. when a user A follows user B, it means user A can get all the messages shared by user B but not vice versa. Due to the restriction on length of tweets sometimes it is difficult to put your idea in fixed characters so, users can share links to other objects on the internet such as article, blogs, video, etc. (termed as artifacts in twitter). As users share what is going around them on twitter and this will be almost instantly [2]. This makes twitter a good source for timely information. Twitter uses the following concept "when you need to know what's going on-in your town or across the globe-get the best of what's happening on twitter". A distinct advantage of applications like Twitter is their ability to distribute data among

O. Gervasi et al. (Eds.): ICCSA 2016, Part V, LNCS 9790, pp. 1–11, 2016.
DOI: 10.1007/978-3-319-42092-9_1

its user through multiple channels. For example, a twitter user can get messages on their phone, through a Facebook application, email, etc. [1].

Event Detection is the process of discovering interesting facts and extracting useful information from this huge source of information. Event detection can be used in various fields such as health epidemics identification, natural disaster detection, trending topic detection and sentiment analysis. The challenges in event detection are related to clustering and classification [3]. Virtually every person can distribute the information regarding the event as the event unfolds. Using the information from such platform can help organizations to acquire actionable knowledge and also provides valuable knowledge.

The goal of this paper is to demonstrate how the popular events can be extracted from tweets posted by the users. Events are the incident that is happening in some place at some point of time. The popular event is the event which receives more attention as compared to other events. This attention is calculated on the basis of frequency of occurrence of tweets related to that topic.

In this paper, we suggest an algorithm to detect popular events, for which the tweets are classified based on their similarity. This similarity will be governed by the keywords used in the tweets. This approach will use nouns used in tweets as their keyword and merge two tweets into one event when their keyword matches above some threshold. The popularity of an event is decided by the number of tweets related to that topic. For this purpose we have used a count of the number of tweets in each and every topic. We also discuss a method to examine the type of a particular event, for example, whether an event is related to politics, entertainment, sports, etc. For this we have classified events in different categories depending on the type of keywords used in tweets about that event. To summarize every event we have made a timeline interface. The timeline will show the frequency of tweets with time. The peak in timeline shows the event is more popular at that point of time as people are more discussing it.

The paper is organized as follows: Sect. 2 describes some related work that has been done till now in this field. Section 3 describes our proposed methodology which we used for different purpose. Section 4 describes the result generated by our method. Section 5 gives the conclusion and future works.

2   Related Work

An event detection algorithm is given in [4] which build a Key Graph using keywords used in tweets. Key Graph contains nodes consisting of keywords and the edges between two nodes will signify the occurrence of two keywords simultaneously. Next, a community detection algorithm is used to cluster different nodes to form the cluster. Inter cluster edges are removed based on betweenness centrality score. Named entities and noun phrases are used as keywords. Using these keywords, they had created the Key Graph.

The problem of online social event monitoring over tweet streams for real applications like crisis management and decision making are addressed in [5]. [5] suggested location-time constrained topic model for associating location with each event. An algorithm for event detection, named as New Event Detection (NED) is given in [6].

NED is consists of two subtasks i.e. Retrospective and Online NED. In Retrospective NED previously unidentified events are detected and in Online NED new events in text stream are detected.

Wavelet transformation is used for event detection in [7]. The problem of identifying events and their user contributed social media documents as a clustering task, where documents have multiple features, associated with domain-specific similarity metrics [8]. A general online clustering framework, suitable for the social media domain is proposed in [9]. Several techniques for learning a combination of the feature-specific similarity metrics are given in [9] that are used to indicate social media document similarity in a general clustering framework. [9] proposed a clustering framework and the similarity metric learning technique is evaluated on two real-world datasets of social media event content.

Location is considered in [10] with every event as incident location and event are strongly connected. The approach in [10] consists of the following steps. First, preprocessing is performed to remove stop words and irrelevant words. Second, clustering is done to automatically group the messages in the event. Finally, a hotspot detection method is performed.

TwitInfo is a platform for exploring Tweets regarding to a particular topic is presented in [11]. The user had to enter the keyword for an event and TwitInfo has provided the message frequency, tweet map, related tweets, popular links and the overall sentiment of the event. The *TwitInfo* user interface contained following thing: The user defined name of the event with keywords in the tweet, Timeline interface with y axis containing the volume of the tweet, Geo location along with that event is displayed on the map, Current tweets of selected event are colored red if the sentiment of the tweet is negative or blue if the sentiment of the tweet is positive and Aggregate sentiment of currently selected event using pie charts.

TwitterMonitor system is presented in [12] that detect the real time events in defined time window. This is done in three steps. In first step bursty keywords are identified, i.e. keywords that are occurring at a very high rate as compared to others. In second step grouping of bursty keyword is done based on their occurrences. In third and last step additional information about the event is collected.

A news processing system for twitter called as *TwitterStand* is presented in [13]. For users, 2000 handpicked seeders are used for collecting tweets. Seeders are mainly newspaper and television stations because they are supposed to publish news. After that junk is separated from news using the naïve Bayes classifier. Online clustering algorithm called leader-follower clustering to cluster the tweets to form events. A statistical method *MABED* (mention-anomaly-based event detection) is proposed in [14]. The whole process of event detection is divided in three steps. In first step detected the events based on mention anomaly. Second, words are selected that best describes each event. After deleted all the duplicated events or merged the duplicate events. Lastly, a list of top k events is generated.

Twitter can be used in important situations which may need current attention [15]. For this [15] considered high profile events, i.e. to national security and two emergency events. *TwiBiNG* (Twitter Bipartite News Generator) is presented in [16]. It has been created to help online journalism. The main part of this platform is to generate two bipartite clusters of user intensions. After that LCS (Longest Common Subsequence) is used along with some user information to separate the useful data from irrelevant data. Using this method would not only generate good news, but also contain less spam.

Event detection and clustering are correlated in [17], which means some clustering methods can be used in event detection. Event detection is similar to aggregated trend changes. Community detection algorithm is applied to find out popular events. Hierarchical clustering of tweets, dynamic cutting and ranking of result cluster is used to obtain the cluster is used in [19]. Bursty word extraction for event detection is used in [20].

Most of the earlier works have considered temporal context of messages, but location information is also an important factor of an event [18]. The Geo referencing used in tweets can be used to detect localized events such as public events, emergency events, etc. User mostly near to event location message more information as compared to others. So these users can serve as human sensors to describe an event.

3 Proposed Methodology

The basic way to detect the events and news from a large amount of data is by using keywords. If two events are similar then up to some extent they contain similar words. This similarity should have some threshold. It means two tweets are only put into one event when they have more same keywords than some threshold.

3.1 Proposed Algorithm

Our proposed algorithm is incremental, i.e. when new tweets enter in the system we don't have to process the whole data again. We just have to process the coming tweet and put it in its proper position. Our algorithm uses unsupervised learning similar to DBSCAN in data mining. Unlike classification we don't know the total number of events that would be created when algorithm stops. The number of events created would solely depend upon data.

The Algorithm is taking tweets one by one and comparing its hash with already formed events. If the hash is same means those two tweets corresponds to the same event. For every tweet we have also stored the number of tweets fall under that event. So put a new tweet along with a previous tweet in the same event and increment the tweet count for that event. Either if hash is not same for new tweet and tweets present in already created events or hash of new tweet entering in system is not known then we have compared the keywords present in the new tweet and tweets present in previous events.

If keywords are matched above the threshold for any event, then we merge this new tweet with that event. Whenever any tweet was merged in already created event the keyword of that tweet is updated in that event. So in this way the event is learning new keywords about that event on addition of new tweets to that event. If no previous event has the same kind of keywords, then make a new event and put tweet in that event along with its keywords. Keyword of this tweet is made as keyword of the newly created event. The above steps are done iteratively until all tweets are put into some already created events or new events are created for them. After this we have distinguished events along with count indicating the number of tweets in that event.

The algorithm is given below. The following symbols are used in the algorithm.

S: set of all tweets along with its keywords and hash which is nonempty
E: set of all events created at any point of time in an algorithm. Initially this set is empty
p: threshold for merging tweet into that event. The value of p can be $0<p<1$
ne: new event to be inserted in the set E
t: tweet in the set S
e: an event from set E
MAX-EV: event in set E with which maximum keyword of t matches. Initially empty

Algorithm for discovering popular events
Input: S   output: E
while S is not empty **do**
    remove an element t from S; S := S \ {t}
    if E is empty
        Create new event ne; E := E ∪ {ne}; add all keywords of t into ne; set count := 1; set hash of t as the hash of ne.
    else   // E is nonempty
      for every event e in set E **do**
        if hash of t and e matches
            put t into event e and count := count+1.
        else
            match keywords of e with that of t.
            if the similarity of keywords > p
                update MAX-EV to max(MAX-EV, e)
      end for
      if MAX-EV is not empty **then**   put t into MAX-EV;
           count :=count+1; add all keywords of t into MAX-EV.
      else   Create new event ne; E := E ∪ {ne}; add all keywords of t into ne; set count := 1; set hash of t as the hash of ne.
end while

4   Implementations and Results

For implementation of our algorithm we have collected our data set from Twitter's developer API. For the data collection we have chosen 14,558 users. We have collected 5,27,548 tweets over a period of 10 months (22 June, 2015 to 25 April,2016). In this section we have shown the snapshots of interface we have created in java. We have also shown the comparison of different events. Comparisons are shown in the form of histogram and in tabular form. Pie chart is also shown for event classification. A First interface that we have created asks the user for value of k i.e. top k events user is interested in. For example, the user is interested in top 5 popular events. After clicking popular event button interface on the frame shows top 5 events. Table 1 show the output when the user clicks on the popular event button.

Table 1. Top k(k = 5) popular events along with tweet count of each event.

S.no.	Event number	Count	Hash
1.	437	134	SecondKashab
2.	558	127	EndTheDisruption
3.	314	126	Baahubali
4.	1033	118	OROP
5.	933	101	SensexDown

On clicking on the tweet button of mainframe second frame appears which allow the user to select one event among top k events. If user is interested in 558th event which corresponds to EndTheDistruption. After user selection it will show the tweets corresponding to that event as shown in Fig. 1. Using the tweets user can get more information about the event. In that frame a button for pie chart is also provided. By clicking on that button user can see the pie chart which will divide the event in different labels. A Label which has the highest percentage would become class of that event.

event_no	tweet_id	Tweet
558	629157551460847616	Congress protests for the third day outside the Parli...
558	629160402601914368	Our Manipur CM, Arunachal CM, Assam CM, who are...
558	629160702486315009	Here is another show of the arrogance of this govern...
558	629162710052147200	Sonia Gandhi, leaders of other oppn parties stage pr...
558	629164591457505280	Our voices are being stifled in the Parliament, the voi...
558	629164749935153153	It is an insult to people of India: Cong VP Rahul Gan...
558	629165479010045952	This Govt is arrogant; this Govt that keeps saying 'sa...
558	629167429550149632	Rajya Sabha adjourned till noon #EndTheDisruption
558	629170690441121792	Rajya Sabha adjourned till 1200 hours after noisy pr...
558	629180914585571328	Rajya Sabha adjourned till 2 PM following Congress...

Fig. 1. Tweets corresponding to the selected event.

We also provided the user a pie chart corresponding to each event to show how tweets in that event are distributed. Larger the percentage of any class means event fall under that class. As a selected event is more falls into category of politics as compared to other classes. Below pie chart is useful when we want to know in which class a particular event will fall (Fig. 2).

Here we compared two events to show how classification of two events will be different. Type of event means under which category an event falls. The category of event is decided by terms used in tweets under that event. There are eight categories for

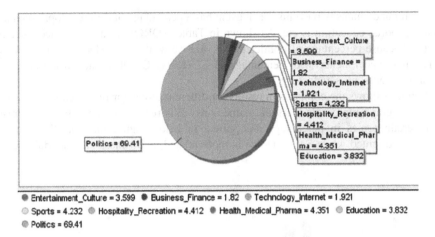

Entertainment_Culture = 3.599   ● Business_Finance = 1.82   Technology_Internet = 1.921
Sports = 4.232   Hospitality_Recreation = 4.412   ● Health_Medical_Pharma = 4.351   Education = 3.832
Politics = 69.41

Fig. 2. Pie chart for classification of tweets corresponding to the EndTheDistruption event. (Color figure online)

every event such as Entertainment and Culture, Business Finance, Politics, Technology and Internet, Sports, Hospitality and Recreation, Health and Medical Pharmacy. For showing this we have used pie chart.

In this section we compare two event named as OROP (One Rank One Pension) and SensexDown. OROP is about ex-servicemen of India is appealing to the Indian government about having one pension after retirement. So there was very much discussion about this event in August 2015. Second event SensexDown is about falling of stocks in Indian share market. Investors are very much tensed about this down fall. Sensex was fallen by 1600 points at that point of time. So OROP seems to be more political event while SensexDown is more business and finance.

Table 2 shows the classification of events in different labels. We have tabulated two events, i.e. OROP and SensexDown. First column contains topic while second and

Table 2. Classification of Events.

Topic\Event ⟶ ↓	OROP	SensexDown
Entertainment_Culture	5.55	5.071
Business_Finance	10.496	37.787
Politics	40.8	10.92
Technology_Internet	5.063	6.016
Sports	8.317	11.096
Hospitality_Recreation	9.395	2.193
Health_Medical_Pharma	5.385	10.624
Education	8.731	2.855
Others	2.434	6.835

third column contains two events. Each event has a percentage of different topic such as politics, sports, education is etc. As shown in Table 2 OROP is more inclined toward politics because percentage of politics part is 40.8 % while SensexDown more related to Business and Finance with percentage of 37.787 %. OROP is an. So the classification of events are satisfactory.

Figure 3 shows the comparison results for different topics for two events. X-axis of the above graph represents the topic name and y-axis represents the value of interest over certain topics in percentage. As shown in below figure that OROP is more politically oriented while SensexDown is more related to Business and entertainment.

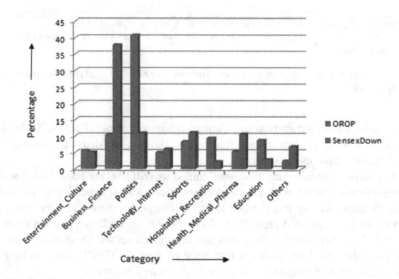

Fig. 3. Comparison of two Events(OROP and SensexDown). (Color figure online)

Timeline interface is used to summarize the event on the time frame. In this subsection we have presented the timeline for OROP event. As shown in Fig. 3 this event was particularly popular in August 2015 and particularly 28th August 2015 is the day when the tweet for this event is maximum as compared to other days. This event has started on 6th August 2015 and proceeded till 18th September. The number of tweets on each typically varies from each other. This event was specifically popular from 25th August to 28th August.

Figure 4 shows the above data in graphical form. In this graph x axis is showing the days while y axis is showing the number of tweets done on that particular day. The peak of the graph shows the time when the event was more popular. Some time graph may contain more than one peak it means the event is going in episodes. Some time less tweets are done by users, but after some real incident the event becomes popular again and more tweets are tweeted by the user.

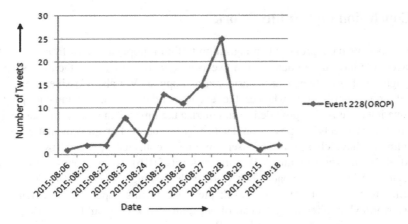

Fig. 4. Timeline interface showing OROP event.

4.1 Discussion

Table 3 shows the comparison between different methods. The comparison is done on the basis in four dimensions, i.e. popular event, Classification of events, Timeline of every event and Geotagging of the event.

As shown in the table that our proposed method is providing all the mentioned features as compared to other existing methods which are providing some of them. To find popular event, our method is better than TwitterMonitor [12]. Since they have used bursty keyword for finding popular events, but we have used set of keywords instead of single word. Timeline and Geotagging is provided in Twitinfo [11] and TwitterStand [13] but these two features are incomparable with our proposed method.

Table 3. Comparison between different Methods.

Features\Method→	Twitinfo[11]	TwitterMonitor[12]	TwitterStand[13]	MABED[14]	Proposed Method
Popular Event	No	Yes	Yes	Yes	Yes
Classification	No	No	No	No	Yes
Timeline	Yes	No	No	No	Yes
Geotagging	Yes	No	Yes	No	Yes

5 Conclusion and Future Work

In this paper we have proposed an algorithm to find k popular events from the twitter messages. We have used noun used in tweets as keywords to cluster the events. Event with high number of tweets are considered as popular. We also classified the events depend upon distribution, whether the event is related to entertainment, politics. Timeline interface is also provided to summarize the time when an event is happening, i.e. when most tweets are posted on Twitter by the users. For implementation of our algorithm we have chosen 14,558 users. We have collected 5, 27,548 tweets over a period of 10 months (22 June, 2015 to 25 April, 2016). Results obtained are quite satisfactory. We have compared our work with existing work and our method out performs over some of them. The work presented in this paper can be extended to various worthwhile directions. As part of our future work we would like to extend this framework by using sentiment analysis. By doing this we can know whether users are giving positive, negative or neutral feedback about an event.

Acknowledgement. The Authors thanks the anonymous reviewers for their valuable suggestions that have helped in improving the paper.

References

1. Krishnamurthy, B., Gill, P., Arlitt, M.: A few Chirps about Twitter. In: Proceedings of the First Workshop on Online Social Networks, pp. 19–24. ACM, Washington (2008)
2. Atefeh, F., Khreich, W.: A survey of techniques for event detection in Twitter. Comput. Intell. **31**(1), 132–164 (2015)
3. Madani, A., Boussaid, O., Zegour, D.E.: What's happening: a survey of tweets event detection. In: Proceedings of the Third International Conference on Communications, Computation, Networks and Technologies (2014)
4. Sayyadi, H., Hurst, M., Maykov, A.: Event detection and tracking in social streams. In: Proceedings of the Third International ICWSM Conference, pp. 311–314. Berlin Heidelberg (2009)
5. Zhou, X., Chen, L.: Event detection over Twitter social media streams. VLDB J. **23**(3), 381–400 (2014)
6. Dou, W., Wang, X., Ribarsky, W., Zhou, M.: Event detection in social media data. In: Proceedings of IEEE VisWeek Workshop on Interactive Visual Text Analytics-Task Driven Analytics of Social Media Content, pp. 971–980 (2012)
7. Becker, H., Naaman, M., Gravano, L.: Learning similarity metrics for event identification in social media. In: Proceedings of the Third ACM International Conference on Web Search and Data Mining, pp. 291–300 (2010)
8. Marcus, A., Bernstein, M.S., Badar, O., Karger, D.R., Madden, S., Miller, R.C.: TwitInfo: aggregating and visualizing microblogs for event exploration. In: Proceedings of the SIGCHI Conference on Human Factors in Computing Systems, pp. 227–236 (2011)
9. Becker, H., Naaman, M., Gravano, L.: Beyond trending topics: real-world event identification on Twitter. In: Proceedings of Fifth International AAAI Conference on Weblogs and Social Media (2011)

10. Mathioudakis, M., Loudas, N.: TwitterMonitor: trend detection over the Twitter stream. In: Proceedings of the 2010 ACM SIGMOD International Conference on Management of data, pp. 1155–1158 (2010)
11. Weng, J., Yao, Y., Leonardi, E., Lee, F.: Event detection in Twitter. In: Proceedings of the 5th International AAAI Conference on Weblogs and Social Media, pp. 401–408 (2011)
12. Unankard, S., Li, X., Sharaf, M.A.: Emerging event detection in social networks with location sensitivity. In: Proceedings of World Wide Web, pp. 1–25 (2014)
13. Sankaranarayanan, J., Samet, H., Teitler, B.E., Lieberman, M.D., Sperling, J.: TwitterStand: news in Tweets. In: Proceedings of the 17th ACM SIGSPATIAL International Conference on Advances in Geographic Information Systems, pp. 42–51 (2009)
14. Guille, A., Favre, C.: Event detection, tracking, and visualization in Twitter: a mention anomaly based-approach. Soc. Netw. Anal. Min. **5:18**, 1–18 (2015)
15. Huges, L., Palen, L.: Twitter adoption and use in mass convergence and emergency events. Int. J. Emerg. Manage. **6**(3–4), 248–260 (2009)
16. Sharma, Y., Bhatia, D., Choudhary, V.K.: TwiBiNG: a bipartite news generator using Twitter. In: Proceedings of the SNOW 2014 Data Challenge, Seoul, Korea (2014)
17. Aggarwal, C.C., Subbian, K.: Event detection in social streams. In: Proceeding of SDM, vol. 12, pp. 624–635 (2012)
18. Abdelhaq, H., Sengstock, C., Gertxz, M.: EvenTweet: online localized event detection from Twitter. In: Proceedings of VLDB Endowment, vol. 6, no. 12, pp. 1326–1329 (2013)
19. Ifrim, G., Shi, B., Brigadir, I.: Event detection in Twitter using aggressive filtering and hierarchical Tweet clustering. In: Proceedings of SNOW WWW Workshop (2014)
20. Wang, X., Zhu, F., Jiang, J., Li, S.: Real time event detection in Twitter. In: Wang, J., Xiong, H., Ishikawa, Y., Xu, J., Zhou, J. (eds.) WAIM 2013. LNCS, vol. 7923, pp. 502–513. Springer, Heidelberg (2013)

Analysis of Users' Interest Based on Tweets

Nimita Mangal[1], Rajdeep Niyogi[1(✉)], and Alfredo Milani[2]

[1] Department of Computer Science and Engineering,
Indian Institute of Technology Roorkee, Roorkee 247667, India
nimitamangal@gmail.com, rajdpfec@iitr.ac.in
[2] Department of Mathematics and Computer Science,
University of Perugia, Perugia 06123, Italy
milani@unipg.it

Abstract. Analysis of tweets would help in designing smart recommendation systems. Analysis of twitter messages is an interesting research area. Sentiment analysis of tweets has been done in some works. Another line of work is the classification of tweets into different categories. However, there are few works that have considered both sentiment analysis and classification to find out users' interest. In this paper, we propose an approach that combines both sentiment analysis and classification. Thus we are able to extract the topic in which users are interested. We have implemented our algorithm using five lakhs of tweets and around one thousand of users. The results are quite encouraging.

Keywords: Sentiment analysis · Twitter user · Social media

1 Introduction

Twitter is one of the popular online social blog where many celebrities post tweets for their fans and also post something related to an event. Twitter is a microblogging service. It is so called by this name because it enables users to send and read a short text message which is known as "tweet". There are 316 million monthly active users on twitter and 500 million tweets are posted per day. We can use these tweets for analyzing the interest of users and get to know the trends going on at any place. Such analysis may help in designing a smart recommendation system.

Several works have been done in the field of social networking, namely, classification of gender, classification of the topic, sentiment analysis of twitter users based on tweets, event detection, community detection, etc. Most of the work on recommendation system is based on network topology. A user's knowledge with social sites could be remarkably improved if other information like demographic attributes and user's personal interest and the interest of other users are considered. Such information allows users to follow a post or a user according to her topic of interest and the user may join a particular community of their own interest.

Moreover, a user may be interested to get recommendations based on her areas of interest. The recommendation first requires to know the user behavior. A person gets information about any event through newspaper, television, social sites or with the people around them. Now, if a person is interested in that event than she may tweet on

© Springer International Publishing Switzerland 2016
O. Gervasi et al. (Eds.): ICCSA 2016, Part V, LNCS 9790, pp. 12–23, 2016.
DOI: 10.1007/978-3-319-42092-9_2

twitter about the event positively or negatively according to her viewpoint. To get this negative and positive viewpoint of the user, sentiment analysis of tweets is necessary. The topic to which a particular tweet belongs is done by categorization and through this we get to know the topic in which the user is interested. By applying both the techniques we can provide better recommendations to users.

In this paper, we make an attempt to come up with a method for analyzing the interest of users based on sentiments (positive, negative or neutral) and the topic to which tweets are related to get the correct positive or negative interest of users. We are particularly interested in users and their tweets to help them to give a better recommendation which they need according to their current interest.

Tweets are collected using the Twitter4j api in Java. Sentiment analysis has been done using Stanford core NLP integrated framework. A core NLP tool pipeline code is run on tweets. Sentiment score is computed based on the words composes and longer phrase. Classification of tweets to which topic it is related has been done using the matching of words with a topic.

The paper is organized as follows: Sect. 2 describes the related work. Section 3 describes our method for analyzing tweets. Section 4 describes the implementation details and results obtained by our method. Conclusion and future work are given in Sect. 5.

2 Related Work

Different methods are proposed for sentiment analysis, finding sentiments in words, sentences, sentiments in topics. Some of these approaches use machine learning, pattern based and natural language processing. Hybrid classifiers are designed in [1] to get better sentiment results. Sentiment analysis of twitter data is studied in [11] and it introduces POS-specific earlier polarity feature and explore the use of tree kernel. Experiments were performed on three models [11]: feature based model uses hundred features only and have the same accuracy as that of unigram model that uses ten thousand features. Kernel tree based model first tokenize the tweet into a tree by separating punctuation mark, exclamatory mark, negation word and emoticon and prior calculate the polarity of word using word-net dictionary. The unigram model is used as a baseline for the experiments.

Two approaches (machine learning and lexical approach) are suggested for sentiment analysis in [3]. First, the machine learning approach which first convert each text into a list of words, consecutive word pairs and consecutive word triplets and then based upon some human coded set of texts 'learn' which of these features tends to associate with sentiment scores to classify the new cases. Second, the lexical approach, uses some grammatical structure of language and some list of words with sentiment scores and polarities is used. The accuracy of both approaches depends on the training set and the score, which is already provided for most of the words.

Sentiment tree bank approach is suggested for sentiment analysis in [2]. The recursive neural network approach computes parent node vectors in bottom-up fashion and use a composition function g and node vector is featuring for that node. An approach for computing sentiment score of short, informal text and sentence that contain phrases within it is suggested in [12].

Many recommender systems provide recommendation using the information based on user profile. A method for user recommendation is suggested in [4] and the method is based on sentiment volume objectivity. User profiling is done and similarity measure is computed between users (similarity measures based on place, sentiments of tweets). A method for friend recommendation is suggested in [5] which uses collaborative filtering and graph structure. Semantic user modeling has been done based on twitter posts in [6]. They suggested a formula for users similarity which is based on topic discussed by the users. A method to predict which political party a twitter user belongs is suggested in [7]. There approach is based on certain characteristic of parties like activity, influence, structure and interaction, context and sentiment and then user classification has been done based on Bayesian classification.

Analysis over user intentions has been done in [8] that are associated at a community level and show how users with similar intentions connect with each other. The task of user classification in social media using machine learning framework is addressed in [9]. User profile features such as followers, friends, username, user-location are collected to know about a user. Tweets of user are collected for judging the behavior of users and to classify users of same types.

Two methods for classification of the Twitter trending topic are proposed in [10] first, based on textual information and the other based on the network structure. In text based model all the hyperlinks are removed from the tweet and then a tokenizer removes stop words and delimited character. Since there is a limitation of 140 characters in a tweet, people use acronyms for words and so a vocabulary is used that has the full form of these words (e.g. BR is used to represent best regard). The network based approach uses a similarity model to find out the trending topic say X. It searches for five topics that are similar to the topic X and finds out the similarity index. Text categorization method is proposed in [14] that uses support vector machines and gives proof both theoretically and logically that svm is well suited for text classification.

Most of the above works are related to sentiments, recommender systems, and trending topic. However, these works do not discuss about a user's interest on the topic being discussed by the users. Our approach is different from others as we compute the interests of a particular user and the users of a certain location by taking their sentiments (positively or negatively inclined) towards certain topics.

If we do only sentiment analysis on tweets, it gives the sentiments of users, whether she tweets positive or negative for any event happened on social media. Based on this, we do not get any idea about a user's interest. If we do only topic categorization then we get to know about the topic on which a user tweeted. It does not provide us the information that the user is positively interested on a certain topic or not. Hence, doing sentiment analysis and topic categorization separately, do not provide us the result for the user's interest. Thus, in this paper, we combine these two techniques that gives better results.

3 Proposed Methodology

Figure 1 shows the basic flow diagram of our method. First, we have collected the tweets of the user for knowing the interest according to her tweet. Sentiment analysis has been done on the tweets that are collected to know the inclination of users, whether she is positively indicated his sentiments over a particular topic or not.

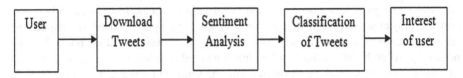

Fig. 1. Overview of our method.

We have used matching of words for classification of tweets to categorize it under a certain label (like sports, politics, entertainment, technology, hospitality, etc.). Finally, interest of a user is obtained that shows the positive inclination of the user towards a certain topic.

3.1 Data Collection

We have collected data for three different problems using Twitter 4j API and implementation results on this data are given in Sect. 4. The first problem deals with the user interest and for this we have collected 2,31,750 tweets of 1,150 users and show their behavior in the form of pie-chart. Tweets for different users were collected for different spans of time period for comparing their interest in different time intervals. The second problem deals with different cities of India in which we have collected around 2,02,578 tweets of 5 different cities of India and show the interested topic at that location. Tweets were collected by taking the value of latitude and longitude of the city. The third problem deals with the comparison of tweets of two countries (India and America). Tweets of the users that act as a bot (bot user is a user whose tweet done automatically by machines and not by persons) like news channel (bbcnews, indiatoday, etc.) are collected. The reason to choose bot user is to get more news about the country and to obtain interest for the country.

3.2 Sentiment Analysis of Tweet

Sentiment analysis is done, using the Stanford coreNLP method. This method is appropriate for short text. One drawback of this method, it is not considering emoticons value and acronym value. To solve this problem, first we check for emoticons and acronyms in the tweet. If it is present we compute the sentiment score accordingly for both. NLP provides some analyzing tools and has some implemented module that tag the words in a sentence, whether they are name of place, people, etc. or belong to noun, verb, and adjective. These analyzing tools include the parser, sentiment analysis, named entity recognizer, open information extraction tools, etc.

We refine the tweet by removing all hashes, @ and extra spaces to make it more readable plain text. A static init method is called that set the properties to get to know what action is needed for an incoming text. In our case we set four properties, tokenize, ssplit, parse and sentiment. Tokenize property breaks the tweet into tokens. The tokenizer saves the offsets of each token from where it starts and ends. Ssplit property splits a sequence of tokens into sentences. Parse property generates the parse tree, based on some grammatical structure and language information to distinguish between phrases, subject and predicate in a sentence. Sentiment property is used to compute the sentiment score of a tweet, a binarized tree form for a tweet based on positivity and negativity. After the init method findsentiment method is called that first make a labeled tree for a given tweet and based on tree find the sentiment score in the range of 0–4. Higher the value of the score represents the positive sentiment of a tweet.

3.3   Classification of Tweet

After sentiment analysis, tweets are classified according to topic to which it is related. The open NLP package is used for classification [13]. This package provides us a tagger file for tagging of sentences. MaxentTagger is a class used for tagging each word in a tweet with its corresponding form, whether it is an adverb, noun, adjective, etc. There are 36 taggers and each word in a tweet belongs to one of these taggers. After tagging a tweet word tagger pair is formed.

Each word tagger pair is compared with ten different categories of topics like entertainment, technology, politics, etc. For comparing word with the topic we are using wordnet similarity module that implements a variety of semantic similarity and relatedness measures which is based on information found in the lexical database WordNet. For using this WordNet similarity we are having WS4J API. For more accurate results we compare these words with the synonyms of topic for example, if we want to compare any word with technology then we compare word with technology, network, industry, etc. A method getSimilarity is available which compares this word with these topics and calculates some relatedness scores and gives a similarity score.

4   Implementation and Results

Below Fig. 2 shows the implemented flow diagram for our system. We have provided an interface to the user in which user provides a screenname (unique name given to each user on twitter) of the user and our backend system calls the download procedure that downloads the tweets of that user and after this sentiment analysis module is called which finds out the sentiment for each tweet and then each tweet is fall under one category positive or negative or neutral. After this classification module is called that runs for the positive and neutral sentiment tweet and gives percentage according to topic to which it belong. It is possible that one tweet belongs to more than one category. The final result for the user about the interested topic is shown in the form of a pie chart.

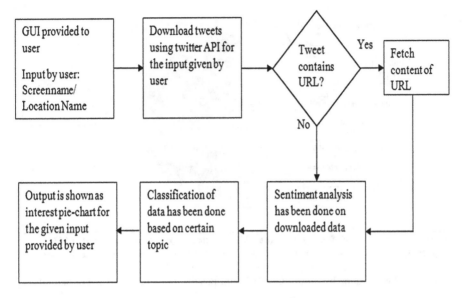

Fig. 2. Implemented flow diagram of our system.

If a tweet contains url then first the content of the link is fetched using some code for content fetching of the url and then on obtaining the plain text we apply sentiment analysis and classification algorithm.

4.1 User Interest

Using the user interface we have obtained the following results. Figure 3 shows the interest pie chart for the tweets done by Shreya Goshal, a popular singer in India, from 1-02-16 to 29-02-16 and this result shows that major topic in which she is interested is entertainment for this period of time. The values shown in the pie chart is in percentage and the entertainment culture topic is having the highest value of the interest that is 49.345 % and second interested topic is a hospitality recreation with 25.199 % value of interest.

Figure 4 shows the interest pie chart for the tweets done by Narendra Modi, Prime Minister of India, and this result shows that major topic in which he is interested are politics and social issues with 22.132 % value of interest. The second interested topic is business finance with 15.996 % value of interest.

4.2 Location Interest

We have collected tweets of different cities of India and show interested topic. Figure 5 represents the bar graph whose x-axis represents the topic name and the y-axis represents the value of interest in percentage. Through figure it has been seen that in Hyderabad most of the people do tweet that belongs to the business finance field with

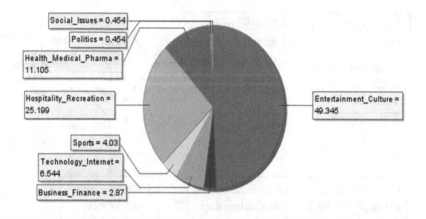

Entertainment_Culture = 49.345 ● Business_Finance = 2.87 ● Technology_Internet = 6.544 ○ Sports = 4.03
Hospitality_Recreation = 25.199 ● Health_Medical_Pharma = 11.105 ○ Politics = 0.454 ● Social_Issues = 0.454

Fig. 3. Tweets done by Shreya Ghoshal with positive interest. (Color figure online)

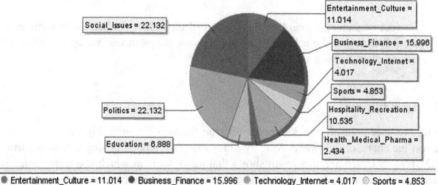

● Entertainment_Culture = 11.014 ● Business_Finance = 15.996 ● Technology_Internet = 4.017 ○ Sports = 4.853
○ Hospitality_Recreation = 10.535 ● Health_Medical_Pharma = 2.434 ○ Education = 6.888 ● Politics = 22.132
● Social_Issues = 22.132

Fig. 4. Tweets done by Narendra Modi with positive interest. (Color figure online)

38.0413 % value of interest and second interested topic is technology with 12.9435 % value of interest.

Table 1 represents the value in percentage of the interest for certain topics that are listed in table for most of the famous cities of India.

Figure 6 shows the comparison results for different topics for five different cities of India. X-axis of the above graph represents the topic name and the y-axis represents the value of interest over certain topics in percentage. Bangalore is one of the cities where most of the multi-national companies are located and most of the business activities take place. The results are showing that among the five cities of India, tweets from

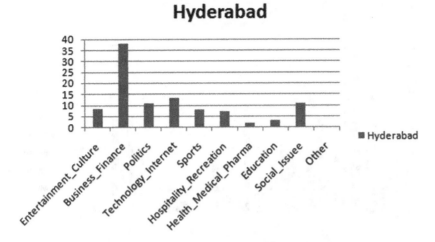

Fig. 5. A topic to which tweets belong to Hyderabad city.

Table 1. Comparison data for different cities.

City \ Topic	Hydera-bad	Delhi	Banga-lore	Chandi-garh	Mum-bai
Entertainment_Culture	8.0628	5.6234	3.9814	27.3675	22.517
Business_Finance	38.0413	22.0722	49.6758	3.3993	22.0722
Politics	10.805	22.5177	10.7765	12.2889	5.6234
Technology_Internet	12.9435	14.3932	16.2475	3.3759	14.3932
Sports	7.9552	4.6632	2.4733	27.367	4.6632
Hospitality_Recreation	6.9499	6.0001	4.0945	12.9485	6.0001
Health_Medical_Pharma	1.4342	0.9744	0.3051	2.5275	0.9744
Education	2.9985	1.2268	1.6645	0.3142	1.2268
Social_Issuee	10.805	22.517	10.7765	10.4095	22.5177
Other	0.0041	0.0107	0.0045	7.62E-04	0.0107

Bangalore are highly related to business-finance. Delhi, the capital of India, is a political hub where many politicians and youth that belong to some non- governmental organization (NGO) reside. From the results we infer that among the five cities, tweets from Delhi are mostly related to politics and social issues. It is useful to have such data, information because it provides us the trend of users at certain locations. Based on the trends of the twitter data, new products may be launched in certain locations.

4.3   Comparison of Data of Different Countries

Table 2 represents the value in percentage of the interest for certain topics that are listed in a table for two countries India and America. We have obtained these values by

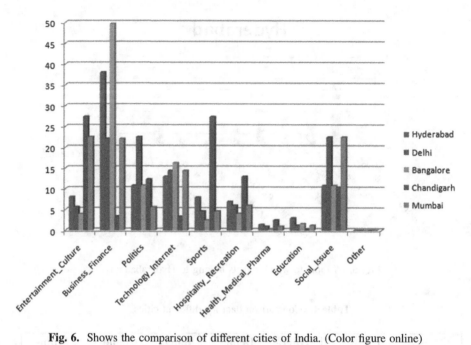

Fig. 6. Shows the comparison of different cities of India. (Color figure online)

collecting the tweets of different bot users of India and America and here we try to get the overall interest of Indian and American users and our analysis shows that most of the tweet done by Indian users are related to politics and social issues with 18.291 % value of interest and then other interested topic is hospitality recreation with 14.393 % value of interest.

Table 2. Comparison data for countries, India and America.

Country \ Topic	India	America
Entertainment_Culture	10.625	15.9
Business_Finance	9.5233	8.3479
Politics	18.291	5.8073
Technology_Internet	5.5157	14.754
Sports	12.85	14.464
Hospitality_Recreation	14.393	23.986
Health_Medical_Pharma	5.6224	7.3929
Education	4.8878	3.5413
Social_Issues	18.291	5.8073
Other	6.90E-05	

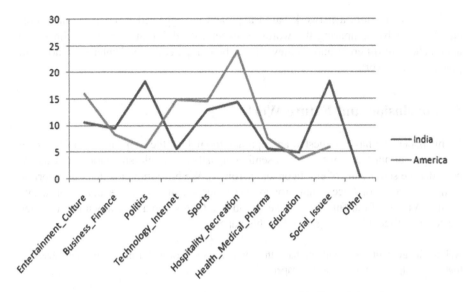

Fig. 7. Shows the comparison between tweets of India and America.

The tweets done by American users are mostly related to hospitality recreation with 23.986 % value of interest and then second major interested topic is entertainment with 15.9 % value of interest. Figure 7 shows the comparison of Indian and American users graphically based on the tweets collected for both the countries.

4.4   Discussion

Table 3 shows the comparison of the proposed approach with other works. Sentiment analysis suggested in [3] is not suitable for short text. Sentiment analysis suggested in [2] is suitable for short text, but they did not consider the sentiment score for the emoticon and acronym. We consider sentiments of emoticon and acronym in our approach. Thus, we obtain better results in tweets.

Table 3. Comparison with other works.

Paper / Features	Sentiment Analysis	Classification	Users' Interest
R. Prabowol [1]	Yes	-	-
R. Socher [2]	Yes	-	-
M. Thelwall [3]	Yes	-	-
K. Lee [10]	-	Yes	-
This Paper	Yes	Yes	Yes

For classification, a network based approach is suggested in [10]; we have done classification by comparing the words of tweet with different categories. Since both approaches are different and so they cannot be compared. Users' interest is not computed in these works.

5 Conclusion and Future Work

In this paper, we have suggested an approach to analyze the user interest based on her tweets. In our approach, we combine sentiment analysis and classification of tweets. We consider the sentiments of emoticon and acronym. We have implemented our algorithm using five lakhs of tweets and around one thousand of users. We obtain promising results. As part of our future work we would like to develop a recommendation system based on the techniques suggested in this paper.

Acknowledgement. The authors thank the anonymous reviewers for their valuable suggestions that have helped in improving the paper.

References

1. Prabowo, R., Thelwall, M.: Sentiment analysis: a combined approach. J. Informetrics **3**(2), 143–157 (2009)
2. Socher, R., Perelygin, A., Wu, J.Y., Chuang, J., Manning, C.D., Ng, A.Y., Potts, C.: Recursive deep models for semantic compositionality over a sentiment Treebank. In: Proceedings of the Conference on Empirical Methods in Natural Language Processing, pp. 1631–1642 (2013)
3. Thelwall, M.: Heart and soul: sentiment strength detection in the social web with SentiStrength. In: Proceedings of the CyberEmotions, pp. 1–14 (2013)
4. Gurini, D.F., Gasparetti, F., Micarelli, A., Sansonetti, G.: A sentiment-based approach to Twitter user recommendation. In: Proceedings of 5th ACM RecSys workshop on Recommender Systems and the Social Web, June 2013
5. Agarwal, V., Bharadwaj, K.K.: A collaborative filtering framework for friends recommendation in social networks based on interaction intensity and adaptive user similarity. J. Soc. Netw. Anal. Min. **3**, 359–379 (2013)
6. Abel, F., Gao, Q., Houben, G.-J., Tao, K.: Semantic enrichment of Twitter posts for user profile construction on the social web. In: Antoniou, G., Grobelnik, M., Simperl, E., Parsia, B., Plexousakis, D., Leenheer, P., Pan, J. (eds.) ESWC 2011, Part II. LNCS, vol. 6644, pp. 375–389. Springer, Heidelberg (2011)
7. Boutet, A., Kim, H., Yoneki, E.: What's in Twitter, I know what parties are popular and who you are supporting now! J. Soc. Netw. Anal. Min. **3**(4), 1379–1391 (2013)
8. Java, A., Song, X., Finin, T., Tseng, B.: Why we Twitter: understanding microblogging usage and communities. In: Proceedings of 9th WebKDD and 1st SNA-KDD Workshop, SanJose, California, USA, pp. 56–65, August 2007
9. Pennacchiotti, M., Popescu, A.: A machine learning approach to Twitter user classification. In: Proceedings of the Fifth ICWSM, pp. 281–288 (2011)

10. Lee, K., Palsetia, D., Narayanan, R., Patwary, M.M.A., Agrawal, A., Choudhary, A.: Twitter trending topic classification. In: Proceedings of 11th IEEE International Conference on Data Mining Workshops, pp. 251–258, December 2011
11. Kiritchenko, S., Zhu, X., Mohammad, S.M.: Sentiment analysis of short informal texts. J. Artif. Intell. Res. **50**(1), 723–762 (2014). USA
12. Agarwal, A., Xie, B., Vovsha, I., Rambow, O., Passonneau, R.: Sentiment analysis of Twitter data. In: Proceeding of LSM 2011 Workshop on Languages in Social Media, pp. 30–38. Association for Computational Linguistics (2011)
13. Manning, D., Christopher, M., Surdeanu, J., Bauer, J., Finkel, S., Bethard, J., McClosky, D.: The stanford CoreNLP natural language processing toolkit. In: Proceedings of the 52nd Annual Meeting of the Association for Computational Linguistics: System Demonstrations, pp. 55–60 (2014)
14. Joachims, T.: Text categorization with support vector machines: learning with many relevant features. In: Proceedings of the 10th European Conference on Machine Learning, London, pp. 137–142 (1998)
15. Tumasjan, A., Sprenger, T.O., Sandner, P.G., Welpe, I.M.: Predicting elections with Twitter: what 140 characters reveal about political sentiment. In: Proceedings of the Fourth International AAAI Conference on Weblogs and Social Media (2010)

User-Friendly Ontology Structure Maintenance Mechanism Targeting Sri Lankan Agriculture Domain

S.W.A.D.M. Samarasinghe[1(✉)], A.I. Walisadeera[2],
and M.D.J.S. Goonetillake[1]

[1] University of Colombo School of Computing, Colombo 07, Sri Lanka
dilinims5@gmail.com, jsg@ucsc.cmb.ac.lk
[2] Department of Computer Science, University of Ruhuna, Matara, Sri Lanka
waindika@cc.ruh.ac.lk

Abstract. As a result of recent enhancements in global knowledge sharing capabilities, knowledge representation and reasoning with ontologies for acquisition of implicit learning is gaining more attention among general communities. Agriculture being a dynamic economic sector in the Sri Lankan context, can massively benefit from such a knowledge repository that can be accessed and maintained by the community. However, existing approaches for ontology maintenance are complex and are designed for users with ontology-engineering expertise. Thus, despite numerous benefits, adoption and diffusion of ontology based knowledge systems within ontology-illiterate agriculture community is significantly hindered. This study investigates means of addressing the said limitation by proposing a user-friendly mechanism to incorporate evolving knowledge into ontologies. Updating the ontology structure while assuring real-time consistency maintenance is considered the prime objective of the approach. Task based evaluation results prove the effectiveness of our approach against the existing work.

Keywords: Knowledge representation · Agriculture ontology · User-friendly ontology structure maintenance · Real-time consistency maintenance

1 Introduction

Knowledge Representation is recently being directed towards new dimensions with the possibility of seamless sharing of global knowledge. The dynamic domains, with their increasing pace of knowledge growth, have urged the need to look for mechanisms to represent existing knowledge as well as incorporate evolving domain knowledge into the representation. A widespread agreement that is followed in achieving this semantic representation of knowledge evolution is to use ontologies populated with concepts [1]. An ontology is a properly structured view of the domain knowledge and serves as a repository of the concepts in the domain allowing knowledge sharing, aggregation, and information retrieval. The inference mechanism based on the evolving ontology system ensures enhancement of knowledge of domain users for decision making requirements.

© Springer International Publishing Switzerland 2016
O. Gervasi et al. (Eds.): ICCSA 2016, Part V, LNCS 9790, pp. 24–39, 2016.
DOI: 10.1007/978-3-319-42092-9_3

As of the Sri Lankan context, agriculture sector stands as a vital economic icon occupying 28.4 % of the total labour force of the country [2]. Stakeholders ranging from farmers to agriculture instructors, consultants, researchers and government authorities are all in need of up-to-date agricultural knowledge for decision making and various other information needs. Signifying the need of such a system, Samansiri and Wanigasundera [3] provide statistics that informational search costs account for 11 % of the total transaction cost of smallholder vegetable farmers in Sri Lanka, which could have been eliminated with a proper knowledge management system. Thus, an ontology based agricultural knowledge management system which maintains up-to-date agriculture knowledge would be an ideal solution to cater to requirements of stakeholders.

With such a system, two ontology components would require to be handled, namely; the structure which represent the concepts/classes and relationships among them, and instances representing individual data pertaining to concepts. For example, vegetables, pests, and relationship among vegetables and pests constitute a part of the agricultural ontology structure. Different vegetables such as carrot, potato and cabbage represent instances of vegetable class. Section 3 elaborates more on the said aspects with technical terminology and further details. Devising a mechanism to update the ontology with instances is relatively easier than maintaining the structure as ensuring consistency, and addressing accurate representation of related concepts and relationships are issues to be dealt with only when updating the ontology structure.

There are several standard ontology editors that are capable of assisting the structure management of such an agricultural knowledge base. However, the major drawback of these is that they are designed only for ontology literate computer engineers and require unique expertise in understanding and handling them. Therefore, Sri Lankan agriculture experts who are unaware of ontology know-how cannot use such advanced systems. As a result, dynamic ontology updating through the information obtained from domain users would require to be carried out by a computer engineer with some standard ontology editor. Yet, the difficulties in obtaining the support of a computer engineer at all times might cause havoc and result in inefficiency and ineffectiveness in maintaining the knowledge base in the long run. Furthermore, if domain users themselves are to handle ontology maintenance with such tools, their complexity, time consumption in learning and simply the annoyance caused by the use of them would most certainly prevent the uptake of the tool by the users [4]. Hence, despite their numerous benefits, ontology based knowledge management systems are not readily adopted within Sri Lankan agriculture domain.

Driven by the said factors, motivation of this research lies behind the issues that have caused the specific need of a convenient mechanism to modify and manage the structure of the agriculture ontology. Devising a framework that can manage the ontology structure while also shielding the underlying ontology framework from the users of the system is investigated in our work. The intension is to propose a mechanism to support acquisition and maintenance of ontology, from an end-user perspective which has often been overlooked in similar approaches, to ultimately ensure knowledge enhancement of agriculture domain users.

The remainder of this paper is organized as follows. Section 2 presents the related work available with regard to ontology structure maintenance. Section 3 elaborates the proposed mechanism for maintaining ontology structure and also addresses the requirement of usability in end-user viewpoint. Section 4 discusses the implementation architecture and technologies used. The evaluation phase is elaborated in Sect. 5 according to different metrics to assess the usability of our approach with the Sri Lankan agriculture domain based ontology. Finally, conclusion and potential future work are presented in Sect. 6 of the paper.

2 Related Work

Here we discuss the existing approaches for ontology maintenance and their limitations in ontology-illiterate end user perspective. The literature identifies several standard workbenches such as Protégé [5], Swoop [6] and OntoEdit [7] which provide a rich set of options for creation, visualization and manipulation of ontologies. However, none of them support ontology validation or real-time consistency maintenance which is checked only at the inference stage. Thus, the user requires having an understanding on underlying logic and ontology terminology to manage the tools. Furthermore, tracing errors back to the origin at the inference point would be a complex and tiresome work as consistency is not maintained throughout.

On contrary to the complex standard tools, several studies have proposed approaches to assist ontology-illiterate users in maintaining knowledge bases. Onto-Web [8] and OntoXpl [9] are both user-friendly exploratory tools designed on a user-friendly basis which support easy navigation, querying and viewing of the ontology without the need of ontology expertise. However, none of the tools possess editing or knowledge inference capabilities which lead to difficulties in updating the ontology with up-to-date knowledge. The work proposed in OWLEasyViz [10] addresses visualization and navigation aspects for ontology-illiterate end users. Nevertheless, ontology editing options are still designed for advance users with knowledge on ontology building. Moving a step ahead, applications such as NavEditOW [11] have ensured more flexibility to end users through provision of comprehensive visualization of the ontologies as well as create, edit and remove options only for ontology instances. The ontology editor which is constructed for Sri Lankan agricultural domain mentioned in [12] is also focused only on instances where it provides a simple web based UI for the end users to populate the ontology with data. Nevertheless, handling the structure of the ontology still needs intervening of computer engineers with sound ontology literacy in both these cases. Thus, we identify a limitation in the existing approaches to cater to user-friendly ontology structure maintenance. In the next sections we elaborate the mechanism we propose to address the said issues and assist ontology-illiterate Sri Lankan agriculture domain experts.

3   Proposed Mechanism for Agriculture Ontology Structure Maintenance

The proposed mechanism is intended to present a design where the end-user of the system is capable of using a web-based user interface to either request a structural modification to be performed or to infer knowledge from the existing concepts and relationships.

A request to modify the ontology would cause the respective change to be updated in the source ontology file along with the propagations that are required to be carried out. An inference request would generate the inferred agriculture ontology file which carries the new knowledge attained from structural changes. Both the scenarios require to ensure consistency of the ontology in order to guarantee accurate knowledge inference. The sub sections below elaborate on the research process followed during the derivation of the mechanism.

3.1 Ontology Preliminaries

In designing the said user-friendly approach to facilitate dynamic update of agriculture ontology, several concepts were studied for making design decisions.

Ontology authoring is conducted based on logical definitions given by Description Logics (DL), which provide means of modeling relationships between concepts or entities in a particular domain [13]. DL identify 3 kinds of entities namely: concepts, roles and individuals, where individuals refer to data in a particular domain, concepts represent the classes or the sets of individuals and roles represent the binary relationships between concepts or individuals. An ontology is categorized into two components based on DL as follows:

T-Box. This defines the terminology or the *structure* of the domain. Out of the 3 entity types described earlier, concepts and roles belong to T-Box [14]. In addition to the atomic concepts and roles, complex descriptions can also be defined under T-Box with DL.

A-Box. It consists of assertions on individual data pertaining to the domain [13]. It comprises of *instances* which have names and assertions to the T-Box concepts and roles.

Although DL is used as the base for building ontologies, a machine readable model is required to use DL for describing the sources in the semantic web. In realizing this requirement, several languages have evolved and the latest W3C[1] recommendation includes Web Ontology Language (OWL) 2 which is being used since 2004 as an extension with added functionality to its initial version OWL 1. OWL offers entities corresponding to DL namely; classes, object properties, data type properties and individuals. It facilitates further capabilities such as transitive, symmetric, and inverse properties, cardinality and value constraints on properties, and constructs such as union, intersection and complement to combine classes. It is based on Extended Markup Language (XML) syntax for modeling ontologies. Three sub languages of OWL exist which are developed with the intention of catering to different types of domain users, namely: OWL Lite, OWL DL and OWL Full [15]. OWL Lite offers only a minimum of expressiveness including classification hierarchy and cardinality. OWL DL has a desirable expressiveness while also ensuring all reasoning tasks to be

---

[1] http://www.w3.org/.

decidable and thus complete in finite time [16]. Complete expressivity is ensured with the OWL Full version, however, with no guarantee that all reasoning tasks are decidable. Therefore, with the work of this paper, OWL DL version is adopted for maximum expressivity and inference of knowledge. The DL variant of our approach confirms to SHOIN, based on OWL version 2 that is utilized with the work.

3.2   Supported Structural Modifications

The work proposed in here is solely concerned with the T-Box or the structure, since there is existing literature as mentioned under related work concentrated on A-Box which can easily be integrated with our work as well. Primitive modifications of T-Box representing 'add' and 'delete' functions, which are the atomic modifications that are possible with any type of ontology, are supported in the said aspect. Since all complex modifications can be replaced with 'add' and 'delete' modifications, primitive modifications are capable of handling all types of ontology modification requirements of end users. The modifications allowed in our approach with respect to classes, object properties, data type properties and their selected axioms are as elaborated in Fig. 1.

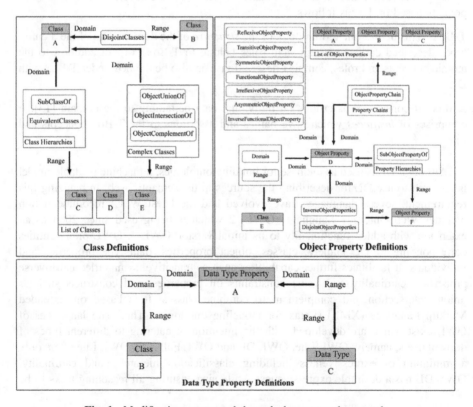

Fig. 1. Modifications supported through the proposed approach

3.3   Propagation Strategy for Modifications

Modifications in ontology elements cannot occur independently. When a particular element is added, it affects the consistency whereas a deletion affects the existence of elements related to deleted element. Hence, in the process of deleting an ontology element, their dependent artifacts need to be addressed to maintain the accuracy and reliability of the ontology. Subsequently, when trying to address these dependencies, it can lead to another issue as ambiguities might arise. Thus, in order to ensure that ontology remains consistent and generates accurate inferences, both the requested modification as well as their implicit dependencies should be catered to. Four types of dependencies are identified over the research process which are responsible for implicit dependencies namely, direct dependencies, indirect dependencies, total dependencies and partial dependencies [17]. Figure 2 illustrates the said dependencies with examples from the agriculture ontology for clarification.

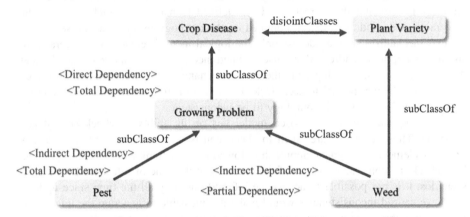

Fig. 2. Dependency types illustrated with class dependencies for class 'Crop Disease'

In addressing the said dependencies, four types of available strategies were ana-lyzed to propagate deletions [18]:

Delete Direct Axioms Strategy: all totally and partially dependent axioms which are directly linked with the deleted element are deleted.
Cascade Strategy: all direct-dependent axioms and totally dependent elements are deleted with their direct-dependent axioms. Partially dependent elements are ignored.
Attach to Parent/Root Strategy: all-direct dependent axioms are deleted and direct-dependent elements are linked to the parent/root element of the deleted element.

In proceeding with the mechanism, *Delete Direct Axioms Strategy* was adopted as the propagation strategy based on the nature of Sri Lankan agriculture domain ontology and the requirement of its end users. The domain users prefer the dependent elements to remain in the ontology as cascading would cause a significant loss to the knowledge base. Thus, only the dependent axioms are deleted and provision is given for the user to perform cascading or attaching to parent if needed. The relevant deletions with regard to implicit dependencies are as follows: deleting a class would implicitly delete direct concept-axiom dependencies and direct concept-instance dependencies, deleting an object/ data type property would implicitly delete direct property-axiom dependencies and property-instance property dependencies. Deleting axioms of classes or properties do not cause implicit dependencies.

3.4 Consistency Maintenance and Knowledge Inference

Three forms of consistency requirements are handled by our work to avoid contradictions in the representation and ensure accurate knowledge inference [19]: Structural Consistency which ensures that the constructs used in ontology representation align with the underlying DL variant, Logical Consistency which checks whether the ontology is semantically correct and User-Defined Consistency which refers to the constraints and conditions that are defined on the ontology by the user himself.

Our mechanism should address only the logical inconsistencies as structural consistency is implicitly addressed and user-defined inconsistencies can be declared under the same categories. With regard to the primitive changes, only addition definitions are capable of inducing inconsistencies. Deleting an axiom from an already consistent set of axioms will not cause the ontology to be inconsistent.

The uniqueness of the proposed model lies on its ability to check consistency real-time. Thus, measures are taken to check consistency at each addition of a new ontology element. The typical approach followed in other frameworks such as protégé, is to check the consistency status of the ontology at the time of inference [5]. In such scenarios, it is not possible to access the inferred ontology all the time since updates that have caused inconsistencies would make it impossible to generate the inferences. Even though they usually provide an error report with all inconsistent axioms, it is quite inconvenient for the user to identify the actual causes of inconsistencies at that stage as there could be many. This has been discussed as an issue in several instances and knowledge engineers have preferred real-time consistency check in place of checking it at inference stage [20].

Tableau based FACT algorithm [21] is adopted in both checking the consistency and generating inferred knowledge. Since our approach ensures that inconsistent modifications are not allowed at each addition of a new ontology object, it is assured that the ontology is consistent at any given moment of time. Therefore, generation of inferred knowledge does not require a prior consistency check to be conducted. The inferences would be generated at the request of the end user directly without interruption through the reasoner.

Based on the described design considerations and explanations, the flow of the ontology structure maintenance mechanism leading to accurate dynamic ontology evolution can be illustrated as shown in the flow chart in Fig. 3.

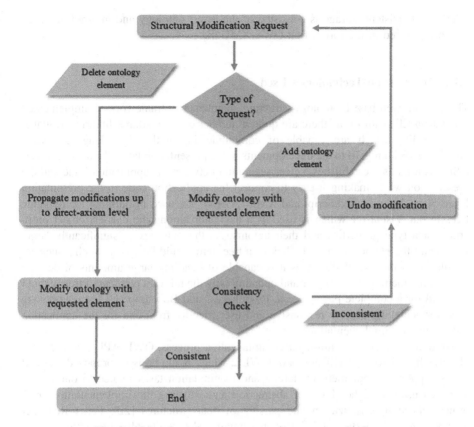

Fig. 3. Flow of ontology structure maintenance mechanism

4 Implementation Overview

From an architectural point of view, the proposed mechanism is implemented as an application composed of several modules. The application is designed to be accessible from a web browser over a network connection, and can be collaboratively maintained by several users. The web-based ontology structure maintenance prototype is mainly comprised of three layers as: Presentation Layer, Semantic Framework Layer and Persistent Storage.

The presentation layer ideally represents the implementation tasks related to web interface which provides flexible options for the users for long term maintenance of the ontology by allowing them to make the required updates. Semantic framework layer is the ground layer for all functionalities handled at presentation layer. It is intended to support manipulation of ontology and inferring tasks by shielding the user from underlying framework. This layer is composed of the main Application Program Interface (API) that communicates between the presentation layer and the OWL source file, and the reasoner which supports in consistency maintenance and knowledge inference. Persistent storage is consisted of the assert ontology and inferred ontology which are stored in the form of an OWL/XML file.

4.1   Overview on Technologies Used

The overall web-based ontology structure maintenance framework is implemented using Java EE platform as there are quite a number of open source libraries available for Java EE, which are capable of communicating with OWL language based ontologies. With regard to the developments in the presentation layer, JavaServer Faces (JSF)[2] is used as it establishes a clear separation between the application logic and the presentation while making it easy to connect the application code to the presentation layer. Asynchronous JavaScript and XML (AJAX)[3] are used in building the dynamic trees in our Visualizer which is used as a navigator among the classes, object properties, data type properties and their axioms. As the ontology is significantly large, loading all the elements on each click of a tree item would be costly. Such issues are handled with the use of AJAX as it is capable of handling large amounts of data by exchanging them with the server and updating only the relevant element of tree without the need for a complete page refresh. PrimeFaces[4] and Bootstrap[5] UI frameworks are used for achieving a more responsive and user-friendly front-end for the application with minimum implementation cost.

In order to facilitate intra-layer communication process, OWL API is used as it is closely aligned with specifications of OWL 2 which is the basis of proposed work. It offers support to implement modelling and manipulation tasks related to ontologies, while ensuring a higher level of abstraction and isolating the implementation from issues related to serialization and parsing of data structures [22]. Pellet Reasoner supports in both A-Box and T-Box reasoning tasks, while also providing several unique features that are important in our work as well. It possesses the capability of reasoning with large number of individuals and uses novel optimization techniques based on partitioning of OWL ontologies and axiom tracing. Its ability for incremental reasoning is much advantageous in our approach especially with real-time consistency maintenance, as it re-computes as little as possible after each update saving performance time up to a great extent [23].

4.2   T-Box Editing and Real-Time Consistency Check

All modifications inclusive of class definitions, object property definitions and data type property definitions, are addressed via suitable GUI components. The uniqueness of our application when compared to the standard ontology maintenance approaches is that, the user is provided with select options instead of unguided input fields in all cases

---

[2] http://www.oracle.com/technetwork/java/javaee/javaserverfaces-139869.html.

[3] http://www.seguetech.com/blog/2013/03/12/what-is-ajax-and-where-is-it-used-in-technology.

[4] http://www.primefaces.org/.

[5] http://getbootstrap.com/2.3.2/.

except in creation of a new element where an input field is a must. The presentation layer identifies the input values as parameters to be forwarded to the semantic framework layer for processing the request. Apart from the convenience it provides for the user, it also has an impact on mitigating inconsistencies which can occur at interface level. User defined consistencies are well maintained as improper axioms would not be added to the ontology due to differently spelled words entered by mistake. Furthermore, as measures are taken to check consistency at each addition of a new ontology element, it allows the user to correct any wrong update instantaneously without having to go through poorly descriptive error reports to understand their mistake at a later stage.

4.3 Navigation and Visualization

Our visualizer helps the user to understand and learn the existing ontology elements and their hierarchy based on sub class definitions and sub object property definitions. Browsing of characteristics pertaining to each ontology element and navigation between the editor and visualizer are facilitated through the interface. Users are provided with options to follow the details of a particular ontology object by expanding a node in the tree. An example of handling and visualizing object properties at interface level is given in Fig. 4.

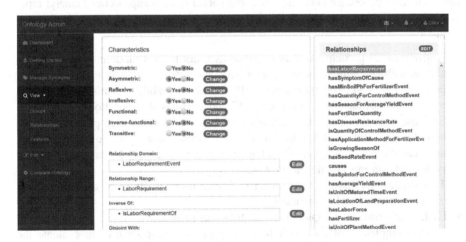

Fig. 4. T-box editing options and visualization of ontology elements

4.4 Feedback, Help and Support

In order to ensure that user does not feel naïve around the uncommon ontology terminology, we have replaced the most commonly used phrases with more understandable terms: Classes as Groups, Object Properties as Relationships, Data type Properties as Features. For further clarifying the meaning and behavior of rest of the terms for axioms, a separate guide is provided along with details on each function, their effect on ontology structure and causes of inconsistencies for each.

Furthermore, feedback is provided on each modification done to the ontology structure, through a notification displaying the effect of the action a user has made. User can view the change in the visualizing pane and how the hierarchy has been affected.

5 Evaluation

A comparison and questionnaire based evaluation was conducted in order to determine the usability of our approach in the end user based perspective. Protégé was chosen as a benchmarking tool and both our system and Protégé were provided for the end users to compare the effectiveness. The reason to have chosen Protégé is that there are no user-friendly mechanisms developed with regard to ontology structure maintenance, and Protégé is the most widely adopted standard ontology editing workbench as of today which is used by computer engineers [24]. Hence, a user based evaluation by comparing against Protégé would be the best methodology to prove the usability and flexibility enhancement entitled by our work for ontology-illiterate domain users.

5.1   Evaluation Methodology

The evaluation procedure constitutes a combination of two approaches namely: repeated measure and task based. Following procedure was adopted in conducting the evaluation:

- A group of 20 participants was recruited as users from a random sample of population with and without ontology expertise, and with at least 5 years of experience in using a general personal computer (PC) to avoid usability results being affected by their lack of familiarity in using a PC and web-based applications.
- Each user was provided with a pre-test questionnaire to assess their prior knowledge on ontologies and related subject area. Also, a brief introduction to ontologies and their capabilities was given for all users along with a 'quick-start-guide' to our system and an introduction to Protégé.
- Two similar task lists were given for each user individually to be carried out with our system and Protégé separately.
- Each user's time to complete the task lists was recorded.
- At the completion of each task list, two post-test questionnaires to evaluate the usability of each tool were given for the users to complete based on their experience with the two tools. Furthermore, a questionnaire to evaluate the user-friendliness of the proposed application interface and a comparison questionnaire on the two tools were also provided for the users to complete at the end.

Two standard questionnaires were used in evaluation process which have been accepted and adopted by a considerable number of published literature in evaluating usability of systems [25]. System Usability Scale (SUS) which was initially developed at Digital Equipment Corp., is used with both tools to compare usability.

Questionnaire for User Interface Satisfaction (QUIS) was used to evaluate the user-friendliness of the proposed application interface.

In carrying out the evaluation process two task lists, A and B were prepared which constituted of a similar structure. The user group was equally divided into two sets (10 each) and one set carried out task list A with our system and task list B with Protégé while the other set carried out task list B with our system and list A with Protégé.

5.2 Statistical Analysis Based on Task Completion Time

Task completion time was recorded for each user performing each task list, in order to test several hypotheses regarding the usability of our proposed approach. Following are the objectives that were intended to evaluate with task completion time analysis:

Is there a difference between the times taken to,

1. complete a task list with our approach for users with and without prior knowledge on ontologies?
2. complete a task list with Protégé for users with and without prior knowledge on ontologies?
3. complete a task list with our editor and Protégé for the users with prior knowledge on ontologies?
4. complete a task list with our editor and Protégé for the users without prior knowledge on ontologies? If yes which approach takes less time for the users?

Figure 5 depicts the time taken for test users to complete the tasks with each tool. Also, from the pre-test scores, 7 users who scored more than 50 % were identified as ontology-literate users while the remaining 13 were categorized into ontology-illiterate test user set. These time and score values are not normally distributed and the measurement scale is ordinal. Hence, non-parametric tests were utilized in analyzing the data set.

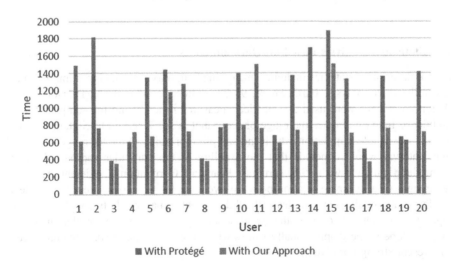

Fig. 5. Time taken to complete task lists (s) (Color figure Online)

Out of the available non-parametric tests, Mann-Whitney U test and Wilcoxon Signed Ranks test were utilized in carrying out the evaluation. Mann-Whitney U test is used when comparing differences between two independent samples, when the dependent variable is not normally distributed and is ordinal or continuous, while Wilcoxon Signed Ranks test is used when comparing two score sets of the same group.

Table 1 gives the results for rejecting null hypothesis (H0) and alternative hypothesis (H1) where, H0 denotes there is no significant difference between the two hypotheses and H1 denotes a difference. From analyzing the results on the hypotheses, it can be deduced that ontology-illiterate end users find Protégé to be much complex than ontology-literate end users and take more time when working with it. Both ontology-literate and illiterate end users take nearly an equal amount of time to work with our approach and it is much less than the time taken by ontology-illiterate end users to work with Protégé and almost equivalent to the time taken by ontology-literate end users to work with Protégé. Therefore, it can be favorably concluded that our approach suits both types of users and it is much helpful for ontology-illiterate end users to efficiently carry out their ontology modification tasks than using Protégé.

Table 1. Statistical analysis results for the time measures

Objective	Parameters		Test	Type	Hypothesis	
					HO	HI
1	With our approach	Prior-knowledge	Mann-Whitney U	Two-tailed	Reject	Retain
		No prior-knowledge				
2	With Protégé	Prior-knowledge	Mann-Whitney U	Two-tailed	Reject	Retain
		No prior-knowledge				
3	With prior-knowledge	Our approach	Wilcoxon Signed Ranks	Two-tailed	Retain	Reject
		Protégé				
4	Without prior-knowledge	Our approach	Wilcoxon Signed Ranks	One-tailed	Reject	Retain
		Protégé				

5.3   Statistical Analysis Based on Questionnaires

General descriptive statistics including mean, median, maximum and minimum values were utilized in analysis. Furthermore, 95 % confidence interval and Pearson's correlation coefficient were used as statistical measures in analyzing the samples. 95 % Confidence Interval denotes the range of score that is 95 % certain to contain the true mean of the whole population which the sample represents. A better range of confidence interval would conclude that the usability of the particular approach is high among the population with 95 % certainty. Pearson's Correlation Coefficient measures the strength of a linear association between two samples. +1 would identify a perfect correlation among the data samples with all data points on a straight line with a positive slope in the graph. −1, on the other hand denotes a perfect inverse correlation with a negative slope in the graph. Usually values which are >+0.7 and <−0.7 are considered to represent strong correlation.

From the descriptive statistics in Table 2, it is clear that our approach has a better usability while Protégé is relatively less usable according to the mean scores among the population for both SUS as well as comparison questionnaires. There is a higher correlation between the SUS scores for Protégé and pre-test scores, which means it was the users with prior-knowledge on ontology who have rated Protégé to be usable, while no such strong correlation is observed with our approach and pre-test score meaning all users irrespective of their prior knowledge have considered our approach to be user-friendly and adoptable.

Individual scores for each question in the questionnaires have suggested that the interface was easy to use, reliable, while task performance was easy to carry out (more than 80 % score). Also, most of them have a positive impression on the reference material provided to learn the tool and the organization of the interface. However, a lesser score was obtained on the aspects such as the support given by the approach for correcting mistakes during structural change performance (52 %) and guidance provided for correcting their mistakes (59 %). These issues would require to be looked into in order to provide maximum benefit of the proposed structure maintenance tool.

Table 2. Descriptive analysis results for questionnaire scores

	SUS with our approach	SUS with protege	Comparison questionnaire	QUIS
Minimum	58	26	61.11	67.65
Maximum	90	68	94.44	85.88
Median	80	43	77.78	74.76
Mean	79.4	45.9	78.89	76.32
Standard deviation	7.79	13.45	9.25	5.13
95 % confidence interval	75.75–83.05	39.61–52.19	74.56–83.22	73.92–78.73
Pearson correlation with pre-test questionnaire score	−0.4136	0.700614	−0.4067	−0.01179
Pearson correlation between SUS score protege vs our approach	−0.01467	–	–	–

5.4 User Comments and Suggestions

Comments were obtained from test users at the end of QUIS questionnaire which pointed out the following recommendations:

- To introduce undo and redo options in order to quickly recover from mistakes. As of now, users require to manually correct mistakes with the 'add' and 'delete' options provided.
- To improve the navigation capability through the ontology elements by supporting hyperlinks for the items inside the view panels as well. Hyperlinks are provided only for elements in hierarchical view in the current system.
- To introduce a check option to clarify if an inconsistency would occur prior to actually adding the ontology element into the source and rejecting if it was inconsistent.

6 Conclusion and Future Work

This study is focused on investigating means of addressing the requirement of Sri Lankan agriculture domain users who are ontology-illiterate, to effectively manage their ontological knowledge base. In assessing the requirement, we proposed a mechanism with flexible means of handling the ontology by hiding the underlying ontology framework from user. Support to modify the ontology T-Box and accurate inference of knowledge based on reasoning techniques were ensured through the design of the model. We also focused on real-time consistency maintenance as a solution to complications at inference stage with traditional approaches, and introduced a unique model to handle inconsistencies at real-time. The model was implemented as a web-based system with a three layer architecture including a presentation layer, semantic framework layer and a persistent storage.

We analyzed the effectiveness of the proposed mechanism through repeated measure, task based user evaluation techniques with an ontology for Sri Lankan agriculture domain. Questionnaire based statistical analysis was also conducted to assess the user preference and to obtain suggestions for future enhancements. The presented results justify the effectiveness of our approach against the existing complex standard tools and prove its usability to be significantly higher than the standard tools.

We intend to expand our mechanism towards other domains as well with several future enhancements, while also considering user comments and suggestions taken. The expressivity and usability of the approach is to be further improved by extending its capabilities to address more complex change definitions apart from the primitive changes supported at this stage. Another aspect is to enhance the mechanism to support a concurrent-collaborative ontology maintenance. Ideally, issues would arise when multiple users request different changes to be performed on same ontology element, and conflict resolution of change requests and control of concurrent access to the system would require to be handled during collaborative maintenance of ontology.

References

1. Jakus, G., Milutinović, V., Omerović, S., Tomažič, S.: Concepts, Ontologies, and Knowledge Representation. Springer, New York (2013)
2. Central Intelligence Agency: The World Factbook. https://www.cia.gov/library/publications/the-world-factbook/fields/2195.html
3. Samansiri, B.A.D., Wanigasundera, W.A.D.P.: Use of information and communication technology (ICT) by extension officers of the tea small holdings development authority of Sri Lanka. Trop. Agric. Res. **25**, 1–16 (2014)
4. Davis, B., Iqbal, A.A., Funk, A., Tablan, V., Bontcheva, K., Cunningham, H., Handschuh, S.: RoundTrip ontology authoring. In: Sheth, A.P., Staab, S., Dean, M., Paolucci, M., Maynard, D., Finin, T., Thirunarayan, K. (eds.) ISWC 2008. LNCS, vol. 5318, pp. 50–65. Springer, Heidelberg (2008)
5. Noy, N.F., Sintek, M., Decker, S., Crubézy, M., Fergerson, R.W., Musen, M.A.: Creating semantic web contents with protege-2000. IEEE Intell. Syst. **16**, 60–71 (2001)

6. Kalyanpur, A., Parsia, B., Sirin, E., Grau, B.C., Hendler, J.: Swoop: a web ontology editing browser. Web Semant. **4**, 144–153 (2006)
7. Sure, Y., Angele, J., Staab, S.: OntoEdit: guiding ontology development by methodology. In: Meersman, R., Tari, Z. (eds.) CoopIS 2002, DOA 2002, and ODBASE 2002. LNCS, vol. 2519, pp. 1205–1222. Springer, Heidelberg (2002)
8. Spyns, P., Oberle, D., Volz, R., Zheng, J., Jarrar, M., Sure, Y., Studer, R., Meersman, R.: OntoWeb - a semantic web community portal. In: Karagiannis, D., Reimer, U. (eds.) PAKM 2002. LNCS (LNAI), vol. 2569, pp. 189–200. Springer, Heidelberg (2002)
9. Haarslev, V., Lu, Y., Shiri, N.: OntoXpl - exploration of owl ontologies. In: IEEE/WIC/ACM International Conference on Web Intelligence, pp. 624–627 (2004)
10. Catenazzi, N., Sommaruga, L., Mazza, R.: User-friendly ontology editing and visualization tools: the OWLeasyViz approach. In: 2009 13th International Conference Information Visualisation, pp. 283–288. IEEE (2009)
11. Bonomi, A., Mosca, A., Palmonari, M., Vizzari, G.: NavEditOW – a system for navigating, editing and querying ontologies through the web. In: Apolloni, B., Howlett, R.J., Jain, L. (eds.) KES 2007, Part III. LNCS (LNAI), vol. 4694, pp. 686–694. Springer, Heidelberg (2007)
12. Walisadeera, A.I., Ginige, A., Wikramanayake, G.N.: Developing a community-based knowledge system: a case study using Sri Lankan agriculture. Int. J. Adv. ICT Emerg. Reg. (ICTer) **8**(3), 1 (2016)
13. Krötzsch, M., Simancik, F., Horrocks, I.: A description logic primer, pp. 1–17 (2012). arXiv Prepr. arXiv:1201.4089
14. Baader, F.: What's new in description logics. Informatik-Spektrum **34**, 434–442 (2011)
15. Plessers, P.: An approach to web-based ontology evolution. Ph.D. thesis, Department of Computer Science, Vrije Universiteit Brussel, Brussel (2006)
16. OWL Web Ontology Language Overview. http://www.w3.org/TR/owl-features/
17. Abgaz, Y.M., Javed, M., Pahl, C.: Dependency analysis in ontology-driven content-based systems. In: Rutkowski, L., Korytkowski, M., Scherer, R., Tadeusiewicz, R., Zadeh, L.A., Zurada, J.M. (eds.) ICAISC 2012, Part II. LNCS, vol. 7268, pp. 3–12. Springer, Heidelberg (2012)
18. Stojanovic, L.: Methods and Tools for Ontology Evolution (2003). http://citeseerx.ist.psu.edu/viewdoc/download?doi=10.1.1.458.5996&rep=rep1&type=pdf
19. Haase, P., Stojanovic, L.: Consistent evolution of OWL ontologies. In: Gómez-Pérez, A., Euzenat, J. (eds.) ESWC 2005. LNCS, vol. 3532, pp. 182–197. Springer, Heidelberg (2005)
20. Vigo, M., Bail, S., Jay, C., Stevens, R.: Overcoming the pitfalls of ontology authoring: strategies and implications for tool design. Int. J. Hum.-Comput. Stud. **72**, 835–845 (2014)
21. Horrocks, I.: Applications of description logics: state of the art and research challenges. In: Dau, F., Mugnier, M.-L., Stumme, G. (eds.) ICCS 2005. LNCS (LNAI), vol. 3596, pp. 78–90. Springer, Heidelberg (2005)
22. Horridge, M., Bechhofer, S.: The OWL API: a Java API for working with OWL 2 ontologies. Semant. Web **2**, 11–21 (2011)
23. Sirin, E., Parsia, B., Grau, B.C., Kalyanpur, A., Katz, Y.: Pellet: A practical OWL-DL reasoner. Web Semant. Sci. Serv. Agents World Wide Web. **5**, 51–53 (2007)
24. Cardoso, J.: The semantic web vision: where are we? IEEE Intell. Syst. **22**, 84–88 (2007)
25. Funk, A., Tablan, V., Bontcheva, K., Cunningham, H., Davis, B., Handschuh, S.: CLOnE: controlled language for ontology editing. In: Aberer, K., et al. (eds.) ASWC 2007 and ISWC 2007. LNCS, vol. 4825, pp. 142–155. Springer, Heidelberg (2007)

A Software Project Management Problem Solved by Firefly Algorithm

Broderick Crawford[1,2,7], Ricardo Soto[1,3,4], Franklin Johnson[1,5(✉)],
Sanjay Misra[6], and Eduardo Olguín[7]

[1] Pontificia Universidad Católica de Valparaíso, Valparaíso, Chile
{broderick.crawford,ricardo.soto}@pucv.cl
[2] Universidad Central de Chile, Santiago, Chile
[3] Universidad Cientifica del Sur, Lima, Peru
[4] Universidad Autónoma de Chile, Santiago, Chile
[5] Universidad de Playa Ancha, Valparaíso, Chile
franklin.johnson@upla.cl
[6] Covenant University, Ogun, Nigeria
sanjay.misra@covenantuniversity.edu.ng
[7] Facultad de Ingeniería y Tecnología, Universidad San Sebastián,
Bellavista 7, Santiago 8420524, Chile
eduardo.olguin@uss.cl

Abstract. In software project management there are several problems to deal, one of those is the Software Project Scheduling Problem (SPSP). This problem requires to assign a set of resources to tasks for a given project, trying to decrease the duration and cost of the whole project. The workers and their skills are the main resources in the project. In this paper we present the SPSP as a combinatorial optimization problem and a novel approach to solve SPSP by a Firefly algorithm. Firefly algorithm is a new metaheuristic based on the behaviour of the firefly. We present the design of the resolution model to solve the SPSP using an algorithm of fireflies and we illustrate some experimental results in order to demonstrate the viability and soundness of our approach.

Keywords: Firefly algorithm · Metaheuristic · Software project scheduling problem · Project management

1 Introduction

In this paper, we present a novel approach to solve the Software Project Scheduling Problem (SPSP) using a Firefly metaheuristic [11]. SPSP consists in determine a schedule for workers to tasks that trying to decrease the duration and cost for the whole project, so that task precedence and resource constraints are satisfied [1]. This is a NP-hard combinatorial problem, being difficult to solve it by a complete search method in a limited amount of time. The main contribution of the paper is to propose a solution to the problem with a Firefly algorithm and be competitive with other techniques. FA is a probabilistic method, inspired

© Springer International Publishing Switzerland 2016
O. Gervasi et al. (Eds.): ICCSA 2016, Part V, LNCS 9790, pp. 40–49, 2016.
DOI: 10.1007/978-3-319-42092-9_4

from the behaviour of natural firefly, recently developed on population based metaheuristic [11,12]. So far, it has been shown that Firefly algorithm is very efficient in dealing with global optimization problems. For a deeper comprehension of review of firefly advances and applications please refer to [10]. Researches on FA for SPSP have not been seen to date.

We illustrate encouraging experimental results where our approach noticeably competes with other well-known optimization methods reported in the literature.

This paper is organized as follows. In Sect. 2 presents the definition of SPSP, in Sect. 3 presents a description FA. In its subsection presents the model and algorithm to solve the SPSP. In Sect. 4 presents the experimental results, the conclusions are outlined in Sect. 5.

2   The Software Project Scheduling Problem

The software project scheduling problem is one of the most common problems in managing software engineering projects [8]. It consists in finding a worker-task schedule for a software project [2,9]. The most important resources involved in SPSP are; the tasks, which is the job needed for completing the project, the employees who work in the tasks, and finally the skills.

Description of Skills: As mentioned above, the skills are the abilities required for completing the tasks, and the employees have all or some of these abilities. These skills can be for example: design expertise, programming expert, leadership, GUI expert. The set of all skills associated with software project is defined as $S = \{s_1, \ldots, s_{|S|}\}$, where s_i is a specific skill and $|S|$ is the number of skills.

Description of Tasks: The tasks are all necessary activities for accomplishing the software project. These activities are for example, analysis, component design, programming, testing. The software project is a sequence of tasks with different precedence among them. Generally, we can use a graph called task-precedence-graph (TPG) to represent the precedence of these tasks [4]. This is a non-cyclic directed graph denoted as $G(V, E)$. The set of tasks is represented by $V = \{t_1, t_2, \ldots, t_{|T|}\}$. The precedence relation of tasks is represented by a set of edges E. An edge $(t_i, t_j) \in E$, means t_i is a direct predecessor task t_j. Consequently, the set of tasks necessary for the project is defined as $T = \{t_1, \ldots, t_{|T|}\}$, where $|T|$ is the maximum number of tasks. Each task has two attributes: t_j^{sk} is a set of skills for the task j. It is a subset of S and corresponds to all necessary skills to complete a task j, t_j^{eff} is a real number and represents the workload of the task j.

Description of Employees: The employees have to be assigned to a task in order to complete the task. The problem is to create a worker-task schedule, where employees are assigned to suitable tasks. The set of employees is defined as $EMP = \{e_1, \ldots, e_{|E|}\}$, where $|E|$ is the number of employees working on the project. Each employee has tree attributes: e_i^{sk} is a set of skills of employee i. $e_i^{sk} \subseteq S$, e_i^{maxd} is the maximum degree of work. It is the ratio between hours for the project and the workday. $e_i^{maxd} \in [0,1]$, if $e_i^{maxd} = 1$ the employee has total dedication to the project, if the employee has a e_i^{max} less that one, in this case is a part-time job, e_i^{rem} is the monthly remuneration of employee i.

2.1   Model Description

The SPSP solution can be represented as a matrix $M = [E \times T]$. The size $|E| \times |T|$ is the dimension of matrix determined by the number of employees and the number of tasks. The elements of the matrix $m_{ij} \in [0,1]$, correspond to real numbers, which represent the degree of dedication of employee i to task j. If $m_{ij} = 0$, the employee i is not assigned to task j. If $m_{ij} = 1$, the employee i works all day in task j.

The solutions generated in this matrix M are feasible if they meet the following constraints. Firstly, all tasks are assigned at least one employee as is presented in Eq. 1. Secondly, the employees assigned to the task j have all the necessary skills to carry out the task, it is presented in Eq. 2.

$$\sum_{i=1}^{|E|} m_{ij} > 0 \ \forall j \in \{1,\ldots,T\} \tag{1}$$

$$t_j^{sk} \subseteq \bigcup_{i|m_{ij} > 0} e_i^{sk} \ \forall j \in \{1,\ldots,T\} \tag{2}$$

We represent in Fig. 1A an example for the precedence tasks TPG and their necessary skills t^{sk} and effort t^{eff}. For the presented example we have a set of employees $EMP = \{e_1, e_2, e_3\}$, and each one of these have a set of skills, maximum degree of dedication, and remuneration. A solution for problem represented in Fig. 1A and C is depicted in Fig. 1B.

First, it should be evaluated the feasibility of the solution, then using the duration of all tasks and cost of the project, we appraise the quality of the solution. We compute the length time for each task as $t_j^{len}, j \in \{1,\ldots,|T|\}$, for this we use matrix M and t_j^{eff} according the following formula:

$$t_j^{len} = \frac{t_j^{eff}}{\sum_{i=1}^{|E|} m_{ij}} \tag{3}$$

Now we can obtain the initialization time t_j^{init} and the termination time t_j^{term} for task j. To calculate these values, we use the precedence relationships, that is described as TPG $G(V, E)$. We must consider tasks without precedence, in this case the initialization time $t_j^{init} = 0$. To calculate the initialization time of tasks with precedence firstly we must calculate the termination time for all previous tasks. In this case t_j^{init} is defined as $t_j^{init} = max\{t_l^{term} \| (t_l, t_j) \in E\}$, the termination time is $t_j^{term} = t_j^{init} + t_j^{len}$.

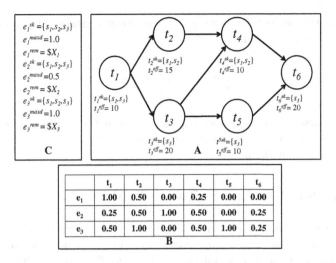

Fig. 1. A: Task precedence graph TPG, B: a possible solution for matrix M, C: employees information

Now we have the initialization time t_j^{init}, the termination time t_j^{term} and the duration t_j^{len} for task j with $j = \{1, \ldots, |T|\}$, that means we can generate a Gantt chart. For calculating the total duration of the project, we use the TPG information. To this end, we just need the termination time of last task. We can calculate it as $p^{len} = max\{t_l^{term} \| \forall l \neq j(t_j, t_l)\}$. For calculating the cost of the whole project, we need firstly to compute each cost associate to task us t_j^{cos} with $j \in \{1, \ldots, |T|\}$, and then the total cost p^{cos} is the sum of costs according to the following formulas:

$$t_j^{cos} = \sum_{i=1}^{|E|} e_i^{rem} m_{ij} t_j^{len} \tag{4}$$

$$p^{cos} = \sum_{j=1}^{|T|} t_j^{cos} \tag{5}$$

The target is to minimize the total duration p^{len} and the total cost p^{cos}. Therefore a fitness function is used, where w^{cos} and w^{len} represent the importance of p^{cos} and p^{len}. Then, the fitness function to minimize is given by $f(x) = (w^{cos}p^{cos} + w^{len}p^{len})$.

An element not considered is the overtime work that may increase the cost and duration associated to a task, consequently increase p^{cos} and p^{len} of the software project. We define the overtime work as e_i^{overw} as all work the employee i less e_i^{maxd} at particular time.

To obtain the project overwork p^{overw}, we must consider all employees. We can use the following formula:

$$p^{overw} = \sum_{i=1}^{|E|} e_i^{overw} \qquad (6)$$

With all variables required, we can determine if the solution is feasible. In this case, it is feasible when the solution can complete all tasks, and there is no overwork, that means the $p^{overw} = 0$.

3   Firefly for Schedule Software Project

Nature-inspired methodologies are among the most powerful algorithms for optimization problems. The Firefly Algorithm (FA) is a novel nature-inspired algorithm originated by the social behaviour of fireflies. By idealizing some of the flashing characteristics of fireflies, a firefly-inspired algorithm was presented in [11, 12]. The pseudo code of the firefly-inspired algorithm was developed using these three idealized rules:

- All fireflies are unisex and are attracted to other fireflies regardless of their sex.
- The degree of the attractiveness of a firefly is proportional to its brightness, and thus for any two flashing fireflies, the one that is less bright will move towards the brighter one. More brightness means less distance between two fireflies. However, if any two flashing fireflies have the same brightness, then they move randomly.
- Finally, the brightness of a firefly is determined by the value of the objective function. For a maximization problem, the brightness of each firefly is proportional to the value of the objective function and vice versa.

As the attractiveness of a firefly is proportional to the light intensity seen by adjacent fireflies, we can now define the variation of attractiveness β with the distance r by:

$$\beta = \beta_0 e^{-\gamma r^2} \qquad (7)$$

where β_0 is the attractiveness at $r = 0$. The distance r_{ij} between two fireflies is determined by:

$$r_{ij} = \sqrt{\sum_{k=1}^{d}(x_k^i - x_k^j)^2} \qquad (8)$$

where x_k^i is the kth component of the spatial coordinate of the ith firefly and d is the number of dimensions. The movement of a firefly i is attracted to another more attractive (brighter) firefly j is determined by:

$$x_i^{t+1} = x_i^t + \beta_0 e^{-\gamma r_{ij}^2}(x_j^t - x_i^t) + \alpha(rand - \frac{1}{2}) \qquad (9)$$

where x_i^t and x_j^t are the current position of the fireflies and x_i^{t+1} is the ith firefly position of the next generation. The second term is due to attraction. The third term introduces randomization, with a being the randomization parameter and *rand* is a random number generated uniformly but distributed between 0 and 1. The value of γ determines the variation of attractiveness, which corresponds to the variation of distance from the communicated firefly. When $\gamma = 0$, there is no variation or the fireflies have constant attractiveness. When $\gamma = 1$, it results in attractiveness being close to zero, which again is equivalent to the complete random search. In general, the value of γ is between $[0, 100]$.

In this implementation each firefly represents a unique solution for the problem. The dimension of each firefly is determined by $|e| * |t| * 3$. The firefly is represented by an array of binary numbers. The FA process it is represented by 7 steps or phases.

Phase 1: Sort Tasks. Each task is ordered according to the TGP which has the projects.

Phase 2: Initialization of Parameters. Input parameters such as the size of the population, the maximum number of cycles, the method of calculating the light intensity (objective function), and the absorption coefficient are received and initialized.

Phase 3: Creating Fireflies. The solution of SPSP is represented by a matrix $M = [E \times T]$, E is the number of employees and T is the number of tasks of the project, each value of the matrix M is a real number between 0 and 1. When $M_{ij} = 0$ the employee i is not assigned to task j, when $M_{ij} = 1$ the employee work all his time in the task. To determine the value of M_{ij} by a firefly we propose to use 3 bits, this allows us to create 8 possible values. To discretize this value, we divide it by seven. That is presented in Fig. 2.

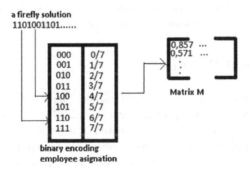

Fig. 2. Representation of a solution by a firefly

Phase 4: Validations. At this stage the fulfilment of all the constraints of the problem is assessed.

Phase 5: Fitness Evaluation. When a firefly reaches a solution the matrix M is generated, then the light intensity is calculated. For SPSP we use the makespan or objective function to minimize.

$$\beta = (W^{cos}p^{cos} + W^{len}p^{len})^{-1} \qquad (10)$$

Phase 6: Update Location. Here the change in the position of fireflies occurs. A firefly produces a change in light intensity based among fireflies position where the lower brightness will approach the more intense. The new position is determined by changing the value using Eq. 9.

Despite working with a binary representation, the result of the new component of the Firefly is a real number. To solve this problem we use a binarization function. This function transform the values between [0,1] as specified in [3] is a function of reaching better solutions faster compared to others. The result of the function with a random number A between 0 and 1. Then compares if A is greater than the random number, is scored 1, otherwise it is assigned 0. Then binarization function $L(xi)$ is as follows.

$$L(x_i) = \frac{e^{2*x_i-1}}{e^{2*x_i+1}} \qquad (11)$$

$$x_i = \begin{cases} 0 & \text{if } L < A \\ 1 & \text{if } L > A \end{cases} \qquad (12)$$

Phase 7: Store Solution. The best solution found so far is stored, and the cycle increases.

Phase 8: Ending. If the maximum number of steps are achieved, the execution is finished and then the best solution is presented. Otherwise go to phase 4.

4   Experimental Results

In this section we present the experimental results. The algorithm was executed 10 trials for each instance and we report the average value from those 10 trials. For the experiments, we use a random instances created by a generator[1]. The instances are labelled as <tasknumber>t<employeesnumber>e<skillsnumber>s. To compare the different results we use the feasible solution in 10 runs, *cost*: average cost of feasible solutions in 10 runs, *duration*: average duration of feasible solutions in 10 runs, to compute the *fitness*: average fitness of feasible solutions in 10 runs.

---

[1] http://tracer.lcc.uma.es/problems/psp/generator.html.

4.1 Comparative Results with Other Techniques

Some results are presented in [9] by Xiao et al., using the similar parameter to our instances. Xiao presents results using Ant Colony System (ACS) and Genetic Algorithms (GA). For the sake of clarity we transform the fitness presented by the author as fitness^{-1} to obtain the same fitness used by us. The comparative results are presented in Table 1.

Table 1. Comparison with other techniques.

Instance	Algorithms	Fitness
$5e - 10t - 5s$	ACS	3.57175
	GA	3.64618
	FA	**3.40654**
$10e - 10t - 5s$	ACS	2.63123
	GA	2.74442
	FA	**2.61583**
$10e - 10t - 10s$	ACS	2.63565
	GA	2.66467
	FA	**2.32065**
$10e - 20t - 5s$	ACS	6.39424
	GA	6.32392
	FA	**6.31515**
$5e - 10t - 10s$	ACS	3.54955
	GA	**3.54255**
	FA	4.24161
$15e - 20t - 5s$	FA	**4.48418**
$10e - 30t - 5s$	FA	**9.63079**
$10e - 30t - 10s$	FA	**8.39779**

From Table 1 we can compare the fitness of the solution. Unfortunately the times of GA and ACS were not informed by the authors. We only compare the fitness values. In this case for the instances with task = 10 always have a better solution compared with ACS and GA. But in the instances with task = 20 and employees = 10 our proposal it is competitive too, only in instance with employee = 5, task = 10 and Skills = 10 the genetic algorithm get a better solution.

Regarding the fitness we can see that Firefly has better results for all instances with task = 10 and task = 20. For instances with task = 30, we do not have a comparative result, because the other author dont shows results for these instances. If we analyse the results based on project cost, we can see that our proposal provides the best results for all instances compared.

5 Conclusion

The paper presents an overview to the resolution of the SPSP using a Firefly algorithm. In the paper it shows the design of a representation of the problem in order to Firefly can solve it, proposing pertinent heuristic information. Furthermore, it is defined a fitness function able to allow optimization of the generated solutions.

The implementation our proposed algorithm was presented, and a series of tests to analyse the convergence to obtain better solutions was conducted. The tests were performed using different numbers of tasks, employees, and skills. The results were compared with other techniques such as Ant Colony System and Genetic Algorithms. The analysis demonstrates that our proposal gives the best results for smaller instances. For more complex instances was more difficult to find solutions, but our solutions always obtained a low cost of the project, in spite of increasing the duration of the whole project.

An interesting research direction to pursue as future work is about the integration of autonomous search in the solving process, which in many cases has demonstrated excellent results [5–7].

Acknowledgments. Broderick Crawford is supported by Grant CONICYT/ FONDECYT/REGULAR/1140897, Ricardo Soto is supported by Grant CONICYT/FONDECYT/REGULAR/1160455, Franklin Johnson is supported by Postgraduate Grant PUCV 2015.

References

1. Alba, E., Chicano, F.: Software project management with gas. Inf. Sci. **177**(11), 2380–2401 (2007)
2. Barreto, A., Barros, M.D.O., Werner, C.M.L.: Staffing a software project: a constraint satisfaction and optimization-based approach. Comput. Oper. Res. **35**(10), 3073–3089 (2008)
3. Chandrasekaran, K., Simon, S.P., Padhy, N.P.: Binary real coded Firefly algorithm for solving unit commitment problem. Inf. Sci. **249**, 67–84 (2013)
4. Chang, C.K., Jiang, H.Y., Di, Y., Zhu, D., Ge, Y.: Time-line based model for software project scheduling with genetic algorithms. Inf. Softw. Technol. **50**(11), 1142–1154 (2008)
5. Crawford, B., Soto, R., Castro, C., Monfroy, E.: Extensible CP-based autonomous search. In: Stephanidis, C. (ed.) Posters, Part I, HCII 2011. CCIS, vol. 173, pp. 561–565. Springer, Heidelberg (2011)
6. Crawford, B., Soto, R., Monfroy, E., Palma, W., Castro, C., Paredes, F.: Parameter tuning of a choice-function based hyperheuristic using particle swarm optimization. Expert Syst. Appl. **40**(5), 1690–1695 (2013)
7. Monfroy, E., Castro, C., Crawford, B., Soto, R., Paredes, F., Figueroa, C.: A reactive and hybrid constraint solver. J. Exp. Theoret. Artif. Intell. **25**(1), 1–22 (2013)
8. Ozdamar, L., Ulusoy, G.: A survey on the resource-constrained project scheduling problem. IIE Trans. **27**(5), 574–586 (1995)

9. Xiao, J., Ao, X.T., Tang, Y.: Solving software project scheduling problems with Ant Colony optimization. Comput. Oper. Res. **40**(1), 33–46 (2013)
10. Yang, X., He, X.: Firefly algorithm: recent advances and applications. CoRR, abs/1308.3898 (2013)
11. Yang, X.-S.: Firefly algorithms for multimodal optimization. In: Watanabe, O., Zeugmann, T. (eds.) SAGA 2009. LNCS, vol. 5792, pp. 169–178. Springer, Heidelberg (2009)
12. Yang, X.S.: Nature-Inspired Optimization Algorithms, 1st edn. Elsevier Science Publishers B. V., Amsterdam (2014)

A Baseline Domain Specific Language Proposal for Model-Driven Web Engineering Code Generation

Zuriel Morales[1]([✉]), Cristina Magaña[1], José Alfonso Aguilar[1],
Aníbal Zaldívar-Colado[1], Carolina Tripp-Barba[1], Sanjay Misra[2],
Omar Garcia[1], and Eduardo Zurita[1]

[1] Señales y Sistemas (SESIS) Facultad de Informática Mazatlán,
Universidad Autónoma de Sinaloa, Mazatlán, Mexico
{zmorales,cmagana,ja.aguilar,azaldivar,ctripp,ogarcia,ezurita}@uas.edu.mx
[2] Department of Computer and Information Sciences,
Covenant University, Ota, Nigeria
ssopam@gmail.com
http://sesis.maz.uasnet.mx

Abstract. It is well-known that Model-Driven Web Engineering requires the development of code-generation tools in order to be adopted outside research field as a complete solution in Web application development industry. Regrettably, a fully-guided methodology supported by a complete code-generation tool that considers a complete development process based on MDA (Model-Driven Architecture) is missing. The idea behind MDA is that requirements are considered (functional and non-functional requirements) from the Computational Independent Model (CIM), to the Platform Specific Model (PSM) passing for the Platform Independent Model (PIM) to generate the source code for the Web application. In our work is presented a baseline DSL (Domain Specific Language) for Web application code-generation considering the basic language used in a small software factory in Mexico. This is an ongoing work which is part of a institutional project in order to build a suite of tools for code-generation for Web application development.

Keywords: MDWE · Metamodel · DSL's

1 Introduction

A model is an abstract representation of a system and a meta-model is an abstract description of a model. The abstraction helps to neglect the less important aspects of a system, while concentrating on favorable parts that are desired to a specific study. Models are used to represent system functionality, Model-Driven Web Engineering (MDWE) has become a success for Web application development because with the use of models it is possible to represent (modeling) the user needs (goals) without neglecting the organizational objectives,

© Springer International Publishing Switzerland 2016
O. Gervasi et al. (Eds.): ICCSA 2016, Part V, LNCS 9790, pp. 50–59, 2016.
DOI: 10.1007/978-3-319-42092-9_5

the software architecture and the business process and from this representation generate the Web application source code. In the last 20 years, several MDWE methods [1] have been emerged for the development of Web applications using models to do it, but only some of them strictly complied with the proposal of the Object Management Group (OMG) for Model-Driven Development named Model-Driven Architecture (MDA) [2]. The basic idea of the use of MDA starts from the Computational Independent Model (CIM), in this first level, the application requirements must be elicited and defined, such that we can generate through model-to-model transformations (M2M) the Platform Independent Model (PIM) to finish in the Platform Specific Model (PSM) with the source code due to model-to-text (M2T) transformation rules. Regrettably, most of the MDWE methods does not provide a complete support tool and only implements MDA from PIM to PSM leaving aside the requirements phase (CIM level) despite this is a critical phase on which the success of the development depends directly [3]. This fact can be seen in their support tools i.e., code generation tools and modeling tools. Although most of them define their own notation for building models such as the navigation, the presentation or the personalization model, we argue that in many cases it is just another notation for the same concepts, i.e. they should be based on a common metamodel for the Web application domain. In addition, tool-supported design and generation is becoming essential in the development process of Web applications due to the increasing size and complexity of such applications, and CASE-tools should be built on a precisely specified metamodel of the modeling constructs used in the design activities, providing more flexibility if modeling requirements change.

Bearing these considerations in mind, this paper presents a first step towards such a baseline DSL (Domain Specific Language), also named metamodel. A metamodel is a model of a model, and metamodeling is the process of generating such metamodels. Metamodeling or meta-modeling is the analysis, construction and development of the frames, rules, constraints, models and theories applicable and useful for modeling a predefined class of problems. As its name implies, this concept applies the notions of meta- and modeling in software engineering and systems engineering. Metamodels are of many types and have diverse applications. In this context, we define a metamodel for code generation based on HTML5 and PHP languages, the state of the implementation of this two languages mixed in a DSL is not present in the most well-known MDWE methods according to [1], i.e., NDT [4], UWE [5] and WebML [6]. Moreover, the tools used or developed for support code generation on each method is analyzed as a previous work in our proposal. It is important to mention that, this is a complex process to decide which techniques will be used in a project, determining, learning, acquiring, discovering the appropriate techniques, so this may negatively influence the development process which results in system failures. Importantly, we do not pretend to establish a full code-generation tool since Web application development is on continuos change.

The rest of the paper is organized as follows: Sect. 2 presents some related work regarding to DSL's and metamodels in Model-Driven Web engineering.

Section 3 describes our baseline DSL for code-generation proposal. Finally, the conclusion and future work is presented in Sect. 4.

2 Background

The advantages of Model-Driven Development (MDD) are well-known, specifically in Model-Driven Web Engineering [7], cost and effort reduced, achieving profits in the project budget, among others. Thus, it is obvious that Web engineering methods are now focus on adapting its approaches in to MDD paradigm, but this transition, until now, has not been easy. This is specially true with regard to tool support for code-generation, in particular for the user interface components by code-generation, controls for the Rich Internet Applications are more complex to build that common controls used in Web 1.0 development. One solution proposed to facilitate this work f are the so called metamodels, also named Domain Spacific Languages (DSL's) in the context of MDD. Next, is described the background regarding the development of DSL's for code generation and user interface modelling.

In [8], the authors proposes a meta-model for abstract web applications that can be mapped to multiple platforms. We extend a UML-based model to support specific features of the Web and Web 2.0 as well as to establish a bridge to functional and usability requirements through use cases and user interface (UI) prototypes. The meta-model also helps avoid a common MDD-related problem caused by name-based dependencies. Finally, mappings to a number of specific web platforms are presented in order to validate the appropriateness of the meta-model as an abstract web model. This is focused in requirements-usability leaving aside code generation. In [9], the authors presents a metamodel for requirements specification, this metamodel is defined in order to be created within NDT methodology. The authors are no using the DSL definition for a User Interface controls in order to be mapped to code. On the other hand, in [10], the work presented by the authors is based on the usability patters applied in Model-Driven Web Engineering, they focuses on the properties for being configured for usability support.

There have been many attempts for adapting Rich Internet Applications controls in Model-Driven Web Engineering for user interface development, but most of the approaches are not supported by a methodology covering Model-Driven Architecture and its three levels (CIM, PIM and PSM) to code-generation, the support tool they offers generates JAVA Web applications such Webratio [11], others look for its integration with another proposals such as [12] and NDT [13], that's the reason why we believe that our first attempt to create a easy to use tool for code generation needs to be carried on based on a basic DSL for easy to use and to learn for software developers not familiarized with code-generation and modeling tools. According to [14], there is a big gap in the learning curve of current software developers regarding to modeling tools for software modeling.

3   The Baseline DSL for Code Generation

This section describes the development of a metamodel for code generation in the context of Model-Driven Web Engineering. The classes defined for each purpouse are described in detail (see Fig. 1).

The Components. The metamodel has six main meta-classes, these are:

1. Website. This is the main class, this metamodel element contains all the web site. It has one attribute named -name- for the web site name.
2. Page. The -Page- class is one of the most important ones because it generates the HTML 5 code for a web page. Its attributes are: -title-, for the page title and -accessibility level- for accesibility according to WCAG 2.0 (as a future work this function will be enabled).
3. DisplayTable. This class is defined for in order to extract data from the database. The attributes are: -title-, for the table title, -table- for the data binding and -edit- and -delete- by attributes.
4. Form. The -Form- class is used to create a HTML form. The attributess defined are: the form -name-, the name of the button, the send -method- (POST or GET) and the table -name- on which the data from the form will be stored or obtained.
5. Database. This class is defined in order to specify the parameters for database connection such as the server name, user and password and the database name. This class allows to select among different connectors.
6. Input. The -Input- class is the main class used to select among different HTML 5 controls. It defines five attributtes used in HTML 5 controls, these are: -id- for the control identifier, -name-, for the control name, -label, for the control label, -attribute- for the control attribute such as -String- and -field type- used to define the data type for the database binding.

Also, seven meta-classes are defined in order to specify HTML 5 controls and attributes for HTML tables, these are: -Text-, Password-, -Radio-, -Combobox-, -Checkbox-, -Option- and -Attributte-. These are basic HTML 5 controls, thats the reason why the explanation of each one is out of the scope of this work, see the reference about HTML 5 on World Wide Web Consortium: https://dev. w3.org/html5/html-author/ for further information. The technology used for the definition of our baseline metamodel is the Eclipse Modeling Project (https:// eclipse.org/modeling/emf/), by means of the implementation of the MOF architecture as a ECORE in Eclipse, we defined an ecore file for the meta-classes definition.

Finally, it is important to mention that this section describes our baseline metamodel for code generation based on HTML 5 and PHP languages, in this sense, the main idea behind this work is to produce a complete model-driven tool for semi-automatic web application generation for a local software factory under open source initiative.

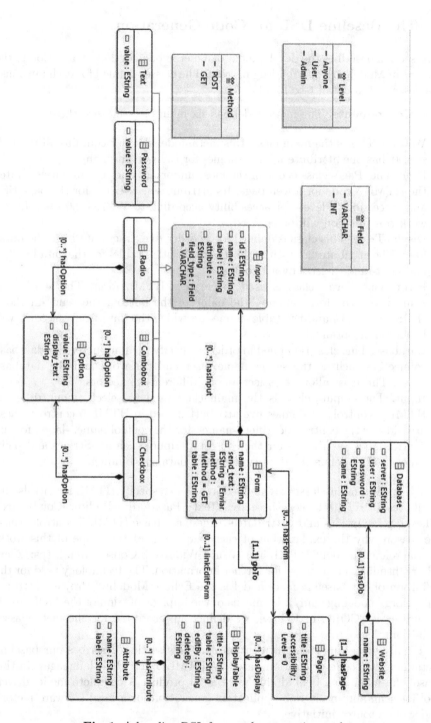

Fig. 1. A baseline DSL for a code-generation tool

Application Example. This subsection describes how the baseline metamodel proposed works. Eclipse Modeling Framework (EMF) describes data models and enables easy code generation from different types of data model artifacts, such as XML Schema, the Rational Rose model, the Ecore model, or Java annotations. In the process of code generation, the EMF generator creates model code, which includes type-safe interfaces and implementation classes for the data model. However, in certain situations, these type-safe interfaces and implementation classes are not required by the application. Instead, data objects are required that could be shared among or processed further by application components. In such situations, Dynamic EMF comes in handy because it allows application developers to manufacture an in-memory core model at run time programmatically, make instances of it dynamically, and access the model instance elements using the EMF reflective API.

Eclipse Modeling Framework provides a metamodel tester named *Dynamic Instance*, this is used to create a model which is conforms to the metamodel. This is the most simple way to create an instance of the metamodel. We need only right click on the root class/model object of our metamodel (in our example, it is the Website class), actually all the classes are only available, for creation, through this root class (the containment property). See Fig. 2.

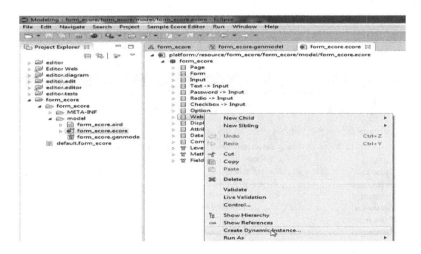

Fig. 2. Create dynamic instance from an ecore file, the main class named *Website*, this is our model.

The next step consist in adding the elements corresponding to the HTML components, the *Dynamic Instance* allows to create a model viewed as tree editor, this model we created represents the final model, see Fig. 3. When we finish the model, we will get a XMI file which we can convert into source code by means of the Acceleo code generation templates (Fig. 4).

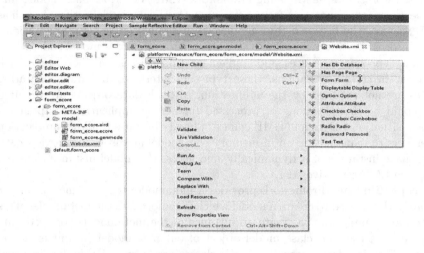

Fig. 3. The HTML elements we can add to the main class (root class) *Website*.

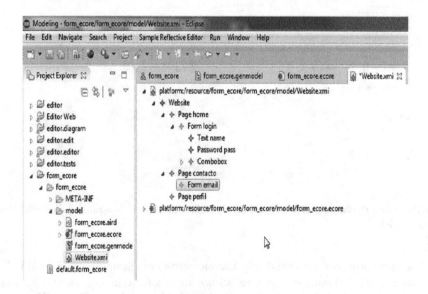

Fig. 4. The final model created from our metamodel.

In Fig. 5 we can see an extract of the Acceleo template for code generation of the model created from our baseline metamodel for Model-Driven Web Engineering code generation.

Fig. 5. The Acceleo template defined for the code generation.

The final source code generated from our baseline metamodel is presented in Fig. 6, the source code is in PHP language, the operation is for delete an elemenent from the database.

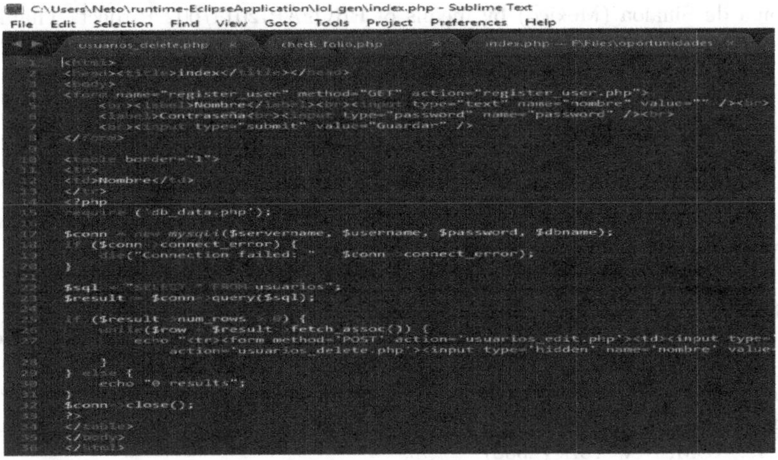

Fig. 6. The PHP code for the delete operation.

4   Conclusions and Future Work

A conclusion that comes out from our DSL is that the use of common models and languages for component representation for code-generation tools is recommended due to this fact will help to the use of the same terminology for user interface modelling, because, a topic we found is the different terminology used by current Model-Driven Web Engineering methods to name their types of components for user interface modeling and code-generation. Therefore, if standard concepts are promoted in Web modelling we can establish a standard form for modeling user interface components and controls. In this respect, although studies have been conducted regarding the benefits of MDD in a development process [7], few refer to the Web engineering domain, among which are the works presented in [10,16], thats the reason why it is necessary to conduct more studies with regard to MDD in WE through empirical studies in order to validate and support the potential application of these methods.

The use of current Web interface techniques for modelling the behavior of the Web application is not common among the current methods for Model-Driven Web Engineering. This leads to possible research lines in which mechanisms to represent the behavior are studied to find a form to standarize its representation, for instance with the new language IFML (Interaction Flow Modeling Language). Furthermore, the notation for building such models should be intuitive and easy to use, and it should have a serialization mechanism that allows integrating the models with other methods.

Acknowledgments. This work has been partially supported by: Universidad Autónoma de Sinaloa (México) by means of PROFAPI2014/002 Project. Thanks to Señales y Sistemas research group for its collaboration. Thanks to the two main authors of this paper for being a exceptional students, you have your thesis!.

References

1. Aguilar, J.A., Garrigós, I., Mazón, J.N., Trujillo, J.: Web engineering approaches for requirement analysis- a systematic literature review. In: Web Information Systems and Technologies (WEBIST), Valencia, Spain, vol. 2, pp. 187–190. SciTePress Digital Library (2010)
2. Brown, A.: Model driven architecture: principles and practice. Soft. Syst. Model. **3**(4), 314–327 (2004)
3. Nuseibeh, B., Easterbrook, S.: Requirements engineering: a roadmap. In: Proceedings of the Conference on the Future of Software Engineering, ICSE 2000, pp. 35–46. ACM, New York (2000)
4. García-García, J.A., Escalona, M.J., Ravel, E., Rossi, G., Urbieta, M.: NDT-merge: a future tool for conciliating software requirements in MDE environments. In: Proceedings of the 14th International Conference on Information Integration and Web-Based Applications and Services, IIWAS 2012, pp. 177–186. ACM, New York (2012)

5. Koch, N., Kraus, A., Hennicker, R.: The authoring process of the UML-based web engineering approach. In: First International Workshop on Web-Oriented Software Technology (2001)
6. Brambilla, M., Fraternali, P.: Large-scale model-driven engineering of web user interaction: the WebML and WebRatio experience. Sci. Comput. Program. **89**, 71–87 (2014)
7. MartíNez, Y., Cachero, C., Meliá, S.: MDD vs. traditional software development: a practitioner's subjective perspective. Inf. Softw. Technol. **55**(2), 189–200 (2013)
8. Fatolahi, A., Some, S.S., Lethbridge, T.C.: A meta-model for model-driven web development. Int. J. Soft. Inform. **6**, 125–162 (2012)
9. Escalona, M.J., Koch, N.: Metamodeling the requirements of web systems. In: Filipe, J., Cordeiro, J., Pedrosa, V. (eds.) Web Information Systems and Technologies. LNCS, vol. 1, pp. 267–280. Springer, Heidelberg (2007)
10. Insfran, E., Fernandez, A.: A systematic review of usability evaluation in web development. In: Hartmann, S., Zhou, X., Kirchberg, M. (eds.) WISE 2008. LNCS, vol. 5176, pp. 81–91. Springer, Heidelberg (2008)
11. Brambilla, M., Butti, S., Fraternali, P.: WebRatio BPM: a tool for designing and deploying business processes on the web. In: Benatallah, B., Casati, F., Kappel, G., Rossi, G. (eds.) ICWE 2010. LNCS, vol. 6189, pp. 415–429. Springer, Heidelberg (2010)
12. Linaje, M., Preciado, J.C., Morales-Chaparro, R., Rodríguez-Echeverría, R., Sánchez-Figueroa, F.: Automatic generation of RIAs using RUX-tool and Webratio. In: Gaedke, M., Grossniklaus, M., Díaz, O. (eds.) ICWE 2009. LNCS, vol. 5648, pp. 501–504. Springer, Heidelberg (2009)
13. Robles Luna, E., Escalona, M.J., Rossi, G.: Modelling the requirements of rich internet applications in WebRe. In: Cordeiro, J., Virvou, M., Shishkov, B. (eds.) ICSOFT 2010. CCIS, vol. 170, pp. 27–41. Springer, Heidelberg (2013)
14. Ceri, S., Brambilla, M., Fraternali, P.: The history of WebML lessons learned from 10 years of model-driven development of web applications. In: Borgida, A.T., Chaudhri, V.K., Giorgini, P., Yu, E.S. (eds.) Conceptual Modeling: Foundations and Applications. LNCS, vol. 5600, pp. 273–292. Springer, Heidelberg (2009)
15. Mishra, D., Mishra, A., Yazici, A.: Successful requirement elicitation by combining requirement engineering techniques. In: First International Conference on the Applications of Digital Information and Web Technologies, ICADIWT 2008, pp. 258–263, August 2008
16. Valderas, P., Pelechano, V.: A survey of requirements specification in model-driven development of web applications. ACM Trans. Web **5**(2), 1–51 (2011)

ArtistRank – Analysis and Comparison of Artists Through the Characterization Data from Different Sources

Felipe Lopes de Melo Faria[1]([✉]), Débora M.B. Paiva[2],
and Álvaro R. Pereira Jr.[1]

[1] Departamento de Computação, Universidade Federal de Ouro Preto,
Ouro Preto, Brazil
felipelmfaria@hotmail.com, alvaro@iceb.ufop.br
[2] Faculdade de Computação, Universidade Federal de Mato Grosso do Sul,
Campo Grande, Brazil
dmbpaiva@gmail.com

Abstract. Understanding how artists, musical styles, and the music itself evolve over time is important when analysing the process of making music. In addition, it can help to study critical music artists that are in the top rankings and why. With the emergence of new ways in which communities are exposed to music, we see the need to reassess how the popularity of an artist is measured. It is noticed that traditional popularity rankings that model artists based only on disk sales and runs on the radio are not sufficient. The presence of numerous web services that allow user interaction in the world of music, whether listening online or watching the life of an artist more closely in digital media, has changed the dynamics of the music industry. Thus, this work proposes a methodology for music artists comparison based on how they perform on different music digital media available on the web. The methodology is also prepared to allow input data from mass media, so that a study case is presented using the TV media in order to properly assess the popularity of artists. Our case study shows that relevant rankings of music artists can be revealed by employing the proposed methodology.

Keywords: Data mining · Rankings · Digital media

1 Introduction

The construction of rankings consists in sorting retrieved results according to a certain criteria [15]. In the music industry, the Billboard magazine is the primary means of popularity rankings measurement of artists in the United States. Their rankings, in part, are based on sales data and executions in radio, and are issued weekly [10].

The volume of music sales is losing its role as an indicator of popularity for artists. Although the sales numbers and the audience may have been widely used to measure the popularity of artists in the 50s and 60s, these variables

© Springer International Publishing Switzerland 2016
O. Gervasi et al. (Eds.): ICCSA 2016, Part V, LNCS 9790, pp. 60–76, 2016.
DOI: 10.1007/978-3-319-42092-9_6

became ineffective as popularity indicators due to the rapid growth of music publication in the digital media [10]. With the emergence of new means through which communities are exposed to music, we see the need to reassess how the popularity of an artist is measured.

With this demand, Next Big Sound[1] has emerged. Considered a revolution in the music industry [6], it is able to analyze the consumption of artists' songs across a wide range of sale channels and digital media, including Spotify[2], Pandora[3], Last.fm[4], Facebook[5], Twitter[6] and Vevo[7]. The company claims it can predict album sales with 20 % accuracy for 85 % of artists, considering the artist's growth in Facebook and other measures. The Next Big Sound, in partnership with Billboard, provides its data to the magazine to create the Social 50 ranking, which shows the 50 most popular artists weekly on Facebook, Pandora, Twitter, Last.fm, Youtube[8] and MySpace[9] [8,13].

Finally, another factor contributing to the popularity of an artist is the analysis of events which are related to the music industry, for example, the release of a song in a television broadcast (TV) and its use on novels and TV series. Being played on a novel opens space in the TV programming, and consequently on radio [7].

A research by Next Big Sound in 2013[10] shows that 91 % of the music bands that exist in the world have not yet been discovered by the public. This is mainly due to the fact that popular digital media focus on playing music from artists who are at the top of the rankings. Approximately, 1 % of the top artists hold 87.3 % of all likes on Facebook and views on Youtube and VEVO.

We note the importance of the digital and popular media to analyze the popularity of artists in the music world. Even though researches on digital media are on wide growth, there is still little material related to the production and distribution of music content in these media. There is also little research related to practices and dissemination strategies used by producers and fans in such environments. In addition, the measurement of the popularity of an artist through the features offered by digital media such as likes, dislikes and comments remains yet unexplored in the context of Information Retrieval [3]. The use of data from various sources in the generation of rankings can generate results, that are different from the current standard.

Although there are rankings in the music industry that make use of digital media, such the Social 50 ranking, it is observed that there is no clear and transparent methodology on how the positions of the rankings are calculated.

---

[1] https://www.nextbigsound.com.
[2] http://www.spotify.com/br.
[3] http://www.pandora.com.
[4] http://www.lastfm.com.br.
[5] http://www.facebook.com.
[6] http://www.twitter.com.
[7] http://www.youtube.com/user/VEVO.
[8] http://www.youtube.com.
[9] http://www.myspace.com.
[10] https://www.nextbigsound.com/industryreport/2013/.

Moreover, it is observed that there is always a tendency that those rankings only consider artists who are emerging, limiting the analysis of the music industry and ignoring, for example, the history of music over the years and how it may influence the music industry today.

The objective of this work is to develop an artists' ranking building methodology from the analysis and data mining often collected from digital media (YouTube, Last.fm, Letras[11], Twitter, Facebook, Vagalume[12], CifraClub[13] and Rdio[14]) and popular TV media, in order to evaluate the popularity of an artist in question. The objective is to analyze the types of rankings that can be developed by studying characteristics and behaviors from the artists in the media. Since different artists may share similar behaviors or have very different profiles in different media, the development of different rankings containing artists that can be comparable is necessary. The proposed methodology generated the implementation of algorithms that produce artists rankings from data of different media.

The following sections of this work are divided as follows. Section 2 lists related work, including important concepts used in this work, as well as articles that relate to the theme ranking and popularity analysis of artists. In Sect. 3, the proposed methodology is presented for the construction of aggregate rankings. In Sect. 4, rankings developed in this work are presented and compared with a ranking built by companie of the phonographic business as a case study. Finally, in Sect. 5, the conclusions from this work are made and future works are presented.

2   Related Work

This section presents works in the scope of construction of artist rankings, using digital media, and also works using the relationship between digital and popular media to build rankings.

Music researchers and sociologists have long been interested in the patterns of music consumption and its relation to socioeconomic status through features in digital media. Thinking about it, the authors Park et al. [14] taking user data from Last.fm and Twitter, designed and evaluated a measure that captures the diversity of musical preferences. The authors used this measure to explore associations between musical diversity and variables such as the level of interest in music. According to the authors, this measure can provide a useful means to study musical preferences and consumption. To that extent, they evaluated that the artists who were in the Top 100 of the Last.fm ranking, for example, in 100 % of the cases, were also present in the ranking of Allmusic[15], used as an artists database for the study.

---

[11] https://www.letras.mus.br.
[12] http://www.vagalume.com.br.
[13] http://www.cifraclub.com.br.
[14] https://www.rdio.com.
[15] https://www.allmusic.com.

Grace [5] proposed an approach that uses text data mining techniques to measure popularity and construct rankings of artists from the analysis of comments from listeners on the social network MySpace, a popular musical media. They demonstrated to have obtained results closer to those that would be generated manually by users (university students) than those generated from the methods used by Billboard magazine.

The authors Bryan and Wang [2] built a ranking of artists based in WhoSampled[16], an information website about the music industry. For this, they used several characteristics of artists and musical genres and adopted the metric PageRank to enable the interpretation and description of patterns of musical influences, trends and music features.

The Next Big Sound, as already explained about its features, built the ranking "Social 50", which takes into account the interaction of users with artists in digital media. MTV also provides rankings that take into account this analysis, in addition to CD sales and radio audience [9].

Influential music companies, such as Sony, are hiring services like those provided by the Next Big Sound. Although the amount of time people spend listening to music has increased because of the number of channels available, many songs can get to the top of the Billboard charts with a modest number of sales. This is because a quarter of people do not pay for the music they listen by using free music services. Thus, these music companies are starting to change to "listen" listeners in order to understand how people want to consume music[17].

Finally, Moss et al. [12] presents a music recommendation service that uses the convergence of media. There are two interesting rankings to be cited. The first takes into account the popularity of the artist in relation to others and the second considers the songs of an artist in relation to the others. Both rankings consider: radio data, be it traditional or web; the ratio of a video of a particular artist that is present in digital media such as Youtube and television broadcasts such as MTV; and information manually provided by experts to categorize artists.

This work also deals with the analysis of data from digital media and TV to build artists rankings. It differs from other studies because it aims to provide a systematic approach for the development of artists rankings from mass and digital media data by providing features to measure the correlation between media, which includes to calculate weights and define metrics. Among the data from the media, we can mention the following: the career age (perennial artists or emerging, for example); categorization of artists by genres (most popular, adult and young people oriented genres, for example); among other features. Therefore, the methodology includes different scenarios with regard to the career characteristics of an artist, once what is observed from the literature are methodologies that do not approach closely the process of the construction of rankings, favoring only certain specific media and do not analyzing all data together or privileging only artists who are at the top for that media (those artists best evaluated in the digital media).

---

[16] http://www.whosampled.com.

[17] http://www.emeraldinsight.com/doi/abs/10.1108/SD-04-2013-0008.

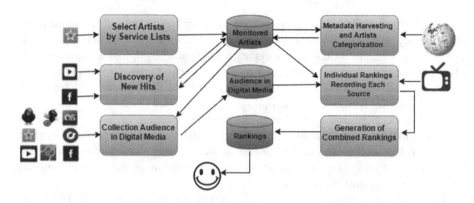

Fig. 1. Construction methodology for an artist ranking

3   Methodology for Artist Ranking

This section presents the construction methodology for an artist ranking. The construction of these rankings is performed using digital media data, such as likes and views, taking into account the profile of the artists in the digital and television media, and content information such as genre, career age and influence in television media. The different types of proposed rankings are presented in Sect. 3.1. Note that the methodology is scalable. The methodology allows to use other digital media different from those used in this work, radio data, a larger or smaller number of artists and different characteristics (career age and musical genres). Figure 1 shows the construction methodology for an artist ranking, which is divided into several steps.

Step 1 in Fig. 1, "Select Artists by Service Lists", has the function of choosing the artists who make up the list of monitored artists and will be used for the construction of the aggregate rankings. This choice occurs by manually choosing artists, taking into account the profile of the artists in this ranking in the monitored digital media. At the end, there is a list of artists to be monitored (1000 in the experiment).

Step 2 in Fig. 1, "Metadata Harvesting and Artists Categorization", has the function of collecting profile information from artists and then categorize them by genre. This collection takes place through the artist profile data obtained in the artist's page on Wikipedia[18]. Then a list containing information of artists profiles is generated. Later, the artists are grouped into categories according to genre. Thus, a list containing information of the categories of the artists is generated [11].

Step 3 in Fig. 1, "Discovery of New Hits", has the function to monitor the list of artists. This monitoring is performed with data from the official pages of the artists on Facebook and Youtube. When keywords related to the music industry are found in the pages of digital media, the data become available to have its

---

[18] http://wikipedia.org.

content audited, and possibly approved as relevant data for the research. With this step, a list of updated information on the artist's career is generated.

Step 4 in Fig. 1, "Collection of Audience in Digital Media", has the function of collecting artists data from digital media, such as likes and views, through the use of APIs or HTML parsers in digital media. Thus, a list containing information of the audience data of the artists in the digital media is generated.

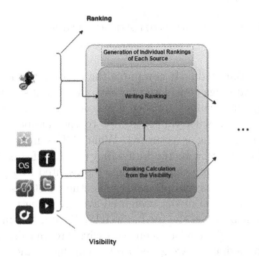

Fig. 2. Construction methodology of individual rankings.

Step 5 in Fig. 1, "Generation of Individual Rankings of Each Source", has the function to generate the individual rankings for each digital media. For better understanding, Fig. 2 shows the generation process of the individual rankings for each source. Audience data, from "Database Audience in Digital Media", are passed to the "Individual Rankings Generation of Each Source", which builds the ranking through the "rank calculation from the Visibility metric" (metric whose calculation is done with the sum of the features of digital media, such as number of likes, views and followers) [1]. It is important to note that the Vagalume data is passed directly for writing on the database because the data used from the base is the own standard ranking.

Finally, step 6 in Fig. 1 "Generation of Combined 'Rankings'", has the function of aggregating the data from the previous step on the digital media. This step receives the rankings of artists generated in the previous step for each digital media and aggregates the data through statistical calculations (using the coefficient of Correlation and Spearman's Rankings [4]). Thus, aggregated rankings of artists are generated.

The algorithm that computed the aggregation is presented in Algorithm 1. RM is an array containing the rankings of each artist in each of the eight digital media (obtained on step 5 above) and $a \in A$ is the set of artists. The algorithm takes as input an RM array that contains the set of artists A with their respective

Data: RM, A
Result: *rank*
initialization;
int *rank* = []
while $a \in A$ **do**
| $rank(a) = p(a)$
end
$rank = Sort(rank(a))$

Algorithm 1. ALGORITHM AGGREGATION RANKINGS

---

positions in each digital media. It computes the average of the positions in the eight digital media rankings considered for each artist and stores it in the array *rank*. Subsequently, these values are sorted in ascending order, and the averages of standardized positions with values from 1 to the number of artists on the base. The aggregate ranking is returned. What differs in each is the average of the positions of each artist in the individual rankings for each digital media (found in step 5) to aggregate these rankings represented by $p(a)$. Averages are made in four different ways: simple average of the positions of artists in the individual rankings of digital media; weighted average using the Spearman rank correlation coefficient between the media; weighted average using the Spearman Ranking Correlation Coefficient between successive days in the rankings collected day history in each media; the weighted average of digital media in the ranking of the Alexa website[19]. We used the Spearman's correlation rank coefficient between the media because it showed better correlation results.

3.1 Developed Rankings

The different rankings generated by this method are described below. They are built with data from digital media (amount of data for each artist on the bases), data concerning the collected metadata and categorization of artists. These data allowed to generate rankings that consider artists divided into several career ages, the ending of the same, or through the TV influence. There is intersection in career age of artists, since, for example, an artist can still be consolidating his/hers career, and may be classified as perennial artist, but at the same time stay contemporary (modern).

Perennial Ranking. The Perennial Ranking is formed by artists who are in evidence in the music industry for over 20 years. The calculation of the position for each artist in each digital media in perennial Ranking was computed by the accumulated data (amount of data) thus represents the total of the entire history of the service; for example, the total number of views on YouTube, or the total number of accesses of lyrics in the Letras. Thus, the artist who has the highest value of the accumulated data, shows up in the top rankings. The other rankings also use the accumulated data.

---

[19] http://www.alexa.com.

Modernity Ranking. The Modernity Ranking is formed by contemporary artists in an attempt to recover artists who are part of the same generation. So it was considered artists that emerged between 7 and 20 years.

Emerging Ranking. The Emerging Ranking is formed by artists who are in career consolidation. Thus representing an attempt to recover artists who are no longer revelations, but can not be considered established artists in the music industry. So it was considered artists that emerged between 2 and 7 years. There is an intersection with artists that were revealed and modern, precisely because artists who fall into this category may be in transition between these two categories.

Memory Ranking. The memory Ranking is formed by artists who are no longer working with the original formation that was successful. An artist may no longer be working for a number of reasons, including having had the band broken up, died or simply had stopped singing.

Revelation Ranking. The Revelation Ranking is formed by artists who emerged recently. So it was considered artists who came up within 3 years.

Trend Ranking. The trend happens when one takes into account changes over time for a large group of individuals. The rankings for each artist in each digital media in Trend Ranking was constructed as follows: we took into consideration the study of the rankings predictions. The study consists in the prediction of rankings positions through regression techniques (K-NN, Linear Regression and Random Forests), using a series of historical data (days of the week containing the positions of the artists in the ranking).

The best results for the metric mean absolute deviation, studied from the techniques mentioned above, were found for the base considering 10 predictors attributes (number of days) using the multiple linear regression technique. Thus, this amount of days is used for the prediction of the positions for the construction of Trend Ranking.

General Ranking. The General Ranking consists of the 1,000 monitored artists. Thus, also considering the amount of data, the objective is to verify both the influence of digital media and the influence of TV media in the construction of the ranking. Moreover, as is considered a complete list of artists, the different genres and career ages of the artists are considered. This ranking, as its name indicates, is intended to take into account all the characteristics considered in the rankings, including the analysis of digital media and TV media.

Summary of Proposed Rankings. The Table 1 of this subsection provides the parameters considered in the proposed rankings. The table consists of the following columns: Ranking, Amount, Period/Age, No. of Artists in the Base. The Ranking column is the type of ranking; the Amount column takes into account the use or not of the digital media data amounts (the amount of data for each digital media for each artist); the Period/Age column is the period considered for analysis (how many years in the artist's career or the amount of time considered in the construction of the rankings); and No. of Artists in Base is the amount of artists that have the characteristics in each ranking considered.

Table 1. Parameters used in the proposed rankings

Ranking	Amount	Period/Age	No. of artists at the base
Perennial	Yes	> 20 years	499
Modernity	Yes	[7, 20] years	392
Emerging	Yes	[2, 7] years	143
Memory	Yes	–	91
Revelation	Yes	≤3 years	72
Trend	Yes	10 days	1000
General	Yes	–	1000

4 Rankings - Study Cases

In the following sections there will be presented some of the main rankings generated for this case study taking into account the characteristics of the artists described above. Not all rankings are presented due to limited space. For the case study, the ranking from December 25, 2014 was considered and 30 artists will be used in each ranking and the percentage analysis will refer to this number. The case study considers international artists, but accessed from the Brazilian versions of the digital media. It is important to say that the methodology is scalable to any nationality.

4.1 Perennial Ranking

This subsection provides the Perennial Ranking, which considers artists with over 20 years of career. Table 2 presents the artists featured in the Perennial Ranking. It brings in the top rankings established artists in the international music scenario (such as The Beatles, Michael Jackson, Elvis Presley, Madonna and The Rolling Stones), artists present in the ranking published in Rolling Stone magazine of the 100 greatest artists of all time [16] and in the top 100 music artists best positioned by Billboard[20], in the edition of 2013 of its fiftieth anniversary.

---

[20] http://www.billboard.com/articles/columns/chart-beat/5557800/
hot-100-55th-anniversary-by-the-numbers-top-100-artists-most-no?page=0%2C0.

It is noticed that several artists present in consecrated rankings (Billboard and Rolling Stone) who consider the best artists positioned over time tend to be artists with a longer career time, are also present in the built Perennial Ranking. The predominant genre in the Perennial Ranking is the Rock, with 15 artists (50 % of the top 30 best ranked artists). This feature is also present on the Hot 100 Billboard ranking of its fiftieth birthday edition, where it is realized that, of the top 20 ranked artists, 10 are of the Rock genre (50 %). The less significant genre is Hip Hop with just one artist.

Table 2. Perennial ranking

Pos	Artist	Genre	Pos	Artist	Genre
1	Red Hot Chili Peppers	Funk Rock	16	Slipknot	Heavy Metal
2	Guns N' Roses	Hard Rock	17	Metallica	Heavy Metal
3	Maroon 5	Pop Rock	18	Black Eyed Peas	Hip Hop
4	Michael Jackson	Pop	19	Iron Maiden	Heavy Metal
5	Bon Jovi	Hard Rock	20	Aerosmith	Hard Rock
6	The Beatles	Psychedelic Rock	21	Evanescence	Alternative Metal
7	Queen	Hard Rock	22	Radiohead	Alternative Rock
8	Foo Fighters	Alternative Rock	23	Elvis Presley	Rock
9	Nickelback	Hard Rock	24	The Rolling Stones	Rock
10	Pink Floyd	Progressive Rock	25	Madonna	Pop
11	Green Day	Punk Rock	26	Mariah Carey	R&B
12	Nirvana	Grunge	27	David Guetta	Electropop
13	Pearl Jam	Alternative Rock	28	Justin Timberlake	Pop
14	Shakira	Latin Pop	29	Black Sabbath	Heavy Metal
15	Oasis	BritPop	30	Christina Aguilera	Soul

4.2 Revelation Ranking

This subsection provides the Revelation Ranking, that shows the ranking of artists who are in evidence at the time of the analysis (in this case, December 2014). It is noteworthy that there are only 13, since those are all the available international artists in the database with this characteristic. Table 3 presents the artists present in the Revelation Ranking, where the most significant genre is Pop, with five artists - 38 %. The less significant genre in this ranking is the Cartoon with only one artist (7.6 %). Interestingly, the artists Sam Smith and Austin Mahone who are present in this ranking, are also present in the Social Rank 50 done by Next Big Sound in partnership with Billboard (week of Dec. 27). This ranking is composed by artists who are currently in evidence in digital media.

4.3 Trend Ranking

This subsection provides the Trend Ranking, which considers the prediction of rankings positions through regression techniques, using series of historical data

Table 3. Revelation ranking.

Pos.	Artist	Genre
1	Lorde	Indie Pop
2	Violetta	Soundtrack
3	Magic!	Reggae
4	Sam Smith	Pop
5	Barbie	Cartoon
6	Austin Mahone	Pop
7	Fifth Harmony	Pop
8	Aliados	Soundtrack
9	The 1975	Indie Rock
10	Banda do Mar	Indie Rock
11	Chet Faker	Electronic Music
12	Whindersson Nunes	Pop
13	Nico &Vinz	Hip Hop

Table 4. Trend ranking

Pos.	Artist	Genre	Pos.	Artist	Genre
1	Meghan Trainor	Pop	16	Violetta	Soundtrack
2	Sam Smith	Pop	17	Filipe Ret	Rap
3	Magic!	Reggae Fusion	18	Projota	Rap
4	Banda do Mar	Indie Rock	19	Joe Cocker	Blues
5	Ariana Grande	Pop	20	One Direction	Dance-pop
6	Mark Ronson	Alternative Rock	21	Imagine Dragons	Indie Rock
7	Iggy Azalea	Hip-Hop	22	Selena Gomez	Pop
8	Ed Sheeran	Folk Music	23	Pharrell Williams	Hip Hop
9	Sia	Downtempo	24	5 Seconds of Summer	Punk Rock
10	Calvin Harris	Electropop	25	Lorde	Indie Pop
11	Haikaiss	Rap	26	Jason Derulo	R&B
12	John Legend	R&B	27	David Guetta	Electropop
13	Charli XCX	Pop Synthpop	28	Austin Mahone	Pop
14	Nick Jonas	Pop	29	Nico &Vinz	Hip-Hop
15	Nicki Minaj	Hip-Hop	30	Taylor Swift	Pop

(days of the week containing the positions of artists in the ranking. Table 4 presents the Trend Ranking.

The Trend Ranking Table 4 is predominantly composed of artists from the Pop genre, featuring 12 artists (40 %). Interestingly, the singer Meghan Trainor, who appears in the first position of the Trend Ranking built in December 2014, won in May 2015 her most relevant awards at the Billboard Music Awards 2015: Billboard Top Music Hot 100 and Billboard Best Digital Music.

This subsection also features some of the rankings visions according to the categories presented in the step "Metadata Harvesting and Artists Categorization".

Regarding visions, Table 5 shows the ranking of the Gospel genre artists. In the ranking, it highlights the singer Mariza, who appears in the first position and despite predominantly being from the Fado genre, is also categorized as Gospel. This artist has been regarded by the British newspaper The Guardian, as a world music star.

Table 5. Trend ranking – Gospel genre.

Pos	Artist
1	Mariza
2	Kari Jobe
3	Johnny Cash
4	David Quinlan
5	Jess Adrin Romero
6	Hillsong United
7	Skillet
8	Casting Crowns
9	Nina Simone
10	Jeremy Camp
11	Aretha Franklin
12	P.O.D

Table 6 presents the ranking of the artists of the Upper Class Category, in which the majority of artists are from the Rock genre (26 artists - 87 %). Interestingly, the Banda do Mar band that is in the first position was nominated for Latin Grammy in 2015 in the category as the best rock album. A curious fact is that the artist John Lennon, deceased former Beatles appeared on the trend of ranking. This can be explained by the anniversary of his death, December 08, close to 25 December 2014, and therefore, within the period considered in the case study.

4.4   General Ranking

This subsection provides the General Ranking, which considers both the analysis of digital media and the TV media. Table 7 presents the artists present in the General Ranking and brings on the top ranking many artists from the Pop genre (14 artist - 47 %), indicating the strong influence of TV media, since most people

Table 6. Trend ranking – Upper class category.

Pos	Artist	Genre	Pos	Artist	Genre
1	Banda do Mar	Rock	16	Of Monsters and Men	Rock
2	Mark Ronson	Rock	17	The Beatles	Rock
3	Joe Cocker	Blues	18	The Black Keys	Blues
4	One Direction	Rock	19	Rod Stewart	Rock
5	Imagine Dragons	Rock	20	Lil Wayne	Rock
6	Selena Gomez	Rock	21	Boyce Avenue	Rock
7	5 Seconds of Summer	Rock	22	Pink Floyd	Rock
8	Onerepublic	Rock	23	The Script	Rock
9	The 1975	Rock	24	Jake Bugg	Rock
10	John Lennon	Rock	25	Led Zeppelin	Rock
11	Coldplay	Rock	26	Paramore	Rock
12	Maroon 5	Rock	27	George Harrison	Rock
13	Bastille	Rock	28	Foster the People	Rock
14	Anselmo Ralph	Blues	29	John Mayer	Blues
15	Arctic Monkeys	Rock	30	Black Veil Brides	Rock

who watch TV, listen music of the Pop genre. Interestingly artists like Adele performed on TV shows in Brazil (the Brazilian Show Fantstico on the Globo broadcast television network). The artist Bruno Mars, present in the ranking, performed at the most watched American television program, the Super Bowl, the final in 2014.

4.5 Relationship Between Constructed Rankings and Industry Rankings Phonographic

This subsection provides a comparative study of an artist ranking constructed in this work with a ranking of artists constructed by Billboard. This study is not intended to show that the rankings generated in this study are better than the rankings provided by Billboard, but just to compare their features. The ranking constructed in this work, which is compared to a ranking provided by an external company, is from December 25, 2014. The best 10 artists from each ranking were considered. The external ranking used is described below:

– the Billboard Hot 100: considering the weekly audience of the songs of the artists on the radio and sales data, both measured by Nielsen Music, and streaming activity data provided by digital media. The ranking consists of 100 songs, with possible repetition of artists. For comparison with the rankings from December 25 built in this paper, it was considered the ranking of the week from December 27, 2014, which includes the day considered in the ranking in this work. The ranking constructed in this paper considers over

Table 7. General ranking.

Pos	Artist	Genre	Pos	Artist	Genre
1	Bruno Mars	Pop	16	Demi Lovato	Pop
2	Beyonc	R&B	17	David Guetta	Electropop
3	Ariana Grande	Pop	18	Katy Perry	Pop
4	Adele	Soul	19	Justin Bieber	Pop
5	Calvin Harris	Electropop	20	Imagine Dragons	Indie Rock
6	Avicii	House Music	21	Led Zeppelin	Hard Rock
7	Avril Lavigne	Pop Rock	22	John Legend	R&B
8	Taylor Swift	Pop	23	Linkin Park	Alternative Rock
9	Ed Sheeran	Folk Music	24	Simple Plan	Pop Punk
10	Arctic Monkeys	Indie Rock	25	Three Days Grace	Alternative Metal
11	One Direction	Dance-pop	26	Sam Smith	Pop
12	Maroon 5	Pop Rock	27	OneRepublic	Pop Rock
13	Coldplay	Alternative Rock	28	Lana Del Rey	English Barroco
14	Bon Jovi	Hard Rock	29	Pink Floyd	Progressive Rock
15	The Beatles	Psychedelic Rock	30	Guns N' Roses	Hard Rock

time positions variation trends like the Top Billboard Hot 100, which considers the weekly variation, is the Trend Ranking. Thus, these two rankings were compared.

Table 8 shows the two rankings to be compared. The table structure consists of: Pos (position of the artist in the specific ranking), Artist, Genre and N/I (artist origin, whether national or international). The rankings are the rankings Billboard Hot 100; and the Trend Ranking (see Table 4 from Sect. 4.3). To compare the rankings, only the song in the best position of each artist was considered. Only the performers present in both bases, Billboard and this work, were considered. This way, there was a total of 72 performers from each base.

It can be seen from Table 8 that rankings share 5 artists in common in their top positions (50 %) and the the predominant genre in both bases is the Pop (accounting for 8 artists in the ranking from Billboard - 80 % and 6 artists in the Trend ranking - 60 %).

It is observed that there is a low similarity between the ranking developed in this work and the ranking from the music industry. It cannot be said that the rankings from this work are more in line with reality than the already existing examples, however, when considering that among their own, there are already several established rankings, such as those constructed by Billboard, it is observed that there is not one way to build rankings that is universally accepted. Having an open methodology from which researchers of the music industry can analyze various parameters, media and artists, shows that the methodology developed in this work is an important input for decision-making in an increasingly dynamic and demanding market.

Table 8. Ranking Billboard Hot 100 and trend ranking.

Ranking Billboard Hot 100				Trend ranking			
Pos	Artist	Genre	N/I	Pos	Artist	Genre	N/I
1	Taylor Swift	Pop	I	1	Meghan Trainor	Pop	I
2	Hozier	Indie Rock	I	2	Sam Smith	Pop	I
3	Mark Ronson	Pop	I	3	Magic!	Reggae	I
4	Meghan Trainor	Pop	I	4	Banda do Mar	Rock	I
5	Sam Smith	Pop	I	5	Ariana Grande	Pop	I
6	Ed Sheeran	Pop	I	6	Mark Ronson	Rock	I
7	Maroon 5	Rock	I	7	Iggy Azalea	Hip Hop	I
8	Ariana Grande	Pop	I	8	Ed Sheeran	Pop	I
9	Nick Jonas	Pop	I	9	Sia	Pop	I
10	Nicki Minaj	Pop	I	10	Calvin Harris	Pop	I

5   Conclusion

Music sales are losing their role as a means of revealing the popularity of artists, due to the way the music is consumed today through downloads on the Internet, with the use of devices for music sharing, online discussion forums, digital media and radio stations on the Internet. Another factor contributing to the popularity of an artist is the analysis of events which are related to the music industry, such as the release of a song in a TV program and the use of his/hers songs in novels and series on TV.

The popularity rankings modeling of artists, due to the current scenario of the way music is consumed, have been directed to this new dynamic, with rankings built by specialized companies in partnership with digital media such as the Social Ranking 50 constructed by the company Next Big Sound in partnership with Billboard. It can be seen that even though the researches on digital media are on wide growth, researchers say there is little material related to the specificities of the production and distribution of music content in digital media, as well as the practices and dissemination strategies used by producers and fans in such environments.

This work's main objective was to develop a construction methodology of artist rankings focusing on the analysis of the influence of digital and television media. With the methodology and algorithms developed, it was possible to reach this goal, as this methodology was marked by satisfactory results achieved by the metrics described above. Several scenarios were considered for the construction of the rankings (achieving one of the objectives of the analytical work of the rankings that could be developed) and then one of the rankings was compared with a consolidated ranking in the music industry constructed by Billboard. Through the case study of the rankings for the day of December 25, 2014, where several rankings were constructed taking into account these various scenarios, it was possible to find interesting results.

With the Trending Ranking, it was observed that the singer Meghan Trainor, first position in the rank on May 2015 won important awards from Billboard. Another ranking, Perennial Ranking, brings in the of top of ranking established artists from the international scene, such as The Beatles and The Rolling Stones. These artists are among the artists most performed over time, according to rankings constructed by Billboard. It is clear, therefore, that the rankings constructed proved to be consistent to those constructed by consolidated companies in the construction of rankings like Billboard. Furthermore, they showed to provide relevant results for the music industry to analyze this market. Finally, another ranking that proved interesting was the General Ranking, where they found several artists who were present in the TV programming such as the artist Adele who performed at the Brazilian Show Fantástico on the channel Globo in 2014 and the singer Bruno Mars who performed at Super Bowl also in 2014 (indicating globalization on access to digital media, once the Super Bowl is broadcasted in the United States). This indicates a strong convergence of media, since the ranking is composed of artists who were present in digital media, on TV or in both.

Regarding the similarity between the rankings of artists constructed on this work and rankings established in the music industry such as Billboard, the lack of observed similarity indicates that no ranking brings an absolute truth, but try to portray in different ways the phonograph scenario that is under constant change. Therefore, this work shows to be relevant as it reflects a systematic methodology that can adapt the rankings for these changes and verify possible erroneous biases.

It is noteworthy that, despite having used eight digital media, the methodology developed in this work is scalable because it proves to be capable of embracing new media, since many digital media have their own APIs that enable easy integration or allow to collect essential audience data for this methodology. In addition, new parameters can be used, such as new values for career intervals, for example.

At this point, it is observed that the division by career age shows to be relevant due to the satisfactory results of correlation of Spearman rankings. In view of this scalability, this approach can be considered for other purposes. For example, it can be used to generate other types of rankings: most sold cars in Brazil in 2015; most popular artists from movies, novels and series; among others.

As future work, the proposal of new rankings is suggested. For example, considering different periods of time. It is also suggested the development of a tool to automate the construction of the rankings, new case studies and the extension of the construction of rankings to other contexts.

Acknowledgements. The authors thank the Federal University of Ouro Preto, Federal University of Mato Grosso do Sul, CAPES and Fundect for supporting the development of this research. Thanks also to the anonymous reviewers for their constructive comments, contributing to a better version of this article.

References

1. Azarite, D.M.R.: Monitoramento e Métricas de Mídias Sociais, 1st edn. DVS, São Paulo (2012)
2. Bryan, N.J., Wang, G.: Musical influence network analysis and rank of sample-based music. In: Proceedings of the International Society for Music Information Retrieval Conference, ISMIR 2011, Florida, Miami, pp. 329–334 (2011)
3. Chelaru, S., Rodriguez, C., Altingovde, I.: How Useful is Social Feedback for Learning to Rank YouTube Videos?, 2nd edn. Springer, Berlin (2009). Proceedings of the World Wide Web, WWW 2013, Hingham, MA, USA, pp. 997–1025 (2013)
4. Cohen, J.: Statistical Power Analysis for the Behavioral Sciences, 2nd edn. Erlbaum, Hillsdale (1988)
5. Grace, J., et al.: Artist ranking through analysis of online community comments. Technical report, IBM (2007)
6. Greenberg, Z.: Moneyball for music: The Rise of Next Big Sound, Forbes, New York, 4 March 2013. http://www.forbes.com/sites/zackomalleygreenburg/2013/02/13/moneyball-for-music-the-rise-of-next-big-sound/
7. Guerrini Jr., I., Vicente, E.: Na trilha do disco: Relatos sobre a indústria fonográfica no Brasil, Rio de Janeiro: E-Papers (2010)
8. Johnston, M.: Moneyball for music: The Rise of Next Big SoundNew Chart Based on Social Networking Data Debuts, Rolling Stone, EUA, 2 December 2010
9. Kaufman, G.: Mtv launches music meter artist-ranking service: New chart will track the popularity of new acts via social media buzz. In MTV News, EUA, 14 December 2010. http://www.mtv.com/news/1654222/mtv-launches-music-meter-artist-ranking-service/
10. Koenigstein, N., Shavitt, Y.: Song ranking based on piracy in peer-to-peer networks. In: Proceedings of the International Society for Music Information Retrieval Conference, ISMIR 2009, Kobe, Japan, pp. 633–638 (2009)
11. Magalhaes, T., Sawaia, J.: Tribos musicais. In: IBOPE Media, São Paulo (2013). http://www.ibope.com.br/pt-br/noticias/Documents/tribos_musicais.pdf/
12. Moss, M., et al.: Multi-input playlist selection. US Patent App. 13/759,540
13. FACEBOOK music like it or not, you may have to 'like' it. Billboard Magazine: United States. **123**(35), 888 (2011)
14. Park, M., Weber, I., Naaman, M.: Understanding musical diversity via online social media. In: Proceedings of the Ninth International AAAI Conference on Web and Social Media, ICWSM 2015, Oxford, UK, pp. 308–317 (2015)
15. Sun, Y., et al.: Personalized ranking for digital libraries based on log analysis. In: Proceedings of the International Workshop on Web Information and Data Management, Napa Valley, USA, pp. 133–140 (2008)
16. Veloso, B.: Os 100 maiores artistas de todos os tempos. In: Rolling Stone, São Paulo (2013)

Solving Manufacturing Cell Design Problems by Using a Dolphin Echolocation Algorithm

Ricardo Soto[1,2,3], Broderick Crawford[1,4,5], César Carrasco[1], Boris Almonacid[1],
Victor Reyes[1(✉)], Ignacio Araya[1], Sanjay Misra[6], and Eduardo Olguín[5]

[1] Pontificia Universidad Católica de Valparaíso, Valparaíso, Chile
{ricardo.soto,broderick.crawford,ignacio.araya}@ucv.cl,
{boris.almonacid.g,victor.reyes.r}@mail.pucv.cl, ccarrascocarre@gmail.com
[2] Universidad Autónoma de Chile, Santiago, Chile
[3] Universidad Científica del Sur, Lima, Peru
[4] Universidad Central de Chile, Santiago, Chile
[5] Facultad de Ingeniería y Tecnología, Universidad San Sebastián, Bellavista 7,
Santiago 8420524, Chile
eduardo.olguin@uss.cl
[6] Covenant University, Ota, Nigeria
ssopam@gmail.com

Abstract. The Manufacturing Cell Design is a problem that consist in organize machines in cells to increase productivity, i.e., minimize the movement of parts for a given product between machines. In order to solve this problem we use a Dolphin Echolocation algorithm, a recent bio-inspired metaheuristic based on a dolphin feature, the *echolocation*. This feature is used by the dolphin to search all around the search space for a target, then the dolphin exploits the surround area in order to find promising solutions. Our approach has been tested by using a set of 10 benchmark instances with several configurations, reaching to optimal values for all of them.

Keywords: Dolphin Echolocation Algorithm · Metaheuristics · Manufacturing cell design problems

1 Introduction

The Manufacturing Cell Design Problem (MCDP) consist in grouping components under the statement: "Similar things should be manufactured in the same way". In order to increase the efficiency and productivity, machines that create products with similar requirements must be close each other. This machine-grouping, known as cell, allow a dedicated and organized production of the differents parts of a given product. Then, the goal of the MCDP is to minimize movements and exchange of material, by finding the best machine-grouping combination.

The research that has been done to solve the cell formation problem has followed two complementary lines, which can be classified in two different groups:

© Springer International Publishing Switzerland 2016
O. Gervasi et al. (Eds.): ICCSA 2016, Part V, LNCS 9790, pp. 77–86, 2016.
DOI: 10.1007/978-3-319-42092-9_7

approximate and complete methods. Most approximate methods focus on the search for an optimal solution in a limited time, however, a global optimal is not guaranteed. On the other hand, complete methods aims to analyze the whole search space to guarantee a global optima but this require a higher cost in both memory and time.

Within the approximate methods, different metaheuristics have been used for cell formation. Aljaber et al. [1] made use of Tabu Search. Lozano et al. [9] presented a Simulated Annealing (SA) approach. Durán et al. [4] combined Particle Swarm Optimization (PSO), which consists of particles that move through a space of solutions and that are accelerated in time, with a data mining technique. Venugopal and Narendran [16] proposed using the Genetic Algorithms (GA), which are based on the genetic process of living organisms. Gupta et al. [5] also used GA, but focusing on a different multi-objective optimization, consisting in the simultaneous minimization of the total number of movements between cells and load variation between them. It is also possible to find hybrid techniques in the problem resolution. Such is the case of Wu et al. [17], who combined SA with GA. James et al. [6] introduced a hybrid solution that combines local search and GA. Nsakanda et al. [10] proposed a solution methodology based on a combination of GA and large-scale optimization techniques. Soto et al. [15], utilized Constraint Programming (CP) and Boolean Satisfiability (SAT) for the resolution of the problem, developing the problem by applying five different solvers, two of which are CP solvers, two SAT solvers and a CP and SAT hybrid.

With regard to the complete methods, preliminary cell formation experiments were developed through the use of linear programming, such as the work of Purcheck [12], and Olivia-López and Purcheck [11]. Kusiak and Chow [8] and Boctor [2] proposed the use of linear-quadratic models. The use of a different paradigm for global optimization stands out, such as Goal Programming (GP), which can be seen as a generalization of linear programming for the manipulation of multiple objective functions. Sankaran and Rodin [13] and, Shafer and Rogers [14] worked on GP.

Lastly, some research has been done combining both approximate and complete methods. Such is the case of Boulif and Atif [3], who combined branch-and-bound techniques with GA.

In this paper, we propose a Dolphin Echolocation Algorithm (DA) [7] to solve the MCDP. The DA is an optimization technique based on the behavior and the echolocation feature of the dolphins. By using a high-frequency click they are able to find a potential prey, all around the search space. Then, incrementally, the dolphins increase its clicks for the purpose of concentrate the search in that region.

The rest of this paper is organized as follows. In Sect. 2 the mathematical model of the MCDP is described and explained in detail. The Dolphin Echolocation Algorithm is presented in Sect. 3. Finally, Sect. 4 illustrates the experimental results that we obtained by using a well-known set of benchmarks instances, followed by conclusions and future work.

2   Manufacturing Cell Design Problem

The MCDP consists in organizing a manufacturing plant or facility into a set of cells, each of them made up of different machines meant to process different parts of a product, that share similar characteristics. The main objective, is to minimize movements and exchange of material between cells, in order to reduce production costs and increase productivity. We represent the processing requirements of machine parts by an incidence zero-one matrix A, known as the *machine-part matrix*. The optimization model is stated as follows. Let:

- M: the number of machines.
- P: the number of parts.
- C: the number of cells.
- i: the index of machines ($i = 1, ..., M$).
- j: the index of parts ($j = 1, ..., P$).
- k: the index of cells ($k = 1, ..., C$),
- M_{max}: the maximum number of machines per cell.
- $A = [a_{ij}]$: the binary machine-part incidence matrix, where:

$$a_{ij} = \begin{cases} 1 & \text{if machine } i \text{ process the part } j \\ 0 & \text{otherwise} \end{cases}$$

- $B = [y_{ik}]$: the binary machine-cell incidence matrix, where:

$$y_{ik} = \begin{cases} 1 & \text{if machine } i \text{ belongs to cell } k \\ 0 & \text{otherwise} \end{cases}$$

- $C = [z_{jk}]$: the binary part-cell incidence matrix, where:

$$z_{jk} = \begin{cases} 1 & \text{if part } j \text{ belongs to cell } k \\ 0 & \text{otherwise} \end{cases}$$

Then, the MCDP is represented by the following mathematical model [2]:

$$\min \sum_{k=1}^{C} \sum_{i=1}^{M} \sum_{j=1}^{P} a_{ij} z_{jk} (1 - y_{i_k}) \tag{1}$$

Subject to the following constraints:

$$\sum_{k=1}^{C} y_{ik} = 1 \quad \forall i \tag{2}$$

$$\sum_{k=1}^{C} z_{jk} = 1 \quad \forall j \tag{3}$$

$$\sum_{i=1}^{M} y_{ik} \leq M_{max} \quad \forall k \tag{4}$$

3   Dolphin Echolocation Metaheuristic

In the animal kingdom, there are some animals that have a particular feature known as *echolocation*, which allow them to find targets. One of these animals is the dolphin, who is able to generate two high frequency clicks. While the first one is used by the dolphins to look for targets, the second one is used to estimate the distance to it. Dolphins can detect targets at ranges varying from a few tens of meters to over a hundred meters. In this section we describe how this algorithm works for solving the MCDP. A more detailed description of the DA can be seen in [7].

3.1   Search Space Ordering

To represent the search space, the first step is to create an ascending or descending ordered matrix. Each machine, represented by j, is defined by a vector A_j. This vector, of lenght LA_j represent all the posibles values that the variable j can take. Then the matrix $Alternatives_{MA:NV}$ is created by putting each of this vectors next to each other, where MA is $Max(LA_j)_{j=1:NV}$; with NV being the number of machines in this particular case.

3.2   Solution Generation

After creating the $Alternatives_{MA:NV}$ matrix, we generate N random solutions. Each solution, represented by L_i is stored in a L_{iN} matrix, so each row will represent a machine-cell matrix defined previously.

3.3   Convergence Factor: Exploration and Exploitation

This metaheuristic has a convergence factor that is defined by the following formula:

$$PP(Loop_i) = PP_1 + (1 - PP_1)\frac{Loop_i^{Power} - 1}{(LoopsNumber)^{Power} - 1} \tag{5}$$

$PP(Loop_i)$ defines the rate of exploration and exploitation of the metaheuristic. For high values, DA will look for promising zones in the search space and low values will imply a local search, instead.

3.4   Metaheuristic Parameters

Input parameters of the DA are:

- $NumberLoops$: Maximum number of iterations.
- R_e: Effective radius of search. This parameter is used to calculate the accumulated fitness of the problem.
- PP_0: This parameter correspond to the convergence factor. It defines the problem convergence.

- *Power*: This parameter defines the convergence curve. This parameter also can be seen as the slope of the curve.
- *NL*: This correspond to the number of locations. Each location will correspond to a potential solution of the problem.

3.5 The Algorithm

The ocean or search space is represented by the matrix $Alternatives_{MA:NV}$. When the dolphin makes a click the random search begins, then when the dolphin receive the echo from a target the L matrix is generated. Using the fitness, the dolphin decides which target is more promising and the dolphin follows that direction. The accumulative fitness is then calculated, according to dolphin rules as follows:

Algorithm 1.

---

for $i = 1$ to the number of locations **do**
    for $j = 1$ to the number of variables **do**
        find the position of $L(ij)$ in the jth column of
        Alternatives matrix and name it as A.
        for $K = -R_e$ to R_e **do**
            $AF_{(A+k)j} \leftarrow \frac{1}{R_e}(R_e - |k|)Fitness(i) + AF_{(A+k)j}$
        end for
    end for
end for

Note that if $A + k$ is not a valid position, i.e. $A + k < 0$ or $A + k > LA_j$, AF must be calculated by using the reflexive property which consist in return to a previous valid position. After the best location is found, the alternatives allocated to the variables of the best location in the AF matrix are set to zero. That is:

Algorithm 2.

---

for $j = 1$: Number of variables **do**
    for $i = 1$: Number of alternatives **do**
        if $i =$ The best location(j) **then**
            $AF_{ij} \leftarrow 0$
        end if
    end for
end for

Then, the probability matrix P is calculated by using the following formula:

$$P_{ij} = \frac{AF_{ij}}{\sum_{i=1}^{LA_j} AF_{ij}} \tag{6}$$

We use a similar procedure with the matrix P as before. This can be seen in the Algorithm 3.

Algorithm 3.

---

for $j = 1$: Number of variables **do**
  for $i = 1$: Number of alternatives **do**
  if $i =$ The best location(j) **then**
    $P_{ij} \leftarrow PP$
  else
    $P_{ij} \leftarrow (1 - PP)P_{ij}$
  end if
  end for
end for

When this values are obtained, the most promising solution is used for exploitation by weighting each solution using the values of the P matrix. After we reach the maximum number of loops, the best solution is selected.

4   Experimental Results

Our approach has been implemented in Java, on an 1.7 GHz CPU Intel Core i3-4005U with 6 GB RAM computer using Windows 10 64 bits and tested out by using a set of 10 incidence matrices [2]. Tests were carried out based on these 10 test problems, using 2 cells with a M_{max} value between 8 and 10, and using 3 cells with a M_{max} value between 6 and 9. The parameters, established based on other problems already solved by this metaheuristic, are defined as follows:

- No. of iterations ($NumberLoops$): 1000
- No. of dolphins (N): 5
- Effective radius (R_e): 1
- Convergence factor (PP_0): 0.1
- Degree of the curve ($Power$): 1

Tables 1 and 2 show detailed information about the results obtained by using DA with the previous configuration. We compare our results with the optimum values reported in [2] (Opt), Simulated Annealing (SA) [2] and with

Table 1. Results of DA using 2 cells

Bench	$M_{max} = 8$				$M_{max} = 9$				$M_{max} = 10$			
	Opt	SA	PSO	DA	Opt	SA	PSO	DA	Opt	SA	PSO	DA
1	11	11	11	11	11	11	11	11	11	11	11	11
2	7	7	7	7	6	6	6	6	4	10	5	4
3	4	5	5	4	4	4	4	4	4	4	5	4
4	14	14	15	14	13	13	13	13	13	13	13	13
5	9	9	10	9	6	6	8	6	6	6	6	6
6	5	5	5	5	3	3	3	3	3	5	3	3
7	7	7	7	7	4	4	5	4	4	4	5	4
8	13	13	14	13	10	20	11	10	8	15	10	8
9	8	13	9	8	8	8	8	8	8	8	8	8
10	8	8	9	8	5	5	8	5	5	5	7	5

Table 2. Results of DA using 2 cells

Pblm	$M_{max} = 11$				$M_{max} = 12$			
	Opt	SA	PSO	DA	Opt	SA	PSO	DA
1	11	11	11	11	11	11	11	11
2	3	4	4	3	3	3	4	3
3	3	4	4	3	1	4	3	1
4	13	13	13	13	13	13	13	13
5	5	7	5	5	4	4	5	4
6	3	3	4	3	2	3	4	2
7	4	4	5	4	4	4	5	4
8	5	11	6	5	5	7	6	5
9	5	8	5	5	5	8	8	5
10	5	5	7	5	5	5	6	5

Particle Swarm Optimization (PSO) [4] as well. Table 3 only shows information of optimal values reported for our metaheuristic and the optimal founded in [2] (Opt). Convergence charts can be seen in Fig. 1.

Experimental results show that the proposed DA provides high quality solutions and good performance using two and three cells, where all the expected optima were achieved.

Table 3. Results of DA using 3 cells

Bench	$MCDPM_{max} = 6$		$M_{max} = 7$		$M_{max} = 8$		$M_{max} = 9$	
	Opt	DA	Opt	DA	Opt	DA	Opt	DA
1	27	27	18	18	11	11	11	11
2	7	7	6	6	6	6	6	6
3	9	9	4	4	4	4	4	4
4	27	27	18	18	14	14	13	13
5	11	11	8	8	8	8	6	6
6	6	6	4	4	4	4	3	3
7	11	11	5	5	5	5	4	4
8	14	14	11	11	11	11	10	10
9	12	12	12	12	8	8	8	8
10	10	10	8	8	8	8	5	5

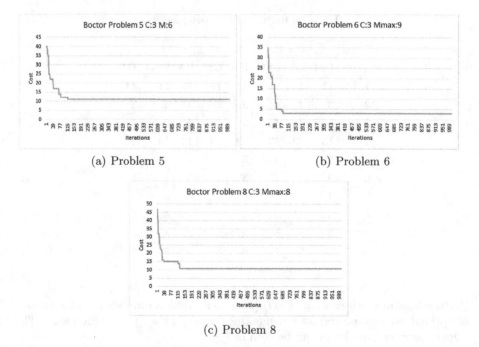

(a) Problem 5   (b) Problem 6

(c) Problem 8

Fig. 1. Convergence plots for benchmarks 5, 6 and 8, with 3 cells and a M_{max} of 6, 9 and 10 respectively

5   Conclusions and Future Work

Manufacturing cell design is a well-known problem, where manufacturing companies seek for a productivity improvement by reducing costs. For this reason, we propose the DA, a new bio-inspired metaheuristic reporting optimum values for all the tested instances. DA has shown high convergence rates at early stages of the search compared to other metaheuristics that has been used for solving the MCDP. This superiority and the capability of *adapting itself* makes DA an interesting metaheuristic to be use in more optimization problems.

As a future work, we plan to use more benchmarks and configurations to test out the robustness of the DA. Also, we plan to do an Autonomous Search integration for setting a possible better parameter configuration.

Acknowledgements. Ricardo Soto is supported by grant CONICYT/FONDECYT/ REGULAR/1160455, Broderick Crawford is supported by grant CONICYT/ FONDECYT/REGULAR/1140897, Victor Reyes is supported by grant INF-PUCV 2015, and Ignacio Araya is supported by grant CONICYT/FONDECYT/REGULAR/ 1160224.

References

1. Aljaber, N., Baek, W., Chen, C.-L.: A tabu search approach to the cell formation problem. Comput. Ind. Eng. **32**(1), 169–185 (1997)
2. Boctor, F.F.: A linear formulation of the machine-part cell formation problem. Int. J. Prod. Res. **29**(2), 343–356 (1991)
3. Boulif, M., Atif, K.: A new branch-&-bound-enhanced genetic algorithm for the manufacturing cell formation problem. Comput. Oper. Res. **33**(8), 2219–2245 (2006)
4. Durán, O., Rodriguez, N., Consalter, L.A.: Collaborative particle swarm optimization with a data mining technique for manufacturing cell design. Expert Syst. Appl. **37**(2), 1563–1567 (2010)
5. Gupta, Y., Gupta, M., Kumar, A., Sundaram, C.: A genetic algorithm-based approach to cell composition and layout design problems. Int. J. Prod. Res. **34**(2), 447–482 (1996)
6. James, T.L., Brown, E.C., Keeling, K.B.: A hybrid grouping genetic algorithm for the cell formation problem. Comput. Oper. Res. **34**(7), 2059–2079 (2007)
7. Kaveh, A., Farhoudi, N.: A new optimization method: dolphin echolocation. Adv. Eng. Softw. **59**, 53–70 (2013)
8. Kusiak, A., Chow, W.S.: Efficient solving of the group technology problem. J. Manuf. Syst. **6**(2), 117–124 (1987)
9. Lozano, S., Adenso-Diaz, B., Eguia, I., Onieva, L., et al.: A one-step tabu search algorithm for manufacturing cell design. J. Oper. Res. Soc. **50**(5), 509–516 (1999)
10. Nsakanda, A.L., Diaby, M., Price, W.L.: Hybrid genetic approach for solving large-scale capacitated cell formation problems with multiple routings. Eur. J. Oper. Res. **171**(3), 1051–1070 (2006)
11. Oliva-Lopez, E., Purcheck, G.F.: Load balancing for group technology planning and control. Int. J. Mach. Tool Des. Res. **19**(4), 259–274 (1979)
12. Purcheck, G.F.K.: A linear-programming method for the combinatorial grouping of an incomplete power set (1975)
13. Sankaran, S., Rodin, E.Y.: Multiple objective decision making approach to cell formation: a goal programming model. Math. Comput. Model. **13**(9), 71–81 (1990)

14. Shafer, S.M., Rogers, D.F.: A goal programming approach to the cell formation problem. J. Oper. Manag. **10**(1), 28–43 (1991)
15. Soto, R., Kjellerstrand, H., Durán, O., Crawford, B., Monfroy, E., Paredes, F.: Cell formation in group technology using constraint programming and boolean satisfiability. Expert Syst. Appl. **39**(13), 11423–11427 (2012)
16. Venugopal, V., Narendran, T.T.: A genetic algorithm approach to the machine-component grouping problem with multiple objectives. Comput. Ind. Eng. **22**(4), 469–480 (1992)
17. Tai-Hsi, W., Chang, C.-C., Chung, S.-H.: A simulated annealing algorithm for manufacturing cell formation problems. Expert Syst. Appl. **34**(3), 1609–1617 (2008)

Cryptanalysis and Improvement User Authentication Scheme for Multi-server Environment

Dongwoo Kang, Jongho Moon, Donghoon Lee, and Dongho Won[✉]

College of Information and Communication Engineering, Sungkyunkwan University,
Seoul, Republic of Korea
{dwkang,jhmoon,dhlee,dhwon}@security.re.kr

Abstract. Because of increasing mobile devices and networks, people who wanted mobile service can access network at anywhere and anytime. User authentication using smartcard is a one of the widely-spread using technique in which server checks the legitimacy of a user between public channel. Currently, the number of user and server is increasing rapidly, user authentication scheme for multi-server environments have been proposed. User authentication scheme for multi-server environments is built more secure and efficient. As schemes are proposed continuously. In 2016, Amin et al. improved both Sood and Li et al.'s schemes and asserted that their scheme is a more secure and efficient for multi-server environment user authentication scheme. However, we discovered that Amin et al.'s scheme still insecure and not suitable to apply real-life application. In this paper, we demonstrate that their scheme is not able to resist several security threats. Finally, we show that our proposed scheme is more secure and provides for more security features.

Keywords: Authentication · Smartcard · Multi-server environment · Session key

1 Introduction

In last few decades, the applications of internet service are rapidly increased. Many users communicate by way of the network service on their own electronic instrument such as smartphone. However, the number of server also rapidly increased, it is very inefficient and hard to control for the user's various identities, passwords for accessing differing remote server repetitively when user used single-server environment. To resolve this problem, there are multi-server environment. In multi-server environment, user logins only once the control server, and then can access diverse different service provide server [1–3].

The most important goal of the user authentication scheme is to design scheme more securely and efficiently. Smartcard is very secure for hold user's own secret information. The eventual aim of the user authentication scheme using smartcard is to verify service access person's legitimacy [4–6]. Since Ford and

© Springer International Publishing Switzerland 2016
O. Gervasi et al. (Eds.): ICCSA 2016, Part V, LNCS 9790, pp. 87–99, 2016.
DOI: 10.1007/978-3-319-42092-9_8

Kaliski [7] proposed password based multi-server authentication protocol, many multi-server environment user authentication scheme were proposed. Jablon [8] improved Ford and Kaliski's scheme direction of do not use public keys. Also, Lin et al. [9] proposed multi-server environment authentication scheme using ElGamal digital signature. In 2007, Hu et al. [10] proposed an efficient password authentication key agreement protocol for multi-server environment which user can access by his/her smartcard. In succession, Tsai [11] proposed smartcard, one-way hash function, nonce based multi-server environment authentication scheme. On the bounce, Liao and Wang [12] proposed a dynamic identity based remote user authentication uses only cryptographic one-way hash function which easy for the implementation. Subsequently, Li et al. [13] proposed a novel remote dynamic ID-based multi-server authenticated key agreement scheme. Nevertheless, Xue et al. [14] pointed that the Li et al.'s scheme is open to attack different types of security threat such as replay attack, masquerade attack and proposed an enhanced scheme. However, Lu et al. [15] demonstrate that the Xue et al.'s scheme also could not resist insider attack, offline password guessing attack. To defeat the security flaw of the Xue et al.'s scheme, Lu et al proposed an robust scheme. Recently, Amin [16] cryptanalyze the Sood [17] scheme and Li et al.'s [18] scheme and Sood et al. scheme is not able to resist offline identity guessing attack and offline password guessing attack and so on, and Li et al. scheme is also not able to resist user impersonation attack, many logged in user's attack. Moreover, Amin et al. proposed more efficient and safety scheme in multi-server environment using smartcard. However, Amin et al. scheme also has security flaws such as offline password guessing attack, offline identitiy guessing attack, session key derived attack.

In this paper, we focus on the security flaws of Amin et al. scheme's safety. After cryptanalyze, we found that, we found their scheme is not able to resist offline password guessing attack, offline identitiy guessing attack, user impersonation attack, and so on. We then proposed more efficient and safety scheme for multi-server environment. Furthermore, we demonstrate that our proposed authentication scheme has more security features and safety compared with Amin et al.'s scheme.

Our paper is organized as following : we review Amin et al.'s scheme in Sect. 2. We then analyse that Amin et al.'s scheme's security weakness in Sect. 3. In Sect. 4, we describe the proposed scheme, and Sect. 5, we show the cryptanalysis of our proposed scheme. Finally, we conclude our article in Sect. 6.

2   Review in Amin et al.'s Scheme

This section, we review the user authentication scheme using hash function proposed by Amin et al. in 2016. Amin et al.'s scheme consists of four phases: Registration, Login, Authentication and Password changing phases which as follows. The notations used in this paper are summarized as Table 1.

Table 1. Notations used in this paper

Symbols	Meaning
CS	Control Server
S_k	k^{th} Service Provide Server
U_i	i^{th} user
ID_i	Identity of U_i
PW_i	Password of U_i
x	Secret key of Control Server
$H(.)$	One-way hash function
SK	Shared secret session key
$\oplus$	Bitwise XOR operation
$\|$	Concatenate operation
T_1, T_2	Current Timestamp

2.1 Registration Phase

When user U_i wants to access server for a service, Service provider registers to control server and User registers to service provider server. Registration phase is described as follows.

Server Registration Phase. In this phase, service provider server S_k chooses identity SID_k and submits to control server by secure channel. After receiving SID_k from S_k, control server computes $P_k = H(SID_k\|x)$ and submits it to the server S_k. After that Server S_k keeps it as secret.

User Registration Phase. When user U_i wants to access server for a service, first, he/she registers with the control server CS. Therefore, U_i chooses his or her identity and password ID_i, PW_i and a random nonce b. After, submits ID_i, $PWR_i = h(PW_i \oplus b)$ to control server via a secure channel. Control server selects random nonce y_i computes $CID_i = h(ID_i \oplus y_i \oplus x)$, $REG_i = h(ID_i\|PWR_i\|CID_i)$, $T_i = h(CID_i\|x)$. Afterwards, Control server issues a smartcard and stores $\{CID_i, REG_i, T_i, y_i, h(.)\}$. After getting smartcard, user U_i stores b into smartcard.

2.2 Login Phase

To start any communication, the user must first login using smart card. The user inserts his/her smart card into card-reader and input ID_i', PW_i' and chooses server identity SID_k. Then, smartcard computes $PWR_i' = h((PW_i' \oplus b)$, $REG_i' = h(ID_i'\|PWR_i'\|CID_i)$ and checks it equals REG_i. If not holds, reject the login. After that, user computes $L_1 = T_i \oplus PWR_i$. Then, smartcard generates random numbers N_1, N_2. Then, further computes $L_2 = N_2 \oplus PWR_i$, $N_3 = N_1 \oplus N_2, L_3 = h(L_1||SID_k||N_1||L_2||N_3)$ and sends login request message $\{CID_i, SID_k, T_i, L_3, L_2, N_3\}$ to control server.

2.3 Authentication Phase

After receiving the login request message from user, the server performs the following sequence of operations to generate same session key. This phase, server computes $A_1 = h(CID_i||x)$ and derives $PWR_i' = T_i \oplus A_1, N_2' = L_2 \oplus PWR_i'$ and $N_1' = N_3 \oplus N_2'$. Further computes $L_3' = h(A_1||SID_k||N_1'||L_2||N_3)$. Then, checks it equals L_3. If not holds, reject the login. After that, Control server generates random numbers N_4 and compute $A_2 = h(SID_k||x), A_3 = A_2 \oplus N_4, N_5 = N_1' \oplus N_4, A_4 = h(A_2||N_4||N_1||CID_i)$. Then control server send the $\{CID_i, A_4, A_3, N_5\}$ to the service provider server S_k of corresponding identity SID_k.

After receiving $\{CID_i, A_4, A_3, N_5\}$ message form the control server, server S_k derives $N_4' = P_k \oplus A_3, N_1' = N_4' \oplus N5$ and computes $A_4' = h(P_k||N_4'||N_1'||CID_i)$. Then checks it equals A_4. If not holds, reject the authentication request. If it holds server S_k generates random nonce N_6 and computes $N_7 = N_1' \oplus N_6, SK = h(SID_k||CID_i||N_6||N_1'), A_5 = h(SK||N_6)$ and sends $\{SID_k, A_5, N_7\}$ to user U_i.

After receiving message from server, the smartcard derives $N_6' = N_7 \oplus N_1$ and compute $SK' = h(SID_k||CID_i||N_6||N_1), A_5' = (SK'||N_6)$, then smartcard checks it equals A_5. If it holds, the mutual authentication are successfully shared by session key SK between user U_i and service provider server S_k.

2.4 Password Change Phase

If user intends to alter his password into an another password, first he/she inputs original identity and password ID_i', PW_i' and smartcard computes $PWR_i' = h(PW_i' \oplus b), REG_i' = h(ID_i'||PWR_i'||CID_i)$ and checks it equals REG_i. If not holds, reject the password change request. If it holds, user inputs his/her new password PW_i^{new}. Smartcard computes $PWR_i^{new} = h(PW_i^{new} \oplus b)$, $REG_i^{new} = h(ID_i'||PWR_i^{new}||CID_i), T_i^{new} = T_i \oplus PWR_i \oplus PWR_i^{new}$. Then, stores REG_i^{new}, T_i^{new} instead of REG_i, T_i.

3 Security Analysis of Amin *et al.*'s Scheme

In this section, we point out security weakness of Amin *et al.*'s Scheme. There are some assumptions are made analysis and design of the scheme.

1. An adversary can eavesdrop all messages communicated between smartcard and remote server with public channel.
2. An adversary can steal smartcard and extract the contents of smartcard by the power consumption attack to the card [19, 20].
3. An adversary can alter, delete or resend the eavesdropped message.

3.1   Offline Password Guessing Attack

Suppose there are outsider attacker U_a, and he/she can steal smartcard and extract information of the smartcard, $< CID_i, REG_i, T_i, y_i, h(.), b >$, by assumption. Then U_a eavesdrops message $\{CID_i, SID_k, T_i, L_3, L_2, N_3\}$. Then U_a can easily do offline password guessing attack by follows:

1. Outsider attacker selects random password PW_i', calculates $PWR_i' = h(PW_i' \oplus b)$.
2. Then, the attacker make $L_3' = ((T_i \oplus PWR_i')||SID_k||(L_2 \oplus PWR_i' \oplus N_3)||L_2||N_3)$
3. If 2)'s result is equal to L_3, the attacker infers that guessing password PW_i' is user U_i's password.
4. Otherwise, outsider attacker selects another password and performs 1 3's step again, until he/she finds password.

3.2   Offline Identity Guessing Attack

Suppose there are outsider attacker U_a, and he/she successes guessing user's password by previous attack, he/she can also guess user's identity by follows:

1. Outsider attacker selects random identity ID_i', calculates $REG_i' = h(ID_i'||h(PW_i \oplus b)||CID_i)$.
2. If 2)'s result is equal to REG_i in the smartcard, the attacker infers that guessing identity ID_i' is user U_i's identity.
3. Otherwise, outsider attacker selects another identity and performs 1,2 step again, until he/she finds identity.

3.3   User Impersonation Attack

To impersonate as a legitimate user, an outsider adversary who successes guessing user's password by previous attack tries to make a login request message. Also, this login request message is authenticated to a server. Under assumption, Amin et al's scheme can not endure user impersonation attack by follows:

1. Outsider attacker selects a random nonce N_1', N_2'.
2. Computes $N_3' = N_1' \oplus N_2', L_2' = N_2' \oplus PWR_i$.
3. Computes $L_3' = h(T_i \oplus PWR_i||SID_k||N_1'||L_2'||N_3')$.
4. Outsider attacker sends $\{CID_i, SID_k, T_i, L_3', L_2', N_3'\}$ to the control server S_k to proof him/herself as a legitimate user.
5. Control server which received login message from outsider attacker, easily proved that received login message is valid.
6. After control server and service provider server's communicated, the outsider adversary successes impersonate as a legitimate user and has valid session key $SK = h(SID_k||CID_i||N_6||N_1')$.

3.4   Replay Attack and Session Key Derived Attack

Replay attack is a type of network attack in which valid data transmission is maliciously repeated. In this scheme, an outsider attacker U_a eavesdrops a message between a user and the server which communicate by public channel. U_a can try to use these messages for starting new communication to a server coming to performs the follows:

1. Outsider attacker U_a eavesdrops a login messages for preparing replay attack $\{CID_i, SID_k, T_i, L_3, L_2, N_3\}$
2. In future, U_a resends a login message to control server.
3. Resent messages can successfully pass server's verification check and control server constructs the authenticated session key $SK = h(SID_k || CID_i || N_6 || N_1)$.

But in this attack, the outsider adversary cannot achieve the session key because he/she doesn't know the N_1 and N_6's value. So, he/she just can disguise the user. But, the adversary who successfully guesses user's password. He/she also can get session key by follows:

1. The outsider adversary U_a calculates $N_1 = L_2 \oplus h(PW_i \oplus b) \oplus N_3$.
2. Then, derives $N_6 = N_1 \oplus N_7$. N_7 is in the message which communicate S_k and smartcard user.
3. Outsider adversary U_a can derive session key $SK = h(SID_k || CID_i || N_6 || N_1)$.

4   Proposed Secure and Efficient Scheme

In previous section, we proved that Amin *et al.*'s scheme are insecure against various threat. To endure these weakness, in this section, we proposed more robust user authentication scheme for multi-server environment using smartcard. The proposed scheme consists of four phases, registration phase, login phase, authentication phase and password change phase. All these phases are discussed as follows:

4.1   Registration Phase

When user U_i wants to access server for a service, Service provider registers to control server and User registers to service provider server. Registration phase is described as follows. Also details procedure of these phase are in Figs. 1 and 2.

Server Registration Phase. In this phase, it is same as Amin *et al.*'s scheme.

User Registration Phase. When user U_i wants to access server for a service, first, he/she registers with the control server CS. Therefore, U_i chooses his or her identity and password ID_i, PW_i and a random nonce b. After, submits ID_i, $PWR_i = h(PW_i \oplus b)$ to control server via a secure channel. Control server selects random nonce N_s computes $\alpha_i = h(X_s \oplus N_s)$, $\beta_i = h(\alpha_i||N_s)$, $T_i = h(PWR_i||ID_i) \oplus \beta_i$, $CID_i = \alpha_i \oplus \beta_i$, $REG_i = T_i \oplus \alpha_i$. Afterwards, Control server issues a smartcard and stores $\{CID_i, REG_i, T_i, N_s, h(.)\}$. After getting smartcard, user U_i stores b into smartcard.

4.2   Login Phase

This phase, user who wants to access the service server S_k, first he/she inserts his/her smartcard into the smart card and inputs ID'_i, PW'_i and chooses server identity SID_k. Then, smartcard computes $h(h(PW_i \oplus b)||ID_i) \oplus CID_i \oplus REG_i$ and checks it equals T_i. If not holds, reject the login. After that, user computes $\beta'_i = T_i \oplus h(PWR_i||ID_i), \alpha'_i = CID_i \oplus \beta'_i$, $L_1 = T_i \oplus \beta'_i$. Then, smartcard generates random numbers N_1, N_2 and timestamp T_1 and further computes $L_2 = N_2 \oplus \alpha_i$, $N_3 = N_1 \oplus N_2$, $L_3 = h(L_1||SID_k||N_1||L_2||T_1)$ and sends login request message $\{CID_i, SID_k, T_i, L_2, L_3, N_3, N_s, T_1\}$ to control server.

4.3   Authentication Phase

This phase, after receiving the login message, control server first verifies whether T_1 is within a tolerable period. If it holds, server then compute $L'_1 = T_i \oplus h(h(X_s \oplus N_s)||N_s), N'_1 = N_3 \oplus h(X_s \oplus N_s), L'_3 = h(L'_1||SID_k||N'_1||L_2||T_1)$. Then, checks it equals L_3. If not holds, reject the login. After that, server generates random numbers N_4, timestamp T_2, Then control server generates random nonce N_4 and computes $L_4 = h(SID_k||x), L_5 = L_4 \oplus N_4, N_5 = N'_1 \oplus N_4, L_6 = h(L_4||N_4||N'_1||CID_i||T_2)$. Then control server sends $\{CID_i, L_6, L_5, N_5, T_2\}$ to the service provide server S_k of the corresponding identity SID_k through public channel.

After receiving messages $\{CID_i, L_5, L_6, N_5, T_2\}$, service provide server first verifies whether T_2 is within a tolerable period. If it holds, service provide server S_k derives $N'_4 = P_k \oplus L_5, N'_1 = N_5 \oplus N'_4$ and computes $L'_6 = h(P_k||N'_4||N'_1||CID_i)$. Then checks it equals L_6. If not holds, reject the authentication request. After that, service provide server generate random nonce N_6, and timestamp T_3 and further computes $N_7 = N'_1 \oplus N_6, SK = h(SID_k||CID_i||N_6||N'_1||T_3), L_7 = h(SK||N_6)$ and sends $\{SID_k, L_7, N_7, T_3\}$ to the user U_i through public channel.

After receiving messages $\{SID_k, L_7, N_7, T_3\}$ from the server. Smartcard first verifies whether T_3 is within a tolerable period. If it holds, smartcard derives $N'_6 = N_7 \oplus N_1, SK' = h(SID_k||CID_i||N'_6||N_1||T_3), L'_7 = h(SK'||N'_6)$. Then, checks it equals L_7. If it holds, the mutual authentication are successfully shared by session key SK between user U_i and service provider server S_k.

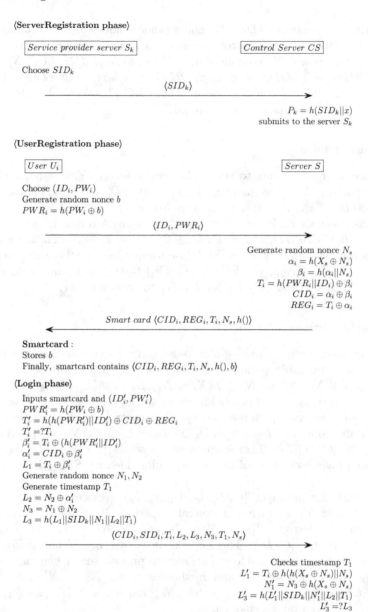

Fig. 1. Registration and Login phase of our proposed scheme

4.4   Password Change Phase

If user intends to alter his password into an another password, first he/she inputs original identity and password ID_i, PW_i and smartcard computes $h((PW_i \oplus b)||ID_i) \oplus CID_i \oplus REG_i$ and checks it equals T_i. If not holds, reject the password change request. If it holds, user inputs his/her new password PW_i^{new}. Smartcard computes $T_i^{new} = T_i \oplus h((PW_i \oplus b)||ID_i) \oplus h((PW_i^{new} \oplus b)||ID_i)$. $REG_i^{new} = REG_i \oplus T_i \oplus T_i^{new}$. Then, stores T_i^{new}, REG_i^{new} instead of T_i, REG_i.

Fig. 2. Authentication and Password change phase of our proposed scheme

5 Security Analysis of Proposed Scheme

In this section, we check that the proposed scheme is safe and can resist various attacks compared with Amin *et al.*'s scheme and other existed scheme.

Theorem 1. *Proposed scheme resists insider attack*

Proof. There is a possibility that a server's insider can directly get the user's password from the registration message. This attack can cause server's insider can guess the user's password. In our proposed scheme, smartcard user sends his/her password information to server in a form of $PWR_i = h(PW_i \oplus b)$ instead of $h(PW_i)$. Thus, server's privileged insider is not able to guess the user generated random nonce b. Thus, the privileged insider can not gain the user's password.

Theorem 2. *Proposed scheme does not need verification table*

Proof. Some user authentication schemes requirement storing some information into verification table in the server. However, our proposed scheme does not need storing the nonce used for login or authentication phase. Also, using verification table can bring about continuing costs of operation problems in server and also some risk about stolen-verifier attack. Our proposed scheme does not require verification table, and so, we know that our scheme prevents overhead, stolen-verifier attack.

Theorem 3. *Proposed scheme supports mutual authentication*

Proof. Mutual authentication is important characteristic of authentication scheme. In our proposed scheme, we have discussed earlier that Sect. 2.3, service provider server, control server and user all check their opposite party's message's valid by checking if derived value L_3, L_6, L_7 is equal to received message, and If not holds, each party reject the phase. If whole these verification are successfully finished, mutual authentication executed correctly.

Theorem 4. *Proposed scheme can detect wrong password quickly*

Proof. Mobile user who uses smartcard sometimes confuses his/her password. Moreover, adversary who want to Denial of Service attack to server in this case, he/she puts wrong identity and password continuously. Our scheme can detect whether input password right or wrong before user sends login message to server, in smartcard. When the mobile user inputs incorrect password PW_i^*, ID_i^* the smartcard calculates $h(PW_i^* \oplus b)||ID_i)$ and compares it $T_i \oplus CID_i \oplus REG_i$. If this condition is not satisfied, smartcard recognizes that user inputs the wrong identity or password and reject login request quickly.

Theorem 5. *Proposed scheme resists offline identity guessing attack*

Proof. According to our proposed scheme, even though attacker can take out all information from smartcard and messages and all communicated message via public channel, but user's identity is protected by server's secret key X_s and it is impossible to guess or extract the server's secret key in our proposed scheme. Therefore, out proposed scheme is secure withstand offline identity guessing attack.

Theorem 6. *Proposed scheme resists offline password guessing attack*

Proof. In accordance with our proposed scheme, only method to guessing user's password is using $T_i = h(h(PW_i^{\oplus}b)||ID_i)$, but we showed in the previous Theorem 5 that outsider attacker can not guess user's identity. Therefore it is also impossible guessing user's identity work up to not guessing user's password.

Theorem 7. *Proposed scheme resists user impersonation attack*

Proof. A violation attacker tries to impersonate legal user so as to do trick opposite party. To start a new session, attacker must make permissible user's L_2, L_3, N_3. However, L_2, L_3 both contains legal user's identity ID_i and password PW_i. We have mentioned earlier that it is infeasible adversary guessing legal user's not only identity but also password. So, attacker can not generate legal user's login message and disguise as a legal user.

Theorem 8. *Proposed scheme resists replay attack*

Proof. Even though an adversary can eavesdrop all transfer messages between user and server by public channel, and using them to make new session or impersonate user. All messages have current timestamp for example T_1, T_2, T_3 and the other party checks the period of delay. Hence, our proposed scheme can resist replay attack.

Theorem 9. *Proposed scheme provides known-key security*

Proof. The contains of session key does not any connection other session keys. Even though a current session key reveals adversary, he/she cannot derived other session keys as ever. Because session key contains randomly generated number N_1 and N_6, also these number are always replacement new value when user processes login phase. Therefore, the proposed scheme can resist known-key-security.

Theorem 10. *Proposed scheme provides a efficient and safe password change phase*

Proof. In our proposed scheme, when the user want to change his/her password, either he/she is legal or illegal user, the smartcard checks old password PW_i^{old}'s legitimacy first. After verify check finished, the user can replace old password to new password PW_i^{new}. All these progresses are not communication with server, only smartcard. Therefore, our proposed password change phase is both efficient and safety.

Theorem 11. *Proposed scheme supports freely password choice*

Proof. In our proposed scheme, in the registration phase, user can choice his/her identity and password freely for remote server access.

Table 2 shows a security features of our proposed scheme and other related schemes. [16–18]

Table 2. Security comparison of the proposed scheme and other related schemes

Features	Amin [16]	Sood [17]	Li *et al.* [18]	Proposed scheme
Theorem 1	O	X	O	O
Theorem 2	O	O	O	O
Theorem 3	O	O	O	O
Theorem 4	O	O	O	O
Theorem 5	X	X	X	O
Theorem 6	X	X	X	O
Theorem 7	X	X	X	O
Theorem 8	X	X	X	O
Theorem 9	X	O	O	O
Theorem 10	O	O	X	O
Theorem 11	O	O	O	O

6   Conclusion

In this paper, we introduced Amin *et al*'s scheme for multi-server environment and showed that Amin *et al*'s scheme has several security flaws such as offline password guessing attack, user impersonate attack, replay attack and so on. Further, we proposed extended scheme which is more efficient and satisfied security feature compared to not only Amin *et al*'s scheme but also other existed scheme. In future, we can adopt this improved scheme with biometric features such as fingerprint, iris to provide more highly safety system.

Acknowledgments. This work was supported by Institute for Information and communications Technology Promotion(IITP)grant funded by the Korea government(MSIP)(No.R0126-15-1111, The Development of Risk-based Authentication Access Control Platform and Compliance Technique for Cloud Security).

References

1. Moon, J., et al.: An improvement of robust biometrics-based authentication and key agreement scheme for multi-server environments using smart cards. PloS one **10**(12), e0145263 (2015)
2. Choi, Y., et al.: Security enhanced anonymous multiserver authenticated key agreement scheme using smart cards and biometrics. Sci. World J. **2014**, 1–15 (2014)
3. Moon, J., et al.: Improvement of biometrics and smart cards-based authentication scheme for multi-server environments. In: Proceedings of the 10th International Conference on Ubiquitous Information Management and Communication. ACM (2016)
4. Jeon, W., Lee, Y., Won, D.: An efficient user authentication scheme with smart cards for wireless communications. Int. J. Secur. Appl. **7**(4), 1–5 (2013)

5. Kyungho, S., DongGuk, H., Dongho, W.: A privacy-protecting authentication scheme for roaming services with smart cards. IEICE Trans. Commun. **95**(5), 1819–1821 (2012)
6. Jung, J., et al.: Cryptanalysis and improvement of efficient password-based user authentication scheme using hash function. In: Proceedings of the 10th International Conference on Ubiquitous Information Management and Communication. ACM (2016)
7. Ford, W., Kaliski Jr., B.S.: Server-assisted generation of a strong secret from a password. In: Proceedings of the IEEE 9th International Workshops on Enabling Technologies: Infrastructure for Collaborative Enterprises, (WET ICE 2000). IEEE (2000)
8. Jablon, D.P.: Password authentication using multiple servers. In: Naccache, D. (ed.) CT-RSA 2001. LNCS, vol. 2020, pp. 344–360. Springer, Heidelberg (2001)
9. Lin, I.-C., Hwang, M.-S., Li, L.-H.: A new remote user authentication scheme for multi-server architecture. Future Gener. Comput. Syst. **19**(1), 13–22 (2003)
10. Hu, L., Niu, X., Yang, Y.: An efficient multi-server password authenticated key agreement scheme using smart cards. In: International Conference on Multimedia and Ubiquitous Engineering, MUE 2007. IEEE (2007)
11. Tsai, J.-L.: Efficient multi-server authentication scheme based on one-way hash function without verification table. Comput. Secur. **27**(3), 115–121 (2008)
12. Liao, Y.-P., Wang, S.-S.: A secure dynamic ID based remote user authentication scheme for multi-server environment. Comput. Stand. Interfaces **31**(1), 24–29 (2009)
13. Li, X., et al.: A novel smart card and dynamic ID based remote user authentication scheme for multi-server environments. Math. Comput. Model. **58**(1), 85–95 (2013)
14. Xue, K., Hong, P., Ma, C.: A lightweight dynamic pseudonym identity based authentication and key agreement protocol without verification tables for multi-server architecture. J. Comput. Syst. Sci. **80**(1), 195–206 (2014)
15. Lu, Y., et al.: A lightweight ID based authentication, key agreement protocol for multiserver architecture. Int. J. Distrib. Sens. Netw. **2015**, 16 (2015)
16. Amin, R.: Cryptanalysis, efficient dynamic ID based remote user authentication scheme in multi-server environment using smart card. Int. J. Netw. Secur. **18**(1), 172–181
17. Sood, S.K.: Dynamic identity based authentication protocol for two-server architecture. J. Inf. Secur. **3**(04), 326 (2012)
18. Li, C.-T., Weng, C.Y., Fan, C.I.: Two-factor user authentication in multi-server networks. Int. J. Secur. Appl. **6**(2), 261–267 (2012)
19. Kocher, P.C., Jaffe, J., Jun, B.: Differential power analysis. In: Wiener, M. (ed.) CRYPTO 1999. LNCS, vol. 1666, pp. 388–397. Springer, Heidelberg (1999)
20. Messerges, T.S., Dabbish, E.A., Sloan, R.H.: Examining smart-card security under the threat of power analysis attacks. IEEE Trans. Comput. **51**(5), 541–552 (2002)

An Operational Semantics for Android Applications

Mohamed A. El-Zawawy[1,2(✉)]

[1] College of Computer and Information Sciences,
Al Imam Mohammad Ibn Saud Islamic University (IMSIU),
Riyadh, Kingdom of Saudi Arabia
[2] Department of Mathematics, Faculty of Science, Cairo University,
Giza 12613, Egypt
maelzawawy@cu.edu.eg

Abstract. The most common operating system for smart phones, which are among the most common forms of computers today, is the Android operating system. The Android applications are executed on Dalvik Virtual Machines (DVMs) which are register-based, rather than stack-based such as Java Virtual Machines (JVMs). The differences between DVMs and JVMs make tools of programm analysis of JVMs are not directly applicable to DVMs. Operational semantics is a main tool to study, verify, and analyze Android applications.

This paper presents an accurate operational semantics *AndroidS* for Android programming. The set of Dalvik instruction considered in the semantics is designed carefully to capture main functionalities of Android programming and to enable the use of semantics to evaluate method of Application analysis. The semantics also simulates the interaction between users and applications during application executions. The semantics also respects constrains of state changes imposed by the life cycle of Android applications.

Keywords: Android platform · Android programming · Operational semantics · Android semantics · Android analysis · *AndroidP* · *AndroidS*

1   Introduction

Examples of smart-devices operating systems are Android and iOS. However Android is the most widely used operating system for smart and embedded devices. The main language to write Android applications is Java (using the rich resource of Java standard libraries). Towards execution, these applications are compiled into Dalvik bytecode [1,22]. Reliability of Android applications is a serious issue, compared to iOS, because digital authentication is not considered before installation. One of the common sources for Android applications is Google Play store. Therefore bad and malicious Android applications can easily find their way to our smart devices. This can result in stealing private data and degrading the performance of the devices [22].

© Springer International Publishing Switzerland 2016
O. Gervasi et al. (Eds.): ICCSA 2016, Part V, LNCS 9790, pp. 100–114, 2016.
DOI: 10.1007/978-3-319-42092-9_9

The wide spread of Android platform as operating system for most mobile devices makes the verification of Android program extremely important. However classical techniques of program analysis [8,9] (applicable to Java programs for example) are not directly applicable to Android applications [18,19,24]. This is so due to the following facts. Dalvik Virtual Machines (DVMs) are the main tools to execute Android applications. There are many discrepancies between Java Virtual Machines (JVMs - executing Java programs) and DVMs. while JVMs are stack-based, DVMs are register-based. Therefore DVMs relies on registers to determine parameters of operations. This role of registers played by operand stacks in JVMs. As a result, the architecture of instruction sets in JVMs and DVMs are not the same. For example while one instruction transfers data in DVMs, many instructions are required to do the same job in JVMs [15].

In this paper, we present a formal operational semantics $AndroidS$ to important components of Android applications including Dalvik bytcode instructions. The semantics is presented on a rich, yet simple, model of Android applications $AndroidP$. The states of semantics are designed to capture accurately the DVMs states during computation. Yet the semantic states are simple to facilitate the further use of semantics to evaluate program analyses developed for Android applications. The proposed semantics considers different issues that are special to Android applications such as gestures (such as clicks and swipes). This means that the semantics, to some extent, simulates the reactions of Android applications to gestures of application users.

The operational semantics $AndroidS$ proposed in this paper can be described as activity centric in the sense that main components of semantic states are activities representations. This is inline with the structure of Android applications which are composed mainly of activities. The proposed semantics $AndroidS$ also takes into consideration and adheres to the constraints of the life cycle of Android activities. This requires constrain the order of executing macros during the run of an application. One important application of the semantics presented in this paper is the verification of the correctness of new method for program analysis [5,20,21,28] of Android Applications.

Compared to most related work such as [15,22], The proposed operational semantics $AndroidS$ has the following advantages:

- The number and the types of Dalvik instructions considered in $AndroidS$ are convenient;
- The semantics captures the semantic of instructions in a simpler formalization;
- More aspects of Android programming are simulated by the semantics.

Motivation. The work in this paper is motivated by the need to an operational semantics for Android programming that

- covers main important Dalvik instructions,
- expresses Android applications interaction with users accurately and simply, and

– expresses simply executions of macros instructions and respects the constraints enforced by the life cycle of Android activities.

Paper Outline

The remaining of the paper is organized as follows. Section 2 presents the syntax of the programming language *AndroidP* used to present the operational semantics introduced in Sect. 3. Section 4 reviews work most related to our proposed semantics. The paper is concluded in Sect. 5.

2 Syntax

The Android programming model, *AndroidP*, presented in this section is an improvement to the syntax models used in [1,15,22].

The grammar of our programming model, *AndroidP*, for Android applications is presented in Fig. 1. The set of types in the model is well-equipped with variety of types including integers, Boolean values, double numbers, classes and arrays. A number of 19 Dalvik instructions are present in the model. Out of the original full set of more than 200 Dalvik instructions, the 19 inductions were selected carefully. Another important type of instructions in *AndroidP* is the set of macros instructions [1,15,22].

Aiming at designing a simple and powerful grammar and to cover main functionalities of Dalivk bytecode, the selection of macros instructions and Dalvik instructions in *AndroidP* was made carefully. The criteria of selecting and designing the macros instructions was to include those used in almost all Android applications and those control the application executions. Hence most analysis techniques of Android applications would study these instructions. Therefore our semantics have good chances to support analysis techniques of Android applications by proving an accurate representations to instructions semantics.

In *AndroidP*, The commands set is a sequence of Dalvik instructions and macros instructions. Methods have the structure (*method m, c, t, t*, com*) where

– m denotes the method name,
– c denotes the class of the method,
– t denotes the return type of the method,
– t^* denotes the sequence of type for method arguments, and
– *textitcom* denotes the method body.

The syntax of classes fields in *AndroidP* has the form (*field f, c, t*) where f is the field name, c is the class owns the field, and t is the field type. Every class in *AndroidP* is an extension of *AppCompatActivity* and includes the following parts:

– c is the class name,
– *app* is the application name owns the class,
– A field *root* of type integer pointing to the view file of the activity owns the class,

$n \in N.$

$l \in \mathcal{L}$ = The set of all memory locations.

$v \in \mathcal{V} = \mathcal{L} \cup$ Integers.

$\delta \in \Delta$ = The set of all labels of program points.

$f \in \mathcal{F}$ = The set of all names of class fields.

$m \in Me$ = The set of all method names.

$c \in C$ = The set of all class names.

$app \in \mathcal{A}$ = The set of all application names.

$t \in$ Types ::= {int, Boolean, ref} $\cup\, C \cup$ array t.

$i \in$ Instructions ::= nop | const i, v | move i, j | goto δ | new-array i, j, t |

array-length i, j | aget i, j, k | aput i, j, k | new-instance i, c |

iget i, j, f | iput i, j, f | invoke-direct $i_1, \ldots, i_n$, method m |

neg-int i, j | add i, j, k | if-eq i, j, k | if-eqz i, j | move-result i |

return-void | return i.

$mc \in$ Macros ::= SetContentView xml | findViewById i, j |

startActivityForResut A | setResult i | finish.

$com \in$ Commands ::= i | mc | skip | Label δ | com_1; com_2.

$M \in$ Methods ::= (method m, c, t, t^*, com).

$F \in$ Fields ::= (field f, c, t).

$C \in$ Classes ::= class $c \lhd$ AppCompatActivity

{app; (field, root, c, int); (field, result, c, int);

(field, finished, c, Boolean); F^*; M^*}.

$App \in$ Applications ::= (Application app, C^*).

Fig. 1. Android$\mathcal{P}$: a model for android applications.

- A field *result* of type integer recording the result of the running an object (activity) of the class,
- A field *finished* of type Boolean to determine whether an object (activity) of the class is finished,
- F^* a sequence of other fields, and
- M^* a sequence of method definitions.

The syntax of an application in *Android$\mathcal{P}$* is of the form (*Application app, C^**) where *app* denotes the application name *app* and C^* denotes a sequence of classes.

Android$\mathcal{P}$ can be thought of as an improvement of grammars in [1,15,22]. The objective of *Android$\mathcal{P}$* is to overcome disadvantages in the programming models in these related works. More specifically, the idea of including a set of macro instructions and the three special fields of classes was inspired by [22]. However, we find the programming model of [22] a very simple one while that of [1,15] a very involved one. Hence there was a need to develop our moderate model. We use the same model $Android\mathcal{P}$ to develop a type system for Android application in [7].

3   Semantics

This section presents an operation semantics [12,14,26,27], $Android\mathcal{S}$, to Android applications using the model $Android\mathcal{P}$ presented in the previous section. Definition 1 presents gradually the states of the semantics.

Definition 1. 1. $Values_\perp = Integers \cup \mathcal{L} \cup \{null\}$.
2. $v \in Values = Values_\perp \cup \{(Values_\perp)^n \mid n \in N\}$.
3. $Registers = N_0 \cup \{MVal, CAct\}$.
4. $R \in Reg\text{-}Val = Registers \rightharpoonup Values$.
5. $Task = Methods \times Reg\text{-}val$.
6. $P \in Partial\text{-}Methods = \{(method\ m, c, l, t, t^*, com) \mid (method\ m, c, t, t^*, com') \in methods \land com\ is\ s\ sub\text{-}command\ of\ com' \land l \in \mathcal{L}\}$.
7. $(P, R) \in Frames = Partial\text{-}Methods \times Reg\text{-}val$.
8. $F_c \in Fields_c = \{(field\ f, c, t) \mid (field\ f, c, t) \in Fields\}$
9. $O_c \in Objects_c = Fields_c \rightharpoonup Values$.
10. $O \in Objects = \{Objects_c \mid c \in \mathcal{C}\}$.
11. $H \in Heaps = Memory\text{-}Locations \rightharpoonup Objects$.
12. $(H\|\Theta) \in Confgs = Heaps \times Frames^*$.
13. The set of different memory locations referenced by members of a sequence of Frames Θ is denoted by $loc(\Theta)$.

A semantic state is composed of a heap and a stack for frames. A heap is a partial map from memory locations into classes objects. A class object is a partial map form the fields of a class to the set of values. The set of values includes integers, locations, the null value, and tuples values. A frame is a pair of partial frame and a register valuation. A register valuation is a partial map from registers, necessary to execute the commands in the associated partial frame, to values. Partial frames has the same syntax and definitions of methods but the command component of Partial frames is part of that of methods. Therefore the partial frames denote partial execution of methods and include the remaining commands to be executed.

Figure 2 presents the first set of semantics rules of the proposed semantics $Android\mathcal{S}$, for $Android\mathcal{P}$. Rule 5 presents the semantics of the instruction new-array i, j, t. The rule assumes a function type-size that calculates the size, m, of a given type. The register j includes the length, n, of the required array. The rule allocates fresh locations $\{l, l+1, \dots, l+m*n-1\}$ in the heap for the required array. The new locations are initialized to 0 in the modified heap H'. Example 1 provides an example of applying Rule 5.

$$\frac{\mathrm{com} = \mathrm{nop};\mathrm{com}' \vee \mathrm{com} = \mathrm{skip};\mathrm{com}' \vee \mathrm{com} = \mathrm{Label}\,\delta;\mathrm{com}'}{\begin{array}{l}(H\|((\mathrm{method}\,m,c,t,t^*,\mathrm{com}),R)::\Theta)\\ \to (H\|((\mathrm{method}\,m,c,t,t^*,\mathrm{com}),R)::\Theta)\end{array}} \quad (1)$$

$$\frac{\mathrm{com} = \mathrm{const}\,i\,v;\mathrm{com}'}{\begin{array}{l}(H\|((\mathrm{method}\,m,c,t,t^*,\mathrm{com}),R)::\Theta)\\ \to (H\|((\mathrm{method}\,m,c,t,t^*,\mathrm{com}'),R[i \mapsto v])::\Theta)\end{array}} \quad (2)$$

$$\frac{\mathrm{com} = \mathrm{move}\,i,j;\mathrm{com}'}{\begin{array}{l}(H\|((\mathrm{method}\,m,c,t,t^*,\mathrm{com}),R)::\Theta)\\ \to (H\|((\mathrm{method}\,m,c,t,t^*,\mathrm{com}'),R[i \mapsto R(j)])::\Theta)\end{array}} \quad (3)$$

$$\frac{\begin{array}{l}\mathrm{com} = \mathrm{goto}\,\delta;\mathrm{com}''\\ (\mathrm{method}\,m,c,t,t^*,\mathrm{com}'''\mathrm{Label}\,\delta\,\mathrm{com}') \in \mathrm{Methods}\end{array}}{\begin{array}{l}(H\|((\mathrm{method}\,m,c,t,t^*,\mathrm{com}),R)::\Theta)\\ \to (H\|((\mathrm{method}\,m,c,t,t^*,\mathrm{com}'),R)::\Theta)\end{array}} \quad (4)$$

$$\frac{\begin{array}{l}\mathrm{com} = \mathrm{new\text{-}array}\,i,j,t;\mathrm{com}' \wedge R(j) = n\\ \mathrm{type\text{-}size}(t) = m \wedge \{l,l+1,\ldots,l+m*n-1\} \in \mathcal{L} \setminus \mathit{dom}(H)\\ H' = H[l \mapsto 0, l+1 \mapsto 0, \ldots, l+m*n-1 \mapsto 0]\end{array}}{\begin{array}{l}(H\|((\mathrm{method}\,m,c,t,t^*,\mathrm{com}),R)::\Theta)\\ \to (H'\|((\mathrm{method}\,m,c,t,t^*,\mathrm{com}'),R[i \mapsto l])::\Theta)\end{array}} \quad (5)$$

$$\frac{\mathrm{com} = \mathrm{array\text{-}length}\,i,j;\mathrm{com}' \wedge R(j) \in \mathit{dom}(H)}{\begin{array}{l}(H\|((\mathrm{method}\,m,c,t,t^*,\mathrm{com}),R)::\Theta)\\ \to (H\|((\mathrm{method}\,m,c,t,t^*,\mathrm{com}'),R[i \mapsto R(j).\mathit{lenght}])::\Theta)\end{array}} \quad (6)$$

$$\frac{\begin{array}{l}\mathrm{com} = \mathrm{aget}\,i,j,k;\mathrm{com}'\\ R(j) \in \mathit{dom}(H) \wedge 0 \le R(k) \le \mathrm{type\text{-}size}(R(j)) - 1\end{array}}{\begin{array}{l}(H\|((\mathrm{method}\,m,c,t,t^*,\mathrm{com}),R)::\Theta)\\ \to (H\|((\mathrm{method}\,m,c,t,t^*,\mathrm{com}'),R[i \mapsto H(R(j)+R(k))])::\Theta)\end{array}} \quad (7)$$

$$\frac{\begin{array}{l}\mathrm{com} = \mathrm{aput}\,i,j,k;\mathrm{com}'\\ R(j) \in \mathit{dom}(H) \wedge 0 \le R(k) \le \mathrm{type\text{-}size}(R(j)) - 1\end{array}}{\begin{array}{l}(H\|((\mathrm{method}\,m,c,t,t^*,\mathrm{com}),R)::\Theta)\\ \to (H[(R(j)+R(k)) \mapsto R(i)\|((\mathrm{method}\,m,c,t,t^*,\mathrm{com}'),R)::\Theta)\end{array}} \quad (8)$$

$$\frac{\mathrm{com} = \mathrm{move\text{-}result}\,i;\mathrm{com}'}{\begin{array}{l}(H\|((\mathrm{method}\,m,c,t,t^*,\mathrm{com}),R)::\Theta)\\ \to (H\|((\mathrm{method}\,m,c,t,t^*,\mathrm{com}'),R[i \mapsto R(\mathrm{MVal})])::\Theta)\end{array}} \quad (9)$$

$$\frac{\mathrm{com} = \mathrm{return\text{-}void};\mathrm{com}''}{\begin{array}{l}(H\|((\mathrm{method}\,m,c,l,\mathrm{void},t^*,\mathrm{com}),R)::(\mathrm{method}\,m',c',l',t',t^{*'},\mathrm{com}'),R')::\Theta)\\ \to (H\|((\mathrm{method}\,m',c',l',t',t^{*'},\mathrm{com}'),R')::\Theta)\end{array}} \quad (10)$$

$$\frac{\mathrm{com} = \mathrm{return}\,i;\mathrm{com}''}{\begin{array}{l}(H\|((\mathrm{method}\,m,c,l,\mathrm{void},t^*,\mathrm{com}),R)::(\mathrm{method}\,m',c',l',t',t^{*'},\mathrm{com}'),R')::\Theta)\\ \to (H\|((\mathrm{method}\,m',c',l',t',t^{*'},\mathrm{com}'),R'[\mathrm{MVal} \mapsto R(i)])::\Theta)\end{array}} \quad (11)$$

Fig. 2. Semantics rules of the operational semantics Android$\mathcal{S}$ (Set 1).

Example 1. *Rule 12 is an example of applying Rule 5.*

$$com = new\text{-}array\ i, j, int;\ com' \land R(j) = 2$$
$$type\text{-}size(t) = 1 \land \{l, l+1\} \in \mathcal{L} \setminus dom(\emptyset)$$
$$\underline{H' = \{l \mapsto 0, l+1 \mapsto 0\}}$$

$$(\emptyset \| ((method\ m, c, t, t^*, com), \{j \mapsto 2\}) :: \Theta)$$
$$\rightarrow (H' \| ((method\ m, c, t, t^*, com'), \{j \mapsto 2, i \mapsto l\}) :: \Theta)$$

(12)

Figure 3 presents the second set of semantics rules of the proposed semantics *AndroidS*, for *AndroidP*. Rules 20 presents the semantics of the instruction *if-eq* i, j, k. The rule simply checks whether the registers i and j have equal values. If this is the case, the control of the execution is transferred to program point whose label is stored in register k.

Example 2. *Rule 24 presents an example of applying Rule 20.*

$$com = if\text{-}eq\ i, j, k;\ com' \land R(i) = R(j)$$

$$(H \| ((method\ m, c, t, t^*, com), \{i \mapsto 1, j \mapsto 1, k \mapsto \text{``}p_1\text{''}\}) :: \Theta)$$
$$\rightarrow (H \| ((method\ m, c, t, t^*, goto\ p_1; com'), \{i \mapsto 1, j \mapsto 1, k \mapsto \text{``}p_1\text{''}\}) :: \Theta)$$

(24)

Figure 4 presents the third set of semantics rules of the proposed semantics *AndroidS*, for *AndroidP*. Rule 29 presents the semantics of the macro insertion *finish*. The rule assumes a function *remove-frames*(Θ, l) that removes from the frame stack Θ all the partial frames whose location is l and returning the remained frame stack Θ'. This is done because the instruction *finish* ends a complete activity with all its methods. The rule also assumes that the frame stack has partial frames belongs to other memory locations (l') rather than l (other activities). The rule also updates the field "finished" of the object at location l into *true*.

Constraints for Method Transitions: To preserve the constraints on method transitions in our semantics forced by the life cycle of activities, the rules for *return-void* (10), *return i* (11), and *finish* (29) are carried out subject to the rules of Fig. 5.

Definition 2 presents necessary terminology to present semantic rules of Fig. 5.

Definition 2. – *An android activity method (**AAM**) is any of the classical life-cycle methods defined in the superclass "AppCompatActivity": {constructor, onStart, onCreate, onRestart, onActivityResult, onResume, onPause, onDestroy, onStop}.*
– *An android user method (**AUM**) is a method that is not an AAM.*
– *An AAM is called freezing AAM (**FAAM**) if it is onPause, onDestroy, or onStop.*

$$com = \textit{new-instance } i, c; com' \wedge R(j) = n$$
$$\textit{type-size}(c) = m \wedge \{l, l+1, \ldots, l+m-1\} \in \mathcal{L} \setminus \textit{dom}(H)$$
$$O_c \in \textit{Objects}_c \wedge H' = H[l, \ldots, l+m-1 \mapsto O_c] \tag{13}$$

$$(H\|((\textit{method } m, c, t, t^*, com), R) :: \Theta)$$
$$\rightarrow (H'\|((\textit{method } m, c, t, t^*, com'), R[i \mapsto l]) :: \Theta)$$

$$com = \textit{iget } i, j, f; com'$$
$$R(j) \in \textit{dom}(H) \wedge \textit{class}(H(R(j))) = c \wedge \textit{field-order}(c, f) = n \tag{14}$$

$$(H\|((\textit{method } m, c, t, t^*, com), R) :: \Theta)$$
$$\rightarrow (H\|((\textit{method } m, c, t, t^*, com'), R[i \mapsto H(R(j) + n)]) :: \Theta)$$

$$com = \textit{iput } i, j, f; com'$$
$$R(j) \in \textit{dom}(H) \wedge \textit{class}(H(R(j))) = c \wedge \textit{field-order}(c, f) = n \tag{15}$$

$$(H\|((\textit{method } m, c, t, t^*, com), R) :: \Theta)$$
$$\rightarrow (H[R(j) + n) \mapsto R(i)\|((\textit{method } m, c, t, t^*, com'), R) :: \Theta)$$

$$com = \textit{invoke-direct } i_1, \ldots, i_n, \textit{method } m'; com''$$
$$R[i_1] \in \textit{dom}(H) \wedge \textit{class}(H(R(i_1))) = c$$
$$\textit{find-method}(c, \textit{method } m') = (\textit{method } m', c', t', t'^*, com')$$
$$R' = \{0 \mapsto R(i_1), \ldots, n-1 \mapsto R(i_n)\} \tag{16}$$

$$(H\|((\textit{method } m, c, t, t^*, com), R) :: \Theta)$$
$$\rightarrow (H\|((\textit{method } m', c', t', t'^*, com'), R') ::$$
$$(\textit{method } m, c, t, t^*, com''), R) :: \Theta)$$

$$com = \textit{neg-int } i, j; com'$$

$$(H\|((\textit{method } m, c, t, t^*, com), R) :: \Theta)$$
$$\rightarrow (H\|((\textit{method } m, c, t, t^*, com'), R[i \mapsto -R(j)]) :: \Theta) \tag{17}$$

$$com = \textit{add } i, j, k; com'$$

$$(H\|((\textit{method } m, c, t, t^*, com), R) :: \Theta)$$
$$\rightarrow (H\|((\textit{method } m, c, t, t^*, com'), R[i \mapsto R(j) + R(k)]) :: \Theta) \tag{18}$$

$$com = \textit{add } i, j, k; com'$$

$$(H\|((\textit{method } m, c, t, t^*, com), R) :: \Theta)$$
$$\rightarrow (H\|((\textit{method } m, c, t, t^*, com'), R[i \mapsto R(j) + R(k)]) :: \Theta) \tag{19}$$

$$com = \textit{if-eq } i, j, k; com' \wedge R(i) = R(j)$$

$$(H\|((\textit{method } m, c, t, t^*, com), R) :: \Theta)$$
$$\rightarrow (H\|((\textit{method } m, c, t, t^*, \textit{goto } R(k); com'), R) :: \Theta) \tag{20}$$

$$com = \textit{if-eq } i, j, k; com' \wedge R(i) \neq R(j)$$

$$(H\|((\textit{method } m, c, t, t^*, com), R) :: \Theta)$$
$$\rightarrow (H\|((\textit{method } m, c, t, t^*, com'), R) :: \Theta) \tag{21}$$

$$com = \textit{if-eq } i, j; com' \wedge R(i) = 0$$

$$(H\|((\textit{method } m, c, t, t^*, com), R) :: \Theta)$$
$$\rightarrow (H\|((\textit{method } m, c, t, t^*, \textit{goto } R(j); com'), R) :: \Theta) \tag{22}$$

$$com = \textit{if-eqz } i, j; com' \wedge R(i) \neq 0$$

$$(H\|((\textit{method } m, c, t, t^*, com), R) :: \Theta)$$
$$\rightarrow (H\|((\textit{method } m, c, t, t^*, com'), R) :: \Theta) \tag{23}$$

Fig. 3. Semantics rules of the operational semantics Android$\mathcal{S}$ (Set 2).

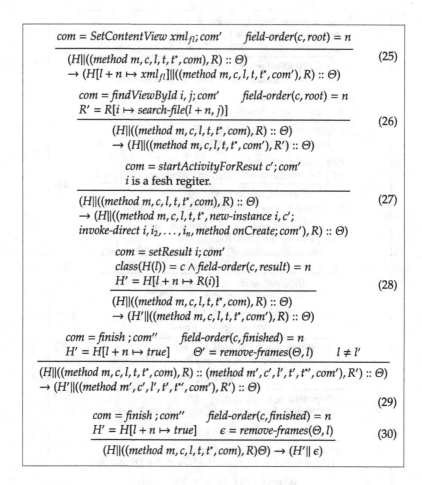

$$\frac{com = SetContentView\ xml_{fl};\, com' \qquad field\text{-}order(c, root) = n}{\begin{array}{c}(H\|((method\ m, c, l, t, t^*, com), R) :: \Theta) \\ \rightarrow (H[l + n \mapsto xml_{fl}]\|((method\ m, c, l, t, t^*, com'), R) :: \Theta)\end{array}} \quad (25)$$

$$\frac{\begin{array}{c}com = findViewById\ i, j;\, com' \qquad field\text{-}order(c, root) = n \\ R' = R[i \mapsto search\text{-}file(l + n, j)]\end{array}}{\begin{array}{c}(H\|((method\ m, c, l, t, t^*, com), R) :: \Theta) \\ \rightarrow (H\|((method\ m, c, l, t, t^*, com'), R') :: \Theta)\end{array}} \quad (26)$$

$$\frac{\begin{array}{c}com = startActivityForResut\ c';\, com' \\ i\ is\ a\ fesh\ regiter.\end{array}}{\begin{array}{c}(H\|((method\ m, c, l, t, t^*, com), R) :: \Theta) \\ \rightarrow (H\|((method\ m, c, l, t, t^*, new\text{-}instance\ i, c'; \\ invoke\text{-}direct\ i, i_2, \ldots, i_n, method\ onCreate;\, com'), R) :: \Theta)\end{array}} \quad (27)$$

$$\frac{\begin{array}{c}com = setResult\ i;\, com' \\ class(H(l)) = c \wedge field\text{-}order(c, result) = n \\ H' = H[l + n \mapsto R(i)]\end{array}}{\begin{array}{c}(H\|((method\ m, c, l, t, t^*, com), R) :: \Theta) \\ \rightarrow (H'\|((method\ m, c, l, t, t^*, com'), R) :: \Theta)\end{array}} \quad (28)$$

$$\frac{\begin{array}{c}com = finish\ ;\, com'' \qquad field\text{-}order(c, finished) = n \\ H' = H[l + n \mapsto true] \qquad \Theta' = remove\text{-}frames(\Theta, l) \qquad l \neq l'\end{array}}{\begin{array}{c}(H\|((method\ m, c, l, t, t^*, com), R) :: (method\ m', c', l', t', t'^*, com'), R') :: \Theta) \\ \rightarrow (H'\|((method\ m', c', l', t', t'^*, com'), R') :: \Theta)\end{array}} \quad (29)$$

$$\frac{\begin{array}{c}com = finish\ ;\, com'' \qquad field\text{-}order(c, finished) = n \\ H' = H[l + n \mapsto true] \qquad \epsilon = remove\text{-}frames(\Theta, l)\end{array}}{(H\|((method\ m, c, l, t, t^*, com), R)\Theta) \rightarrow (H'\|\,\epsilon)} \quad (30)$$

Fig. 4. Semantics rules of the operational semantics Android$\mathcal{S}$ (Set 3).

$$\frac{m\ in\ an\ AUM \qquad l = l' \qquad |loc(\Theta)| > 1}{m' = onPause} \quad (31)$$

$$\frac{field\text{-}order(c, finished) = n \qquad H(l + n) = true \\ m \in \{AUM, onPause, onStop\}}{m\ is\ a\ FAAM} \quad (32)$$

Fig. 5. Semantics rules of the operational semantics Android$\mathcal{S}$ (Set 4): semantics rules for constraints of method transitions.

Semantics Rules for Interaction with Users: Figure 6 presents the fifth set of semantics rules of the proposed semantics $AndroidS$, for $AndroidP$. These rules enables capturing the concept of interaction between users and applications.

Rule 33 captures the semantics of hitting the back button while the activity holds the focus. Rule 34 simulates the screen-orientation change of a mobile device. In this case, a configuration update, according to the rule details, happens. The active activity executes its onDestroy callback and hence it is destroyed and a new activity of the same class is created. Rule 35 captures the scenario when a new activity has become active and hence replaced a previously active activity.

$$\frac{field\text{-}order(c,finished) = n \qquad H' = H[l + n \mapsto true] \qquad \epsilon = remove\text{-}frames(\Theta, l)}{(H\|((method\ m,c,l,t,t^*,skip),R)\Theta) \to (H'\|((method\ m,c,l,t,t^*,skip),R)\Theta))} \tag{33}$$

$$\frac{\begin{array}{c} type\text{-}size(c) = m \wedge \{l',l'+1,\ldots,l'+m-1\} \in \mathcal{L} \setminus dom(H) \\ O_c \in Objects_c \wedge H' = H[l',\ldots,l'+m-1 \mapsto O_c] \\ \Theta' = remove\text{-}frames(\Theta, l) \end{array}}{\begin{array}{c} (H\|((method\ onDestory,c,l,t,t^*,skip),R) :: \Theta) \\ \to (H\|((method\ Constrcutor,c,l',t,t^*,com),\emptyset) :: \Theta') \end{array}} \tag{34}$$

$$\frac{}{\begin{array}{c} (H\|((method\ onResume,c,l_r,t_r,t^*_r,com_r),R_r) \\ :: (method\ onPause,c',l_p,t_p,t^*_p,com_p),R_p) :: \Theta) \\ \to (H\|((method\ onDestory,c',l_d,t_d,t^*_d,com_d),R_d) \\ :: (method\ onPause,c,l_p,t_p,t^*_p,com_p),R_p) :: \Theta) \end{array}} \tag{35}$$

Fig. 6. Semantics rules of the operational semantics $AndroidS$ (Set 5): semantic rules for interaction with users.

Table 1 presents a comparison between the proposed semantics $AndroidS$ and most related semantics of [15, 22].

Definition 3. – A well-structured (according to Definition 1) semantic state (H, Θ) is final if Θ is empty.
– A well-structured (according to Definition 1) semantic state (H, Θ) is disabled if it not a match to the pre-state of any of the rules in the semantics due to a syntax error in command of the partial frame at the top of the stack Θ.

The following theorem is proved by induction on the application structure according to $AndroidP$.

Theorem 1. If (H, Θ) is a well-structured (according to Definition 1) semantic state, then either:

– it is final, or
– it is disabled, or
– there exists (H', Θ') such that $(H, \Theta) \to (H', \Theta')$.

Table 1. A comparison between the proposed semantics and most related semantics

	Payets and Spoto's semantics [22]	Erik's and Henrik's semantics [15]	*AndroidS*
Semantic states complexity	Relatively complex	Relatively complex	Moderate complexity
Semantic rules complexity	Relatively complex	Relatively complex	Moderate complexity
Dalvik instructions treated	Little number	Too many	Moderate
Activity life-cycle	Treated by the semantics	Ignored by the semantics	Treated by the semantics
Convenience of use in program analysis	Convenient	Very hard to use	Convenient
Accuracy of semantic rules	Moderate accuracy	Very accurate	Very accurate
Macros instructions common in Android Apps	Treated by the semantics	Ignored by the semantics	Treated by the semantics

4   Literature Review

One direction of semantics applications for Android applications is the use of semantics to reveal security breaches of Android applications [3,4,6,10,11,33]. It is well known fact that Android applications create static program-analysis challenges. Android semantics can help to ease these challenges. The other main direction of applications for Android semantics is the development of program analyses using the Android semantics [13,23,25,31,32,34].

In [29,30], a semantics for a language of Dalvik bytecode with reflection characteristics was introduced. Also for the same langauge, an advanced control-flow analysis was presented. Concepts like reflection, dynamic dispatch, and exceptions were considered in this study. Typically a model syntax for a programming-language is supposed to be simple enough to facilitate studying program-analysis problems using the syntax. However the syntax should be powerful enough to express the important features of the concerned programming language. The Android syntax utilized in [29,30] is powerful but not simple enough. This is not helpful when using the syntax of [29,30] to develop analysis for Android applications. The syntax presented in our paper overcomes this problem.

In [22], an operational semantics for important concepts of Android applications was introduced including some commands of Dalvik bytecode. Also the techniques for inter-activity communications were partially modeled in [22]. The syntax presented in [22] for Android activities is simple but not powerful enough as it considers only a set of 12 instructions out of the used 218 Dalvik byte-code instructions [2]. For example instructions dealing with arrays, like "new-array", are not considered; which are covered in our semantics. However "new-array" instruction was reported in [30] to be used in 95.47 % of a sample of

1700 applications. Also the instruction *nop* which wastes a cycle of commutation is not included in [22]. The *nop* is important for many static analyses as it can be used to replace a block of unnecessary instructions towards optimizing the code.

In [10], semantic ideas were used to formulate a method, Apposcopy, to recognize a common Android malware (in the from of a class). Private information can be stolen by this malware. Apposcopy is based on the following two facts. Semantic features of the malware are captured using signatures formulated using a high-level language. The malware signature is searched for in applications using tools of static analyses [10]. In [3], a mechanism for monitoring running applications in the form of a dynamic verification mechanism was introduced.

In [4] and on a typed model of the Android programming simulating Android applications, an operational semantics was introduced. The objective of this work is to study the applications security focusing on data-flow security features. The semantics and the type system formulated important concepts necessary to study security issues of Android applications.

Concepts of asynchronous event-based and multi-threading programming are coupled in programming models for Android applications. This of course results in the ability to produce efficient applications. However sophisticated programming errors can easily result from arbitrary scheduling of asynchronous tasks and unexpected interleavings of thread commands [17]. A semantics for the concurrency in Android models of programming was presented in [17]. The semantic model was used to build a runtime race detection method using the concept of the "happens-before" relationship.

In [34], to monitor effective actions of Android applications VetDroid, a dynamic program analysis was presented. The focus of VetDroid is the genuineness use of permissions given to the applications. Hence VetDroid modeled in a systematic fashion the permission use; this included modeling the resources usage by the applications. Of course this analysis has the direct application of revealing security breaches.

Another program analysis for Android applications, DroidSafe, was introduce in [13] to statically trace information flow in order to discover possible exposure of information. DroidSafe can be thought of as a mix of static analysis model and an Android runtime system. This resulted in wide range of application of DroidSafe. The concept of stubs, (a method for analyzing code when its semantics is not determined only by Java) was utilized to enables the mixing mentioned above.

In [32] the problem of recognizing implicit information flow (IIF), was tackled using concepts of program analysis for Android bytecode. The analysis also aimed at stop security threats. The applied program analysis was that of control-migrate-oriented semantics. The analysis managed to recognize many types of IIFs. The exploitability of the IIFs was also studied using proof-of-concepts (PoCs) [16].

In [31], it is explained that classical techniques for the analysis of control-flow can not be work for Android applications. This is due to the event-driven and framework-based natures of Android applications. Also in [31], the user-event-driven parts are studied together with its groups of macros. This was done both for application code and Android framework. This study provides representations for these concepts taking into consideration event-handler and life-cycle macros. The form of representations are static analysis that is context-sensitive.

5   Conclusion

This paper presented an operational semantics Android$S$ to Android applications. The semantics have serval advantages over related work. The sets of Dalvik and macros instructions of considered in the semantics is well-designed. The semantics rules are simple yet accurate. The semantics expresses contains on states transition due to life-cycle of activities which are the main components of Android applications. The semantics also expresses the reaction of Android applications to users gestures. One important direction of future work is to show in details how the semantics can be used to express any of the common static analyses of Android applications.

References

1. Dalvik bytecode. https://source.android.com/devices/tech/dalvik/dalvik-bytecode.html. Accessed 1 Feb 2016
2. Dalvik docs mirror. http://www.milk.com/kodebase/dalvik-docs-mirror/. Accessed Feb 2016
3. Bauer, A., Küster, J.-C., Vegliach, G.: Runtime verification meets android security. In: Goodloe, A.E., Person, S. (eds.) NFM 2012. LNCS, vol. 7226, pp. 174–180. Springer, Heidelberg (2012)
4. Chaudhuri, A.: Language-based security on android. In: Proceedings of ACM SIG-PLAN 4th Workshop on Programming Languages and Analysis for Security, pp. 1–7. ACM (2009)
5. Cousot, P.: Semantic foundations of program analysis. In: Muchnick, S.S., Jones, N. (eds.) Program Flow Analysis: Theory and Application. Prentice Hall, Englewood Cliffs (1981)
6. Crussell, J., Gibler, C., Chen, H.: Scalable semantics-based detection of similar android applications. In: Proceedings of ESORICS, vol. 13. Citeseer (2013)
7. El-Zawawy, M.A.: A type system for Android applications. In: Computational Science and Its Applications-ICCSA (2016)
8. El-Zawawy, M.A.: Abstraction analysis and certified flow and context sensitive points-to relation for distributed programs. In: Murgante, B., Gervasi, O., Misra, S., Nedjah, N., Rocha, A.M.A.C., Taniar, D., Apduhan, B.O. (eds.) ICCSA 2012, Part IV. LNCS, vol. 7336, pp. 83–99. Springer, Heidelberg (2012)
9. El-Zawawy, M.A.: Novel designs for memory checkers using semantics and digital sequential circuits. In: Murgante, B., Gervasi, O., Misra, S., Nedjah, N., Rocha, A.M.A.C., Taniar, D., Apduhan, B.O. (eds.) ICCSA 2015. LNCS, vol. 9158, pp. 597–611. Springer, Heidelberg (2012)
10. Feng, Y., Anand, S., Dillig, I., Aiken, A.: Apposcopy: semantics-based detection of android malware through static analysis. In: Proceedings of 22nd ACM SIGSOFT International Symposium on Foundations of Software Engineering, pp. 576–587. ACM (2014)

11. Fuchs, A.P., Chaudhuri, A., Foster, J.S.: Scandroid: automated security certification of android (2009)
12. Gelfond, M., Lifschitz, V.: The stable model semantics for logic programming. In: ICLP/SLP, vol. 88, pp. 1070–1080 (1988)
13. Gordon, M.I., Kim, D., Perkins, J.H., Gilham, L., Nguyen, N., Rinard, M.C.: Information flow analysis of android applications in droidsafe. In: NDSS (2015)
14. Heim, I., Kratzer, A.: Semantics in Generative Grammar, vol. 13. Blackwell, Oxford (1998)
15. Karlsen, H.S., Wognsen, E.R.: Static analysis of Dalvik bytecode and reflection in android. Master's thesis, Aalborg University, June 2012
16. Yuanyang, H.: Proof-of-concepts of distributed detection with privacy (2014)
17. Maiya, P., Kanade, A., Majumdar, R.: Race detection for android applications. In: ACM SIGPLAN Notices, vol. 49, pp. 316–325. ACM (2014)
18. Mednieks, Z., Laird Dornin, B., Nakamura, M.: Programming Android. O'Reilly Media, Inc., Sebastopol (2012)
19. Milette, G., Stroud, A.: Professional Android Sensor Programming. Wiley, New York (2012)
20. Newcomer, K.E., Hatry, H.P., Wholey, J.S.: Handbook of Practical Program Evaluation. Wiley, New York (2015)
21. Nielson, F., Nielson, H.R., Hankin, C.: Principles of Program Analysis. Springer, Heidelberg (2015)
22. Payet, É., Spoto, F.: An operational semantics for android activities. In: Chin, W.-N., Hage, J. (eds.) Proceedings of ACM SIGPLAN 2014 Workshop on Partial Evaluation and Program Manipulation, PEPM, San Diego, California, USA, pp. 121–132. ACM, 20–21 January 2014
23. Rasthofer, S., Arzt, S., Miltenberger, M., Bodden, E.: Harvesting runtime data in android applications for identifying malware and enhancing code analysis. Technical report, TUD-CS-2015-0031, EC SPRIDE (2015)
24. Rogers, R., Lombardo, J., Mednieks, Z., Meike, B.: Android Application Development: Programming with the Google SDK. O'Reilly Media Inc., Sebastopol (2009)
25. Rountev, A., Yan, D.: Static reference analysis for GUI objects in android software. In Proceedings of Annual IEEE/ACM International Symposium on Code Generation and Optimization, p. 143. ACM (2014)
26. Schmidt, D.A.: Denotational semantics. A methodology for language development (1997)
27. Stoy, J.E.: Denotational Semantics: The Scott-Strachey Approach to Programming Language Theory. MIT Press, Cambridge (1977)
28. Wegbreit, B.: Mechanical program analysis. Commun. ACM 18(9), 528–539 (1975)
29. Wognsen, E.R., Karlsen, H.S.: Static analysis of Dalvik bytecode and reflection in android. Master's thesis, Department of Computer Science, Aalborg University, Aalborg, Denmark (2012)
30. Wognsen, E.R., Karlsen, H.S., Olesen, M.C., Hansen, R.R.: Formalisation and analysis of Dalvik bytecode. Sci. Comput. Program. 92, 25–55 (2014)
31. Yang, S., Yan, D., Haowei, W., Wang, Y., Rountev, A.: Static control-flow analysis of user-driven callbacks in android applications. In: Proceedings of 37th International Conference on Software Engineering, vol. 1, pp. 89–99. IEEE Press (2015)
32. You, W., Liang, B., Li, J., Shi, W., Zhang, X.: Android implicit information flow demystified. In: Proceedings of 10th ACM Symposium on Information, Computer and Communications Security, pp. 585–590. ACM (2015)

33. Zhang, M., Duan, Y., Yin, H., Zhao, Z.: Semantics-aware android Malware classification using weighted contextual API dependency graphs. In: Proceedings of 2014 ACM SIGSAC Conference on Computer and Communications Security, pp. 1105–1116. ACM (2014)
34. Zhang, Y., Yang, M., Bingquan, X., Yang, Z., Guofei, G., Peng Ning, X., Wang, S., Zang, B.: Vetting undesirable behaviors in android apps with permission use analysis. In: Proceedings of 2013 ACM SIGSAC Conference on Computer and Communications Security, pp. 611–622. ACM (2013)

A Type System for Android Applications

Mohamed A. El-Zawawy[1,2]([⊠])

[1] College of Computer and Information Sciences,
Al Imam Mohammad Ibn Saud Islamic University (IMSIU),
Riyadh, Kingdom of Saudi Arabia
[2] Department of Mathematics, Faculty of Science,
Cairo University, Giza 12613, Egypt
maelzawawy@cu.edu.eg

Abstract. The most common form of computers today is the hand held form and specially smart phones and tablets. The platform for the majority of smart phones is the Android operating system. Therefore analyzing and verifying the Android applications become an issue of extreme importance. However classical techniques of static analysis are not applicable directly to Android applications. This is so because typically Android applications are developed using Java code that is translated into Java bytecode and executed using Dalvik virtual machine. Also developing new techniques for static analysis of Android applications is an involved problem due to many facts such as the concept of gestures used in Android interaction with users.

One of the main tools of analyzing and verifying programs is types systems [11]. This paper presents a new type system, Android$\mathcal{T}$, which is specific to Android applications. The proposed type system ensures the absence of errors like "method is undefined". Also the type system is applicable to check the soundness of static analyses of Android applications.

Keywords: Android applications · Type systems · Android activities · Android applications analysis · Typed programming languages · Android$\mathcal{T}$ · Android$\mathcal{P}$

1 Introduction

The growing highest percentage of today's number of computers comes from hand held devices; tablets and smart phones. The most common operating systems for these computers is Android. For hand held devices, several properties of Android OS have led it to become the main platform [26]. The number of produced hand held computers in 2015 exceeded billion devices. This number is much more than the number of produced iOS-MacOS-Windows machines [2,4,33]. The nature of computing has greatly been affected, developed, and changed due to the great dominance of tablets and smart phones. The widespread also of these devices made their security, accuracy, reliability important issues for both developers and users. Software engineering is the branch

© Springer International Publishing Switzerland 2016
O. Gervasi et al. (Eds.): ICCSA 2016, Part V, LNCS 9790, pp. 115–128, 2016.
DOI: 10.1007/978-3-319-42092-9_10

of computer science that is most responsible for taking care of these issues by producing a variety of tools for verifying, analyzing, optimizing, and correcting Android applications. More precisely, techniques and strategies of static analyses are the main tools to approach, take care, and treat these issues. However existing techniques of static and dynamic analyses do not fit well for studying Android applications [20,22,29,32]. This is so as Android applications are developed mainly using Java which gets translated into Dalvik bytecode [26,33].

There are many challengers concerning the analysis of Android applications. The capabilities of hardware are increasing (memory capacity and processor speed). This results in better ability to produce complex and advanced Android applications. This of course results in more difficulty in studying these applications. The concepts of event-driven and framework-based of Android platform

$n \in N$.

$l \in \mathcal{L}$ = *The set of all memory locations.*

$v \in \mathcal{V} = \mathcal{L} \cup Integers$.

$\delta \in \Delta$ = *The set of all labels of program points.*

$f \in \mathcal{F}$ = *The set of all names of class fields.*

$m \in Me$ = *The set of all method names.*

$c \in C$ = *The set of all class names.*

$app \in \mathcal{A}$ = *The set of all application names.*

$t \in types ::= \{int, Boolean, ref\ t, double\} \cup C \cup array\ t$.

$in \in Instructions ::= nop \mid const\ i, v \mid move\ i, j \mid goto\ \delta \mid new\text{-}array\ i, j, t \mid$
$array\text{-}length\ i, j \mid aget\ i, j, k \mid aput\ i, j, k \mid new\text{-}instance\ i, c \mid$
$iget\ i, j, f \mid iput\ i, j, f \mid invoke\text{-}direct\ i_1, \ldots, i_n, method\ m \mid$
$neg\text{-}int\ i, j \mid add\ i, j, k \mid if\text{-}eq\ i, j, k \mid if\text{-}eqz\ i, j \mid move\text{-}result\ i \mid$
$return\text{-}void \mid return\ i$.

$mc \in Macros ::= SetContentView\ xml \mid findViewById\ i, j \mid$
$startActivityForResut\ A \mid setResult\ i \mid finish$.

$com \in Commands ::= in \mid mc \mid skip \mid Label\ \delta \mid com_1; com_2$.

$M \in Methods ::= (method\ m, c, t, t^*, com)$.

$F \in Fields ::= (field\ f, c, t)$.

$C \in Classes ::= class\ c \lhd AppCompatActivity$
$\{app; (field, root, c, int); (field, result, c, int);$
$(field, finished, c, Boolean); F^*; M^*\}$.

$App \in Applications ::= (Application\ app, C^*)$.

Fig. 1. Android$\mathcal{P}$: a model for Android applications.

contributes to the difficulty of studding Android applications. In Android applications, gestures (such as swipe and click) are main tools of the user to use the applications [5,15]. The Android applications are responsible for capturing user gestures and act accordingly. This includes recognizing accurate parameters like screen pixel and event type. The interaction between users and applications requires executing macros methods which are part of the Android framework. Taking care of these issues and details in a precise way makes it difficult to produce an efficient and accurate static analysis [12] for Android applications [26,33].

In this paper, we present *AndroidT*, a simple type system [7,18,30] for Android applications (considering activity composition). Our proposed type system *AndroidT* ensures that the Android applications that are typeable in the type systems do not cause method is undefined errors. The proposed type system treats macros that are executed during interaction with users via gestures. The problem of developing such type systems is involved due to the fact that macro instructions related to the life cycle of Android activities have to be considered by the type system. Also the fact that the composition of many activities in different states of execution is allowed in Android applications complicates the development of a type system for Android applications. A key idea in developing the type system is to let the typing environment composed of two parts; one part reports types of registers and memory locations and the other part links current activities with memory locations.

There are many applications to type type system developed in this paper. Besides that the type system guarantees the absence of specific errors, the type system can be used as a verification tool for Android applications. The type system can be used to ensure the convenience of any modifications done to Android applications by techniques of program analysis [9,24,25,31]. Although the proposed type system is simple, it is powerful enough as it is designed on a rich model of Android programming, *AndroidP*, that include main Dalvik bytecode operations. For example the array-operations are included in our programming model and considered in our type system. Another advantage of the proposed type system is that the used set of types is rich and specific to Android applications. For example the type "reference to view" ref_v is defined and used in the type system to capture types of registers hosting addresses of views composing Android activities.

Motivation. The work in this paper is motivated by the need to a type system that

- is specific to Android applications and macro instructions related to the life cycle of Android activities,
- ensures the absence of errors like "method is undefined", and
- is applicable to check the soundness of static alanyses of Android applications.

Paper Outline

The rest of the paper is organized as follows. Section 2 presents the langauge model, *AndroidP*, used to develop the type system. The detailed type system, *AndroidT*, is presented in Sect. 3. Section 4 presents a review to the most related up-to-date work. The paper is concluded in Sect. 5.

2   Langauge Model

Figure 1 presents the syntax of our programming model, *AndroidP*, for Android applications. The model has a rich set of types including integers, Boolean values, double numbers, classes, and arrays. The model uses a set of 19 Dalvik instructions. These inductions were selected carefully out of the full set which includes more than 200 instructions. The other main component in *AndroidP* is a set of macros instructions [1,14,27].

The selection of Dalvik instructions and macros instructions in *AndroidP* was made to cover main functionalities of Dalivk bytecode in a simple way. The macros included in the syntax are those used in almost all Android applications and that controls the application execution. Therefore the instructions are expected to be the focus of most analysis techniques of Android applications. This will help our semantics to serve these techniques by proving a mathematical background to the analysis techniques.

The set of commands in *AndroidP* is a sequence of instructions and macros. Every method has the form (*method m, c, t, t*, com*) where

- m is the method name,
- c is the class owns the method,
- t is the return type of the method,
- t^* is the sequence of type for method arguments, and
- *textitcom* is the body of the method.

Each class filed in *AndroidP* has the syntax (*field f, c, t*) where c denotes the class having the field, f denotes the field name, and t denotes type of the field. Android classes in *AndroidP* are extensions of *AppCompatActivity* and hence have the following components:

- c denotes the class name,
- *app* denotes the name of the application having the class,
- A special field *root* of type integer to record the address of view file of the activity having the class,
- A special field *result* of type integer to keep the result of the executing an object (activity) of the class,
- A special field *finished* of type Boolean to inform if an object (activity) of the class is done,
- F^* denotes a sequence of more fields, and
- M^* denotes a sequence of method structures.

$$\frac{n \in integers}{\Pi, \Gamma : n \vdash int} \qquad \frac{l \in \mathcal{L} \wedge l\ points\ to\ value\ of\ type\ \tau}{\Pi, \Gamma : l \vdash ref\ \tau} \tag{1}$$

$$\frac{\{i \mapsto \tau\} \subseteq \Gamma}{\Pi, \Gamma : i \vdash \tau} \qquad \frac{\{l \mapsto \tau\} \subseteq \Gamma}{\Pi, \Gamma : l \vdash \tau} \tag{2}$$

$$\frac{}{\Pi, \Gamma : nop \vdash OK} \tag{3}$$

$$\frac{\Pi, \Gamma : v \vdash \tau}{\Pi, \Gamma : const\ i, v \vdash \tau} \tag{4}$$

$$\frac{\Pi, \Gamma : j \vdash \tau}{\Pi, \Gamma : move\ i, j \vdash \tau} \tag{5}$$

$$\frac{}{\Pi, \Gamma : goto\ \delta \vdash OK} \tag{6}$$

$$\frac{\Pi, \Gamma : i \vdash ref\ array\ \tau \qquad \Pi, \Gamma : j \vdash int}{\Pi, \Gamma : new\text{-}array\ i, j, \tau \vdash ref\ array\ \tau} \tag{7}$$

$$\frac{\Pi, \Gamma : i \vdash int \qquad \Pi, \Gamma : j \vdash ref\ array\ t}{\Pi, \Gamma : array\text{-}length\ i, j \vdash int} \tag{8}$$

$$\frac{\Pi, \Gamma : i \vdash \tau \qquad \Pi, \Gamma : j \vdash ref\ array\ \tau \qquad \Pi, \Gamma : k \vdash int}{\Pi, \Gamma : aget\ i, j, k \vdash \tau} \tag{9}$$

$$\frac{\Pi, \Gamma : i \vdash \tau \qquad \Pi, \Gamma : j \vdash ref\ array\ \tau \qquad \Pi, \Gamma : k \vdash int}{\Pi, \Gamma : aput\ i, j, k \vdash \tau} \tag{10}$$

$$\frac{class\text{-}size(c) = n \qquad \Pi, \Gamma : i \vdash ref\ c \\ \{\{l, l+1, \ldots, l+n-1\} \mapsto fields(c)\} \subseteq \Gamma \\ (c, l) \in \Pi}{\Pi, \Gamma : new\text{-}instance\ i, c \vdash ref\ c} \tag{11}$$

$$\frac{\Pi, \Gamma : i \vdash \tau \qquad \Pi, \Gamma : j \vdash ref\ c \qquad type(c, f) = \tau}{\Pi, \Gamma : iget\ i, j, f \vdash \tau} \tag{12}$$

$$\frac{\Pi, \Gamma : i \vdash \tau \qquad \Pi, \Gamma : j \vdash ref\ c \qquad type(c, f) = \tau}{\Pi, \Gamma : iput\ i, j, f \vdash \tau} \tag{13}$$

Fig. 2. Inference rules of the type system Android$\mathcal{T}$ (Set 1).

Finally an application in *AndroidP* has the form (*Application app, C**) which includes a name *app* and a sequence of classes C^*.

Our model can be realized as a combination model of the models used in [1, 14, 27] to overcome disadvantages in the related models. The idea of adapting the set of macro instructions and the three special fields of classes was followed from [27]. We believe the langauge model in [27] is very simple while that in [1, 14] is very complex and our model is moderate; in between the two models.

We use the same model *AndroidP* to develop an operational semantics for Android application in [10].

3 Type Systems

This section presents a new type system, *AndroidT*, for *AndroidP*. The type system *AndroidT* is meant to guarantee the soundness of types in Android applications; guaranteeing the absence of dynamic type-errors such as method-not-found and field-not-found. For *AndroidP*, our proposed type system also supports typing macro instruction related to activity state methods like onCreate and onDestroy.

Definition 1 introduces the typing judgments and their components of *AndroidT*.

Definition 1. – *The set of judgment types, T, used in the typing judgments are the set of types defined in the langauge syntax AndroidP plus view-reference type (ref_v), the file-reference type (ref_f), the void type, and the method types:*

$$\tau \in T = types \cup \{ref_v, ref_f, void, (\tau_1, \ldots, \tau_n) \to \tau\}.$$

– *A type environment, Γ, is a partial map from $\mathcal{L} \cup \mathcal{R}$ to types.*
– *A class environment, Π, is a stack of pairs of activity classes and memory locations.*
– *A type judgement for $e \in \{commands, Methods, Fields, Classes\}$, has one of the following forms:*

$$\Pi, \Gamma : e \vdash \tau \qquad\qquad \Pi, \Gamma : e \vdash OK.$$

Definition 2 introduces the sub-typing relation defined on the set T.

Definition 2. *We let $\Subset$ denotes the least reflexive transitive closure of $\sqsubseteq$ on the set types where:*

1. *$Boolean \sqsubseteq int \sqsubseteq double$.*
2. *$\forall \tau \in T. \ \tau \sqsubseteq \tau$.*
3. *$\forall \tau, \tau' \in T. \ \tau \sqsubseteq \tau' \Rightarrow arrary \ t \sqsubseteq arrary \ t'$.*
4. *$\forall \tau, \tau' \in types. \ \tau \sqsubseteq \tau' \Rightarrow ref \ \tau \sqsubseteq ref \ \tau'$.*
5. *$\forall \{\tau_1, \ldots, \tau_n, \tau, \tau_1', \ldots, \tau_n', \tau'\} \subseteq types. \ (\forall i. \ \tau_i \sqsubseteq \tau_i') \wedge \tau \sqsubseteq \tau' \Rightarrow (\tau_1, \ldots, \tau_n \to \tau) \sqsubseteq (\tau_1', \ldots, \tau_n' \to \tau')$.*
6. *$\forall c \in \mathcal{C}. \ c \sqsubseteq AppCompatActivity$.*

Definition 2 guarantees the sub-typing relationship is defined conveniently over arrays types, references types, and methods types. The definition also makes it explicit that all activities (as defined classes) in Android applications are extensions for the class *AppCompatActivity*. The proof of Lemma 1 is straightforward. However it is important for the sub-typing relationship to be a partial order.

Lemma 1. *The binary relation $\Subset$ is a partial order.*

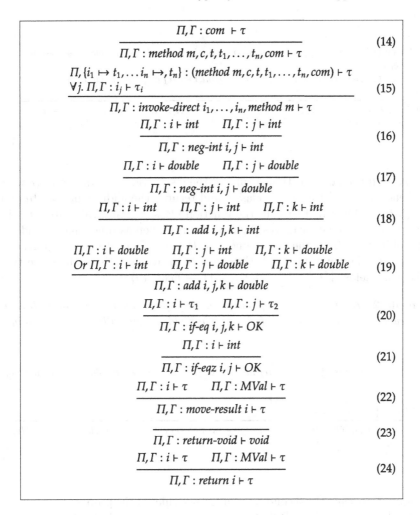

$$\frac{\Pi,\Gamma : com \vdash \tau}{\Pi,\Gamma : method\ m,c,t,t_1,\dots,t_n, com \vdash \tau} \quad (14)$$

$$\frac{\Pi,\{i_1 \mapsto t_1,\dots i_n \mapsto, t_n\} : (method\ m,c,t,t_1,\dots,t_n, com) \vdash \tau \quad \forall j.\ \Pi,\Gamma : i_j \vdash \tau_i}{\Pi,\Gamma : invoke\text{-}direct\ i_1,\dots,i_n, method\ m \vdash \tau} \quad (15)$$

$$\frac{\Pi,\Gamma : i \vdash int \qquad \Pi,\Gamma : j \vdash int}{\Pi,\Gamma : neg\text{-}int\ i,j \vdash int} \quad (16)$$

$$\frac{\Pi,\Gamma : i \vdash double \qquad \Pi,\Gamma : j \vdash double}{\Pi,\Gamma : neg\text{-}int\ i,j \vdash double} \quad (17)$$

$$\frac{\Pi,\Gamma : i \vdash int \qquad \Pi,\Gamma : j \vdash int \qquad \Pi,\Gamma : k \vdash int}{\Pi,\Gamma : add\ i,j,k \vdash int} \quad (18)$$

$$\frac{\Pi,\Gamma : i \vdash double \quad \Pi,\Gamma : j \vdash int \quad \Pi,\Gamma : k \vdash double}{Or\ \Pi,\Gamma : i \vdash int \quad \Pi,\Gamma : j \vdash double \quad \Pi,\Gamma : k \vdash double}{\Pi,\Gamma : add\ i,j,k \vdash double} \quad (19)$$

$$\frac{\Pi,\Gamma : i \vdash \tau_1 \qquad \Pi,\Gamma : j \vdash \tau_2}{\Pi,\Gamma : if\text{-}eq\ i,j,k \vdash OK} \quad (20)$$

$$\frac{\Pi,\Gamma : i \vdash int}{\Pi,\Gamma : if\text{-}eqz\ i,j \vdash OK} \quad (21)$$

$$\frac{\Pi,\Gamma : i \vdash \tau \qquad \Pi,\Gamma : MVal \vdash \tau}{\Pi,\Gamma : move\text{-}result\ i \vdash \tau} \quad (22)$$

$$\frac{}{\Pi,\Gamma : return\text{-}void \vdash void} \quad (23)$$

$$\frac{\Pi,\Gamma : i \vdash \tau \qquad \Pi,\Gamma : MVal \vdash \tau}{\Pi,\Gamma : return\ i \vdash \tau} \quad (24)$$

Fig. 3. Inference rules of the type system Android$\mathcal{T}$ (Set 2).

Figure 2 presents the first set of inference rules of the proposed type system *AndroidT*, for *AndroidP*. This figure presents the typing rules for Dalvik bytecdoe inductions. Rule 7 presents the inference rule for the instruction *new-array* i,j,τ that establishes a new array of type τ and of length j and returns the address of the array to the register i. Therefore the rule requires i to be of type *ref* τ, and j to be of type *int*. In this case, the rule infers that the instruction is to type *ref* τ.

Example 1. *Rule 25 provides an example for applying Rule 7.*

$$\{(A, l_A)::(B, l_B)::(C, l_C)\}, \{i \mapsto ref\ int, j \mapsto int, l \mapsto ref\ c, l_C \mapsto double, l_C + 1 \mapsto double\}$$
$$: i \vdash ref\ int$$
$$\{(A, l_A)::(B, l_B)::(C, l_C)\}, \{i \mapsto ref\ int, j \mapsto int, l \mapsto ref\ c, l_C \mapsto double, l_C + 1 \mapsto double\}$$
$$: j \vdash int$$

$$\{(A, l_A)::(B, l_B)::(C, l_C)\}, \{i \mapsto ref\ int, j \mapsto int, l \mapsto ref\ c, l_C \mapsto double, l_C + 1 \mapsto double\} :$$
$$new\text{-}array\ i, j, int \vdash ref\ int$$

$$(25)$$

Rule 11 formulates the typing rule for the instruction *new-instance* i, c which establishes a new object of the class c and returns the address of the object to the register i. The rule first calls the function *class-size* that calculates the size, n, of the class in memory unites (bytes). The rule then assumes that the address of the object is l from the environment $(c, l) \in \Pi$. The rules also requires that according to the environment Γ the memory locations to have the same types as the class fields; $\{\{l, l + 1, \ldots, l + n - 1\} \mapsto fields(c)\} \subseteq \Gamma$.

Example 2. *Inference rule 26 presents an example of the application of Rule 11. We assume that the class c is of size 2. The fields of the calls are of type double.*

$$class\text{-}size(c) = 2$$
$$\{(A, l_A)::(B, l_B)::(C, l_C)\}, \{i \mapsto ref\ int, j \mapsto int, l \mapsto ref\ c, l_C \mapsto double, l_C + 1 \mapsto double\}$$
$$: k \vdash ref\ c$$
$$\{\{l_C \mapsto double, l_C + 1 \mapsto double\} \subseteq \Gamma$$
$$(C, l_C) \in \{(A, l_A)::(B, l_B)::(C, l_C)\}$$

$$\{(A, l_A)::(B, l_B)::(C, l_C)\}, \{i \mapsto ref\ int, j \mapsto int, l \mapsto ref\ c, l_C \mapsto double, l_C + 1 \mapsto double\} :$$
$$new\text{-}instance\ k, c \vdash ref\ c$$

$$(26)$$

Figure 3 presents the second set of inference rules of the proposed type system *AndroidT*, for *AndroidP*. Rule 20 presents the typing rule for the instruction *if-eq* i, j, k which checks if the registers i and j are equal. If this is the case, then the control goes to location k. The rule requires that i and j are just typeable. If so then the instruction is *OK*.

Example 3. *Inference rule 34 presents an example of the application of Rule 20.*

$$\{(A, l_A)::(B, l_B)::(C, l_C)\}, \{i \mapsto ref\ int, j \mapsto int, l \mapsto ref\ c, l_C \mapsto double, l_C + 1 \mapsto double\}$$
$$: i \vdash ref\ int$$
$$\{(A, l_A)::(B, l_B)::(C, l_C)\}, \{i \mapsto ref\ int, j \mapsto int, l \mapsto ref\ c, l_C \mapsto double, l_C + 1 \mapsto double\}$$
$$: j \vdash int$$

$$\{(A, l_A)::(B, l_B)::(C, l_C)\}, \{i \mapsto ref\ int, j \mapsto int, l \mapsto ref\ c, l_C \mapsto double, l_C + 1 \mapsto double\} :$$
$$if\text{-}eq\ i, j, k \vdash OK$$

$$(34)$$

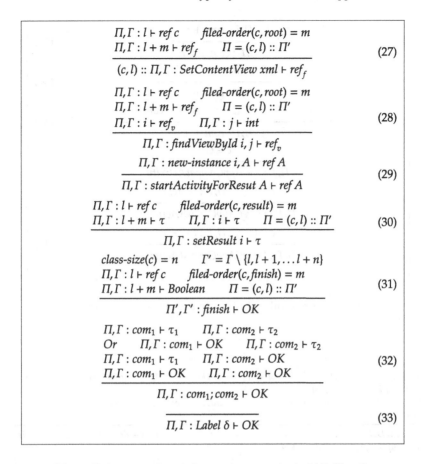

$$\frac{\Pi, \Gamma : l \vdash ref\ c \quad filed\text{-}order(c, root) = m}{\Pi, \Gamma : l + m \vdash ref_f \quad \Pi = (c, l) :: \Pi'} \tag{27}$$
$$(c, l) :: \Pi, \Gamma : SetContentView\ xml \vdash ref_f$$

$$\frac{\Pi, \Gamma : l \vdash ref\ c \quad filed\text{-}order(c, root) = m}{\Pi, \Gamma : l + m \vdash ref_f \quad \Pi = (c, l) :: \Pi'}{\Pi, \Gamma : i \vdash ref_v \quad \Pi, \Gamma : j \vdash int} \tag{28}$$
$$\Pi, \Gamma : findViewById\ i, j \vdash ref_v$$

$$\frac{\Pi, \Gamma : new\text{-}instance\ i, A \vdash ref\ A}{\Pi, \Gamma : startActivityForResut\ A \vdash ref\ A} \tag{29}$$

$$\frac{\Pi, \Gamma : l \vdash ref\ c \quad filed\text{-}order(c, result) = m}{\Pi, \Gamma : l + m \vdash \tau \quad \Pi, \Gamma : i \vdash \tau \quad \Pi = (c, l) :: \Pi'} \tag{30}$$
$$\Pi, \Gamma : setResult\ i \vdash \tau$$

$$\frac{class\text{-}size(c) = n \quad \Gamma' = \Gamma \setminus \{l, l+1, \dots l+n\}}{\Pi, \Gamma : l \vdash ref\ c \quad filed\text{-}order(c, finish) = m}{\Pi, \Gamma : l + m \vdash Boolean \quad \Pi = (c, l) :: \Pi'} \tag{31}$$
$$\Pi', \Gamma' : finish \vdash OK$$

$$\frac{\Pi, \Gamma : com_1 \vdash \tau_1 \quad \Pi, \Gamma : com_2 \vdash \tau_2}{Or \quad \Pi, \Gamma : com_1 \vdash OK \quad \Pi, \Gamma : com_2 \vdash \tau_2}{\Pi, \Gamma : com_1 \vdash \tau_1 \quad \Pi, \Gamma : com_2 \vdash OK}{\Pi, \Gamma : com_1 \vdash OK \quad \Pi, \Gamma : com_2 \vdash OK} \tag{32}$$
$$\Pi, \Gamma : com_1; com_2 \vdash OK$$

$$\frac{}{\Pi, \Gamma : Label\ \delta \vdash OK} \tag{33}$$

Fig. 4. Inference rules of the type system Android$T$ (Set 3).

Figure 4 presents the third set of inference rules of the proposed type system *AndroidT*, for *AndroidP*. Rule 30 introduces the typing rule for the macro instruction *setResulti* that sets the content of the register i into the field *Result* of the currently active object known from the first element of the environment Π; $\Pi = (c, l) :: \Pi'$. The rule calls the method *field-order* to determine the order of the field, m, *Result* in the class c; *filed-order*$(c, result) = m$. The rule also requires the register i and the location $l + m$ to have equal type τ. If this is the case, then the rule concludes that the type of the instruction is τ.

Figure 5 presents the fourth set of inference rules of the proposed type system *AndroidT*, for *AndroidP*. Rule 37 says that assigning a type to a command in an environment, $\Pi', \Gamma' : com \vdash \tau$, guarantees assigning the same type to the command in a bigger environment.

Example 4. *Inference rule 41 presents an example of the application of Rule 32 on results of Examples 1 and 3.*

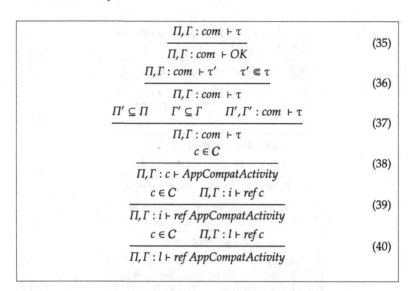

Fig. 5. Inference rules of the type system Android$\mathcal{T}$ (Set 4).

$$\frac{\{(A, l_A)::(B, l_B)::(C, l_C)\}, \{i \mapsto \textit{ref int}, j \mapsto \textit{int}, l \mapsto \textit{ref c}, l_C \mapsto \textit{double}, l_C + 1 \mapsto \textit{double}\} :}{\textit{new-array } i, j, \textit{int} \vdash \textit{ref int}}$$
$$\{(A, l_A)::(B, l_B)::(C, l_C)\}, \{i \mapsto \textit{ref int}, j \mapsto \textit{int}, l \mapsto \textit{ref c}, l_C \mapsto \textit{double}, l_C + 1 \mapsto \textit{double}\} :$$
$$\textit{if-eq } i, j, k \vdash \textit{OK}$$

$$\{(A, l_A)::(B, l_B)::(C, l_C)\}, \{i \mapsto \textit{ref int}, j \mapsto \textit{int}, l \mapsto \textit{ref c}, l_C \mapsto \textit{double}, l_C + 1 \mapsto \textit{double}\} :$$
$$\textit{new-array } i, j, \textit{int}; \textit{if-eq } i, j, k \vdash \textit{OK}$$

$$(41)$$

It is not hard (using structure induction) to prove the following fact about the type system *AndroidT*.

Theorem 1. *Every well-structured command of AndroidP is typeable in AndroidT.*

4   Literature Review

The focus of a big program-analysis audience has been attracted by the common use of Android mobile devices. Methods such as dynamic monitoring, information-flow analysis, modifying the Android system have been applied to reveal, study, fix essential problems and subtle structure flaws of Android applications [6].

Type systems have been used as one of the main tools to study programs in general and Android programs in particular. There are a variety of analyses and studies that can be carried out for Android applications using type systems [13, 23, 26]. Security of Android programs is a famous example of an issue that can be analyzed using type systems [6, 16, 17, 19, 28].

In [19] for Android applications, a security type system was presented to achieve the analysis of static data-flow. This analysis has the direct application of revealing privacy leaks. The idea of the analysis is to check the code of Android applications against privacy protocol using the type system [19].

In [6], concepts of type systems were utilized to analyze Android applications. The Android API for inter-component communication was reasoned for using a theoretical calculus. This paper also presented a type system to stop privilege. The idea in this paper is that well-typed components of applications are guaranteed to be attacks-protected. This work was implemented to produce a type checker, Lintent [6].

A technique, Dare, for covering Android applications into Java classes was presented in [26]. The conversion included the introduction of a new intermediate form for Applications. The concept of typing inference was also utilized in this paper in the form of solving strong constraints. A few number of inference rules was used to convert the full set of DVM opcodes (more than 200) [3]. This paper provides a way to treat unverifiable Dalvik bytecode [26].

In [28], it is claimed that application-centric rather than resource-centric models of permissions for resources (like GPS, camera, and Internet connection) in Android applications would better serve both users and developers. For example a permission for the use of camera has to be application justified (for paper scanning for example) and another permission for the Internet access can be granted but for downloading from a specific server for example. It was also claimed in [28] that Android is already equipped with required mechanisms to adapt a precise and efficient structures of policies that are application-centric.

Proved sound against the concept of non-interference, a type system for DEX bytecode, an operational semantics for Dalvik VM were introduced in [13]. The aim of this work was to verify characteristics of non-interference for DEX bytecode. Moreover this paper introduced an abstract conversion (that is non-interference preserving) of Java bytecode into DEX bytecode.

In [17] it is claimed that although the already paid high cost in applying the process of application approval, some mobile application stores let Malware sneaks to our mobile devices. It is obvious that application stores must adapt strong mechanisms guarantee the genuineness (not malicious) of their applications. In [17], a verification technique for achieving this task is presented.

Targeting Android applications, a type system for taint analysis was introduced in [16]. The type system, DFlow, is classified as data-flow and context-sensitive. This type system was associated with an analysis, DroidInfer, for type inference. The direct application of the type system and its analysis is to reveal privacy leaks. Also in [16] methods for treating Android characteristics such as inter-component communication, macros, and many entry program points were presented. DFlow and its analysis also are quipped with methods to report errors using CFL-reachability.

In [21], it was noted that methods that control access to private information such as location and contacts in mobile applications are not strong enough. In this paper, declassification policies, that control the reveal of private information by the interaction of users with constrains of mobile applications, were presented. Relying on appropriate event sequences in the application execution, the policies were independently formalized from the application implementation details.

In [8] a new programming langauge was introduced to facilitate expressing Android local policies and Android Inter-Process Communication logic. This can be realized as a try to effectively and efficiently formalize the application and verification of policies. The syntax of the proposed langauge is equipped with a scope operator used to force the application of local policies to specific parts of the code. Such techniques is also applicable to web services and object-oriented programming [8].

5   Conclusion

This paper presented a type system, $AndroidT$, for Android applications. The type system was developed using a rich model for Android programming, $AndroidP$. The model includes a completed set of Dalvik instructions and a set of important android macros instructions. The type system $AndroidT$ can be used to prove that an Android application if free of type errors such as "method is undefined". The type system can also be used to prove the adequacy and correctness of new techniques of program analysis for Android applications. The set of types used in the type system is rich enough to host types such as reference to view which is the main component of activity which is the main component of Android applications.

References

1. Dalvik   bytecode.   https://source.android.com/devices/tech/dalvik/dalvik-bytecode.html. Accessed 1 Feb 2016
2. Dalvik docs mirror. http://www.statista.com/topics/840/smartphones/. Accessed Feb 2016
3. Dalvik   docs   mirror.   http://www.milk.com/kodebase/dalvik-docs-mirror/. Accessed Feb 2016
4. Gartner, Inc.: Worldwide Traditional PC, Tablet, Ultramobile and Mobile Phone-shipments. www.gartner.com/newsroom/id/2692318
5. Amalfitano, D., Fasolino, A.R., Tramontana, P., De Carmine, S., Memon, A.M.: Using GUI ripping for automated testing of android applications. In: Proceedings of 27th IEEE/ACM International Conference on Automated Software Engineering, pp. 258–261. ACM (2012)
6. Bugliesi, M., Calzavara, S., Spanò, A.: Lintent: towards security type-checking of android applications. In: Beyer, D., Boreale, M. (eds.) FORTE 2013 and FMOODS 2013. LNCS, vol. 7892, pp. 289–304. Springer, Heidelberg (2013)
7. Cardelli, L.: Type systems. ACM Comput. Surv. **28**(1), 263–264 (1996)

8. Costa, G.: Securing Android with local policies. In: Bodei, C., Ferrari, G.-L., Priami, C. (eds.) Programming Languages with Applications to Biology and Security: Essays Dedicated to Pierpaolo Degano on the Occasion of His 65th Birthday. LNCS, vol. 9465, pp. 202–218. Springer, Heidelberg (2015). doi:10.1007/978-3-319-25527-9_14

9. Cousot, P.: Semantic foundations of program analysis. In: Muchnick, S., Jones, N. (eds.) Program Flow Analysis : Theory and Applications. Prentice Hall, Englewood Cliffs (1981)

10. El-Zawawy, M.A.: An operational semantics for Android applications. In: Computational Science and Its Applications-ICCSA (2016)

11. El-Zawawy, M.A.: Heap slicing using type systems. In: Murgante, B., Gervasi, O., Misra, S., Nedjah, N., Rocha, A.M.A.C., Taniar, D., Apduhan, B.O. (eds.) ICCSA 2012, Part III. LNCS, vol. 7335, pp. 592–606. Springer, Heidelberg (2012)

12. El-Zawawy, M.A.: Recognition of logically related regions based heap abstraction. J. Egypt. Math. Soci. 20(2), 64–71 (2012)

13. Gunadi, H.: Formal certification of non-interferent Android bytecode (DEX bytecode). In: 2015 20th International Conference on Engineering of Complex Computer Systems (ICECCS), pp. 202–205. IEEE (2015)

14. Karlsen, H.S., Wognsen, E.R.: Static analysis of Dalvik bytecode and reflection in Android. Master's thesis, Aalborg University, June 2012

15. Cuixiong, H., Neamtiu, I.: Automating gui testing for Android applications. In: Proceedings of 6th International Workshop on Automation of Software Test, pp. 77–83. ACM (2011)

16. Huang, W., Dong, Y., Milanova, A., Dolby, J.: Scalable and precise taint analysis for Android. In: Proceedings of 2015 International Symposium on Software Testing and Analysis, pp. 106–117. ACM (2015)

17. Just, R., Ernst, M.D., Millstein, S.: Collaborative verification of information flow for a high-assurance app store. Softw. Eng. Manag. 77 (2015)

18. Machiry, A., Tahiliani, R., Naik, M.: Dynodroid: an input generation system for Android apps. In: Proceedings of 2013 9th Joint Meeting on Foundations of Software Engineering, pp. 224–234. ACM (2013)

19. Mann, C., Starostin, A.: A framework for static detection of privacy leaks in android applications. In: Proceedings of 27th Annual ACM Symposium on Applied Computing, pp. 1457–1462. ACM (2012)

20. Mednieks, Z., Dornin, L., Meike, G.B., Nakamura, M.: Programming Android. O'Reilly Media Inc, Sebastopol (2012)

21. Micinski, K., Fetter-Degges, J., Jeon, J., Foster, J.S., Clarkson, M.R.: Checking interaction-based declassification policies for Android using symbolic execution. In: Pernul, G., Ryan, P.Y.A., Weippl, E. (eds.) ESORICS 2015. LNCS, vol. 9327, pp. 520–538. Springer, Heidelberg (2015). doi:10.1007/978-3-319-24177-7_26

22. Milette, G., Stroud, A.: Professional Android Sensor Programming. Wiley, New York (2012)

23. Mohr, M., Graf, J., Hecker, M.: JoDroid: adding android support to a static information flow control tool. In: Software Engineering (Workshops), pp. 140–145 (2015)

24. Newcomer, K.E., Hatry, H.P., Wholey, J.S.: Handbook of Practical Program Evaluation. Wiley, New York (2015)

25. Nielson, F., Nielson, H.R., Hankin, C.: Principles of Program Analysis. Springer, Heidelberg (2015)

26. Octeau, D., Jha, S., McDaniel, P.: Retargeting Android applications to Java bytecode. In: Proceedings of ACM SIGSOFT 20th International Symposium on the Foundations of Software Engineering, p. 6. ACM (2012)

27. Payet, É., Spoto, F.: An operational semantics for android activities. In: Chin, W.-N., Hage, J. (eds.) Proceedings of ACM SIGPLAN 2014 Workshop on Partial Evaluation and Program Manipulation, PEPM, San Diego, California, USA, pp. 121–132. ACM, 20–21 January 2014

28. Reddy, N., Jeon, J., Vaughan, J., Millstein, T., Foster, J.: Application-centric security policies on unmodified Android. UCLA Computer Science Department, Technical report 110017 (2011)

29. Rogers, R., Lombardo, J., Mednieks, Z., Meike, B.: Android Application Development: Programming with the Google SDK. O'Reilly Media Inc., Sebastopol (2009)

30. Schwartzbach, M.I., Palsberg, J.: Object-Oriented Type Systems. Wiley, New York (1994)

31. Wegbreit, B.: Mechanical program analysis. Commun. ACM **18**(9), 528–539 (1975)

32. Rubin, X., Saïdi, H., Anderson, R.: Aurasium: practical policy enforcement for android applications. Paper presented at the Part of 21st USENIX Security Symposium (USENIX Security 12), pp. 539–552 (2012)

33. Yang, S.: Static analyses of GUI behavior in Android applications. Ph.D. thesis, The Ohio State University (2015)

A Weed Colonization Inspired Algorithm for the Weighted Set Cover Problem

Broderick Crawford[1,2,3], Ricardo Soto[1,4,5], Ismael Fuenzalida Legüe[1(✉)],
Sanjay Misra[6], and Eduardo Olguín[2]

[1] Pontificia Universidad Católica de Valparaíso, 2362807 Valparaíso, Chile
ifuenzalida17@gmail.com
[2] Facultad de Ingeniería y Tecnología, Universidad San Sebastián,
Bellavista 7, 8420524 Santiago, Chile
[3] Universidad Central de Chile, 8370178 Santiago, Chile
[4] Universidad Autónoma de Chile, 7500138 Santiago, Chile
[5] Universidad Científica del Sur, Lima 18, Lima, Peru
[6] Covenant University, Ogun 110001, Nigeria

Abstract. The Weighted Set Cover Problem (SCP) is a popular optimization problem that has been applied to different industrial applications, including scheduling, manufacturing, service planning and location problems. It consists in to find low cost solutions covering a set of requirements or needs. In this paper, we solve the SCP using a recent nature inspired algorithm: Invasive Weed Optimization (IWO). IWO imitates the invasive behavior of real weeds: natural reproduction and selection where the best weed has more chance of reproduction. We test our approach using known ORLIB test problems for the SCP. The computational results show that the IWO metaheuristic can find very good results.

Keywords: Invasive Weed Optimization · Set covering problem · Combinatorial optimization

1 Introduction

Nature inspired algorithms are metaheuristics that mimics the nature for solving optimization problems. In the past years, numerous research efforts has been concentrated in this area. The main motivation to use metaheuristics is because former methods to solve optimization problems require enormous computational efforts, which tend to fail as the problem size increases.

Bio inspired stochastic optimization algorithms are computationally efficient alternatives to the deterministic approaches. Metaheuristics are based on the iterative improvement of either a population of solutions (as in swarm or evolutive algorithms [8,11,14]) or a single solution (as in hill-climbing or tabu search [23]) and mostly employ randomization and local search to solve a given optimization problem.

In this paper we use Invasive Weed Optimization, this metaheuristic was proposed in year 2006 by Mehrabian, A., and Lucas, C., and it is based on the

© Springer International Publishing Switzerland 2016
O. Gervasi et al. (Eds.): ICCSA 2016, Part V, LNCS 9790, pp. 129–138, 2016.
DOI: 10.1007/978-3-319-42092-9_11

behavior of invasive weeds seeking to imitate its robustness and facility with which some herbs have front the hardness environment, the playability of the herbs and natural selection of them [19].

The problem to be solved is the Weighted Set Covering Problem (SCP), it is a classic problem which belongs to the category NP-Complex [17]. In general, in this kind of optimization problems the goal is to find a set of solutions that are able to meet the constraints of the problem minimizing the solution cost.

It should also be noted that at present the vast majority of metaheuristics appearing in literature achieved be close of the optimum for each of the instances of SCP especially, when problems have hundreds of rows and thousands of columns [5]. However, when problems grow up exponentially, and they have thousands rows and millions of columns algorithms are approaching the 1 % of the optimum solutions in a reasonable computational performance [6].

Recent metaheuristics used to solve the SCP are: Teaching-learning-based optimization algorithm [8], Artificial bee colony algorithm [10], Cultural algorithms [11], Binary firefly algorithm [12], Binary fruit fly optimization algorithm [14] and Shuffled Frog Leaping Algorithm [13].

This paper is organized as follows. Section 2 explains the SCP. Sections 3 and 4 explain IWO and binary IWO. Section 4 introduces the operating parameters used to configure IWO and the results obtained solving ORLIB SCP instances [3]. Finally in Sect. 6, we conclude.

2   The Weighted Set Cover Problem

The Weighted Set Cover Problem is one of representative combinatorial optimization problems, which has many practical applications. SCP consists of a set of variables that have a relationship together, and by a objective function are able to minimize cost allocation. It is a classic problem that belongs to the category NP-Complex [17]. In the case of this problem in specific, the goal is to search variables assignment to the lowest possible cost. That is, it seeks to cover all the needs (rows) with the lowest cost (columns).

Such as mentioned in the previous paragraph, we can mention that the representation of the problem is a matrix assignments (MxN). Where M represents the needs that must be cover and N columns variables to assign. The assignment matrix is based on a series of restrictions that must be satisfied to be considered a "workable solution" [17].

The SCP has many applications in industry and in real life: facility location [16], ship scheduling [2], production planning [1], crew scheduling [7,18], vehicle routing [4,27], musical composition [22], information retrieval [30], erritory design [15], sector location [20,21,25] or fire companies [28].

As shown in the application examples mentioned, the problem can be applied in different circumstances of decision making. Where more information have these decisions, it will help to improve the quantitative and qualitative performance of the assets of the company, that could be used in a better way, thus improving their performance and quality of service.

To land the explanation of the problem it is necessary to explain the mathematical formulation. This will be explained in a better way by using formulas and mathematical notation helping to expose a more didactic way the complexity of the problem and its characteristics. Thus achieving a better understanding and comprehension of the problem. The mathematical model of the SCP is:

$$Minimize \qquad Z = \sum_{j=1}^{n} c_j x_j \qquad (1)$$

Subject to:

$$\sum_{j=1}^{n} a_{ij} x_j \geq 1 \qquad \forall\, i \in \{1, 2, 3, ..., m\} \qquad (2)$$

$$x_j \in \{0, 1\} \qquad (3)$$

Consequently, the Eq. (1) represents the objective function of the problem. This function allows to know the *fitness* of the solution evaluated. Where c_j represents the cost of the $j - th$ column and x_j is the decision variable, this variable determines whether a column is activate or not. Equation (2) represent the constraints: each one row should be cover by at least one column. Where a_{ij} is an element of the MxN matrix such elements can only have values 0 and 1. Finally, Eq. (3) represents the values that can take the decision variables: 1 or 0, where 1 represents if the column is active and 0 otherwise [17].

3   Invasive Weed Optimization

Invasive Weed Optimization is based on how the invasive weeds behave in when colonize. An invasive weed is a type of plant that grows without being desired by people. In general, the term invasive or undesirable, is used in agriculture, and it is used for herbs that are a threat to the crop plants.

In IWO a weed represents a point in the search space of solutions and seeds represent other points explored [26].

Considerations for a more detailed explanation: D as the problem dimension, P_{int} as the initial size of the colony of herbs, P_{max} as the maximum size of the colony where $1 \leq P_{int} \leq P_{max}$ and W^P as a set of herbs, where each weed represents a point in the search space. Importantly, for calculating the *fitness* of each weed it is used the objective function defined in the problem. Which is as follows $F : R^D \rightarrow R$ [26].

There are some distinctive properties of IWO in comparison with other metaheuristics: *Way of reproduction*, *Spatial dispersal* and *Competitive exclusion* are its main stages. In the following subsections we show their main characteristics.

3.1   Initialization

The IWO process begins with initializing a population. Given the generation G, we proceed to create a weed population size P_{int}, which is randomly generated and the weeds W_i^P are uniformly distributed (W_i^P $(U(X_{min}, X_{max})^D)$).

Where X_{min} and X_{max}, are defined according to the type of problem to be implemented [26]. For the SCP, these values are determined by 0 and 1 [26].

3.2 Reproduction

In each iteration, each weed W_i^P of the current population, are reproduced from seed. The amount of seeds for each weed W_i^P, is given by S_{num}, this number depends on the fitness [26]. Where best fitness has the evaluated weed, the greater the amount of seeds for may have to breed [26].

$$S_{sum} = S_{min} + (\frac{F(W_i^P) - F_{worse} F_{best} - F_{worse}}{)}(S_{max} - S_{min}) \qquad (4)$$

where S_{max} and S_{min} represent the maximum and minimum allowed by weed W_i^P [26]. All seeds S_{sum} are distributed in space and close to the father weed, that is, starting these solutions is created a neighborhood of solutions [26].

3.3 Spatial Distribution

As explained in the previous section, the seeds are distributed in the search space and in this way in, generating new solutions looking to the best for the problem [26]. To achieve this, we should be consider a way to achieve the correct distribution of seeds for this, the use of the normal distribution [26].

$$S_j = W_i^P + N(0, \sigma_G)^D \qquad (1 \leq j \leq S_{num}) \qquad (5)$$

where σ_G represents the standard deviation, which will be calculated as follows:

$$\sigma_G = \sigma_{final} + \frac{(N_{iter} - G)^{\sigma_{mod}}}{(N_{iter})^{\sigma_{mod}}}(\sigma_{init} - \sigma_{final}) \qquad (6)$$

where N_{iter} represents the maximum number of iterations, σ_{mod} represented nonlinear index modulation, σ_{init} and σ_{final} is parameters input.

3.4 Competitive Exclusion

At this stage, we proceed to verify the amount of herbs and seeds created by the algorithm not exceeding the maximum permitted W_{max}, it proceeds to make a pruning the worst weed. This, in order to let the herbs with better results to own the best opportunities to breed and find the best solution to the problem [26].

4 Binary Invasive Weed Optimization

Most of the IWO applications have been solving continuous problems. Our implementation of IWO is aimed to solve combinatorial problems in a binary search space such as SCP. The Binary Invasive Weed Optimization (BIWO) is a variation of the main algorithm, it has some modifications to the main algorithm [26]:

Instead of working with real domains R^D for solutions, BIWO works in a binary search space $B^D \in \{0, 1\}$. Therefore, the objective function also undergoes changes in its definition. Consequently, the objective function is:

$$F : B^D \rightarrow R$$

In the phase *Distribution Spatial*, the formula for the distribution of seed undergoes the following:

$$S_j = N(W_i^P, \sigma_G)^D \qquad (1 \leq j \leq S_{num}) \tag{7}$$

In the new formulation we propose that a seed is the assignment of a father weed to the seed; but a bit change which is determined by the calculate of normal distribution and it will determine that so close is the seed of the father weed W_i^P [26]. In turn, the positive part of the normal distribution is used, which will imply that the number of bits that will change will diminish with each iteration on seeds and weed, belonging the population. It is explained because the algorithm is sensitive to changes, across of calculating the standard deviation, which directly impacts on the calculation of the normal distribution and therefore in

---

Algorithm 1. Binary Invasive Weed Optimization

1: Generate initial random population of W^P weeds (Stage-I)
2: **for** $iter \ni 1..MaxIter$ **do**
3:    Calculate maximum and minimum fitness in the population
4:    **for** $w_i^P \ni W^P$ **do**
5:       Determine number of seeds w_i^P, corresponding to its fitness (Stage-II)
6:       $NewWeed$ = Use Neighborhood generation algorithm (Stage-III)
7:       Add $NewWeed$ to the W^P
8:    **end for**
9: **end for**
10: **if** $W^P.Size > W^P.SizeMax$ **then**
11:    Remove Weeds with worst Fitness (Stage-IV)
12:    Sort the population of weed with smaller fitness
13: **end if**

Algorithm 2. Neighborhood generation algorithm

Require: Weed W_i^P and σ_G
1: $Nchange_{bits} = N^+(0, \sigma_G)$
2: $Change_{probality} = \dfrac{Nchange_{bits}}{ProblemDimension}$
3: $Seed$ = Weed W_i^P
4: **for** $d \ni 1..D$ **do**
5:    $Random_{number} = U(0,1)$
6:    **if** $Random_{number} \leq Change_{probality}$ **then**
7:       $Seed_d = \neg Seed_d$
8:    **end if**
9: **end for**
10: **return** $Seed$

the mutation of solutions. Importantly, that the mutation of solutions is similar as it is used in genetic algorithms.

The above process can be understood better through the Algorithms 1 and 2:

5   Experiments and Results

BIWO has a number of parameters that are required to tune.

5.1   Configuring the Algorithm

BIWO was configured before performing the experiments. To this end and start-ing from default values, a parameter of the algorithm is selected to be tuned. Then, independent runs are performed for each configuration of the parameter, considering a reduced stop condition. Next, the configuration which provides the best performance on average is selected, assuming the hypervolume metric as quality indicator. Next, another parameter is selected so long as all of them are fixed. At following we show the values considered for the best configuration obtained:

- Number of generations 30.
- Number of iterations (N_{iter}) 400.
- Initial amount of weed (P_{init}) 100.
- Maximum number of weed (P_{max}) 20.
- Minimum number of seed (S_{min}) 20.
- Maximum number of seed (S_{max}) 80.
- σ_{init} = Problem Dimension.
- σ_{final} = 1.
- σ_{mod} = 3.

It is necessary to mention that the BIWO was executed on a computer with the following characteristics:

- Operative System. Microsoft Windows 8.1.
- Memory Ram: 6 Gb.
- CPU: Intel Core i5 2.60.

5.2   Computational Results

The following table presents the results obtained for the ORLIB instances of SCP. We should mention that instances used in our experiments were preprocessed deleting redundant columns. This delete redundant column process is explain in [24, 29].

Table 1. Results of experiments

Results					
Instances	Optimum	Best results	Worse results	Average	RPD (Best results)
scp41	429	429	443	432,2	0 %
scp42	512	512	535	519,57	0 %
scp43	516	516	550	526,4	0 %
scp44	494	494	530	503,07	0 %
scp45	512	512	528	518,3	0 %
scp46	560	560	574	563,5	0 %
scp47	430	430	444	434,37	0 %
scp48	492	492	505	496,57	0 %
scp49	641	649	675	661,83	1,25 %
scp51	253	253	275	259,1	0 %
scp52	302	302	324	310,63	0 %
scp53	226	226	231	228,93	0 %
scp54	242	242	247	244,13	0 %
scp55	211	211	219	215,63	0 %
scp56	213	213	222	215,83	0 %
scp57	293	293	303	295,56	0 %
scp58	288	288	300	292,47	0 %
scp59	279	279	289	281,13	0 %
scp61	138	142	151	144,2	2,90 %
scp62	146	146	159	150,56	0 %
scp63	145	145	157	151,1	0 %
scp64	131	131	135	132,96	0 %
scp65	161	161	169	165,37	0 %
scpa1	253	254	266	257,93	0,40 %
scpa2	252	256	266	260,9	1,59 %
scpa3	232	233	244	237,4	0,43 %
scpa4	234	236	245	241,07	0,85 %
scpa5	236	236	240	237,9	0 %
scpb1	69	69	77	72,4	0 %
scpb2	76	77	85	80,63	1,32 %
scpb3	80	80	86	82	0 %
scpb4	79	80	87	86,23	1,27 %
scpb5	72	72	77	72,7	0 %
scpc1	227	229	237	232,33	0,88 %
scpc2	219	221	231	224,83	0,91 %
scpc3	243	250	262	255,23	2,88 %
scpc4	219	219	237	227,83	0 %
scpc5	215	215	229	220,77	0 %

The table is structured as: First column is the instance name, Second shows best value known, Third column is best result obtained; Fourth is worse result obtained; Fifth is average of the results obtain by each instance and the Sixth column is Relative Percentage Deviation (RPD) [9], it is calculate as:

$$RPD = \frac{(Z - Z_{opt})}{Z_{opt}} * 100 \tag{8}$$

Where Z is our best result and Z_{opt} is the optimum known by instance Table 1.

6 Conclusion

In this paper, we solve the Weighted Set Cover Problem using a Binary Invasive Weed Optimization algorithm. After performing the experiments solving ORLIB instances of SCP, our main conclusions is in relation with the parameter tuning, we recommend a tuning of parameters by group of instances because, this will allow generate custom tests and results better for each group of instances.

In order to improve the performance we are working in the incorporation of an elitist method for mutation. It is expected that the incorporation of elitism can improve performance in the search for good solutions in terms of quality and further making improvements in execution times.

The results obtained with BIWO show very good results in almost all ORLIB instances tested.

Acknowledgements. The author Broderick Crawford is supported by grant CONICYT/FONDE-CYT/REGULAR/1140897 and Ricardo Soto is supported by grant CONICYT/FONDECYT/REGULAR/1160455.

References

1. Adulyasak, Y., Cordeau, J.-F., Jans, R.: The production routing problem: a review of formulations and solution algorithms. Comput. Oper. Res. **55**, 141–152 (2015)
2. Bai, R., Xue, N., Chen, J., Roberts, G.W.: A set-covering model for a bidirectional multi-shift full truckload vehicle routing problem. Transp. Res. Part B: Methodol. **79**, 134–148 (2015)
3. Beasley, J.E.: OR-Library: distributing test problems by electronic mail. J. Oper. Res. Soc. **41**(11), 1069–1072 (1990)
4. Cacchiani, V., Hemmelmayr, V.C., Tricoire, F.: A set-covering based heuristic algorithm for the periodic vehicle routing problem. Discrete Appl. Math. **163**, 53–64 (2014)
5. Caprara, A., Fischetti, M., Toth, P.: A heuristic method for the set covering problem. Oper. Res. **47**(5), 730–743 (1999)
6. Caprara, A., Toth, P., Fischetti, M.: Algorithms for the set covering problem. Ann. Oper. Res. **98**(1–4), 353–371 (2000)
7. Chen, S., Shen, Y.: An improved column generation algorithm for crew scheduling problems. J. Inf. Comput. Sci. **10**(1), 175–183 (2013)

8. Crawford, B., Soto, R., Aballay, F., Misra, S., Johnson, F., Paredes, F.: A teaching-learning-based optimization algorithm for solving set covering problems. In: Gervasi, O., Murgante, B., Misra, S., Gavrilova, M.L., Rocha, A.M.A.C., Torre, C., Taniar, D., Apduhan, B.O. (eds.) ICCSA 2015. LNCS, vol. 9158, pp. 421–430. Springer, Heidelberg (2015)

9. Crawford, B., Soto, R., Cuesta, R., Paredes, F.: Application of the artificial bee colony algorithm for solving the set covering problem. Scientific World J. **2014**, 8 (2014)

10. Crawford, B., Soto, R., Cuesta, R., Paredes, F.: Using the bee colony optimization method to solve the weighted set covering problem. In: Stephanidis, C. (ed.) HCI 2014, Part I. CCIS, vol. 434, pp. 493–497. Springer, Heidelberg (2014)

11. Crawford, B., Soto, R., Monfroy, E.: Cultural algorithms for the set covering problem. In: Tan, Y., Shi, Y., Mo, H. (eds.) ICSI 2013, Part II. LNCS, vol. 7929, pp. 27–34. Springer, Heidelberg (2013)

12. Crawford, B., Soto, R., Olivares-Suárez, M., Palma, W., Paredes, F., Olguin, E., Norero, E.: A binary coded firefly algorithm that solves the set covering problem. Rom. J. Inf. Sci. Technol. **17**(3), 252–264 (2014)

13. Crawford, B., Soto, R., Peña, C., Riquelme-Leiva, M., Torres-Rojas, C., Johnson, F., Paredes, F.: Binarization methods for shuffled frog leaping algorithms that solve set covering problems. In: Silhavy, R., Senkerik, R., Oplatkova, Z.K., Prokopova, Z., Silhavy, P. (eds.) CSOC2015. AISC, vol. 349, pp. 317–326. Springer, Heidelberg (2015)

14. Crawford, B., Soto, R., Torres-Rojas, C., Peña, C., Riquelme-Leiva, M., Misra, S., Johnson, F., Paredes, F.: A binary fruit fly optimization algorithm to solve the set covering problem. In: Gervasi, O., Murgante, B., Misra, S., Gavrilova, M.L., Rocha, A.M.A.C., Torre, C., Taniar, D., Apduhan, B.O. (eds.) ICCSA 2015. LNCS, vol. 9158, pp. 411–420. Springer, Heidelberg (2015)

15. Elizondo-Amaya, M.G., RÃos-Mercado, R.Z., DÃaz, J.A.: A dual bounding scheme for a territory design problem. Comput. Oper. Res. **44**, 193–205 (2014)

16. Farahani, R.Z., Asgari, N., Heidari, N., Hosseininia, M., Goh, M.: Covering problems in facility location: a review. Comput. Ind. Eng. **62**(1), 368–407 (2012)

17. Feo, T.A., Resende, M.G.: A probabilistic heuristic for a computationally difficult set covering problem. Oper. Res. Lett. **8**(2), 67–71 (1989)

18. Juette, S., Thonemann, U.W.: Divide-and-price: a decomposition algorithm for solving large railway crew scheduling problems. Eur. J. Oper. Res. **219**(2), 214–223 (2012)

19. Mehrabian, A.R., Lucas, C.: A novel numerical optimization algorithm inspired from weed colonization. Ecol. Inform. **1**(4), 355–366 (2006)

20. Revelle, C., Marks, D., Liebman, J.C.: An analysis of private and public sector location models. Manag. Sci. **16**(11), 692–707 (1970)

21. Schreuder, J.A.: Application of a location model to fire stations in rotterdam. Eur. J. Oper. Res. **6**(2), 212–219 (1981)

22. Simeone, B., Nouno, G., Mezzadri, M., Lari, I.: A boolean theory of signatures for tonal scales. Discrete Appl. Math. **165**, 283–294 (2014)

23. Soto, R., Crawford, B., Galleguillos, C., Paredes, F., Norero, E.: A hybrid alldifferent-Tabu search algorithm for solving sudoku puzzles. Comput. Int. Neurosci. **2015**, 286354:1–286354:10 (2015)

24. Soto, R., Crawford, B., Muñoz, A., Johnson, F., Paredes, F.: Pre-processing, repairing and transfer functions can help binary electromagnetism-like algorithms. In: Silhavy, R., Senkerik, R., Oplatkova, Z.K., Prokopova, Z., Silhavy, P. (eds.) Artificial Intelligence Perspectives and Applications. AISC, vol. 347, pp. 89–97. Springer, Heidelberg (2015)
25. Toregas, C., Swain, R., ReVelle, C., Bergman, L.: The location of emergency service facilities. Oper. Res. **19**(6), 1363–1373 (1971)
26. Veenhuis, C.: Binary invasive weed optimization. In: 2010 Second World Congress on Nature and Biologically Inspired Computing (NaBIC), pp. 449–454. IEEE (2010)
27. Vidal, T., Crainic, T.G., Gendreau, M., Prins, C.: Heuristics for multi-attribute vehicle routing problems: a survey and synthesis. Eur. J. Oper. Res. **231**(1), 1–21 (2013)
28. Walker, W.: Using the set-covering problem to assign fire companies to fire houses. Oper. Res. **22**, 275–277 (1974)
29. Xu, Y., Kochenberger, G., Wang, H.: Pre-processing method with surrogate constraint algorithm for the set covering problem
30. Zhang, J., Wei, Q., Chen, G.: A heuristic approach for λ-representative information retrieval from large-scale data. Inf. Sci. **277**, 825–841 (2014)

Software Architecture and Software Quality

Michal Žemlička[1]([⊠]) and Jaroslav Král[2]

[1] The University of Finance and Administration,
Estonská 500, 101 00 Praha 10, Czech Republic
michal.zemlicka@post.cz
[2] Masaryk University, Faculty of Informatics,
Botanická 68a, 602 00 Brno, Czech Republic
kral@fi.muni.cz

Abstract. Software quality is a crucial but partly subjective concept. Assessment of quality of software systems is typically a two-stage process consisting of the evaluation od related quality aspects and assessment of the quality of the software. If the software architecture discussed in the paper is used, the evaluation of many commonly considered aspects is increased and therefore the assessment of the system quality is enhanced. We show that some known but neglected aspects as well as some new architecture related ones ought to be considered.

Keywords: Software confederations · Document-oriented communication · Hierarchical system composition · Business-oriented information systems · Autonomous services and systems

1 Introduction

We will discuss software quality aspects from the point of view preferable for small software firms developing software for small to medium-sized customers and for user experts involved in software development, maintenance, and use.

Software quality is a subjective concept. It depends on the requirements, aims, and needs of involved people. The quality is assessed. The evaluator need not be an IT expert, it can be a user. The assessment as a rule uses assessments of a set of quality aspects. The aspects can be chosen from a list of quality aspects being generally useful for the assessment of quality [6,14].

The outputs of the assessment of an aspect are expressed by values of fuzzy metrics. The fuzzy metric of enterprise size can have, for example, the values *small*, *medium*, *large*, and *giant*. This metric is an indicator of the enterprise type.

The values of the metrics are ordered. An aspect A is positively related to an aspect B if there is (or we believe it) a positive correlation between them in the sense that increase of a chosen metric of A tends to increase a chosen metric of B (compare [10]). A and B are tightly connected if A is positively correlated with B and if B is positively correlated with A.

One of the aims of software system quality assessment and analysis is to find relevant quality aspect A positively correlated with the quality of the software

© Springer International Publishing Switzerland 2016
O. Gervasi et al. (Eds.): ICCSA 2016, Part V, LNCS 9790, pp. 139–155, 2016.
DOI: 10.1007/978-3-319-42092-9_12

system. It then can make sense to look for a way for enhancement of A (i.e. to increase its metrics). Such a metric can be included into a list of aspects to be considered during the quality assessment. We will discuss generally considered aspects first. We will find new aspects becoming meaningful due to the use of a specific software architecture.

The assessments are difficult to be fully formalized. Partial formalization is sometimes possible. Then the quality assessment process can be standardized (ISO). A standardized quality aspect is then called quality characterization. It is a quite complex task; see the number and size of the ISO/IEC standards with numbers 250nn. We show below that many aspects are substantially enhanced and new aspects ought to be considered if a variant of service-oriented architecture called software confederation is used.

2   Quality Aspects

2.1   Characteristics (Aspects) from ISO/IEC 25010:25010

The international standard ISO/IEC 25010:2011 Systems and software engineering – Systems and software Quality Requirements and Evaluation (SQuaRE) – System and software quality models deals with the following characteristics (formalized aspects) and sub-characteristics of software quality, see Table 1.

Table 1. Characteristics and sub-characteristics of software quality according ISO/IEC 25010:2011

Functional suitability	Reliability
Functional completeness	Availability
Functional correctness	Fault tolerance
Functional appropriateness	Recoverability
Performance efficiency	Security
Time behaviour	Confidentiality
Resource utilization	Integrity
Capacity	Non-repudiation
	Accountability;
	Authenticity
Compatibility	Maintainability
Co-existence	Modularity
Interoperability	Reusability
Usability	Analysability
Appropriateness recognizability	Modifiability
Learnability	Testability
Operability	Portability
User error protection	Adaptability
User interface aesthetics	Installability
Accessibility	Replaceability

2.2   Quality Aspects Enhancible by Software Architectures

There are software quality aspects not mentioned in ISO/IEC 25010:2011 or in literature. We suppose that some of them are important for overall quality of software systems, especially of information systems[1]. Let us give the most important examples:

Openness – weakening dependency on a single vendor (compare Vendor Lock-In Antipattern [3]).

Quick (re)configurability (opportunity to change the system behavior very quickly) – if something unexpected happens (accident, system failure, natural disaster, partner failure), we need to react in the situation instantly (we cannot wait till an agreement with a consulting company and the system vendor is made and until the change is designed and implemented – typically within weeks or months).

Agility of development process and maintenance

Autonomy of (coarse-grained) components and systems – The parts could be independently developed and often also used. It is, according our experience, a key aspect.

Incrementality of the specification, development, deployment, use, and maintenance – Is it possible to specify, develop, deploy, update, maintain, and use it part-by-part or only entire at once?

Transparency – The user can understand the behavior of the system as a whole as well as of its parts (components) and is able to be involved in system design, maintenance, and use.

Hierarchical composability – Ability to adopt hierarchical structures.

Close relation to the structure of the organization and its agendas – We can take advantage from closer responsibility for the system and its parts.

Business orientation – Are the structure of the system as well as the interfaces business oriented (in contrary to currently popular programmer-oriented APIs and layered system structure)? We will show that the majority of the characteristics from Table 1 and all the quality aspects from Sect. 2.2 can be substantially enhanced if a proper software architecture is used.

Document-based communication – The system internal and external communication is based on exchange of human readable documents.

3   Software Architectures

Large software systems must be constructed as a composition (a network) of components being autonomous as much as possible. The main reasons for it are:

– Large monolithic system cannot be developed quickly enough as its development time cannot be for a given system shortened under some quite large limit whatever is the size of a development team [2,9]. It causes that such a system is not tested enough and that it is permanently obsolete. Their quality gradually deteriorates. It must be periodically redeveloped/rewritten.

---

[1] Cmp. https://en.wikipedia.org/wiki/List_of_system_quality_attributes.

- Monolithic systems are not good for the incremental development in the large, i.e. when increments are large and have business document oriented interfaces.
- Legacy systems and applications must be reused. It is difficult to implement if the system is monolithic.
- Many forms of agility of development as well as business processes are very difficult to implement.
- There can be difficulties to enable flexible collaboration of business partners.

A good solution is a system having service-oriented architecture (SOA) integrating large application services wrapped by wrappers as services. The system portals are wrapped similarly. The wrappers communicate with each other using (digitalized business) documents. We call such a SOA *software confederations*.

Good business processes are difficult to implement unless they are controlled and supported by digitalized documents. The document-oriented paradigm is very different from the currently very popular object-oriented one. It could seem to be not important. It is, however, crucial. People used to use object-oriented thinking are very often unable to accept document oriented thinking. Partial explanation is that object-orientation is good for coding and coders whereas document orientation is mainly business oriented (and therefore easier to be understood by business people).

Object-oriented design is well-applicable for development of not too complex systems. When talking about large and complex systems, object-oriented attitude is often applied in the cases of large and complex systems. It makes the systems rather tightly bound. It holds not only for monolithic systems, but very often also for distributed ones as well as web-service based ones. The interfaces are typically based on method invocation (local or remote one) and logically it is still imperative programming with fine-grained parts and fine-grained interfaces.

Object-oriented development is well supported by many tools, libraries, and frameworks. It simplifies the development and modification of systems being not too complex. Even in this case there are problems to meet user needs. Practice shows that for upper levels of larger systems the object orientation is a handicap. According to our observations, the people used to use object-oriented attitudes are not only unable to accept service-oriented[2] and document-oriented paradigms, they even hate it. There are multiple reasons causing it. The probably most important one is that there is no smooth and easy way to achieve (using object orientation) a high level of autonomy of large components.

4   Cathedral, Bazaar, or Town

Raymond in [17] has described two approaches to development and use: It can be built as a cathedral or a bazaar. We suppose that there is a third way: to build it as a town.

---

[2] We exclude the SOA applying the antipattern Service as a Class.

Cathedrals need not be a proper choice as they require great resources and too long development times. The solution is to construct the system as a town – a network of communicating cathedrals (and small churches as well). The communication is supported by architectural (infrastructure) tools. Let us take inspiration from real world and distinguish software designers, software architects, and software urbanists. We can even distinguish multiple levels of software architectures and architecting.

It is in some sense similar to build a single family house or a simple software system. In both cases entire project could be kept in a single mind. When developing a village or a middle-sized or middle-complexity software system, we get to the limits of a single mind (it, however, depends on the problem as well as on the mind). The proposed solution enables separation of concerns. The cathedrals can be developed autonomously (as autonomous units). The communication and integration tools can be built gradually.

It is clear, that if we develop of a town or a complex application, then on the top level we must use approaches and tools different from the ones used for design individual houses or small components. At the top levels there are usually set up basic rules, assigned basic functionality to various parts, and set up some limitations. Basic rules, limitations, and functionalities of parts or components are set up at the topmost levels.

At lower level the entities (e.g. housing blocks or streets) can be built if they follow agreed rules and limitations. At the same time it should be enough freedom of selection of proper solutions. Similarly, it is reasonable to assign individual logical entities (system parts) to individual designers (architects, urbanists).

Towns can outlive centuries and they need not get obsolete. They can be continuously reconstructed or adapted, they did not get obsolete. Town-like software can have similar properties. It can be updated continuously for a very long time.

It is possible to build a town using software confederations.

Software towns are the best choice for systems being not small. Bazaars and cathedrals [17] have specific properties. Cathedrals (very large monoliths) become gradually obsolete and must be therefore periodically rewritten. Their development costs and durations are as a rule overcome. Cathedrals can be built by very large vendors only. Even the largest vendors are unable to meet/fulfill quality requirements like adaptability, openness, and simple and quick maintainability and prevent gradual obsolescence.

5   Software Confederations

In this section we describe a service-oriented software architecture different from the solutions of Erl [4,5], OASIS [13], or OpenGroup [16] – *Software Confederations* [7]. Although the services in a software confederation could be available on the web (they can be web services) they are no Web Services [18] in the sense of OASIS or W3C (https://www.w3.org/2002/ws/).

The notion of *Software confederations* describes software architecture covering the following features:

- The system is a virtual peer-to-peer network of cooperating autonomous software entities (services).
- The communication of every two services can be agreed and optimized according to the purpose of the communication. It should be inspired by the communication of the real-world counterparts of the services.
- The overall structure of the system can be hierarchical and similar to the structure of the organizational structure of users. It should reflect formal organizations (enterprises, offices) as well as informal organizations (alliances, supply chains, business partner groups).
- Services can be either encapsulated applications or (architectural) services supporting communication and with help of politics providing system architecture.

Software confederations are a variant of service-oriented architectures (SOA, [4,7,13]). An overview of main SOA variants and their comparison is discussed in [11].

The concept of software confederations is based on authors' decades lasting practical experience with structured systems, control systems, and other architectures and system types as well as on research, teaching and consulting. It allows a deeper involvement of users as well as agility of them during software design, development, testing, operation, and maintenance. It is a very important aspect of software (information) system quality.

The main idea is, that if an organization is successful, it has its own know-how that should be not only preserved but also supported by the system. It is therefore preferable to apply the existing know-how of the organization if possible. We will see how it can be achieved. It implies the tendency to enable the exchange of digitalized business documents.

Inter-human communication tends to be declarative, business-problem-oriented, and coarse-grained. Formal inter-human communication (based on exchange of forms, legal or business documents, etc.) is usually asynchronous. It is usually as precise as necessary and as vague as possible. Years or decades of use of the real-world service interfaces usually caused that they are well tuned.

The application of such interfaces (or using interfaces very similar to them) has many advantages like system transparency, flexibility, or information hiding.

It is not common for communication inside classic applications and information systems. The issue is solved in confederations using tools and turns described below.

5.1 Encapsulation/Wrapping

Basic functionality of confederations is provided by services being typically wrapped legacy or third-party systems supporting the basic business operations. It is possible and often also reasonable that such systems are supported by people executing some operations.

The issue is that existing applications are usually equipped with interface being imperative and fine-grained. We need to transform the interfaces into the asynchronous document-oriented form first.

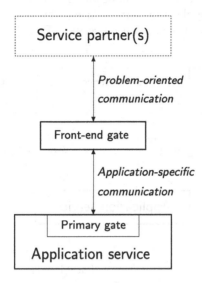

Fig. 1. Application service with front-end gate

Other issue is that many applications are not designed to communicate asynchronously or are even not equipped with any network interface. It is good to wrap the applications twice (see Fig. 1):

1. We extend an application with a message-passing ability (*primary gate*) if necessary. We call the resulting system *application service.*
2. The application service is further wrapped by a special service transforming the interface from being application specific to be problem specific.

The first transformation is usually done with a wrapper tightly bound to the wrapped software making it accessible from the rest of the system. The second transformation is reasonable to design in a partner-specific way. It needs not be so tightly bound to the application. It is reasonable to design it as an autonomous service – *front-end gate* – that maps sequences of coarse-grained declarative messages or documents in one language to sequences of (fine-grained, imperative, and application-oriented) messages for primary gate of the service. The message transformation is very similar to the transformations made by compilers, so the techniques and tools are already available and known.

Front-end gates transform also the messages sent by the application service to problem-specific form. The decompiling techniques can be used here.

It is necessary to use the structure from Fig. 1 as one logical whole. It can be achieved if we introduce a limitation (a policy) stating that any communication to as well as from the application service must pass the front-end gate.

The proposed solution can be easily enriched if we establish the policy changing the partner services. This concept could be further generalized in different ways – especially if architectural services are introduced.

The concept described in Fig. 1 can be enriched if we allow that there is more front-end gates for an application service. It enables that different partners can use different interfaces to the application service (Fig. 2).

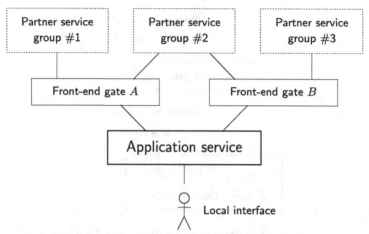

Partner services from groups 1 and 3 use their specific interface. Partners from group 2 can access both interfaces – they can e.g. play both roles. For local support and precise (expert) handling a local interface can be used. Any communication path to and from the application service must pass just one front-end gate.

Fig. 2. Application service with its interfaces

Then we have services providing basic functionality of the system equipped with interfaces matching the needs of the communication partners and understandable to the users providing the agenda and to their partners. Let us compose the services into a system.

5.2   Towards a Town

The system structure is implemented using various types of architectural services [12]. The first type of them – front-end gates – has been already described. They allow keeping most changes of the service implementation hidden from the rest of the system. It is often possible to change the application providing the capability by another application providing it without influencing the rest of the system. It significantly simplifies maintenance (it reduces change propagation) and increases system flexibility.

System Interfaces. Many systems have one interface for employees and one for customers. Some systems are moreover equipped by further interfaces for business partners.

The interfaces (portals) can be again services. A system can have several portals. There are following reasons for that:

- Every portal can contain only the functionality necessary for the expected group of users. The interface supported by the portal can be optimized for given group of users.
- It simplifies its design, implementation, and maintenance.
- Portals act for typically human users as front-end gates of the entire system.
- It is better also for system security: if there is no support for strong functions in the interface available from the internet, potential hack of that interface gives the attacker less opportunities for further influence on the system.
- Technically complex portals can be built for specific group of people. They can be maintained independently. As they are not available only for a very limited group of users, it increases system security. An example is given below.
- If a portal is down, it is still possible that other portals are available. The systems are therefore available for at least some users even if the main portal is under attack (DoS or destruction).

Structuring the System. If an organization grows, it could be necessary to change its structure to be hierarchical. In other cases it could be meaningful to decompose the enterprise as well as the system into parts. Such transformation could be necessary for developers as well as for users:

- If the system structure mirrors the organizational structure, the users can feel responsibility for the part of the system supporting their organizational subunit as well as for its data.
- The organizational decomposition usually follows the rule that the number of direct subunits and supported agendas is kept relatively low (about ten). It is close to the psychological limitation for understanding system structure described in [15].

Composition of Services. The services can form groups called *composite services*. The composite services are for other services (the ones outside the group) represented by a special architecture service – *head of composite service* (HCS) concentrating all communication to and from the group. Such group can be viewed as a single service if necessary. It also behaves this way. Creation of hierarchical system structure using composite services and their heads is shown in Figs. 3 and 4.

As any other service in the system, also a composite service represented by its head can be equipped by front-end gates to provide required interfaces of the service.

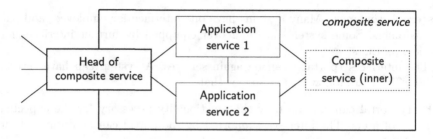

Fig. 3. Composite service – logical view

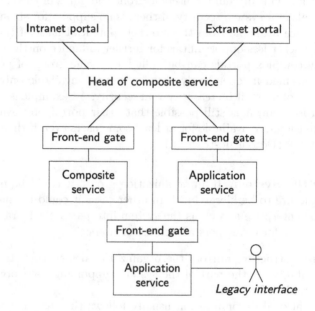

Fig. 4. A very simple confederation

It is possible that a composite service communicates differently with different partners. Moreover, no one from the services inside the composite service needs not to understand the formats required by the partners. It again enhances flexibility of the system.

Supporting Business Processes. Key role in processing agendas of the organization is played by business processes. There are multiple opportunities for implementation of business processes in software:

1. **Hardcoding business processes into application code.** It is fast but it is very laborious and time-consuming to change it. It is moreover necessary to ask vendor for every change of business processes implemented this way.
2. **Describing business process using Business Process Execution Language (BPEL, [1]) or other business process execution language.** It potentially allows modifying business processes without changing application

code. (Yes, one can argue that BPEL code is part of the application code but let us expect that such code is at other logical and technical level, so it can be handled separately.) Here again, it is possible to recognize wide spectrum of various business process execution languages. Some can be highly declarative and can expect high expertise of people executing it, other can be very imperative and describe any movement in detail. Currently, most tools support business process execution languages that are close to the last case: the languages are technical (require some programming skill instead of problem domain expertise) and imperative (do this and now). Such solutions are in fact another module implemented by the software vendor. There is no practical chance that the change could be done by usual business process owners.

3. **Encoding business processes into the documents over them the business processes should keep control.** It is relatively flexible (the business process is written to the document on its entering into the system). It is then hard to track responsibility (how the process owner could influence an already running business process?). Similarly, when some problems arise, it is complicated to stop or hold execution of influenced business processes.

4. **Rule-based control over business processes.** It is hard to develop, there is lack of responsibility for the process (its execution is driven by the predefined rules, the owner has no chance to influence it). The system is very sensitive to completeness and correctness of the rule set.

The discussed architecture enables support for multiple business process execution languages. It is important as most organizations have various business process types. It is quite common that some business processes are performed frequently and are highly optimized and some ones are executed with high variability or rarely. They expect higher problem domain expertise from their executors. Such activities are typical for development of prototypes, for larger structural changes, for business process restructuring, and other cases.

The control of business processes is let on special kind of services – *process managers* [8]. Business process manager controls given business process. It supports process owner (typically through a specific portal) with data on the business process and accepts instructions for the process control (like hold, stop, ...). It is therefore reasonable if the process manager supports the business process execution language that is corresponding to the business process type and to the skills of the process owner. The process owner then can react intuitively and very quickly to any issue if necessary.

When talking about business processes in structured organizations, we can have business processes working across inter-organizational or even organizational borders. It is reasonable, if such business process is split into parts:

- individual parts of handling activities within corresponding units,
- coordination part managing the coordination of individual units (e.g. monitoring overall limits).

Integration of Autonomous Parts. Individual agendas (or individual organizational subunits) could be of various nature. They then require different behavior and different requirements on cooperating people, their work, or communication. The agendas then can have really different requirements on the information systems supporting them. It can happen that at least for some of the information systems different development paradigms could be optimal. Then it is reasonable if such part can be really loosely coupled.

In manufacturing individual workplaces are cooperating together. The critical places (bottlenecks) are equipped with buffers (even in just-in-time manufacturing case there are such places – smaller than usual but they usually are there).

Other buffers may be of technological use: Some partial products are used later in pairs or m-tuples for larger m. In other cases it is reasonable if some partial products wait a bit until applied (they must dry, change temperature, ...) or it is reasonable to wait for more critical parts to be prepared.

It is, we propose to build buffers allowing to relax and simplify communication between parts of the system. We can take inspiration from structured systems [19] where there are elements processing data and elements storing data.

We call the services playing the role of a buffer or data store *data store services* or *data stores* for short.

They can be used to bridge the gap between interactive and batch parts (technical reason) of the system or for collection of data for some specific purposes (problem-domain reason). They can also support some complex communication schemata and strengthen autonomy of the connected parts.

It simplifies asynchronous communication (there can be more requests waiting for processing), better load handling, higher independence of the systems (partner systems then could use different communication style).

5.3   Access Points

The services in a confederation can use various communication means. The basic one is asynchronous message (document) passing.

A quite important practical issue is that services often should be available only to partners that are trusted and equipped with a valid agreement of cooperation. This issue can be solved using specific services being and called *access points*. Every service interface can be either accessed directly (in the case of trusted services) or through access points (agreed service partners). The access point is expected to be created or generated when an agreement is signed and can be destroyed or locked when the agreement is terminated or broken.

It is possible (and in some situations reasonable) to limit communication to messages or documents in a specific language or set of languages. Such checking is expected to be used primarily on external access points.

It can restrict accepted messages to specified format only. The language checking can be quite precise – it can restrict not only language constructs but also used identifiers and values. Such policy can increase system security.

On the other hand, keeping all such firewalls up to date can be a non-trivial task. It is necessary to find a boundary between updatable firewalls or replacement of hard-coded firewalls.

6 Confederations and Software Quality Aspects

A properly implemented confederation substantially enhances the majority of almost all quality aspects (characteristics) from the list discussed in ISO/IEC 25010:2011. The level of achievable enhancements of the aspects is shown in Table 2. The list of aspects from Table 2 does not contain some aspects relevant for business document driven SOA (confederations). Let us discuss some enhancements from Table 2. The enhancement can be achieved only if the system is properly designed. Some enhancements are very simple to be achieved in SOA. For example: maintenance is supported by simple user involvement, decomposi-

Table 2. Characteristics and sub-characteristics of software quality according ISO 25010:2011 that could be significantly improved by the use of software confederations

+	Functional suitability	?+	Reliability
+	Functional completeness	?+	Availability
+	Functional correctness	(+)	Fault tolerance
+	Functional appropriateness	++	Recoverability
	Performance efficiency	?+	Security
(+)	Time behavior	+	Confidentiality
?	Resource utilization	+	Integrity
?	Capacity	?-	Non-repudiation
		?+	Accountability;
		?+	Authenticity
++	Compatibility	++	Maintainability
+	Co-existence	++	Modularity
+	Interoperability	++	Reusability
+	Usability	++	Analyzability
+	Appropriateness recognizability	++	Modifiability
+	Learnability	++	Testability
+	Operability	++	Portability
+	User error protection	++	Adaptability
-	User interface aesthetics	+	Installability
?	Accessibility	++	Replaceability

++ strong positive influence ? unclear
+ positive influence   ?- probably negative effect
?+ probably positive effect (+) multiple factor effects

tion into coarse-grained components, and moreover by the fact that the system uses business-oriented principles.

The crucial (mainly new) quality aspects tightly related to confederations are:

1. autonomy of components/services
2. the systems are (business) document driven
3. agility of development processes as well as of business processes
4. possibility to use systems as autonomous components (it is crucial for in- and out-sourcing)

The relations between crucial aspects related to confederations are shown in Fig. 5.

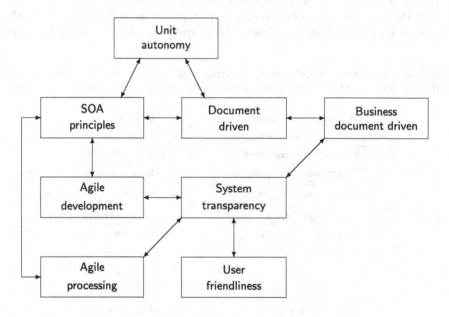

Fig. 5. Mutual stimulation of software quality aspects

Confederation can enhance also some aspects being not listed in ISO/IEC 25010:2011 or even in https://en.wikipedia.org/wiki/List_of_system_quality_attributes. Some of the aspects have sense if and only if the confederations are used. Very important examples are:

– economical aspects of development:
  • development costs reduction,
  • project duration reduction,
  • incremental development and implementation;
– managerial aspects:
  • easy supervising of development and use,
  • easy user involvement,

- possibility to reuse and to insource and outsource,
- transparent correlation between the activities of processes and business reality;
- agility together with understandability and ability to control the processes:
  - hierarchy composition,
  - new dimension of several aspects, especially of usability and composability;
- user interface can be implemented as a set of services; it enables flexible interfaces for different partners according to their skills.

7 Open Issues and Further Research and Development

Although principles of software confederations are known for decades and confederations themselves are used developed also more than ten years, there are still open issues waiting for further research as well as design and development work:

- What is the optimal combination of object-oriented and service-oriented attitudes.
- Application of document management systems as a confederative middleware could lead to some disadvantages of Enterprise Service Bus.
- What formalisms could be applied and where it is better not to formalize.
- Advantages and disadvantages of virtualization in confederations.
- Tools simplifying the development of architectural services.
- Challenges and threats of cloud techniques for service-oriented attitudes.
- Confederation design patterns for the systems working as intelligent data collection tools.

8 Conclusions

We have discussed aspects of software quality with respect to used software architecture. It has been demonstrated on a specific service-oriented architecture – on software confederation. The architecture is based on long-time practical experience of the authors with development of software for small to medium enterprises. Last years we solved practical cases where confederations were combined with existing systems.

We have shown that the proposed software-oriented architecture (we compared it to towns) can very substantially increase the quality of business systems. It is open whether this attitude can be applied in e-government. It is clear (technically as well as from outsider software engineer's point of view) that confederations should be the best attitude for e-government. Confederations are not applied in e-government yet. The technical reason for that could be complex paths of collections of documents.

Acknowledgement. The paper has supported by the Institutional support for long-term strategic development of the research organization University of Finance and Administration.

References

1. Andrews, T., Curbera, F., Dholakia, H., Goland, Y., Klein, J., Leymann, F., Liu, K., Roller, D., Smith, D., Thatte, S., Trickovic, I., Weerawarana, S.: Specification: business process execution language for web services version 1.1 (2003)
2. Boehm, B.W.: Software Engineering Economics. Prentice-Hall, Englewood Cliffs (1981)
3. Brown, W.J., Malveau, R.C., McCormick, H.W., Mowbray, T.J.: AntiPatterns: Refactoring Software, Architectures, and Projects in Crisis. Wiley, New York (1998)
4. Erl, T.: Service-Oriented Architecture: Concepts, Technology, and Design. Prentice Hall PTR, Upper SaddleRiver (2005)
5. Erl, T.: SOA: Principles of Service Design. Prentice Hall Pearson Education, Upper SaddleRiver (2008)
6. International Organization for Standardization, International Electrotechnical Commission: ISO/IEC 25010:2011 systems and software engineering - systems and software quality requirements and evaluation (SQuaRE) - system and software quality models (2011). https://www.iso.org/obp/ui/#iso:std:iso-iec:25010:ed-1:v1: en
7. Král, J., Žemlička, M.: Software confederations - an architecture for global systems and global management. In: Kamel, S. (ed.) Managing Globally with Information Technology, pp. 57–81. Idea Group Publishing, Hershey (2003)
8. Král, J., Žemlička, M.: Implementation of business processes in service-oriented systems. In: 2005 IEEE International Conference on Services Computing (SCC 2005), vol. 2, pp. 115–122. IEEE Computer Society (2005)
9. Král, J., Žemlička, M.: Inaccessible area and effort consumption dynamics. In: Dosch, W., Lee, R., Tuma, P., Coupaye, T. (eds.) Proceedings of 6th International Conference on Software Engineering Research, Management and Applications (SERA 2008), pp. 229–234. IEEE CS Press, Los Alamitos (2008)
10. Král, J., Žemlička, M.: Popular SOA antipatterns. In: Dini, P., Gentzsch, W., Geraci, P., Lorenz, P., Singh, K. (eds.) 2009 Computation World: Future Computing Service Computation Cognitive Adaptive Content Patterns, pp. 271–276. IEEE Computer Society, Los Alamitos (2009)
11. Král, J., Žemlička, M.: SOA Worlds. In: Quintela Varajão, J.E., Cruz-Cunha, M.M., Putnik, G.D., Trigo, A. (eds.) CENTERIS 2010. CCIS, vol. 110, pp. 10–19. Springer, Heidelberg (2010)
12. Král, J., Žemlička, M.: Support of service systems by advanced SOA. In: Lytras, M.D., Ruan, D., Tennyson, R.D., Ordonez De Pablos, P., García Peñalvo, F.J., Rusu, L. (eds.) WSKS 2011. CCIS, vol. 278, pp. 78–88. Springer, Heidelberg (2013)
13. MacKenzie, C.M., Laskey, K., McCabe, F., Brown, P.F., Metz, R.: Reference model for service-oriented architecture 1.0, OASIS standard, 12 October 2006. http://docs.oasis-open.org/soa-rm/v1.0/
14. Microsoft: Chapter 16: Quality attributes. In: Microsoft Application Architecture Guide, pp. 191–204. Microsoft Press, October 2009. https://msdn.microsoft.com/en-us/library/ee658094.aspx
15. Miller, G.A.: The magical number seven, plus or minus two: some limits on our capacity for processing information. Psychol. Rev. **63**, 81–97 (1956)
16. Open Group: Open Group standard SOA reference architecture, November 2011. https://www2.opengroup.org/ogsys/jsp/publications/PublicationDetails.jsp?publicationid=12490

17. Raymond, E.S.: The cathedral and the bazaar. First Monday **3**(3) (1998). http://www.firstmonday.org/ojs/index.php/fm/issue/view/90

18. Weerawarana, S., Curbera, F., Leymann, F., Ferguson, T.S.D.F.: Web Services Platform Architecture. Prentice Hall PTR, Upper Saddle River (2005)

19. Yourdon, E.: Modern Structured Analysis, 2nd edn. Prentice-Hall, Englewood Cliffs (1988)

Multi-hop Localization Method
Based on Tribes Algorithm

Alan Oliveira de Sá[1(✉)], Nadia Nedjah[2], Luiza de Macedo Mourelle[3],
and Leandro dos Santos Coelho[4,5]

[1] Center of Electronics, Communications and Information Technology,
Admiral Wandenkolk Instruction Center, Brazilian Navy, Rio de Janeiro, Brazil
alan.oliveira.sa@gmail.com
[2] Department of Electronics Engineering and Telecommunication,
Engineering Faculty, State University of Rio de Janeiro, Rio de Janeiro, Brazil
nadia@eng.uerj.br
[3] Department of System Engineering and Computation, Engineering Faculty,
State University of Rio de Janeiro, Rio de Janeiro, Brazil
ldmm@eng.uerj.br
[4] Department of Electrical Engineering, Federal University of Parana (UFPR),
Curitiba, PR, Brazil
[5] Industrial and Systems Engineering Graduate Program,
Pontifical Catholic University of Parana (PUCPR), Curitiba, PR, Brazil

Abstract. In many applications of Swarm Robotic Systems (SRS) or
Wireless Sensor Networks (WSN), it is necessary to know the position of
its devices. A straightforward solution should be endowing each device,
i.e. a robot or a sensor, with a Global Positioning System (GPS) receiver.
However, this solution is often not feasible due to hardware limitations of
the devices, or even environmental constraints present in its operational
area. In the search for alternatives to the GPS, some multi-hop localization methods have been proposed. In this paper, it is proposed a novel
multi-hop localization method based on Tribes algorithm. The results,
obtained through simulations, shows that the algorithm is effective in
solving the localization problem, achieving errors of the order of 0.01 *m.u.*
in an area of 100×100 *m.u.*. The performance of the algorithm was also
compared with a previous PSO-based localization algorithm. The results
indicate that the proposed algorithm obtained a better performance than
the PSO-based, in terms of errors in the estimated positions.

1 Introduction

In some distributed systems, such as swarm robotic systems (SRS) or wireless
sensor networks (WSN), it is often required that each device, *i.e.* a robot or
a sensor, be aware of its own position. In a SRS, for example, the positioning
information of a robot is needed to perform auto-formation tasks, where each
robot must position itself within a pre-defined formation, or even self-healing
tasks, where the robots reorganize themselves in order to restore the swarm
formation that eventually has been undone [1]. In the same way, in a WSN,

© Springer International Publishing Switzerland 2016
O. Gervasi et al. (Eds.): ICCSA 2016, Part V, LNCS 9790, pp. 156–170, 2016.
DOI: 10.1007/978-3-319-42092-9_13

the information of the position of each sensor is needed to make it possible to locate the measured events. The environmental monitoring of rivers, lakes and forests [2,3], or even nuclear waste stored underwater [4], are examples where it is necessary to know the sensor's location. Additionally, in a WSN, the sensor's position information is often needed to allow the execution of location based routing protocols [5], such as the MECN (Minimum Energy Communication Network) [6], SMECN (Small Minimum Energy Communication Network) [7], GAF (Geographic Adaptive Fidelity) [8] and GEAR (Geographic and energy-aware routing) [9]. In this routing protocols, the location information is used to find the path with the lowest energy cost, and also allows the data requests to be sent only to specific regions of the sensed area, preventing unnecessary transmissions to the other areas [5].

A straightforward solution to assess the positions of the robots of a SRS, or the sensors of a WSN, should be endowing each device with a Global Positioning System (GPS). However, this solution if often not feasible due to some typical characteristics of these devices, such as: the reduced size, the limited embedded power source and the low cost. Furthermore, some of its operational environments, such as indoor ou underwater, affects the reception of the GPS localization signals. In this sense, some other techniques have been developed, to provide alternatives to the GPS.

In many localization algorithms, classified as range-based, the unknown nodes, *i.e.* robots or sensors that do not know their own position, estimate their positions based on: the distances measured between the network nodes; and the coordinates of reference nodes, also called anchors, whose positions are previously known. In this systems, the most common techniques to measure the distances are based either on the received signal strength (RSS), the propagation time of the signal, or comparing the propagation time of two signals with different propagation speeds [19].

Once the aforementioned distance measurement techniques are based on signals propagation, it is necessary to consider a limit for such distance measurements. In the simple case, where all the anchors are within the distance measurement limit, the measures are made directly and the localization process is done in one-hop [10]. However, in cases where one or more anchors are out of the distance measurement limit, as the problem addressed in this paper, the localization process is performed in multiple hops [11,20]. The advantage of the multi-hop approach is that it requires less anchor nodes distributed in the network than in the one-hop case. On the other hand, the muti-hop strategy usually requires more computational resources. According to [20], the multi-hop localization algorithms, in general, can be designed and described in a standard three stage approach:

1. Determine the distances between each unknown node and each anchor, in multiple hops.
2. Compute a rough position of each unknown node, based on the distances measured in stage 1.
3. Refine the position of each node using the position and distance informations of its neighboring nodes.

Considering the multi-hop localization problem, recent solutions based on bio-inspired optimization techniques have presented promising results [11,12,19]. In this sense, targeting a more accurate solution for the multi-hop localization problem, it is proposed in this paper a localization algorithm based on the Tribes optimization algorithm [13,14]. The performance of the proposed algorithm is evaluated through a set of simulations and compared with the results obtained by the algorithm proposed in [12], which is based on the Particle Swarm Optimization (PSO) algorithm [15,16].

The rest of this paper is organized as follows: First, in Sect. 2, we present some related works. Then, in Sect. 3, we briefly describe the Tribes algorithm. After that, in Sect. 4, we explain the proposed localization algorithm. Then, in Sect. 5, we present and compare the obtained results. Finally, in Sect. 6, we present the conclusions and some possibilities for future work.

2   Related Works

The hardware and environmental constraints often present in SRSs or WSNs, as mentioned in Sect. 1, have motivated the search for more efficient localization algorithms, as alternatives to the GPS.

One possible, and simple, strategy is to estimate the position of each unknown node based on the distances measured directly to a set of anchors, that are within the node's sensor range. Considering this one-hop strategy, it is proposed in [18] a method based on Genetic Algorithms (GA) for problems without obstacles and without noise. Another one-hop GA-based localization algorithm is presented in [21], where mobile anchors are used to locate static unknown nodes. In [10], the authors propose another one-hop localization algorithm based on the Backtracking Search Optimization Algorithm (BSA), and compare its performance with a GA-based algorithm. The BSA-based algorithm achieved better accuracy than the GA-based algorithm, with lower computational cost. In general, a drawback of these one-hop localization algorithms is that they often require more anchors than the multi-hop localization algorithms.

Another strategy consists in perform the localization process in a multi-hop fashion, where the unknown nodes does not need to have the anchors reachable in only one hop. In this case, the anchors can be deployed in a smaller density than in the one-hop strategy. In general, these multi-hop localization algorithms follow the three stage approach proposed by [20] and presented in Sect. 1.

In [19], the author introduce the Swarm-Intelligent Localization (SIL) algorithm, which uses the PSO in its second and third stages to solve multi-hop localization problems in static WSNs. In [22], it is demonstrated the ability of SIL to locate mobile nodes in WSNs. Another multi-hop localization algorithm is presented in [12] where, targeting accuracy and reduced computational cost, the authors propose an algorithm that is based on the Min-Max [20] method and the PSO.

In this context, targeting a better accuracy for the multi-hop localization problem, this work introduces a new Tribes-based localization algorithm, which the performance is compared with the PSO-based algorithm proposed in [12].

3   Tribes Optimization Algorithm

The Tribes optimization algorithm was proposed by Clerc [13, 14], as a variant of the classical PSO, aiming a more robust optimization algorithm. It is a parameter free algorithm, in which the population's size and topology evolves during its execution, in response to the performance of the population.

Once the Tribes algorithm is a variant of the PSO, its individual is referred as particle and its population is referred as swarm. The particle is a possible solution that moves exploring a multidimensional search space. The fitness of a particle is the value that the objective function returns for the particle coordinates. The swarm is a set of possible solutions, *i.e.* a set of particles.

Each particle j has informers q, which are other particles that can share informations with j. Typically, the shared informations are: the best position ever visited by q; and its fitness in such position.

A tribe is a sub-swarm where each particle has all other particles of the tribe as its informers. Each particle belongs to only one tribe. In this algorithm the tribes concept is not related to a spatial proximity of the particles, but can be regarded as a "cultural vicinity", where the neighborhood is defined based on the links stablished to share information. A particle j may also have one or more external informers. An external informer is a member of another tribe, different from the tribe of j. Thus, the information exchange between tribes occur through links established with their external informers.

The number of tribes and the number of particles contained in each tribe changes during the algorithm execution. To determine the changes in a tribe, as well as the particle's displacement, all particles of a tribe are firstly classified as:

- Neutral if: $f(P_j[k]) \geq f(P_j[k-1])$
- Good if: $f(P_j[k]) < f(P_j[k-1])$
- Excellent if: $f(P_j[k]) < f(P_j[k-1]) \land f(P_j[k]) < f(P_j[k-2])$
- The worst of tribe T if: $\forall P_b \in T, (P_j[k]) \geq (P_b[k])$
- The best of tribe T if: $\forall P_b \in T, (P_j[k]) \leq (P_b[k])$

wherein f is the objective function, P_j is the position of the particle j, k is the iteration number and P_b is the position of a generic particle contained in tribe T.

A tribe T containing a total of m particles, of which g particles are classified as "good", is considered:

- A good tribe if: $rand(0, m) < g$
- A bad tribe if: $rand(0, m) \geq g$

In each topology update, the tribes considered "good" have their worst particle removed. Once a good tribe already has good problem solutions, the reduction of the population of the tribe is done to save computational resources. On the other hand, the "bad" tribes generate new particles to increase the possibility to discover better solutions.

The particles displacement in Tribes algorithm, follows two possible strategies: the "simple pivot"; or the "noisy pivot". The "simple pivot" strategy is used to update the positions of particles classified as "excellent". In this strategy, two positions are taken into account: B_j, which is the best position so far visited by a given particle j; and B_q, which is the best position so far visited by its informer q. Then, two hyperspheres of radius $|B_j - B_q|$ are established, centered in B_j and B_q. After that, the new position $P_j[k + 1]$ of particle j is computed as in (1):

$$P_j[k + 1] = w_1 H_j + w_2 H_q, \tag{1}$$

wherein w_1 and w_2 are weights proportional to the relative fitness of B_j and B_q, respectively, and H_j and H_q are randomly generated points in the two hyperspheres centered in B_j and B_q, respectively.

On the other hand, the "noisy pivot" strategy is used to compute the positions of particles classified as "good" or "neutral". The procedure adopted in this strategy is the same as that used in the "simple pivot" strategy, but a random noise is added to the obtained position. This noise allows eventual explorations beyond those hyperspheres.

It is worth mentioning that the topology updates in the tribes, does not happen at each iteration, since some iterations are necessary to let the information propagate through the swarm before each update. Thus, the topology updates are executed every $C/2$ iterations, where C is the dynamically changing number of links in the swarm.

A summarized description of Tribes is presented in Algorithm 1. In this paper, the stopping criterion used is the maximum number of evaluations of the objective function.

Algoritmo 1. Tribes Algorithm

>   **begin**
>   >   Initialize the swarm;
>   >   **repeat**
>   >   >   Evaluate each particle of the swarm;
>   >   >   Move the swarm using the "simple pivot" and "noisy pivot" strategies;
>   >   >   Update the swarm topology after $C/2$ iterations;
>   >   **until** *Stopping criterion is met*;
>   **end**

---

4   The Proposed Localization Algorithm

The localization algorithm herein proposed can be used in problems with nodes distributed in either two or three dimensions. However, for the sake of clarity, but without loss of generality, the algorithm is described focusing the two dimensions case. It is considered that, in most cases, the unknown nodes cannot directly measure their distance to the anchors, characterizing a multi-hop localization problem. As a premise, all nodes are static, and the formulation does not take into account distance measurement errors.

The algorithm follows the standard three stage approach proposed by [20] and described in Sect. 1. Thus, in order to provide a clear understanding of the algorithm, the underlying details of each STAGES-I, II and III are presented in Sects. 4.1, 4.2 and 4.3, respectively. Finally, in Sect. 4.4, the complete algorithm is presented.

4.1   Distance to Anchors

The first stage (STAGE-I) of the algorithm is intended to allow each unknown node to estimate its distance to anchors, through multiple hops. For this purpose, the Sum-Dist [17] algorithm is used. The Sum-Dist starts with a message originated at each anchor, disseminating its coordinates. The message also contains the information of the total distance that it travelled so far, which the anchor initially sets to 0. While the massage is forwarded from node to node, the distance of the last hop is added to the total distance traveled by the message. Each node stores and retransmits a received message if and only if, the total distance traveled by the message is the smallest known so far, to a given reference node. Thus, at the end of the execution of the Sum-Dist, each unknown node i will be aware of the coordinates $(\hat{x}_r, \hat{y}_r)$ of each anchor r and the length of the shortest path $l_{i,r}$ that exists between them.

It is worth mentioning that the shortest path existing between an unknown node i and an anchor r, $l_{i,r}$, is not necessarily the actual distance between them. In a network with nodes randomly deployed, for example, it is common to have a misalignment between the nodes that compose the path from i to r, resulting in a $l_{i,r}$ greater than the actual distance between i and r. The inaccuracy of $l_{i,r}$ may impact the initial coarse position computed in STAGE-II, as described in Sect. 4.2. However, this error is mitigated during the refinement process performed by Tribes algorithm, during STAGE-III, as described in Sect. 4.3.

4.2   Initial Node Position

The second stage (STAGE-II) of the algorithm aims to compute the initial coarse position of each unknown node i, which is obtained through the Min-Max method [17,20]. Firstly, each unknown node i creates regions, denominated by *bounding boxes* $B_{i,r}$, whose boundaries are computed by adding and subtracting $l_{i,r}$ from the coordinates $(\hat{x}_r, \hat{y}_r)$ of the respective anchor r, in accordance with (2):

$$B_{i,r} : [\hat{x}_r - l_{i,r}, \hat{y}_r - l_{i,r}] \times [\hat{x}_r + l_{i,r}, \hat{y}_r + l_{i,r}]. \qquad (2)$$

Then, an area S_i, that corresponds to the intersection of all of these $B_{i,r}$ regions, is computed as defined in (3). Finally, the initial position u_i of the unknown node i is computed as the center point of S_i.

$$S_i = \bigcap_{\forall r \in R_i} B_{i,r} : [\max_{\forall r \in R_i} (\hat{x}_r - l_{i,r}), \max_{\forall r \in R_i} (\hat{y}_r - l_{i,r})] \times$$
$$[\min_{\forall r \in R_i} (\hat{x}_r + l_{i,r}), \min_{\forall r \in R_i} (\hat{y}_r + l_{i,r})] \tag{3}$$

It is noteworthy that the Min-Max method comprehends a small number of low complexity operations, including the process to compute S_i area, that is done by simple searches for a minimum/maximum value in sets with R_i elements, wherein R_i is the total number of anchor discovered by node i.

A graphical representation of the Min-Max computation is shown in Fig. 1, where the position of an unknown node i is estimated based on three anchors r_1, r_2 and r_3

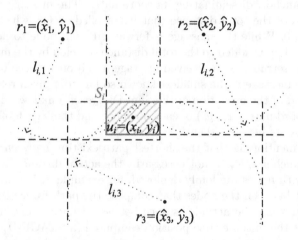

Fig. 1. Graphical illustration of the Min-Max method

4.3   Node Position Refinement

The third stage (STAGE-III) targets the improvement of the accuracy of the unknown nodes positions. During this stage, each unknown node i iteratively reevaluates its position based on the distances measured to its neighbors and their positions, which are also updated at each iteration. Since the nodes are static, and thus the distances measured between them remain constant, the estimated positions of the unknown nodes tend to be gradually adjusted in order to, finally and hopefully, converge to their actual positions. To achieve such convergence, the Tribes algorithm is used in this stage to minimize an objective function that is composed by three main terms, that are described in the remainder of this Section.

The first term gathers the contributions of all one-hop neighbor n of an unknown node i. Node n, which can either be an anchor or unknown node, is considered an one-hop neighbor if and only if it is within the distance measurement threshold L from node i. So, the distance error between the estimated positions of nodes i and n is defined in (4) [19]:

$$g_{i,n} = (d_{i,n} - ||p_n - p_i||), \tag{4}$$

wherein $d_{i,n}$ is the measured distance between unknown node i and its neighbor node n, p_n is the position of n and p_i is the estimated position of node i.

Thus, the first term of the objective function of an unknown node i is computed by the sum of the square of all distance errors $g_{i,n}$ divided by a confidence factor that represents how precise is the knowledge passed by each neighbor n, depending on how close it is to a reference node. This first term is defined in (5):

$$f_1 = \sum_{n \in V_i} \frac{g_{i,n}^2}{\zeta_n}, \tag{5}$$

wherein V_i is the set of all one-hop neighbors of node i and ζ_n is the confidence factor assigned to the one-hop neighbor n. The technique to compute ζ_n is further detailed in this Section.

The second term of the objective function, takes into account the contribution of the nodes w that are two hops away from the unknown node i. It considers that the distance between the nodes i and w is always greater than the distance measurement threshold L, but smaller than $2L$ [19]. Based on this statement, the contribution of each node w to the second term of the objective function is computed as in (6):

$$h_{i,w} = \max(0, L - ||p_w - p_i||, ||p_w - p_i|| - 2L)^2, \tag{6}$$

where p_w is the estimated position of node w. Therefore, if p_i is such that $L \le ||p_w - p_i|| \le 2L$, then 0 is added to the objective function. However, if $||p_w - p_i|| < L$, then the square of the distance between p_i and the circle of radius L, centered at p_w is added to the objective function. Otherwise, if $2L < ||p_w - p_i||$, then the square of the distance between p_i and the circle of radius $2L$, centered at p_w is added to the objective function.

Thus, considering the contribution of all two-hop neighbors of node i, the second term of the objective function is computed as in (7):

$$f_2 = \sum_{w \in W_i} \frac{h_{i,w}}{\zeta_w}, \tag{7}$$

wherein W_i is the set of two-hop neighbors of node i and ζ_w is the confidence factor of node w.

In both (5) and (7), the confidence factor is intended to give more importance to the contribution informed by the nodes that tend to have more accurate positions. The criteria used to compute the confidence factor considers that the nodes closer to the anchors, in general, have more accurate positions [19]. Based on this statement, the confidence factor is computed according to (8):

$$\zeta_n = \sum_{r \in R_n} s_{n,r}{}^2 \qquad (8)$$

wherein R_n is the set composed by the three closest anchors to n, in terms of total number of hops, and $s_{n,r}$ is the number of hops between n and the anchor r. The same computation is also done to obtain ζ_w, by simply replacing n by w.

The third term of the objective function is intended to lead to possible solutions within the S_i area, that is defined by the Min-max technique during STAGE-II of this algorithm (see Sect. 4.2). This third therm is computed as in (9):

$$f_3 = \begin{cases} 0 & \text{if } p_i \in S_i; \\ 1000 + (1000\|p_i - u_i\|)^2 & \text{if } p_i \notin S_i. \end{cases} \qquad (9)$$

wherein u_i is the center of S_i. Thus, if a possible solution is inside S_i, then 0 is added to the objective function. However, if it is out of the S_i area, then a value proportional to the square of the distance between p_i and u_i is added to the fitness function.

Finally, the objective function f that must be minimized is then given by (10). Note that, without distance measurements errors, the global minimum value of f is 0.

$$f = f_1 + f_2 + f_3 \qquad (10)$$

The described objective function allows for each node to estimate its position locally, based on its one-hop and two-hop neighbors. Thereby, the nodes tend to update their estimated coordinates simultaneously, in a collaborative process. With this behavior, the convergence to the zero positioning error, for all nodes, follows from the achievement of the objective of each node, that is expected to be correctly positioned with respect to its neighbors.

4.4   Algorithm of the Proposed Method

The complete algorithm is represented in Fig. 2. It includes the three main stages presented in Sects. 4.1, 4.2 and 4.3. It is worth mentioning that the algorithm is executed simultaneously by each node, causing the emergence of the solution from the distributed processing. In Fig. 2, Δ defines the number of cycles performed during the third stage while μ defines the number of evaluations of the objective function performed by Tribes in each of these cycles.

5   Results

The Tribes-based localization algorithm was evaluated through a set of simulations performed in MATLAB. All networks were created in an area of

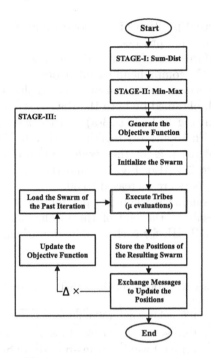

Fig. 2. The complete algorithm

100×100 measurement units ($m.u.$). In this area, 100, then 150, and then 200 unknown nodes were randomly placed, using an uniform distribution. Furthermore, also using an uniform distribution, 10, then 20, and then 30 anchors were randomly placed in the mentioned area. The distance measurement threshold L was set to 20 $m.u.$ For each combination of quantity of anchors and reference nodes, 10 scenarios were generated, totalizing 90 different simulations.

The performance of the Tribes-based algorithm, herein proposed, was compared with the PSO-based algorithm proposed in [12]. That PSO-based algorithm also uses the Sum-Dist and Min-Max method in STAGES-I and II, respectively, and the parameters of the PSO, executed in STAGE-III of that algorithm, were the same as in [12]. The PSO-based algorithm was configured with 100 particles and ran 10 iterations at each STAGE-III cycle.

The STAGE-III of both algorithms was executed through 40 cycles, $i.e.$ $\Delta = 40$. The number of evaluations of the objective function performed by Tribes at each cycle of STAGE-III was $\mu = 1000$. This number was chosen to establish a fair comparison with the PSO-based algorithm, matching the same number of evaluations, per STAGE-III cycle, that the PSO-based algorithm executes in 10 iterations with a swarm of 100 particles.

The positioning error of an unknown node i at any time of the localization process was computed as described in (11):

$$PE_i = ||p_i^{real} - p_i^{comp}||, \tag{11}$$

wherein p_i^{real} is the actual position of the node i and p_i^{comp} it is the position computed by the optimization process.

To evaluate the performance of the algorithm, it was computed the Median Positioning Error (MPE), comprehending all nodes of all 10 scenarios generated for each specific network configuration. The results of the MPE for each cycle of the STAGE-III, obtained by both the PSO-based and our Tribes-Based algorithm, are shown in Fig. 3. In Figs. 3(a) and (b) are presented the results achieved in networks with 10 anchors together with 100, 150 and 200 unknown nodes. Figures 3(c) and (d) present the results obtained when the number of anchors are increased to 20. Also, Fig. 3(e) and (f) shows the results achieved when the number of anchors are increased to 30.

Table 1 presents a summary where its possible to see the algorithm that achieved the better result at the end of the localization process, *i.e.* at the end of 40 cycles of STAGE-III, for each combination of number of anchors and unknown nodes.

Table 1. Summary with the algorithms that achieved better results

	100 unknown	150 unknown	200 unknown
10 anchors	Tribes-based	Tribes-based	Tribes-based
20 anchors	Tribes-based	Tribes-based	Tribes-based
30 anchors	Tribes-based	Tribes-based	PSO-based

From Fig. 3, it is possible to observe that in the worst case, *i.e.* in networks with 10 anchors and 100 unknown nodes, the Tribes-based algorithm achieved a MPE of the order of 1 $m.u.$ In the best case, *i.e.* when there are 30 anchors and 150 unknown nodes, the proposed algorithm reached the MPE of order of 10^{-2} $m.u.$, in an area of 100×100 $m.u.$ The obtained MPEs indicates the effectiveness of the proposed algorithm. Note that the accuracy of the proposed algorithm tends to increase with the number of anchors present in the network.

It is also possible to observe from Fig. 3 and Table 1 that, in almost all network configurations, the Tribes-based algorithm was the one with better performance, achieving MPEs lower than the PSO-based algorithm, except in the simulations with 200 unknown nodes and 30 anchors. The tendency of the graphics also indicates that the PSO-based algorithm, in most cases, stagnated its MPE before the end of the algorithm, while the MPE of the Tribes-based algorithm, in general, continues with a decreasing tendency. This tendency of reduction of the MPE, that is still present at the end of the Tribes-based algorithm, is another advantage that must be highlighted. The better performance of the Tribes-based algorithm, can be explained by its strategy to create new particles and tribes when the optimization stagnates. When the optimization process stagnates, the tribes tend to be classified as "bad". As explained in Sect. 3, the "bad" tribes generate new particles, enhancing the exploration of the search space and, thus, increase the possibility to discover better solutions. This behavior explains not

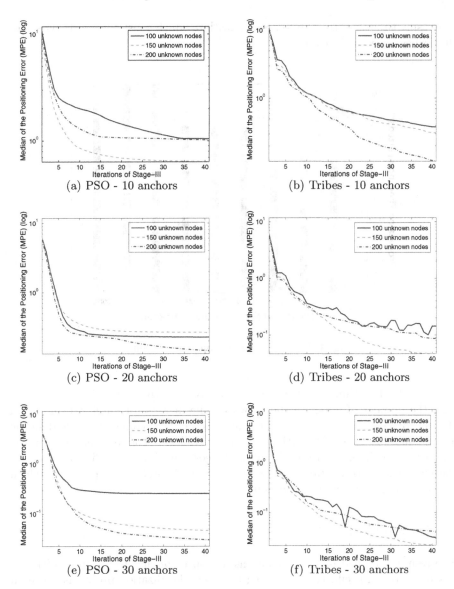

Fig. 3. Evolution of the localization process performed by the PSO-based [12] and the Tribes-based algorithms, with different numbers of anchors and unknown nodes (Color figure online).

only the better final results achieved by the Tribes-based algorithm, but also explains the tendency of reduction of the MPE still present at the end of the localization process.

Figure 4 shows an assessment of the processing time of both algorithms. It is possible to verify that the Tribes-based algorithm requires 10 to 14 times more processing time than the PSO-based algorithm, which constitutes a drawback of the Tribes-based algorithm.

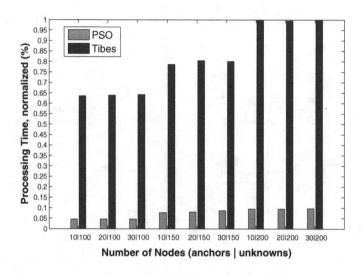

Fig. 4. Processing time (Color figure online)

In both algorithms, it is possible to verify that the processing time grows with the number of nodes in the network. The increase of number of nodes in the network causes the growth of the average network connectivity, and so the average number of neighbors that a node i may have. The augmentation of the number of neighbors of a node i, raises the computation of the first and second therms of the objective function, according with (5) and (7), respectively, which explains the growth of the processing time.

6   Conclusion

The results obtained in the simulations demonstrate that the proposed Tribes-based algorithm is effective in solving multi-hop localization problems, since, in the worst case, the achieved MPE was of the order of 1 $m.u.$ and, in the best case, it was of the order of 10^{-2} $m.u.$ in an area of 100×100 $m.u.$

Based on the results, we conclude that the Tribes-based algorithm herein proposed, in general, has a better performance than the PSO-based algorithm, proposed in [12]. Besides the lower MPE achieved by the Tribes-based algorithm, its MPE decreasing tendency is still present at the end of the algorithm execution, which is another advantage when compared with the PSO-based algorithm that, in most cases, stagnates its MPE before the end of the algorithm. Thus, the use of this Tribes-based algorithm is encouraged for applications requiring lower MPEs, which can possibly be achieved by simply increasing the number of

Δ cycles of STAGE-III. A drawback of this algorithm is its high computational cost, when compared with the PSO-based algorithm.

Another conclusion is that the MPE decreases as the number of anchors are incremented. Thus, this may be another alternative to reduce the MPE, without increasing the Δ of STAGE-III. It suggests that, to achieve lower MPEs, we may have a tradeoff between the computational cost, caused by a higher Δ, and the number of anchors in the network.

For future work, we propose the investigation of the behavior of the algorithm in problems with noise in the distance measurements, also applying the algorithm in real world experiments. It is also desirable to extend the algorithm for problems involving node mobility.

Acknowledgement. We thank the State of Rio de Janeiro (FAPERJ, http://www.faperj.br) for funding this study.

References

1. Rubenstein, M.: Self-assembly and self-healing for robotic collectives. Thesis, University of Southern California, California, USA (2009)
2. Şahin, E.: Swarm robotics: from sources of inspiration to domains of application. In: Şahin, E., Spears, W.M. (eds.) Swarm Robotics 2004. LNCS, vol. 3342, pp. 10–20. Springer, Heidelberg (2005)
3. Rodic, A., Katie, D., Mester, G.: Ambient intelligent robot-sensor networks for environmental surveillance and remote sensing. In: 7th International Symposium on IEEE Intelligent Systems and Informatics, SISY 2009, Subotica, Srvia, pp. 39–44 (2009)
4. Nawaz, S., Hussain, M., Watson, S., Trigoni, N., Green, P.N.: An underwater robotic network for monitoring nuclear waste storage pools. In: Hailes, S., Sicari, S., Roussos, G. (eds.) S-CUBE 2009. LNICST, vol. 24, pp. 236–255. Springer, Heidelberg (2010)
5. Akkaya, K., Younis, M.: A survey on routing protocols for wireless sensor networks. Ad Hoc Netw. **3**(3), 325–349 (2005). Elsevier
6. Rrodolpu, V., Meng, T.H.: Minimum energy mobile wireless networks. IEEE J. Sel. Areas Commun. **17**(8), 1333–1344 (1999)
7. Li, L., Halpern, J.Y.: Minimum-energy mobile wireless networks revisited. In: IEEE International Conference on Communications, ICC 2001, Helsinki, Finlândia, vol. 1, pp. 278–283. IEEE (2001)
8. Xu, Y., Heideman, J., Estrin, D.: Geography-informed energy conservation for ad hoc routing. In: Proceedings of the 7th Annual International Conference on Mobile Computing and Networking, Rome, Italy, pp. 70–84. ACM (2001)
9. Tu, Y., Govindan, R., Estrin, D.: Geographical and energy aware routing: a recursive data dissemination protocol for wireless sensor networks, Los Angeles, CA, USA (2001)
10. de Sá, A.O., Nedjah, N., de Macedo Mourelle, L.: Genetic and backtracking search optimization algorithms applied to localization problems. In: Murgante, B., Misra, S., Rocha, A.M.A.C., Torre, C., Rocha, J.G., Falcão, M.I., Taniar, D., Apduhan, B.O., Gervasi, O. (eds.) ICCSA 2014, Part V. LNCS, vol. 8583, pp. 738–746. Springer, Heidelberg (2014)

11. de Sá, A.O., Nedjah, N., de Macedo Mourelle, L.: Distributed efficient localization in swarm robotic systems using swarm intelligence algorithms. Neurocomputing, Elsevier, p. aceito para publicao (2015)
12. de Sá, A.O., Nedjah, N., de Macedo Mourelle, L.: Distributed efficient localization in swarm robotics using min-max and particle swarmoptimization. In: Proceedings of The Latin American Congress on Computational Intelligence, LA-CCI, San Carlos de Bariloche, Argentina, vol. 1, pp. 7–12 (2014)
13. Clerc, M.: TRIBES-un exemple d'optimisation par essaim particulaire sans parametres de contrle. Optimisation par Essaim Particulaire (OEP 2003), Paris, France 64 (2003)
14. Clerc, M.: Stagnation analysis in particle swarm optimization or what happens when nothing happens. Department of Computer Science, University Essex, Technical report CSM-460 (2006)
15. Kennedy, J., Eberhart, R.: Particle swarm optimization. In: Proceedings of 1995 IEEE International Conference on Neural Networks, Perth, WA, USA, vol. 4, pp. 1942–1948. IEEE (1995)
16. Shi, Y., Eberhart, R.: A modified particle swarm optimizer. In: The 1998 IEEE International Conference on Evolutionary Computation Proceedings, IEEE World Congress on Computational Intelligence, Anchorage, AK, USA, pp. 69–73. IEEE (1998)
17. Savvides, A., Park, H., Srivastava, M.B.: The bits and flops of the n-hop multilateration primitive for node localization problems. In: Proceedings of the 1st ACM International Workshop on Wireless Sensor Networks and Applications, Atlanta, GA, USA, pp. 112–121. ACM (2002)
18. Sun, W., Su, X.: Wireless sensor network node localization based on genetic algorithm. In: 3rd International Conference on Communication Software and Networks, pp. 316–319. IEEE (2011)
19. Ekberg, P.: Swarm-intelligent localization. Thesis, Uppsala Universitet, Uppsala, Sweden (2009)
20. Langendoen, K., Reijers, N.: Distributed localization algorithms. In: Zurawski, R. (ed.) Embedded Systems Handbook, pp. 1–23. CRC Press, Boca Raton (2005)
21. Huanxiang, J., Yong, W., Xiaoling, T.: Localization algorithm for mobile anchor node based on genetic algorithm in wireless sensor network. In: International Conference on Intelligent Computing and Integrated Systems, pp. 40–44. IEEE (2010)
22. Ekberg, P., Ngai, E.C.: A distributed swarm-intelligent localization for sensor networks with mobile nodes. In: 7th International Wireless Communications and Mobile Computing Conference, pp. 83–88 (2011)

Developing Tasty Calorie Restricted Diets Using a Differential Evolution Algorithm

João Gabriel Rocha Silva[1], Iago Augusto Carvalho[2],
Michelli Marlane Silva Loureiro[1], Vinícus da Fonseca Vieira[1],
and Carolina Ribeiro Xavier[1(✉)]

[1] Universidade Federal de São João del Rei, São João del Rei, Brazil
carolinaxavier@ufsj.edu.br
[2] Universidade Federal de Minas Gerais, Belo Horizonte, Brazil

Abstract. The classical diet problem seeks a diet that respects the indicated nutritional restrictions at a person with the minimal cost. This work presents a variation of this problem, that aims to minimize the number of ingested calories, instead of the financial cost. It aims to generate tasty and hypocaloric diets that also respect the indicated nutritional restrictions. In order to obtain a good diet, this work proposes a Mixed Integer Linear Programming formulation and a Differential Evolution algorithm that solves the proposed formulation. Computational experiments show that it is possible to obtain tasty diets constrained in the number of calories that respect the nutritional restrictions of a person.

Keywords: Differential evolution · Diet problem · Evolutionary Algorithms · Calories · Obesity

1 Introduction

Metaheuristics are generic algorithms commonly used to solve optimization problems. Evolutionary Algorithms (EA) are metaheuristics based on the idea of the natural evolution of Darwin, that can be easily applied to problems that arise in several areas of the human knowledge. They are based on a set of solutions, called population. Each solution of the population is said to be an individual. The optimization process of EA are based on probabilistic transitions between two or more individuals of a population, instead the deterministic rule of classical optimization algorithms.

When solving linear or non-linear problems, one of the most prominent EA is the Differential Evolution algorithm (DE). In DE, as in others EA, the evolutionary process can be initiated from a population generated at random. At each generation, individuals suffer recombinations and mutations. Then, the most adapted individuals are selected to go to the next generation. The concept of population, mutation, and recombination turns DE into a fast, robust and stable algorithm, that can easily escape from local optima and achieve good solutions with a reasoning time.

© Springer International Publishing Switzerland 2016
O. Gervasi et al. (Eds.): ICCSA 2016, Part V, LNCS 9790, pp. 171–186, 2016.
DOI: 10.1007/978-3-319-42092-9_14

DE can be easily adapted to a great number of optimization problems. One of the most classical optimization problem is the Diet Problem (DP), introduced by George Stigler in [17]. DP aims to build a diet for a 70 Kg man that respects the minimal quantity of nutrients recommended by the North American National Research Council in 1943, and has the minimum financial cost.

This work proposes a modified version of DP, denominated Caloric Restricted Diet Problem (CRDP). Instead the classic objective function that minimizes the diet financial cost, CRDP aims to build a diet that minimizes the number of ingested calories, while respecting the minimal quantity of nutrients. Moreover, this work is concerned to provide a tasty diet that can be easily adopted by a great number of individuals. We present a Mixed Integer Linear Programming (MILP) formulation for CRDP, and solve it through a DE algorithm. We also present how to develop a real life daily diet with low number of calories composed by six meals.

The remainder of this work is organized as follows. The background and motivation of this work is presented in Sect. 2. CRDP is formally defined in Sect. 3. A Differential Evolution algorithm for CRDP is presented in Sect. 4. Computational experiments are developed and discussed in Sect. 5. Finally, concluding remarks are drawn in the last section.

2   Background and Motivation

2.1   The Diet Problem

DP is a classic linear optimization problem introduced in [17]. It aims to select a subset of foods, from a large set of 77 elements, a subset of they that should be eaten on a daily basis in order to produce healthy diet, taking in account nine different nutrients, and has the minimal cost [8]. Many variations of this problem can be found in Literature. As example, we can cite a fuzzy-logic DP [13], a multi-objective variant of DP [7], and a a robust optimization approach for DP [2].

Let F be the set of available foods. Moreover, let N be the set of nutrients considered at the problem. The LP formulation is defined with decision variables $p_i \geq 0$ that represent the amount of food $i \in F$ to be consumed. Moreover, c_i represents the cost per portion of food $i \in F$, m_{ij} represents the amount of nutrient $j \in N$ that is contained at food $i \in F$, and b_j is the minimal requirement of the nutrient $j \in N$. The corresponding LP formulation is defined by Eqs. 1-3.

$$min \sum_i c_i p_i \tag{1}$$

$$\sum_i m_{ij} p_i \geq b_j, \forall j \in \{0, \dots N\} \tag{2}$$

$$p_i \geq 0, \forall i \in \{0, \dots, F\} \tag{3}$$

The objective function (1) aims to minimize the cost of the aliments that are included on the diet. Inequalities (2) maintain all the necessary nutrients at a minimal level. Equation (3) defines the domain of the variable p.

The main objective of the classical DP is to minimize the financial cost of a diet that provides all necessary nutrients. However, DP is only focused on healthy individuals. It is not concerned with individuals that have overweight. Moreover, the original solution provided in [17] lacks of food variety and palatability, thus being hard to follow the proposed diet. As the number of individuals with overweight has increased in the past 20 years [23], the development of computational methods to generate diets that are healthy and tasty at the same time is needed.

2.2 Diet, Calories, and Heath

It is notable that a diet that contains a high amount of calories is correlated with weight gain, both in humans [21] and animals [11]. More than simple affecting the body weight, it can greatly affects the health of an individual, being responsible for a great number of chronic diseases [22].

It was shown that a balanced diet, together with physical exercises, can help an individual to lose weight and improve it's health [16]. However, it is not easy to develop a diet with low calories amount that provides all necessary daily basis nutrients, such as proteins, zinc, and iron, among others. Moreover, the palatability of a meal is affected by the amount of calories present at the consumed foods, such that high amount of calories is directly related to a tastier food [21]. Thus, one of the main challenges for nutritionists all around the world is to develop a diet restricted in the number of calories that provides the necessary nutrients, and that is, at the same time, healthy and tasty.

It is recommended for an individual to daily consume from 2000 to 2500 calories. Caloric restricted diets generally uses 800 kilocalories as a lower bound for the daily calories intake. In order to develop a healthy diet, this caloric restricted diet also need to offer all the nutrients necessary for an individual. Moreover, it need to provide tasty foods, such that it will be easier for an individual to follow the recommended meals.

2.3 Evolutionary Algorithms

There exists many EA variants and implementations. However, all of they rely on one basic idea, back from the natural selection theory: given an initial population, the selection is realized at each generation such that the most fitted individuals have a greater chance to survive from one generation to another. Thus, it is possible to achieve a population most adapted to its environment.

A pseudo-code of a general EA is presented at Algorithm 1. The initial population is generated, commonly with individuals generated at random, at line 1. Next, at line 2, each individual have it fitness evaluated. The evolutionary process occurs from line 4 to 10, while a given stopping criterion is not met.

First, at line 5, some individuals are selected as parents. Next, the selected individuals are recombined in order to generate a new individual, at line 6. These new individuals go through a mutation process at line 7. Next, their fitness function are evaluated again at line 8. Finally, the most fitted individuals are selected to continue to the new generation at line 9.

```
 1 begin
 2    initialize the population with individual;
 3    evaluate each individual;
 4    while stopping criterion not met do
 5       select the parents;
 6       cross the selected parents;
 7       mutate the individuals;
 8       evaluate the offspring;
 9       evolute the most fitted individuals to survive at next generation;
10    end
11 end
```

Algorithm 1. Pseudo-code of an general EA

2.4 Differential Evolution

The Differential Evolution (DE) algorithm is an instance of an EA. It was first presented in [18], with the objective to solve the polynomial adjust problem of Chebychev. DE is widely applied to non-linear optimization problems. According to [3], the choose of DE despite other EA is based on the following facts.

- DE is efficient to solve problems with discontinuously functions, because it does not require information about the derivate of a function.
- DE can run properly with a small population.
- DE can easily escape from local optima, due its efficiently mutation operator.
- DE input and output parameters can be manipulated as floating point number without any additional manipulation.

As DE was developed to handle real parameters optimization, it can be easily adapted to a large number of areas. In Literature, it is possible to find DE applications in engineering [10], chemistry [1], biology [19], finances [9], among others [5,14]. In order to apply DE to a problem, it is just necessary to model the input of the problem as an input for DE.

DE follows the five main evolution stages of an EA, as presented in Algorithm 1. However, in DE, the evolution stages occur in a different order. First, the parents are selected. Next, occurs the mutation, recombination and evaluation phases, in order. Finally, the most fitted individuals are selected to go to the next generation.

Initialization. This stage initializes a population with $|X|$ individuals. Each individual $x_i \in X$ is a d-dimensional vector, such that each position of the vector (also called a gene) represent one characteristic of that individual. Figure 1 shows, for an arbitrary problem, a representation of an individual with 6 genes. It is possible to see that each gene has a real value, thus representing one parameter or variable of the problem.

2.1	2.8	0.5	1.3	2.3	1.7

Fig. 1. Example of an individual with 6 genes

Mutation. Mutation is the process of perturbation of the population. Mutation produces new individuals, denominated trial individuals, from the addition of an aleatory individual γ from the population to the difference between two other aleatory individuals α and β from the same population. This process utilizes a factor of perturbation $F \in [0;2]$ that ponderate the mutation operator. Let x_α, x_β, and x_γ be three distinct individuals that belongs to the population X at generation g. A trial individual V, that belongs to the generation $g+1$ can be generated as show in Eq. (4).

$$V^{g+1} = x_\gamma^g + F(x_\alpha^g - x_\beta^g) \tag{4}$$

Crossing. The crossing operator is used in order to increase the diversity of the trial individuals V generated at the mutation process. Consider that, for each target individual $x_s^g, s \in X, s \neq \alpha \neq \beta \neq \gamma$ a trial individual V^{g+1} is generated from mutation. Then, an individual U^{g+1}, denominated offspring, is generated as shown in Eq. (5).

$$U_i^{g+1} = \begin{cases} V_i^{g+1}, & \text{if } r_i \leq Cf \\ x_{s,i}^g, & \text{if } r_i > Cf \end{cases}, \quad \forall i \in \{0,\dots,N\}, r_i \in [0;1] \tag{5}$$

where $U_i^{g+1}, V_i^{g+1}, x_{s,i}^{g+1}$ are the i-th gene of the individuals U^{g+1}, V^{g+1}, and x_s^g, respectively. Value r_i is a number generated at random, at each iteration of the algorithm. Let $Cf \in [0;1]$ represents the possibility of the offspring to inherit the genes from the trial individual V_i^{g+1}. When $Cf = 1$, the offspring will be equal to the trial individual V^{g+1}. At the other hand, when $Cf = 0$, the offspring will be equal to the target vector x_s^g. If $0 < Cf < 1$, the offspring can receive genes from both V^{g+1} and x_s^g.

Evolution. The evolution is the last stage of the evolutionary process. At evolution stage of generation g, the individuals that will be at the generation $g+1$ are selected. If the generated offspring fitness is better than the target vector offspring, then the target vector is replaced at the population by the offspring. Otherwise, the population remains the same. This process occurs for all offspring generated at each generation.

3   Caloric Restricted Diet Problem

The Caloric Restricted Diet Problem (CRDP) aims to develop caloric restricted diets that provides all necessary nutrients for a daily basis intake. Moreover, it is also concerned to provide tasty meals, that can be easily accepted by a patient. Thus, one can lose weight by a healthy way, while consuming a great diversity of meals.

In Literature, it is possible to find a greater number of variations for the number of recommended calories into a hypocaloric diet. The work [15] suggest an initial diet with 1200 kilocalories (Kcal) for overweight women. Moreover, the work [4] suggests that diets from 1000 to 1200 Kcal can improve the health of a set of obese individuals within 25 days. Thus, this work adopt as objective to develop diets such that $Kcal \approx 1200$. CRDP objetive function is stated as Eq. (6).

$$min \quad |1200 - Kcal| \tag{6}$$

where $Kcal$ denotes the number of calories in the diet. As this work is interested to obtain a diet with approximate 1200 $Kcal$, the objective function use the modulus of the result to minimize the distance between the number of calories at the formulated diet and the recommended number of calories. Let F be the set of available foods. The value of $Kcal$ is calculated as Eq. (7).

$$Kcal = \sum_i Kcal_i p_i y_i \tag{7}$$

where $Kcal_i$ denotes the amount of $Kcal$ contained in one portion of the product $i \in F$. The value $p_i \in [0.5; 3]$ corresponds to the number portions of each product $i \in F$, and y_i is a binary variable such that $y_i = 1$ if the product i is part of the diet, and $y_i = 0$ otherwise, for all $i \in F$. We consider 100 g of a food, or 100 mL of a beverage as a portion. The value of p_i is limited at the interval $[0.5; 3]$ in order to avoid a greatly quantity of some product to appear, thus dominating the diet, or to consume a minor portion of some product, that may be impracticable at some real situation (imagine to cook only 10 g of fish at dinner, for example).

Besides the number of calories of the diet, CRDP takes into account the amount of proteins (pt), carbohydrates (c), sodium (Na), dietary fibers (f), calcium (Ca), magnesium (Mg), manganese (Mn), phosphor (P), iron (Fe), and zinc (Zn). Thus, CRDP have ten constraints, as shown in Table 1. Each couple of columns display the name of the nutrient and its necessary daily intake, respectively. Then, a set of constraints is inserted at the problem, as shown in Eq. (8).

$$\sum_i m_{ij} p_i y_i \geq b_j \tag{8}$$

Table 1. Daily recommended nutrients intake [20]

Nutrient	amount (g)	Nutrient	amount (mg)	Nutrient	amount (mg)
pt	≥ 75	Ca	≥ 1000	Mg	≥ 260
c	≥ 300	Mn	≥ 2.3	P	≥ 700
f	≥ 25	Fe	≥ 14	Zn	≥ 7
Na	≤ 2.4				

3.1 Food Database

This work consider the Brazilian Table of Food Composition (TACO) (Tabela Brasileira de Composição de Alimentos, in portuguese) [12] as the information source about the nutritional data of foods. TACO was elaborated at the University of Campinas, Brazil. It contains data from a large number of foods and beverages, with a quantitative description about 25 of their nutrients and properties. Each data represents a portion of 100 g (or 100 mL, when appropriated) of a product. Moreover, TACO displays information about foods in a variety of states (frozen, cooked, with or without salt, for instance).

TACO presents each product in more than one presentation. For instance, it presents uncooked meat, uncooked meath with salt, and cooked meat. Thus, only subset of the products of TACO (that can be served in a meal) are selected to be in F. Moreover, TACO presents 25 different information about each product. This work considers only the most important of then, thus considering 11 from 25 nutrients in the problem. The selected nutrients are displayed at the odd columns of Table 1.

The subset of selected products were classified into nine different categories, as shown in Table 2. The first column displays the class of the product. The

Table 2. Different classification of products from TACO

Product	Symbol	#	Information
Beverages	B	21	Beverages, except natural juices and alcoholic beverages
Juices	J	11	Natural fruits juice
Fruits	F	62	Fruits, in general
Lacteal	L	19	Products derived from milk
Carbohydrates 1	C1	21	Snacks, as bread, cookie, and cracker
Carbohydrates 2	C2	12	Main meals carbohydrates, as rice, potato, and cassava
Grains	G	12	Leguminous foods, such as lentils and beans
Vegetables	V	41	Vegetables, in general
Proteins	P	95	High proteic foods, as meat, chicken, and eggs

second column displays the symbol that represents the class. The third column displays the number of products at each class. The last column displays some extra information about the products contained in each class.

4   Differential Evolution for CRDP

4.1   Solution Representation

In order to represent correctly the diet of the people, an individual, also called a solution of DE, needs to represent the different meals that exists during a day. Thus, a solution is represented as a combination of the most common types of products consumed in each meal. As products are divided in categories, it is possible to build a diet that has a great variety of products every day.

Figure 2 shows the diversity of products in each meal. This work considers 6 different meals at a day, namely the breakfast, two snacks, lunch, dinner, and supper. Breakfast is the first meal by the morning. Snacks are small meals that can exists between two major meals. Supper is the last meal, after the dinner and before sleep, that is common to a great number of individuals.

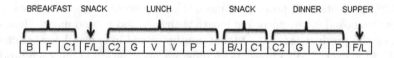

Fig. 2. Representation of a daily diet based on their food classes

Based at Fig. 2, a solution for CRDP is expressed as shown in Fig. 3. A solution consists in two vectors of 17 genes each, that aggregates 6 different meals, as expressed by Fig. 2. The first vector contains the quantity of portions p_i of each product $i \in F$ to be consumed. The second vector contains a bijective function $f : \mathbb{N} \mapsto F$ that maps each number as a product.

0.6	2.1	2.9	1.2	1	0.8	2	2.7	1.9	1.1	0.6	1	1.6	1.1	1.7	2	0.5
13	52	16	17	19	7	33	12	84	10	18	6	16	11	31	9	41

Fig. 3. Representation of an individual

4.2   Penalty Function

When the constraint is not violated, the CRPD employs the objective function (Eq. (6)) as the fitness function. However, when constraints are violated, a mechanism to handle these constraints must be adopted. It is worth mentioning that the DE method does not deal with the constraints properly and, thus, some technique must be adopted to deal with this issue. Hence, the fitness function becomes the objective function plus a penalization term [6]. This term is defined in accordance with Eq. (9).

$$min \quad |Kcal - 1200| + \sum_j \left(\frac{\left| \sum_{i \in F} (m_{ij} p_i y_i) - b_j \right|}{b_j} \right) \cdot M, \quad \forall j \in N \quad (9)$$

where m_{ij} denotes the amount of the nutrient $j \in N$ at product $i \in F$, M is a big constant number, used as a penalty term for violated constraints. Both p_i, y_i, and b_j are the same as denoted at Eq. (7). One can see that there is no penalty for the value of y_i. The limits of y_i are handled by DE, such that a value out of the interval $[0.5; 3]$ is rounded to the closest value inside the interval.

4.3   Population Initialization, Mutation, Crossing, and Evolution

The population is initialized at random. For each gene of an individual $x_i \in X$, a product is selected at random, always respecting the group that it belongs. Next, a random value $[0.5; 3]$ is assigned to each product.

At each generation, $|X|$ new offsprings are generated through the mutation and crossing operators. The mutation operator is the same as described at Subsect. 2.4. The crossing operator uses Eq. (5) to produce new values for the first vector of the individual. The number of the product at the trial vector is selected at random among the chosen individuals α, β, and γ. Figure 4 shows an example of the crossing operator between individuals with 3 genes. It can be see that the product of the trial vector is one of the products of individuals α, β, or γ. The offspring products, as the products from the trial vector, are random chosen between the target and the trial vector.

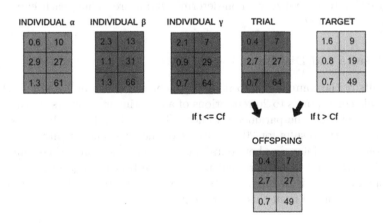

Fig. 4. Illustration of a mutation operation and crossing at the DE

As the number of offsprings is the same of the number of individuals at the current generation, the evolution process compares one offspring with one individual from the current population. Then, the individual with the worst value of fitness is discarded, and the other individual is included at the population. This procedure ensures that a generation $g + 1$ will be equal or better than generation g.

5 Computational Experiments

Computational experiments have been performed on a single core of an Intel Core i5 CPU 5200U with 2.2 GHz clock and 4 GB of RAM memory, running Linux operating system. DE was implemented in C and compiled with GNU GCC version 4.7.3.

5.1 Parameter Tuning

DE has three different parameters, namely the perturbation factor, the crossing factor, and the population size. The perturbation factor F varies in the interval $[0; 2]$. The crossing factor is a percent value, and varies in the interval $[; 1]$. The population size $|X|$ is an integer number greater than 2. In order to tune the algorithm, a complete factorial experiment was developed with their three parameters, in order to choose the best value for each one.

The value of the perturbation factor F was varied in the set $\{0.3, 0.8, 1.3, 1.8\}$. The crossing factor Cf was varied in the set $\{0.4, 0.6, 0.8\}$. Finally, the population size $|X|$ was varied in the set $\{10, 50, 100\}$. Using these values, a complete factorial experiment was performed in order to evaluate DE parameters. As DE is not a deterministic algorithm, it is necessary to execute each experiment repeatedly, in order to minimize its non-determinist nature. Thus, each experiment was repeated 20 times.

5.2 Results and Discussion

The results of the complete factorial experiment are shown in Table 3. Each row of the table corresponds to 20 repetitions of a experiment. The first, second, and third columns show the parameter values. The fourth, fifth, sixth, and seventh columns show the results for 20 executions of the experiments with the defined parameter values. There is a horizontal line that split the results according to the population size, and the experiment that has the best average fitness value for each population size is highlighted. As DE ran with up to 2 s, the computational time was disregarded from this experiment.

Table 3. Results for the complete factorial experiment

Parameters			Fitness			
Population size	F	Cf	Average	Best	Worst	Standard deviation
10	0.3	0.4	1665.81	1544.91	1820.98	75.96
10	0.3	0.6	1761.16	1552.56	2166.99	142.26
10	0.3	0.8	1869.79	1685.96	2393.99	153.65
10	0.8	0.4	1631.13	1524.63	1778.55	67.24
10	0.8	0.6	1659.09	1495.48	1915.12	99.23
10	0.8	0.8	1675.09	1545.51	1846.58	89.65
10	**1.3**	**0.4**	**1630.83**	**1512.67**	**1793.14**	**71.12**
10	1.3	0.6	1633.73	1528.59	1740.91	66.66
10	1.3	0.8	1696.47	1586.12	1847.05	82.50
10	1.8	0.4	1650.69	1537.55	1782.75	64.80
10	1.8	0.6	1667.26	1520.58	1849.50	91.60
10	1.8	0.8	1749.45	1611.98	1911.00	83.72
50	**0.3**	**0.4**	**1507.89**	**1456.03**	**1579.59**	**26.23**
50	0.3	0.6	1508.93	1449.48	1608.58	36.26
50	0.3	0.8	1522.09	1449.70	1578.57	38.45
50	0.8	0.4	1519.45	1461.62	1570.28	28.50
50	0.8	0.6	1514.38	1479.55	1558.03	22.77
50	0.8	0.8	1522.01	1434.66	1596.34	42.71
50	1.3	0.4	1530.11	1488.79	1578.24	26.67
50	1.3	0.6	1518.93	1470.49	1567.46	25.05
50	1.3	0.8	1528.54	1478.57	1579.48	28.65
50	1.8	0.4	1517.64	1483.21	1588.33	24.99
50	1.8	0.6	1527.42	1485.89	1594.77	28.71
50	1.8	0.8	1531.60	1469.17	1632.94	43.39
100	0.3	0.4	1491.56	1450.89	1528.48	20.88
100	**0.3**	**0.6**	**1479.08**	**1430.06**	**1531.20**	**24.66**
100	0.3	0.8	1484.84	1442.14	1567.03	31.84
100	0.8	0.4	1496.23	1459.74	1527.54	17.44
100	0.8	0.6	1487.97	1435.62	1520.59	19.29
100	0.8	0.8	1493.18	1452.27	1532.95	26.63
100	1.3	0.4	1507.45	1461.57	1546.01	21.19
100	1.3	0.6	1499.52	1441.36	1564.64	28.28
100	1.3	0.8	1509.47	1474.18	1546.89	22.49
100	1.8	0.4	1511.40	1483.83	1541.32	14.18
100	1.8	0.6	1503.08	1471.72	1551.44	23.44
100	1.8	0.8	1507.61	1468.57	1558.26	25.49

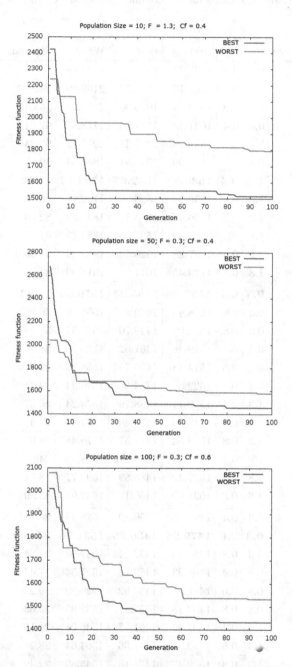

Fig. 5. Convergence curve for the best and the worst executions, for each set of highlighted parameters (Color figure online)

Table 4. Violations of the diet without constraint

Nutrient	pt	c	f	Ca	Mg	Mn	P	Fe	Na	Zn
Constraint	≥ 75	≥ 300	≥ 25	≥ 1000	≥ 260	≥ 2.3	≥ 700	≥ 14	≤ 2400	≥ 7
Reached	62.7	174.4	22.4	736.5	253.0	1.84	1050.5	10.0	819.0	5.8
Violation (%)	16.4	41.8	10.2	23.6	2.69	20.0	0.0	28.5	0.0	17.4

In Table 3, one can see that the population size is positive correlated with the solution value. There is no one other apparent correlation between the perturbation factor F and the crossing factor Cf with the solution value. Based on this experiment, in order to solve CRDP, the best set of parameter for DE is $F = 0.3, Cf = 0.6$, and $|X| = 100$.

Figure 5 shows a detailed convergence curve for the three highlighted parameter set in Table 3. One can see a difference of almost 300 $Kcal$ between the best and worst results with a population size $|X| = 10$. Moreover, as the population size increases, the algorithm convergence curve also increases. It shows that the solutions diversity has an important role in DE, preventing it to be stuck at a local minima prematurely.

In order to show the importance of the constraints, a DE with the best set of parameters was executed another 20 times at CRDP without the 10 nutrients constraints. Thus, DE aims only to minimize the caloric cost of the diet, without considering the basic nutrient needs of a person. Table 4 summarizes the results of this experiment. It shows that a diet that just only minimizes the number of calories can be very dangerous to the health, as 8 of the 10 nutritional requirements are not satisfied.

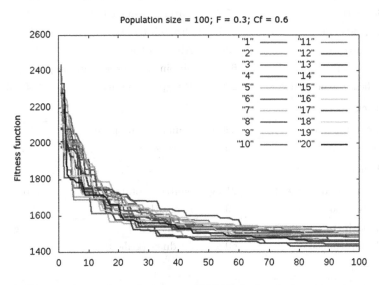

Fig. 6. Evolution of 20 independent runs of DE with the the best set of parameters

Figure 6 shows the convergence curve of 20 independent executions of DE with the best set of parameters found. One can see that all the convergence curves are close to each other. Thus, it indicates that DE is a stable method to solve CRDP.

5.3   Example Diets

Table 5 shows two possible caloric restricted diets generated though DE. One can see the diversity of foods and beverages among diets. DE can be executed many times to solve CRDP, thus generating different diets at each execution. Thus, a weekly diet can be generated in seconds. The great diversity of foods

Table 5. Example of two possible diets

Diet 1		Diet 2	
1464.50 Kcal		1490.36 Kcal	
Breakfast			
Nonfat plain yogurt	190 ml	Coconut water	170 ml
Banana Pacovan (raw)	285 g	Plum (raw)	265 g
Morning cereal, corn flavor	50 g	Morning cereal, corn flavor	115 g
Snack			
White guava with peel	50 g	Pear (raw)	80 g
Lunch			
Potatoes (baked)	145 g	Sweet potato (baked)	70 g
Beans (baked)	95 g	Carioca beans (cooked)	130 g
Alfavaca (raw)	210 g	Broccoli (cooked)	270 g
Tomato (raw)	230 g	Alfavaca (raw)	275 g
Corvina fish (cooked)	65 g	Sirloin tri tip roast (grilled)	50 g
Galician lemon juice	220 g	Rangpur lime juice	280 ml
Snack			
Tangerine juice	220 ml	Rangpur lime juice	85 ml
Green corn	50 g	Green corn	50 g
Dinner			
Sweet potato (baked)	210 g	Sautéed potatoes	100 g
Lentil (cooked)	50 g	Beans (baked)	80 g
Spinach (raw)	275 g	Celery (raw)	125 g
Shrimp (boiled)	60 g	Sardines (grilled)	60 g
Supper			
Pitanga (raw)	225 g	Pear (raw)	225 g

and beverages makes generated diets easy to follow, such that the palate will not be tired with repetitive meals.

6   Conclusions and Future Works

In this work, the Caloric Restricted Diet Problem (CRDP) was proposed. A Mixed Integer Linear Programming formulation was proposed, so as a Differential Evolution (DE) algorithm that solves the proposed formulation. Computational experiments with the proposed algorithm shows that it is possible to generate tasty and health diets with a reduced number of calories. Another contribution of this work is to model meals as a set of different foods and beverages classes. This modeling guarantee that generated meals are closer to the reality.

The key point to solve CRDP is to generate a great diversity of solutions. Thus, various diets can be easily generated through DE. Deterministic methods, such as branch-and-bound or mathematical programming approaches, can not be properly used to solve CRDP, as they always generate the same diet.

Although CRDP is not a hard problem, from the computational point of view, it can be a hard problem in practice. It can takes an expensive amount of time to generate a weekly diet, with a great variety of meals, as the ones generated though DE. Thus, this work can be applied to real situations. It can be used by practitioners to generate a great amount of diets, in a small amount of time.

As future works, it is proposed to generate different types of diets. This model and algorithm can be used to generate an uncountable number of diets, from high-protein diets, for those who are interested in muscle mass gain, to balanced vegan diets. Another future work is to develop a multi-objective version of CRDP, that minimizes, besides the number of calories, the financial cost or the sodium intake, while maximizes the amount of dietary fibers and proteins, for example.

Acknowledgments. This work was partially supported by the Brazilian National Council for Scientific and Technological Development (CNPq), the Foundation for Support of Research of the State of Minas Gerais, Brazil (FAPEMIG), and Coordination for the Improvement of Higher Education Personnel, Brazil (CAPES).

References

1. Babu, B., Angira, R.: Modified differential evolution (mde) for optimization of non-linear chemical processes. Comput. Chem. Eng. **30**(6), 989–1002 (2006)
2. Bas, E.: A robust optimization approach to diet problem with overall glycemic load as objective function. Appl. Math. Model. **38**(19), 4926–4940 (2014)
3. Cheng, S.L., Hwang, C.: Optimal approximation of linear systems by a differential evolution algorithm. IEEE Trans. Syst. Man Cybernetics Part A: Syst. Hum. **31**(6), 698–707 (2001)
4. Crampes, F., Marceron, M., Beauville, M., Riviere, D., Garrigues, M., Berlan, M., Lafontan, M.: Platelet alpha 2-adrenoceptors and adrenergic adipose tissue responsiveness after moderate hypocaloric diet in obese subjects. Int. J. Obes. **13**(1), 99–110 (1988)

5. Das, S., Suganthan, P.N.: Differential evolution: a survey of the state-of-the-art. IEEE Trans. Evol. Comput. **15**(1), 4–31 (2011)
6. Datta, R., Deb, K.: Evolutionary Constrained Optimization. Springer, Heidelberg (2015)
7. Gallenti, G.: The use of computer for the analysis of input demand in farm management: a multicriteria approach to the diet problem. In: First European Conference for Information Technology in Agriculture (1997)
8. Garille, S.G., Gass, S.I.: Stigler's diet problem revisited. Oper. Res. **49**(1), 1–13 (2001)
9. Hachicha, N., Jarboui, B., Siarry, P.: A fuzzy logic control using a differential evolution algorithm aimed at modelling the financial market dynamics. Inf. Sci. **181**(1), 79–91 (2011)
10. Kim, H.K., Chong, J.K., Park, K.Y., Lowther, D.A.: Differential evolution strategy for constrained global optimization and application to practical engineering problems. IEEE Trans. Magn. **43**(4), 1565–1568 (2007)
11. Kohsaka, A., Laposky, A.D., Ramsey, K.M., Estrada, C., Joshu, C., Kobayashi, Y., Turek, F.W., Bass, J.: High-fat diet disrupts behavioral and molecular circadian rhythms in mice. Cell Metab. **6**(5), 414–421 (2007)
12. Lima, D.M., Padovani, R.M., Rodriguez-Amaya, D.B., Farfán, J.A., Nonato, C.T., Lima, M.T.d., Salay, E., Colugnati, F.A.B., Galeazzi, M.A.M.: Tabela brasileira de composição de alimentos - taco (2011). http://www.unicamp.br/nepa/taco/tabela.php?ativo=tabela Acessed 16 03 2016
13. Mamat, M., Rokhayati, Y., Mohamad, N., Mohd, I.: Optimizing human diet problem with fuzzy price using fuzzy linear programming approach. Pak. J. Nutr. **10**(6), 594–598 (2011)
14. Neri, F., Tirronen, V.: Recent advances in differential evolution: a survey and experimental analysis. Artif. Intell. Rev. **33**(1–2), 61–106 (2010)
15. Pasquali, R., Gambineri, A., Biscotti, D., Vicennati, V., Gagliardi, L., Colitta, D., Fiorini, S., Cognigni, G.E., Filicori, M., Morselli-Labate, A.M.: Effect of long-term treatment with metformin added to hypocaloric diet on body composition, fat distribution, and androgen and insulin levels in abdominally obese women with and without the polycystic ovary syndrome. J. Clin. Endocrinol. Metab. **85**(8), 2767–2774 (2000)
16. Stefanick, M.L., Mackey, S., Sheehan, M., Ellsworth, N., Haskell, W.L., Wood, P.D.: Effects of diet and exercise in men and postmenopausal women with low levels of hdl cholesterol and high levels of ldl cholesterol. N. Engl. J. Med. **339**(1), 12–20 (1998)
17. Stigler, G.J.: The cost of subsistence. J. Farm Econ. **27**(2), 303–314 (1945)
18. Storn, R., Price, K.: Differential evolution-a simple and efficient heuristic for global optimization over continuous spaces. J. Glob. Optim. **11**(4), 341–359 (1997)
19. Thomsen, R.: Flexible ligand docking using differential evolution. In: The 2003 Congress on Evolutionary Computation, 2003, CEC 2003, vol. 4, pp. 2354–2361. IEEE (2003)
20. de Vigilância Sanitária (ANVISA), A.N.: Resolution rdc number 360. Diário Oficial da União (2003). http://goo.gl/PZAChm. Acessed 16 03 2016
21. Warwick, Z.S., Schiffman, S.S.: Role of dietary fat in calorie intake and weight gain. Neurosci. Biobehav. Rev. **16**(4), 585–596 (1992)
22. Who, J., Consultation, F.E.: Diet, nutrition and the prevention of chronic diseases. World Health Organ Technical report Ser 916(i-viii) (2003)
23. Yang, L., Colditz, G.A.: Prevalence of overweight and obesity in the United States, 2007–2012. JAMA Intern. Med. **175**(8), 1412–1413 (2015)

Using Classification Methods to Reinforce the Impact of Social Factors on Software Success

Eudisley Anjos[1,2(✉)], Jansepetrus Brasileiro[2], Danielle Silva[2],
and Mário Zenha-Rela[1]

[1] CISUC, Centre for Informatics and Systems, University of Coimbra,
Coimbra, Portugal
{eudisley,mzrela}@dei.uc.pt

[2] CI, Informatics Center, Federal University of Paraíba, Santa Rita, Brazil
{eudisley,danielle}@ci.ufpb.br, jansebp@eng.ci.ufpb.br

Abstract. Define the way of success for software can be an arduous task, especially when dealing with OSS projects. In this case, it is extremely difficult to have control over all stages of the development process. Many researchers have approached ways to identify aspects, whether social or technical, that have some impact on the success or failure of software. Despite the large number of results found, there is still no consensus among which types of attributes have a greater success impact. Thus, after identifying technical and socio-technical factors that influence the success of OSS using data-mining techniques in about 20.000 projects data from GitHub, this study aims to compare them in order to identify those which most influence in determining the success of an OSS project. The results show that it is possible to identify the status (active or dormant) in more than 90 % of the cases based, mainly, in social attributes of the project.

1 Introduction

Technological advancement and market needs have resulted in constant modifications of software systems. These changes and the unsuitability to replace existing systems, demand that they are highly maintainable. This is one of the reasons why maintenance is one of the most expensive phases in software development [8,28]. It constitutes between 60 % to 80 % of the total effort spent on the software product [10,15]. Thus, the quality of a system is not defined only by ensuring the functions for which the system was designed are fit for purpose, but also by ensuring a high level of understanding, modifiability and maintainability. A system with these characteristics allows quick and easy changes in the long term.

As the system evolves, it is important that the maintainability of the system is sustained at an acceptable level. Many authors cite the laws of software evolution of Lehman [18] to demonstrate what happens when the system changes, addressing actions that could be taken to avoid a decline in maintainability. However, ensuring a high level of system maintenance of architectural projects

© Springer International Publishing Switzerland 2016
O. Gervasi et al. (Eds.): ICCSA 2016, Part V, LNCS 9790, pp. 187–200, 2016.
DOI: 10.1007/978-3-319-42092-9_15

is still a very complex task. There is a lack of architectural support to ensure that the modifications allow high maintainable projects, leading it to success.

In Open Source Software this evaluation procedure is even more difficult since there is a lack of a means for controlling several variables. Despite this, FLOSS projects have increasingly featured more prominently among programmers and their success is well known. Several systems have achieved high acceptability in the market evolving to very maintainable and stable versions [7]. All the FLOSS projects have managed to achieve success and a level of high quality and they are evolving very quickly, although this depends on their ability to attract and retain developers [33]. However, not all FLOSS achieve a high quality and success [20].

Most of the work spent on software evolution focuses on practices relating to managing code. However, code management to assure a high level of quality is not an easy undertaking. The amount of inactive/dormant software is undoubtedly higher compared to the number of active projects [1].

Several metrics of code have been proposed to attempt to ensure that the system remains as maintainable as possible. Nonetheless, many of these metrics seem not to be mutually consistent [31] and there is no concise standardization to support researchers and practitioners in deciding which metric should be used [27]. The same process happens to socio-technical metrics. Due to this reason, the constant monitoring of these metrics are not enough to ensure a high level of maintenance leading the projects to success.

The work presented in this paper shows a comparison between active and dormant systems using different metrics divided in two groups: metrics of code and socio-technical parameters. The main objective is to understand which group can better define the activeness of systems' projects directing the efforts of software engineers towards improving system maintainability. For this, the results of two main experiments involving around 20.000 systems of 10 different programming languages are presented in this paper.

2   State of Art

The body of knowledge of software engineering about software success grown up and several researches have tried to address the main causes of project's abandonment. Most of work found in literature analysing FLOSS success impact has concentrated efforts on quality attributes related to modularity [11]. The core measures approached in these works are software complexity such as in [5], composability or decomposability as in [19]; coupling and cohesion [3] and so forth. Such works have a focus on the product and therefore the use of code metrics facilitates the longitudinal evaluation of projects behaviour.

On the other hand, some works are concerned with processing characteristics with regard to systems development. Some authors use the term 'Network Structure' to refer to social characteristics such as: collaborators, as approached in [34,40], the social integration within the project team, as in [32,36] and external parameters such as: acceptability and popularity [26]. Since these parameters are more subjective, it is more difficult to measure them automatically. It explains why many works take few projects as subject.

In [29] the authors investigated the effects of external characteristics on the success of OSS projects, as measured by the number of subscribers and developers working on the project. The authors found a relationship between the language in which the application is written or the platform in which it is being developed and the numbers of subscribers and developers. On the basis of their findings, OSS projects that develop software compatible with Windows/UNIX operating systems, as well as software written in C or its derivative languages, tend to experience larger increase in the number of subscribers and usually attract more developers than the projects without these features.

There is another branch of studies which examines many different aspects of a project to determine whether they can be classified as successful or not. In [14], for example, the authors collected data from 1025 OSS projects in a longitudinal study. The results show that the extent of a project's operating systems, the range of translated languages, programming languages, and project age, achieve better outcomes for the OSS projects with regard to market penetration and the attraction of human resources, which they view as an indication of success.

The use of data mining techniques to validate software engineering hypotheses regarding software success is approached in many works. In [37] it is shown that the success of an OSS project can be predicted by just considering its first 9 months of development data with a K-Means clustering predictor at relatively high confidence. Similarly, in [25] several clustering algorithms are applied on a set of projects to predict success, based on defect count. Their findings indicates that K-means algorithms are more efficient than other clustering techniques in terms of processing time, efficiency and it is reasonably scalable.

We also found other studies where authors applied further concepts of machine learning to predict how healthy a project were in a given moment of time. In [22] the authors claim they used machine learning to create a models that can determine, with some accuracy, the stage a software project has reached. To create this model they took into account time-invariant variables such as type of license, the operating system the software runs on, and the programming language in which the source code is written. They also considered time-variant variables such as project activity (number of files, bug fixes and patches released), user activity (number of downloads) and developer activity (number of developers per project). The authors validated their model using two performance measures: the success rate of classifying an OSS project in its exact stage and the success rate over classifying a project in an immediately inferior or superior stage than its actual one. In all cases they obtain an accuracy of above 70 % with "one away" classification (a classification which differs by one) and about 40 % accuracy with an exact classification.

As presented in Table 1, recent researches have worked with social and technical attributes analysing them as factors of success of FLOSS projects. Nevertheless, most of work are guided by different perspectives. Some of them approach the success as product using measures of technical attributes, for example complexity of code. Others, approaches the number of subscribers and developers as main factors of success. Thus, it is still not so clear which group of attributes should be monitored in way to define the activeness of systems projects, increasing the possibility of success along a FLOSS project life. This is one of the main motivations for the studies presented in this paper. In way to understand better the table, consider the following abbreviations: attributes evaluated (A), evaluation techniques (B), repository used: SourceForge or GitHub (C), is the evaluation longitudinal? - yes or no (D), and the next line shows the findings of the research.

Table 1. Researches approaching software success

	A	B	C	D
[11]	Modularity	Static model	SourceForge	Yes
Findings: The relationship between software modularity and software quality depends on project characteristics and how software quality is measured				
[5, 10, 19]	Complexity	Static model	SourceForge	Yes
Findings: The results showed that inactive FLOSS projects do not seem to be able to keep up with the extra work required to control the systems complexity, presenting a different behaviour of the successful active FLOSS projects				
[17]	Composability or decomposability	Static model	SourceForge	No
Findings: The results indicated that both design architectures and designer network structures are important to FLOSS project outcomes				
[3]	Coupling and Cohesion	Static model	SourceForge	No
Findings: The observations contradict the common assumption that software quality attributes (aka non-functional requirements) are mostly determined at the architectural level				
[27]	Time-invariant[a] and time-variant[b]	Hierarchy Bayes model	SourceForge	Yes
Findings: The results focus on subscribers and developers as measures of OSS project success. Also it was found a correlation between developers on subscribers is positive and significant				
[32, 35]	Network structure and time-variant	Static model	GitHub	No
Findings: The results show that understanding how developers and projects are actually related to each other on a social coding site is the first step towards building tools to aid social programmers				
[13]	Time-invariant[c]	Static model	SourceForge	Yes
Findings: The results showed that FLOSS projects compatible with more operating systems; translated into more languages; built in a widely used programming language; and have been running for a long time relative to other projects, are more likely to be successful				
[34]	Time-variant[d]	Machine learning	SourceForge	Yes
Findings: The results define a model that predict the success of a given OSS project by just examining its first 9 month development data				

[a] License type, operating system, programming language
[b] Project age, monthly number of developers, monthly number of subscribers
[c] Operating systems, languages of translation, programming languages, and project age
[d] Defect count, outdegree (measure of interactions between actors)

3   Methodology

Machine Learning algorithms are employed in many practical data mining applications. They are used to find and describe the structural patterns present in data, helping researchers to explain it and make new predictions from it. It is possible to make new previsions of what will happen in new situations, based on data that describes what happened in past. Thus, it is possible to use these algorithms to anticipate the behaviour of FLOSS systems, starting from a set of metadata.

One of the way to reach this purpose is using classification models (classifier). The classifier maps instances to classes and allows inferring an aspect of data (class) from other aspects, called attributes. This paper aims to find a way of describing a project status (Active or Dormant) based on project's metadata and predict the status from this perspective. In way to test our hypothesis we performed data collection and several tests, confirming the feasibility of our predictions. The following subsections explains the process of data collection and data analysis for success prediction.

3.1   Data Collection

In order to perform the classification tasks we extracted data from public projects hosted on GitHub repository. The API provided by GitHub made it easier to sample the projects using the following criteria: active projects were updated in last two years and dormant projects were not updated in the same period. The projects were downloaded automatically using a script that make requests and store the data in a database scheme. All collected metadata were used as input to the classifiers, with purpose of inferring the status (active or dormant). The types of metadata are described in the Table 2. This step resulted in a set of 19,994 projects where: 9,995 of them dormants and 9,999 actives. The projects used 10 different programming languages: Python, C, JavaScript, C++, Objective-C, C#, Java, PHP, Shell and Ruby. In the section results we explain this choices.

3.2   Success Predictions

The viability of project success prediction was attested by performing some classification tests with popular machine learning algorithms in Weka program, that contains a collection of these algorithms, as well as preprocessing tools, allowing users to compare different methods and identify the most appropriate methods to the problem. To this end, we used the Explorer interface that allows the use of all Weka resources easily. Each instance of the dataset is represented by the set of attributes (metadata) and the associated class (active or inactive). We choose six different classifiers, belonging to different classes: IBk, from the group of lazy classifiers; J48, member of the classification trees; Simple Logistic, part of mathematical functions; Decision Table, from rule based classifiers; Multilayer Perceptron, from neural network.

Table 2. Metadata used to classify projects by status

Metadata	Description
Language	Programming language used to code the project
Size	Project size in bytes
Number of contributors	Developers who contributed to the project
Stars	Project rating by users
Subscribers	Number of subscribed users in the project
Forks	Number of times that the project was forked
Number of commits	Number of commits in the project
Issues	Number of open issues
Has downloads	Has anyone downloaded the project?
Has page	The project has web page?
Has Wiki	The project has Wiki page?

- **Simple logistic** is a kind of regression algorithm that fits a logistic multinomial regression model. In each iteration, it adds one Simple Linear Regression model per class into the logistic regression model. Logistic regression is a powerful statistical way of modelling a binomial outcome with one or multiple explanatory variables. One of the main advantages of this model is the use of more than one dependent variable, also providing a quantified value for the strength of the association. However you need a large sample size and the definition of the variables must to be carefully planned in way to obtain precise results.
- **J48** decision tree is an open-source implementation available on Weka of C4.5 algorithm developed by Quinlan [23]. Both are an evolution from ID3 (Iterative Dichotomiser 3) and use divide and conquers approach to growing decision trees [21]. J48 is a classification algorithm based on the standardized measure of lost information when a set A is used to approximate B, called information gain. The attribute used to divide the set is the higher value of information gain.
- **Bagging**, also known as Bootstrap Aggregating it creates separate samples of training dataset and classify each sample combining the results (by average or voting). Each trained classifier have a different focus and perspective on the problem, increasing the accuracy of the results. The advantage of Bagging is reduce variance and avoid overfitting[1]. One of the biggest advantages of bagging is the capacity of handling small data sets. Also, this algorithm is less sensitive to noises or outliers.
- **IBk** (Instance Based) is a Weka implementation of k-NN (Nearest Neighbour) Algorithm. IBk generates a prediction for a test instance just-in-time differing for other algorithms which build a model. The IBk algorithm uses a distance

---

[1] It occurs when the data model fits too much to the statistical model causing deviations by measuring errors.

measure using k-close instances to predict a class for a new instance. So, it computes similarities between the selected instance and training instances to make a decision. The use of cross validation allows this algorithms to select appropriate values of k.

– **Decision Table** has two main components: a schema and a body. Schema is a set of features and body is labelled instances defined by the features in the schema [17]. When applied to unlabelled instances, the classifier searches the decision table using the features in the schema to found which class the instance will be classified. Decision tables has been largely used in software engineering area. They are easy to understand and allows developers to work from the same information or to successfully produce system's specifications.

– **Multilayer Perceptron** is an artificial neural network model that use layers between input and output and create flows between them. It is a logistic classifier that uses a learnt non-linear transformation, projecting input date into a space where it becomes linearly separable. It is known as a supervised network because it requires a desired output in order to learn.

In all tests, the division between training and test set was performed by cross-validation, with 10 folds. In this approach the method is executed 10 times, using the proportion of 90 % to train the classifier and 10 % to test it. In each execution, different segments are used to training and testing. Cross-validation is considered as a highly effective method for automatic model selection [2]. A more detailed explanation about how cross-validation works is beyond the scope of this work, although it can be easily found in the literature for further research. The important thing here is the fact that this option led us to more accurate results than other more common, such as the use of the training set.

4 Results and Discussions

The purpose of this study is to identify factors that have led to the success of the FLOSS projects. The understanding of these factors can enable architectural decisions to be made that improve the systems level of maintainability. Before we turn to the experiments conducted here, several underlying assumptions were analysed and tested. The findings led to key decisions and opened up avenues of inquiry and new research areas, which are currently being addressed in other works. The results were obtained during two main experiments both guided by previous studies of our research team. The details of each experiments are shown following.

4.1 Experiments - Phase 01

The initial experiments were guided by previous results obtained by our team and available in [3–5,13,30]. The results showed that systems can be very strongly influenced by social parameters, leading the necessity of compare the results of metrics of code with measures of socio-technical factors. The systems adopted for this stage were downloaded either from GitHub as SourceForge. The use of these two main repositories was necessary to obtain as much social information as possible from the systems to carry out a complete assessment of all the characteristics.

Based on the obtained information, we applied various statistical analysis algorithms to compare the data. The initial comparisons included all the methods from Table 3. In all the comparisons, the main objective was to analyse which factors are most relevant to be considered in architectural enhancements to ensure the success of the projects. At this stage, 20 Java projects were tested. The code metrics were applied using a tool called SonarQube and the social metrics obtained directly from the repositories API.

Table 3. Methods used to compare the results

Correlation	Pearson	Kendall	Spearman
Classification	J48	Simple logistic	IBk (K = 3)
	Bagging	Decision table	Multilayer perceptron
Correlation	Factor analysis		

These initial results are showed in Table 4 and evidence that the projects success was greatly influenced by social parameters and that the code measures (particularly complexity) were practically irrelevant when compared with the others. This supports the theory of some researchers that the FLOSS systems are naturally more modular [6] and therefore code attributes may not be a crucial factor in determining the success or abandonment of the project.

It is clear that some social metrics had a great impact while the results of the metrics of code do not have the same relevance. Thus, the results followed the

Table 4. Attribute-based considerations using classification methods

Socio-technical factor	Consideration	Code metrics	Consideration
Forks	0.9612	Files	0.0000
Subscribers	0.9612	Duplicated lines density	0.0000
Stars	0.9611	Functions	0.0000
Number of contributors	0.5716	Complexity by file	0.0000
Size	0.3739	McCabe complexity	0.0000
Number of commits	0.3217	Comment lines density	0.0000
Issues	0.3204	Directories	0.0000
Has page	0.1938	NLOC	0.0000
Has Wiki	0.0328	Violations	0.0000
Has downloads	0.0276	TLOC	0.0000

initial hypothesis that social factors impact more on success than code. However, as the initial experiments only took into account a small number of OO codes, the results could be influenced by the programming paradigm or the restricted group of 20 projects. Thus, the scope of the experiments was expanded to a bigger number of target projects as well as the programming paradigms used. For this reason it was necessary to semi-automatize the entire process to get the projects information and analysis. This led the authors to determine a second stage of experiments described as following.

4.2   Experiments - Phase 02

Due to the requirement to automate part of the analysis process and the necessity to restrict the evaluated factors it was necessary to chose one of the two repositories used during the first phase of experiments. Although the GitHub has a large project database that is publicly available, there are some difficulties in using it for data mining, as demonstrated in [16]. As a result, many academic researchers end up by choosing SourceForge. This explains the large amount of items in our literature review, and is what led us to choose SourceForge first as a repository for analysis. However, Weiss [38] found some problems with SourceForge, such as its change of categorization or high rate of abandonment. Furthermore, Rainer and Gale [24] analysed the quality of SourceForge data and advised caution in the use of this data due to the large number of abandoned projects. On the other hand, few studies have highlighted the quality of the GitHub data, and there is still a lack of information about what kind of methodology can be employed for it. Thus was necessary to define our own acquisition method.

In 2013, SourceForge experienced a popularity problem and as a result included advertisement buttons inside the project's websites which made the users confused and reduced its usability. In response, many developers removed the download option from their project's website. At that time, SourceForge had 3 million registered users (700,000 less than today) while GitHub had approximately 2.5 million users in early 2013, reaching 3 million users on April 2013 and almost 5 million (with 10 million repositories) in December. As well as this, extracted data from Alexa.com shows the increasing popularity of GitHub in recent years, as SourceForge continues to decline. Furthermore, GitHub provides an access speed about 64 % greater than SourceForge (1.117 s and 2.155 s). Today, GitHub is among the top one hundred most visited Web sites (89°), while the SourceForge is at 272°.

One of the reasons for the extensive use of SourceForge in academic studies is the availability of free databases aimed at research, such as: FLOSSmole, from Syracuse University (discontinued today), and the SourceForge Research Data Archive (SRDA) from University of Notre Dame. This database has been backing up instances of data from SourceForge for academic purposes for many years. Today, some attempts have been made to make GitHub available for big databases involving projects such as GHTorrent and GitHub Archive. However due to restrictions in its use, we had to develop our own application to obtain the data necessary for our research. Thus, the choice of Github was threefold: the rising importance of this repository, the lack of academic work and the peculiarities of socio-technical factors (some of them absent in other repositories).

In total, 19.994 projects were selected (9,995 considered dormant and 9,999 active) from 10 different programming languages: Java, Python, C, Shell, C++, C #, Objective-C, JavaScript, Ruby and PHP. During the experiments, the attributes previously tested were refined, remaining only the most important ones. Furthermore, the number and complexity of the classification algorithms were increased, since they enabled to obtain the best analysis. The results provided greater support to the validation of the initial hypothesis - that the socio-technical parameters are a decisive factor in ensuring architectural enhancements and if controlled may lead the project to success.

The Table 5 shows the Accuracy, True Positives (TP) and False Positives (FP) of the classification algorithms for the dataset. Accuracy is defined by the percentage of correctly classified instances, true positives are related to instances that are correctly classified as a given class in the same way and, false positives are instances falsely classified as a given class.

Despite the high degree of accuracy of the algorithms, some attributes have been shown to have little weight in the classification, such as: complexity and programming language (as shown in Table 6). Our attention was drawn to the

Table 5. Accuracy, true positive and false positive for classification methods

	Simple logistic	J48	Bagging	IBk (K = 3)	Multilayer perceptron	Decision table
Accuracy	92.56 %	96.25 %	96.62 %	93.72 %	92.86 %	95.05 %
TP rate active projects	91.00 %	98.80 %	99.50 %	96.00 %	93.50 %	99.10 %
FP rate active projects	5.90 %	6.30 %	6.30 %	8.60 %	7.80 %	9.00 %
TP rate dormant projects	94.10 %	93.70 %	93.70 %	91.40 %	92.20 %	91.00 %
FP rate dormant projects	9.00 %	1.20 %	0.50 %	4.00 %	6.50 %	0.90 %

Table 6. Importance of attributes

Attribute	Importance
Stars	0.52531051
Subscribers	0.50602662
Forks	0.39342657
Number of contributors	0.35564825
Issues	0.25643352
Has pages	0.01838373
Has downloads	0.00001036
Language	0.00000109
Cyclomatic complexity	0.00000000

fact that the programming language was not assigned a big weight in the importance table. This information contradicts several other results found in some studies, such as [12, 14].

5 Conclusions and Future Works

The increasing trend towards the use and development of open-source projects has led to a growing body of research on exploring qualities of open-source development [39]. The wide range of methodologies and techniques imply a lack of consistency in this area. This results in difficulties in maintaining consistent measurements or being able to conduct an analysis that is adaptable to different projects and environments [35]. Hence, an analysis of the factors that can affect software architecture, and thus maintainability, can lead to improvements in software development.

The main objective of this paper was to help software engineers to improve system maintainability. The results showed an analysis using machine learning algorithms about how two groups of metrics (social and code) affects the activeness of systems' projects. This goal was obtained using data extracted from public projects available on GitHub and SourceForge and analysing them using tools such as: SonarQube and Weka.

In the first phase, 20 Java projects were analysed showing the importance of social aspects the their impact on the prediction of project success while the results of the code metrics do not have the same relevance. However, as Weka works with machine learning algorithms, the use of few projects may provide insufficient data to the analysis, biasing the results and leading to a not realistic analysis. Besides the use of one programming paradigm makes impossible to generalize the results Also, it is known that, as much as analysing FLOSS projects by code metrics is a complex task, many studies have shown the importance of these metrics to assess maintainability.

The results shown in Table 4 indicates a relevance tending to 0 % for code metrics, this value can be explained by how Weka analyses the informations. In other words, the result does not show the relevance of code metrics isolated to indicate success of systems' project, but in comparison of socio-technical aspects. This proves our initial hypothesis that social factors impact more on success than code. As a result, it was reasonable to consider that it makes more sense analyze these projects through socio-technical parameters instead of code metrics.

Driven by this result, in order to validate our results, the second phase used data from almost 20,000 projects of 10 different programming languages, from different paradigms. This scenario aims to identify if the previous results can be generalized for a bigger and more heterogene FLOSS projects database. After the analysis of all these data, using six different classifiers, belonging to different classes and working with "ten-fold cross-validation" technique, available in Weka, we found again, that socio-technical aspects have a huge impact on the prediction of success for FLOSS projects. Using the analysis proposed here, it was possible to identify the correct status (active or dormant) in more than 90 % of the cases, reaching up to 96 % in some cases. This proves that, besides the fact that code metrics can have a great importance on the success prediction of systems' projects [9], socio-technical parameters are also able to achieve this goal with a bigger accuracy.

Besides the interesting results obtained here, our study has limitations that motivates future works. Although this study used a large database with 20,000 projects, it is still unclear if the greater the amount of projects, the higher is the reliability. Due to this we are performing a bigger analysis using a bigger data base (around 160,000 projects). Also it is necessary that such database be the most heterogeneous as possible, especially with regard to new paradigms and new programming languages.

Even obtaining a great result of success or failure prediction, it is necessary to make the comparison with metrics of code for a larger database. As the analysis of code used only 20 projects of the Java programming language, it is possible that this sample set does not represent the majority of the project universe. Only then it is possible to say whether the use of social aspects is more feasible than code and how much more feasible it is, in comparative terms.

Finally, it is essential to understand which of the dormant projects reached, in fact, a level of maturity where the changes are not required. Although some works have approached the conditions of software maturity, it is still difficult to have an automatic process to select these projects. Our team is currently approaching this subject for future analysis.

References

1. Brito e Abreu, F., Pereira, G., Sousa, P.: A coupling-guided cluster analysis approach to reengineer the modularity of object-oriented systems. In: Proceedings of Conference on Software Maintenance and Reengineering, CSMR 2000, pp. 13–22. IEEE Computer Society, Washington, DC (2000)
2. Lee, M.S., Moore, A.: Efficient algorithms for minimizing cross validation error. In: Cohen, W.W., Hirsh, H. (eds.) Proceedings of 11th International Confonference on Machine Learning, pp. 190–198. Morgan Kaufmann (1994)
3. Anjos, E., Castor, F., Zenha-Rela, M.: Comparing software architecture descriptions and raw source-code: a statistical analysis of maintainability metrics. In: Murgante, B., Misra, S., Carlini, M., Torre, C.M., Nguyen, H.-Q., Taniar, D., Apduhan, B.O., Gervasi, O. (eds.) ICCSA 2013, Part III. LNCS, vol. 7973, pp. 199–213. Springer, Heidelberg (2013)
4. de L. Lima, P.A., da C. C. Franco Fraga, G., dos Anjos, E.G., da Silva, D.R.D.: Systematic mapping studies in modularity in IT courses. In: Gervasi, O., Murgante, B., Misra, S., Gavrilova, M.L., Rocha, A.M.A.C., Torre, C., Taniar, D., Apduhan, B.O. (eds.) ICCSA 2015. LNCS, vol. 9159, pp. 132–146. Springer, Heidelberg (2015)
5. Anjos, E., Grigorio, F., Brito, D., Zenha-Rela, M.: On systems project abandonment: an analysis of complexity during development and evolution of floss systems. In: 6TH IEEE International Conference on Adaptive Science and Technology, ICAST 2014, Covenant University, Nigeria, 29–31 October 2014
6. Baldwin, C.: Modularity and Organizations. International Encyclopedia of the Social & Behavioral Sciences, 2nd edn, pp. 718–723. Elsevier, Amsterdam (2015)

7. Bass, L., Clements, P., Kazman, R.: Software Architecture in Practice, 2nd edn. Addison-Wesley Professional, Boston (2003)
8. Benaroch, M.: Primary drivers of software maintenance cost studied using longitudinal data. In: Proceedings of International Conference on Information Systems, ICIS 2013, Milano, Italy, 15–18 December 2013
9. Blondeau, V., Anquetil, N., Ducasse, S., Cresson, S., Croisy, P.: Software metrics to predict the health of a project? An assessment in a major IT company. In: Proceedings of International Workshop on Smalltalk Technologies, IWST 2015, Brescia, Italy, pp. 9:1–9:8, 15–16 July 2015
10. Bode, S.: On the role of evolvability for architectural design. In: Fischer, S., Maehle, E., Reischuk, R. (eds.) GI Jahrestagung. LNI, vol. 154, pp. 3256–3263. Springer, Heidelberg (2009)
11. Conley, C.A.: Design for quality: the case of open source software development. Ph.D. thesis, AAI3340360 (2008)
12. Emanuel, A.W.R., Wardoyo, R., Istiyanto, J.E., Mustofa, K.: Success factors of OSS projects from sourceforge using datamining association rule. In: 2010 International Conference on Distributed Framework and Applications (DFmA), pp. 1–8, August 2010
13. Anjos, F.G.E., Brito, M.Z.-R.D.: Using statistical analysis of floss systems complexity to understand software inactivity. Covenant J. Inform. Commun. Technol. (CJICT) 2, 1–28 (2014)
14. Ghapanchi, A.H., Tavana, M.: A longitudinal study of the impact of OSS project characteristics on positive outcomes. Inf. Syst. Manag. (2014)
15. Glass, R.L.: Software Engineering: Facts and Fallacies. Addison-Wesley Longman Publishing Co. Inc., Boston (2002)
16. Kalliamvakou, E., Gousios, G., Blincoe, K., Singer, L., German, D.M., Damian, D.: The promises and perils of mining GitHub. In: Proceedings of 11th Working Conference on Mining Software Repositories, MSR 2014, pp. 92–101. ACM, New York (2014)
17. Kohavi, R.: The power of decision tables. In: Lavrač, Nada, Wrobel, Stefan (eds.) ECML 1995. LNCS, vol. 912, pp. 174–189. Springer, Heidelberg (1995)
18. Lehman, M.M.: Programs, cities, students, limits to growth? In: Gries, D. (ed.) Programming Methodology. TMCS, pp. 42–69. Springer, New York (1978)
19. Liu, X.: Design architecture, developer networks and performance of open source software projects. Ph.D. thesis, AAI3323131, Boston, MA, USA (2008)
20. Michlmayr, M.: Hunt, F., Probert, D.: Quality practices and problems in free software projects. pp. 24–28 (2005)
21. Mathuria, M., Bhargava, N., Sharma, G.: Decision tree analysis on J48 algorithm for data mining. Int. J. Adv. Res. Comput. Sci. Softw. Eng. 3(6), 1114–1119 (2013)
22. Piggot, J., Amrit, C.: How healthy is my project? Open source project attributes as indicators of success. In: Petrinja, E., Succi, G., El Ioini, N., Sillitti, A. (eds.) OSS 2013. IFIP AICT, vol. 404, pp. 30–44. Springer, Heidelberg (2013)
23. Quinlan, J.R.: C4.5: Programs for Machine Learning. Morgan Kaufmann Publishers Inc., San Francisco (1993)
24. Rainer, A., Gale, S.: Evaluating the quality and quantity of data on open source software projects. pp. 29–36 (2005)
25. Ramaswamy, V., Suma, V., Pushphavathi, T.P.: An approach to predict software project success by cascading clustering and classification. In: International Conference on Software Engineering and Mobile Application Modelling and Development (ICSEMA 2012), pp. 1–8, December 2012

26. Rotaru, O.P., Dobre. M.: Reusability metrics for software components. In: The 3rd ACS/IEEE International Conference on Computer Systems and Applications, p. 24 (2005)
27. Saraiva, J.: A roadmap for software maintainability measurement. In: Proceedings of 2013 International Conference on Software Engineering, ICSE 2013, pp. 1453–1455. IEEE Press, Piscataway (2013)
28. Schneidewind, N.F.: The state of software maintenance. IEEE Trans. Softw. Eng. 13, 303–310 (1987)
29. Sen, R., Singh, S.S., Borle, S.: Open source software success: measures and analysis. Decis. Support Syst. 52(2), 364–372 (2012)
30. Siebra, B., Anjos, E., Rolim, G.: Study on the social impact on software architecture through metrics of modularity. In: Murgante, B., et al. (eds.) ICCSA 2014, Part V. LNCS, vol. 8583, pp. 618–632. Springer, Heidelberg (2014)
31. Sjøberg, D.I., Anda, B., Mockus, A.: Questioning software maintenance metrics: a comparative case study. In: Proceedings of ACM-IEEE International Symposium on Empirical Software Engineering and Measurement, ESEM 2012, pp. 107–110. ACM, New York (2012)
32. Ståhl, D., Bosch, J.: Modeling continuous integration practice differences in industry software development. J. Syst. Softw. 87, 48–59 (2014)
33. Stewart, K.J., Ammeter, A.P., Maruping, L.M.: Impacts of license choice and organizational sponsorship on user interest and development activity in open source software projects. Inf. Syst. Res. 17(2), 126–144 (2006)
34. Thung, F., Bissyande, T.F., Lo, D., Jiang, L.: Network structure of social coding in GitHub. In: 2013 17th European Conference on Software Maintenance and Reengineering (CSMR), pp. 323–326, March 2013
35. Vlas, R.E.: A requirements-based exploration of open-source software development projects - towards a Natural language processing software analysis framework. Ph.D. thesis, AAI3518925, Atlanta, GA, USA (2012)
36. Walz, D.B., Elam, J.J., Curtis, B.: Inside a software design team: knowledge acquisition, sharing, and integration. Commun. ACM 36(10), 63–77 (1993)
37. Wang, Y.: Prediction of Success in Open Source Software Development. University of California, Davis (2007)
38. Weiss, D.: Quantitative analysis of open source projects on SourceForge. In: OSS2005: Open Source Systems, pp. 140–147 (2005)
39. Vlas, R., Robinson, W.: Requirements evolution and project success: an analysis of SourceForge projects. In: AMCIS 2015 Proceedings, Systems Analysis and Design (SIGSAND) (2015)
40. Zanetti, M.S., Sarigöl, E., Scholtes, I., Tessone, C.J., Schweitzer, F.: A quantitative study of social organization in open source software communities. CoRR, abs/1208.4289 (2012)

When to Re-staff a Late Project – An E-CARGO Approach

Haibin Zhu[1,3(✉)], Dongning Liu[2], Xianjun Zhu[3], Yu Zhu[4],
Shaohua Teng[2], and Xianzhong Zhou[3]

[1] Department of Computer Science and Mathematics, Nipissing University,
North Bay, ON, Canada
haibinz@nipissingu.ca
[2] School of Computer Science and Technology, Guangdong University
of Technology, Guangzhou, Guangdong, China
[3] School of Management and Engineering, Nanjing University, Nanjing, China
[4] Faculty of Mathematics, University of Waterloo, Waterloo, ON, Canada

Abstract. Brooks' law is popular in software development. It has been used as a reference for managing software projects for over four decades. However, not enough investigations express this law in a quantitative way that can provide specific project recommendations at critical times in the development process. This paper offers a quantitative way based on our research vehicle of Role-Based Collaboration and the related Environments-Classes, Agents, Roles, Groups, and Objects (E-CARGO) model. The proposed approach is verified by simulations, experiments and a case study. The results produce insights into Brooks' law, and quantitatively present the applicable scope of the law. This contribution is believed to be significant because it provides a quantitative measurement useful in the development of a software project.

Keywords: Brooks' law · Quantitative method · Role-Based Collaboration (RBC) · E-CARGO · Role assignment · Coordination factors

1 Introduction

Brooks' law is still frequently cited [3–5, 7, 9] in software engineering and project management after four decades. This law is stated as "adding manpower to a late software project makes it later" [5]. However, the acceptance of Brooks' law largely depends on intuitive reasoning, i.e., adding more people brings in increased communication costs thus delaying the project. This intuition may not obtain enough solid support from a real-world project. On the other hand, after decades' effort, many new methodologies and technologies are continuously developed [12] to facilitate software development. Innovative styles of software development, such as, agile development,

This work was supported in part by Natural Sciences and Engineering Research Council, Canada (NSERC) under grant RGPIN262075-201, National Natural Science Foundation of China (NSFC) (No. 71171107), and Guangzhou City Foreign Cooperation Special Program (No. 7421255280918).

© Springer International Publishing Switzerland 2016
O. Gervasi et al. (Eds.): ICCSA 2016, Part V, LNCS 9790, pp. 201–217, 2016.
DOI: 10.1007/978-3-319-42092-9_16

cloud computing, outsourcing, and crowdsourcing, make software development much different from that was 40 years ago. The production rate of software teams is improved gradually [12]. Adding human power to some types of software projects indeed increases the production rate. This situation makes it imminent to review Brooks' law in consideration of new types of software development.

Even though simulation research related to Brooks' law has been conducted, no clear conclusions are well accepted. This situation simply expresses the challenge of this problem. That is to say, there are too many factors or aspects that affect the productivity of software development. These factors make the problem so difficult that it is hard for software engineers, developers and managers to model and quantitatively specify this problem.

This paper contributes a consistent quantification of the Brook's law in the sense of coordination by introducing factors of coordination, i.e., the performance of a person (agent) on a position (role) may decrease due to coordination.

The key considerations include:

(1) Some tasks in software development can be performed by a single person while others require coordination; and
(2) Some tasks are well delineated while others are not.

With consideration of coordination factors on role performance, the evaluation model is built with the support of the Role-Based Collaboration (RBC) methodology and its model Environments-Classes, Agents, Roles, Groups, and Objects (E-CARGO). With the E-CARGO model, this paper addresses the Brooks' law from a quantitative perspective and contributes to the formalization of Brooks' law. With this formalized problem as a tool, Brooks' law is revisited quantitatively by simulations.

RBC is an emerging and promising computational methodology that uses roles as a primary underlying mechanism to facilitate collaborative activities [18–22]. With the continuous research effort in RBC, traditional problems in software development can be formally represented and specified. As a result, such problems can be solved with appropriate methods [18, 22]. Role assignment is important and has been revealed as a complex task through the RBC process. Group Role Assignment (GRA) aims at finding an optimal assignment from roles to agents with the agent evaluation result [22] and largely affects the efficiency of collaboration and the degree of satisfaction among members involved in RBC. However, there are many constraints in role assignment, and GRA can become very complex under various constraints. Coordination factors are such an example.

In this paper, we assume that *an agent's performance on a role can be affected by two factors, i.e.,* the parallelism of a role, and the coordination efforts required by the role. With appropriate numerical values for these factors, we can quantitatively model the problems raised by Brooks' law. In this way, we may balance the trade-off between the tasks that can be partitioned [2] and those are not [15] by using the E-CARGO model.

This paper is organized as follows. Section 2 discusses the related work. Section 3 condenses the E-CARGO model. Section 4 defines and specifies the proposed problem by introducing related concepts and definitions. Section 5 proposes the solution to the formalized problem of Brooks' law, verifies the situations that are able to break

Brooks' law, and prove the practicability of the proposed solution by experiments. Section 6 provides a case study. The paper concludes by pointing out topics for future work in Sect. 7.

2 Related Work

Brooks' Law has been cited extensively [1, 3, 4, 7, 9, 12, 13, 15, 16] in the past decades. The law is also investigated by many researchers. However, there are only a few quantitative investigations.

Abdel-Hamid [1] proposes a simulation approach to software project staffing based on system dynamics. The model he applied in simulations includes four subsystems, i.e., human resource management (H), Software production (S), Controlling (C) and Planning (P). The subsystems are connected by the parameters of workforce available, workforce needed, progress status, tasks completed, schedule and effort remaining. His simulation indicates that "*adding manpower to a late software project does always cause a negative net cumulative contribution to the project, and thus, does not always make the project later.*" He also presents that Brooks' law only holds when the time parameter is less than or equal to 30 working days in his simulation. In Abdel-Hamid's simulations, there are assumptions that are arguable. First, their model assumes that development tasks can be partitioned. Second, project managers continuously add new people if they find that they need to. Third, the people in a team are not differentiated.

Berken [3] argues that the Brooks' law belongs to the category of "oversimplification." He lists many exceptions of the law. His qualitative arguments have prompted us to investigate a new quantitative approach.

Building on Brooks' law, Blackburn *et al.* [4] conduct an empirical study from a dataset of 117 software development projects conducted in Finland. They conclude that complexity increases the maximum team size in software development projects, and that maximum team size decreases software development productivity. The logical complexity of software development is abstracted in our proposed approach as coordination factors.

To curb the limits of the two arguable assumptions of Abdel-Hamid, Hsia *et al.* [1, 9] introduces the sequential constraint into the model applied by Abdel-Hamid. They conclude that "*adding people to a late project will always increase its cost, but the project may not always be late*". They also state an optimal time range for adding people without delaying a project through assumptions made for their simulations. Similar to [1], they ignore the differences among team members, i.e., all the team members are engineers.

Weinberg discusses Brooks' Law from the viewpoint of system dynamics [17]. He states that the effect of Brooks' Law is caused by coordination and training effort. More coordination means more extra work. The training effort offered by the experienced staff affects their production. His discussion activates our proposed quantifying method of the coordination factors in this paper.

In project management, some quantitative research has been done. Chen and Zhang [6] propose a model for software project planning. They develop an approach with an event-based scheduler (EBS) and an ant colony optimization (ACO) algorithm.

The basic idea of the EBS is to adjust the allocation of employees at events and keep the allocation unchanged at nonevents. With this strategy, the proposed method enables the modeling of resource conflict and task preemption and preserves the flexibility in human resource allocation. Their work does not mention Brooks' law but informs us some hints in our work to design our model.

Our previous work on GRA [22] makes it possible for us to model Brooks' Law in a quantitative way. Combine the mentioned idea with our previous work, the proposed approach is established.

3   Revised E-CARGO Model

With the E-CARGO model [21], a system Σ can be described as a 9-tuple $\Sigma ::= <C, O, \mathcal{A}, \mathcal{M}, \mathcal{R}, \mathcal{E}, \mathcal{G}, s_0, \mathcal{H}>$, where C is a set of classes, O is a set of objects, $\mathcal{A}$ is a set of agents, $\mathcal{M}$ is a set of messages, $\mathcal{R}$ is a set of roles, $\mathcal{E}$ is a set of environments, $\mathcal{G}$ is a set roups, s_0 is the initial state of the system, and $\mathcal{H}$ is a set of users. In such a system, $\mathcal{A}$ and $\mathcal{H}$, $\mathcal{E}$ and $\mathcal{G}$ are tightly-coupled sets. A human user and his/her agent perform a role together. Every group should work in an environment. An environment regulates a group.

To be self-contained and concise, the definitions of the components of E-CARGO [18–22] are described and simplified as follows to address the proposed problems: *Classes, objects, and Messages* are ignored in this paper, because they are not highly relevant to the proposed problems. Many items in the major components, such as *roles* and *agents*, of the original E-CARGO model are simplified.

To formally state the related concepts, we use some special notations in definitions and algorithm descriptions. If S is a set, $|S|$ is its cardinality. If v is a variable, $v \in |S|$ denotes that v may take an element of S as its value. If a and b are objects, "$a ::= b$" denotes that a is defined as b; "$a := b$" denotes that a is assigned with b; $a.b$ denotes b of a or a's b; and $\{a, b, ...\}$ denotes a set of enumerated elements of $a, b,$ and others. If a and b are integers, a/b is the integer quotient; $a\%b$ is the remainder of the division of a/b (e.g., $12/5 = 2$ and $12 \%5 = 2$). If a and b are real numbers, $[a, b]$ and $[a, b)$ denote the set of all the real numbers between a and b, where the former includes both a and b but the latter includes a but not b. If $\mathcal{Y}$ is a vector, $\mathcal{Y}[i]$ denotes the element at its ith position. If $\mathcal{Y}$ is a matrix, $\mathcal{Y}[i, j]$ denotes the element at the intersection of row i and column j in $\mathcal{Y}$, $\mathcal{Y}[a{:}b, c{:}d]$ is a matrix taking out of $\mathcal{Y}$ the rows of a to b and columns c to d inclusively. We use $\mathcal{N}$ to denote the set of positive integers, i.e., $\{1, 2, 3, ...\}$.

Definition 1: A *role* [18–22] is defined as $r ::= <id, \circledR>$ where,

- id is the *id*entification of the role; and
- $\circledR$ is the *r*equirement for its players.

Definition 2: An *agent* [18–22], as a role player, is defined as $a ::= <id, \mathbb{Q}>$ where

- id is the identification of a; and
- $\mathbb{Q}$ is the abilities (or *q*ualifications).

Note: The term *agent* can be any entity that is described in the way of the definition, such as, human being, software agent, machine, or commodity.

Definition 3: An *environment* [18–22] is defined as $e::=<id, \mathcal{R}_e, \text{\textcircled{S}}, \mathcal{B}>$ where

- id is the identification of the environment;
- $\mathcal{R}_e$ is a finite set of *r*oles (in *e*);
- \text{\textcircled{S}} is the shared object for $\mathcal{R}_e$; and
- $\mathcal{B}$ is a finite set of tuples consisting of roles and their ranges, i.e., $<r, q>$, where $r \in \mathcal{R}_e$. The role range (also called cardinalities) q (It is symbolized from requirements, because *r* and *e* have been used.) is expressed by $<l, u>$ and tells how many agents must (*l, i.e.,* lower bound) and may (*u*, upper bound) play *r* in this environment.

Definition 4: A *group* [18–22] is defined as $g::= <id, e, \mathcal{A}_g, \mathcal{J}>$ where

- id is the identification of the group;
- e is an environment for the group to work in;
- $\mathcal{A}_g$ is a finite set of agents (in *g*); and
- $\mathcal{J}$ is a finite set of tuples of agents and roles, i.e., $\mathcal{J} = \{<a, r> \mid a \in \mathcal{A}_g, r \in e.\mathcal{R}_e\}$.

Definition 5: For a group *g*, a tuple $<a, r>$ of *g*.$\mathcal{J}$ is called a *role assignment*, also called *agent assignment* [18–22].

In formalizing role assignment problems, only agents and roles are emphasized. In the following discussions, current agents or roles [18–22] are our focus. Environments and groups are simplified into vectors and matrices, respectively. Compared with the definitions in [18–22], Definitions 1–4 are simplified. Furthermore, we use non-negative integers m $(=|\mathcal{A}|)$ to express the size of the agent set $\mathcal{A}$, n $(=|\mathcal{R}|)$ the size of the role set $\mathcal{R}$, $i, i', i'', ..., i_1, i_2, ...$ the indices of agents, and $j, j', j'', ..., j_1, j_2, ...$ the indices of roles.

Definition 6: A *lower role range vector L* [18–22] is a vector of the lower bound of the ranges of roles in environment *e* of group *g*. Suppose that roles in *g.e* are numbered as $j \in \mathcal{N}$ $(0 \leq j < n)$ and $\mathcal{B}[j]$ means the tuple for role *j*, then $L[j] = g.e.\mathcal{B}[j].q.l$. The role range vector is denoted as $L[j] \in \mathcal{N}$.

Definition 7: An *upper role range vector U* [18, 21, 22] is a vector of the *upper* bound of the ranges of roles in environment *e* of group *g*. Suppose that roles in *g.e* are numbered as $j \in \mathcal{N}$ $(0 \leq j < n)$ and $\mathcal{B}[j]$ means the tuple for role *j*, then $U[j] = g.e.\mathcal{B}[j].q.u$. The role range vector is denoted as $U[j] \in \mathcal{N}$.

Note: *L* and *U* are valuable components in the E-CARGO model. They reveal many challenges in role assignment and the process of RBC. Without these components, we would not discover the problems discussed in this paper.

Definition 8: A *qualification matrix Q* [18, 21, 22] is an $m \times n$ matrix, where $Q[i, j] \in [0,1]$ expresses the qualification value of agent $i \in \mathcal{N}$ $(0 \leq i < m)$ for role $j \in \mathcal{N}$ $(0 \leq j < n)$. $Q[i, j] = 0$ means the lowest and 1 the highest.

Note that, a Q matrix can be obtained by comparing all the Ⓠs of agents with all the Ⓢs of roles. This process is called agent evaluation [21, 22]. In this paper, we suppose that Q is obtained through conventional quantitative evaluation methods [18–22]. For example, a qualification value can be considered individual performance in software development, i.e., if $Q[i, j] = 0.87$, we can translate it into that agent i plays role j with the production rate of $0.87*10 = 8.7$ Lines of Code (LOC) per hour.

$$
\begin{bmatrix}
0.18 & 0.82 & 0.29 & 0.01 \\
0.35 & 0.80 & 0.58 & 0.35 \\
0.84 & 0.85 & 0.86 & 0.36 \\
0.96 & 0.51 & 0.45 & 0.64 \\
0.22 & 0.33 & 0.68 & 0.33 \\
0.96 & 0.50 & 0.10 & 0.73 \\
0.25 & 0.18 & 0.23 & 0.39 \\
0.56 & 0.35 & 0.80 & 0.62 \\
0.49 & 0.09 & 0.33 & 0.58 \\
0.38 & 0.54 & 0.72 & 0.20 \\
0.91 & 0.31 & 0.34 & 0.15 \\
0.85 & 0.34 & 0.43 & 0.18 \\
0.44 & 0.06 & 0.66 & 0.37
\end{bmatrix}
\quad
\begin{bmatrix}
0 & 1 & 0 & 0 \\
0 & 1 & 0 & 0 \\
0 & 0 & 1 & 0 \\
0 & 0 & 0 & 1 \\
0 & 0 & 1 & 0 \\
0 & 0 & 0 & 1 \\
0 & 0 & 0 & 0 \\
0 & 0 & 1 & 0 \\
0 & 0 & 0 & 0 \\
0 & 0 & 1 & 0 \\
1 & 0 & 0 & 0 \\
0 & 0 & 0 & 0 \\
0 & 0 & 0 & 0
\end{bmatrix}
$$

(a)   (b)

Fig. 1. A qualification matrix Q and an assignment matrix T.

Definition 9: A *role assignment matrix* T [18–22] is defined as an $m \times n$ matrix, where $T[i, j] \in \{0,1\}$ ($0 \leq i < m, 0 \leq j < n$) expresses if agent i is assigned to role j (i.e., $<ai, \eta> \in g.\mathcal{J}$) or not (i.e., $<ai, \eta> \notin g.\mathcal{J}$). $T[i, j] = 1$ means yes and 0 no.

Definition 10: *The group performance σ of group g* [18, 21, 22] is defined as the sum of the assigned agents' qualifications, i.e.,

$$
\sigma = \sum_{i=0}^{m-1} \sum_{j=0}^{n-1} Q[i,j] \times T[i,j]
$$

Note that, *group* is an important concept of the E-CARGO model. Challenges related to role assignment are discovered by investigating this concept. For example, with e, we establish the concept of L; with $\mathcal{J}$ of an e, we discover the role assignment problem by introducing a new concept Q; with L and Q, we discover complex role assignment problems by introducing constraints to forming $\mathcal{J}$. *The proposed approach in this paper is discovered from role assignment with considering the coordination factors in forming $\mathcal{J}$.*

Definition 11: Role j is *workable* [18, 21, 22] in group g if it is assigned with enough agents, i.e., $\sum_{i=0}^{m-1} T[i,j] \geq L[j]$.

Definition 12: T is *workable* [18, 21, 22] if each role j is workable, i.e., $\forall (0 \leq j < n) \sum_{i=0}^{m-1} T[i,j] \geq L[j]$. Group g is *workable* if T is workable.

From the above definitions, group g can be expressed by a Q, an L, and a T. In the following discussions, we assume that $m \geq \sum_{i=0}^{n-1} L[j]$ if we do not clearly state special cases. For example, Fig. 1(a) is a qualification matrix for Table 2. Figure 1(b) is an assignment matrix that makes the group (Table 2) work with vector $L = [1, 2, 4, 2]$ in Table 1. The sum of the assigned values is 6.96.

Definition 13: Given Q, and L, the *Group Role Assignment (GRA)* problem [18, 21, 22] is to find a matrix T to

$$\sigma_1 = \max\{ \sum_{i=0}^{m-1} \sum_{j=0}^{n-1} Q[i,j] \times T[i,j] \}$$

subject to

$$T[i,j] \in \{0,1\} \qquad (0 \leq i < m, 0 \leq j < n) \tag{1}$$

$$\sum_{i=0}^{m-1} T[i,j] = L[j] \quad (0 \leq j < n) \tag{2}$$

$$\sum_{i=0}^{n-1} T[i,j] \leq 1 \quad (0 \leq i < m) \tag{3}$$

Where, constraint (1) tells that an agent can only be assigned or not, (2) makes the group workable, and (3) means that each agent can only be assigned to one role.

We use T^* to express the T that satisfies Definition 13.

Note that, GRA seems a general assignment problem, but it is a different one because of constraint (2). Our previous work provides an efficient solution to GRA by adapting the Kuhn-Munkres algorithm [11, 14] that is used to solve general assignment problems [11].

4 Formalization of the Problem

In fact, Definitions 9–13 relate to the state of a group after role assignment that is highly dependent on various conditions. Thus, we need to specify some constraints when we conduct the assignment.

In fact, the actual working efficiency of a person is different from the individual performance (or qualification) due to many constraints in a team. Here, we borrow the analysis method of Amdahl's law that is used in analyzing the speedup of parallel processors.

Amdahl's law [2, 8] is stated as $S_P = \frac{P}{P \times f_s + f_p}$ (Fig. 2), where, P is the number of processors, S_P is the speedup of P processors, f_s is the serial fraction of code, and f_p the parallel fraction of code, and $f_s + f_p = 100\,\%$.

Definition 14: The performance of role j is the sum of all the qualifications of agents assigned to j, i.e., $\sum\limits_{i=0}^{m-1} Q[i,j] \times T[i,j]$ $(0 \leq j < n)$.

Definition 15: The vector of parallelism of roles P_r is an n-dimensional vector, where $P_r[j] \in [0, 1]$ $(0 \leq j < n)$ expresses the parallelism of role j, i.e., the portion that all the assigned agents can work in parallel without coordination. At the same time, $1 - P_r[j] \in [0, 1]$ $(0 \leq j < n)$ is the part of role j that needs all the assigned agents to coordinate.

For example, we may set $P_r = [0.43, 0.35, 0.86, 0.57]$ for problem in Fig. 1.

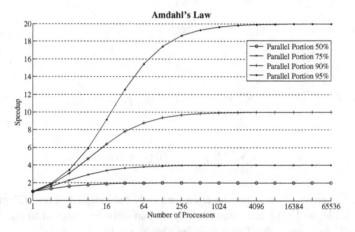

Fig. 2. Amdahl's Law.

Definition 16: The vector of coordination of roles C_r is an n-dimensional vector, where $C_r[j] \in [0, 1]$ $(0 \leq j < n)$ expresses the changing rate of an agent on role j.

For example, we may set $C_r = [0.6, 0.5, 0.6, 0.8]$ for the problem in Fig. 1. Because $Q[1, 2] = 0.8$, the real performance under coordination is $Q[1, 2] \times C_r[2] = 0.4$.

Definition 17: The adjusted performance of role j is the sum of all the qualifications of agents assigned to role j, i.e.,

$$(1 - P_r[j]) \times (C_r[j])^{L[j]-1} \times \sum_{i=0}^{m-1} Q[i,j] \times T[i,j]$$

$$+ P_r[j] \times \sum_{i=0}^{m-1} Q[i,j] \times T[i,j] \qquad (0 \le j < n).$$

Note that, Definition 17 "$L[j] - 1$" expresses that the coordination factors are effective when $L[j] > 1$. This definition supposes that the more agents are added, the less is the increase of performance.

Definition 18: Given m, n, Q, L, P_r, and T^*, the actual *group performance* is

$$\sigma_2 = \sum_{j=0}^{n-1} \begin{array}{l} ((1 - P_r[j]) \times (C_r[j])^{L[j]-1} \times \sum_{i=0}^{m-1} Q[i,j] \times T^*[i,j] + \\ P_r[j] \times \sum_{i=0}^{m-1} Q[i,j] \times T^*[i,j]) \end{array} \qquad (4)$$

subject to (1), (2), and (3).

From Definition 18, we assume that the coordination factors affect the performance during the non-parallel part of performing a role. Evidently, $\sigma_2 \le \sigma_1$.

Definition 19: The completion rate μ is the percentage of the overall time of the project. $\mu \in [0, 1]$ tells how much of the total time has passed.

Definition 20: The tuple $<m, n, Q, L, P_r, C_r, \mu, T>$ is called the state of group g, denoted by Π.

Definition 21: Suppose that the current state of group g organized with GRA is $\Pi = <m, n, Q, L, P_r, C_r, \mu, T>$. The **Problem of Brooks' law** is that given m', Q', find a new state $\Pi' = <m', n, Q', L', P_r, C_r, \mu, T'>$, where $m' > m$, Q' is an $m' \times n$ matrix ($Q = Q'[0:m-1, 0:n-1]$), find a new $m' \times n$ assignment matrix T' and a new vector L', to make

$$\sigma_3 \times (1-\mu) + \sigma_2 \times \mu > \sigma_1, \text{ where}$$

$$\sigma_3 = \max\{ \sum_{j=0}^{n-1} \begin{array}{l} ((1 - P_r[j]) \times (C_r[j])^{L'[j]-1} \times \sum_{i=0}^{m'-1} Q'[i,j] \times T'[i,j] + \\ P_r[j] \times \sum_{i=0}^{m'-1} Q'[i,j] \times T'[i,j]) \end{array} \} \qquad (5)$$

subject to:

$$T'[i,j] \in \{0,1\} \qquad (0 \le i < m', 0 \le j < n) \qquad (6)$$

$$L'[j] \in \{1, 2, \ldots, m'\} \quad (0 \le j < n) \tag{7}$$

$$\sum_{i=0}^{m'-1} T'[i,j] = L'[j] \quad (0 \le j < n) \tag{8}$$

$$\sum_{i=0}^{n-1} T'[i,j] \le 1 \; (0 \le i < m') \tag{9}$$

$$\sum_{j=0}^{n-1} L'[j] \le m' \tag{10}$$

$$L[j] \le L'[j] \le U[j] \; (0 \le j < n) \tag{11}$$

Evidently, the problem of Brooks' law is a Non-Linear Programming (NLP) problem, and the complexity of this problem is very high. From our previous work, we can obtain σ_1 and σ_2 directly from the GRA algorithm. However, the GRA algorithm cannot be directly used to get σ_3, because $L'[j]$ $(0 \le j < n)$ is a variable, and new constraints (7), (10), and (11) are added. We at first hope to use the IBM ILOG CPLEX Optimization package [10] to solve this problem but fail, because it does not support the solutions to NLP problems.

5 A Solution Based on GRA

5.1 The Solution

If L' is determined, the problem of Brooks' law defined in Definition 21 may become a GRA problem [22] if L' satisfies (10). We may use the enumeration of Ls to obtain σ_3 based on the GRA solution. The number of L's should be $\prod_{j=0}^{n-1} (U[j] - L[j] + 1)$. If $U[j] - L[j] + 1$ is a constant, i.e., $k > 1$, then the complexity is k^n. This tells the complexity of the **Problem of Brooks' law**.

Our solution is to use the exhaustive search method to enumerate all the possible L's that satisfy the constraints in the problem; and then call the GRA algorithm [18, 21, 22] with each L' to get a T and the corresponding value of σ_1, and finally get the maximum σ_1 to obtain σ_3. Note that, if there were no the GRA algorithm within a polynomial time, the algorithm would be much more complex.

The algorithm can be described in a Java-like format as follows, where "$a := b$" means that a is assigned with the value of b:

Input:
Q– the current m×n qualification matrix
L – the n-dimensional lower range vector;
U – the n-dimensional upper range vector;
P_r – the n-dimensional upper range vector;
C_r– the n-dimensional upper range vector; and
Q' – the new m'×n qualification matrix.

Output:
T' - the new assignment matrix;
L' – the new lower range vector; and
σ_3 - the group performance with L'.

Algorithm BrooksLaw() {
    LL:=All the possible L's;
    k:=the length of LL;
    $\sigma_3:=0$; *T':={0}; L':={0};*
    for *(0≤i≤k){*

    if $(\sum_{j=0}^{n-1} LL[i][j] \leq m')$ {

    $Q''[i, j] := ((1 - P_r[j]) \times (C_r[j])^{LL[i][j]} + P_r[j]) \times Q'[i, j]$;

    v:=**GRA**$(m', n, LL[i], Q'', T)$;
    if $(v > \sigma_3)$ {
       $\sigma_3:=v$; *T':=T; L':= LL[i];*
    }
    }
    }
    return σ_3;*//Note: the results are also in L' and T'.*
}

5.2   Simulations

Because the proposed problem is related to many parameters, i.e., m, m', n, Q, P_r, C_r, μ, L, and U, there are many variations. The solution in 5.1 can be applied in different ways. In this paper, we concentrate on the effects of C_r on μ.

We set $m = 13$, $m' = 20$, $n = 5$, $P_r = [0.8\ 0.6\ 0.7\ 0.6]$, $L = [2\ 3\ 5\ 4\ 3]$, and $U = [2\ 3\ 8\ 7]$ as determined numbers.

Then, we randomly create 100 tuples of <Q, Q'>. In each case of Qs, we randomly create a C_r, then we determine the largest μ that satisfies $\sigma_3 \times (1-\mu) + \sigma_2 \times \mu \geq \sigma_1$.

To create a C_r, we investigate the following cases: (1) $0 \leq C_r[j] < 0.2$; (2) $0.2 \leq C_r[j] < 0.4$;   (3) $0.4 \leq C_r[j] < 0.6$;   (4) $0.6 \leq C_r[j] < 0.8$;   and (5) $0.8 \leq C_r[j] \leq 1.0$.

Table 1. shows the relationships between the ranges of C_rs and μs. This simulation reveals the situation for an agile team ($m = 13$, $m' = 20$) with positions that possess good parallelisms ($P_r[j] = 0.6$ to 0.8, $0 \leq j < 5$). We note an exception of the correlation between μs and C_rs, i.e., the 2nd row and the 3rd row of Table 1. This shows that the team performance is related to many factors. It is reasonable because the Q matrices are

Table 1. Confirmed relationships by simulation between the ranges of C_rs and μs ($m = 13$, $m' = 20$, $n = 5$, $P_r = [0.8\ 0.6\ 0.7\ 0.6]$, $L = [1\ 2\ 4\ 2]$, and $U = [2\ 3\ 8\ 7]$).

C_r	μ
[0.0, 0.2)	[0.38, 0.60]
[0.2, 0.4)	[0.34, 0.61]
[0.4, 0.6)	[0.45, 0.64]
[0.6, 0.8)	[0.58, 0.77]
[0.8, 1.0)	[0.75, 0.96]

randomly created. It also shows that the significance of the proposed approach in engineering applications, i.e., every case may be different. Figure 3 shows the scattered μs for each 100 cases for $C_r \in [0.4, 0.6)$.

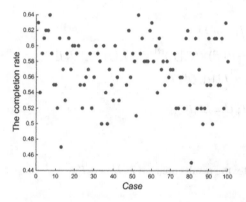

Fig. 3. Simulation of random 100 cases ($0.4 \leq C_r[j] < 0.6$).

5.3 Performances

To check the practicability of the proposed method, we conduct some experiments on the time used to solve a question.

We set $m = 10, 20, 30$, $m' = m + 4$, the incremental step of 3 to ≤ 50, $n = 3, 5$, ..., $\leq m$. Under each tuple $<m, m', n>$, we use random 100 tuples of $<Q, Q', L, U, P_r, C_r>$ to compute μs, we collect the maximum, minimum, and average time used in each tuple $<m, m', n>$. To make the case meaningful, we create $L[j] \leq m/n$, and $L[j] \leq U[j] \leq m'/n$.

In the experiments, we use the platform shown in Table 2. To save space, we only show the maximum and average times used in one experiment in Figs. 4 and 5. From the figures, we find that the proposed approach is practical to relatively large agile teams (m is up to 50). Also from the figures, we do notice the fluctuations of the meshes. The drops and fluctuations are reasonable because (1) the number of available Ls drops significantly when n becomes big enough, and (2) some specific groups can be solved quickly by the GRA algorithm [20].

Table 2. Configurations of the test platform

Hardware	
CPU	Intel core i7-4650 M @1.7 and 2.5 GHz
MEM	8 GB
HDD	WD WD3200BEKT @7200 rpm
Software	
OS	Windows 7 Enterprise
Eclipse Standard/SDK	4.4.0.20140612-0500
Java	JRE7

Maximum times used in different cases

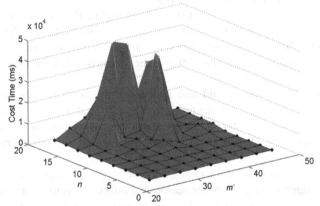

Fig. 4. Maximum times used in different cases ($m = 20$, $m' = m+4$, …, $m + 30$, $n = 3, 5,…, 13$).

Average times used in different cases

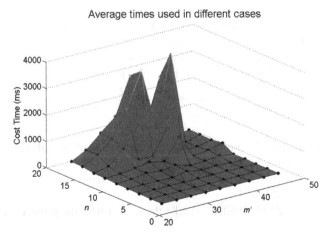

Fig. 5. Average times used in different cases ($m = 20$, $m' = m+4$, …, $m + 30$, $n = 3, 5,…, 13$).

6 Case Study

Let us conduct a case study with a demonstrative scenario. In company Z, five months ago, Ann, the Chief Executive Officer (CEO), signed a 10-month contract the value of which is a million dollars. This project was estimated as 67 Thousands (Kilo) Lines of Code (KLOC). She has been recently reported that it seems likely that the project could not be delivered by the planned date. She is an experienced CEO in the software industry and well acquainted with Brooks' law. She also knows that Brooks' law does not fit all situations. She tries to determine if her team can break the law. Therefore, she asks Bob, the human resource officer, to investigate the coordination issues in the current team and possible people who may join the team. The final aim is to assure if they can add other people to the team to make the due date or inform the client that they need to extend the delivery date of the project.

Bob participated in the team formation by optimizing the team performance (i.e., production with Kilo Lines Of Codes (KLOCs) in a month) with GRA (Definition 13) and thought that the project should be done on time (Table 4). Because the team is organized as an agile team, he could estimate the productivity of the group. The initial estimation is as the following:

The assignment (Table 4, **bold** and underlined numbers) is done by GRA, the maximum of group performance (the productivity) is 6.96, that can be transformed into LOC as 6.96 * 10KLOCs = 69.6KLOC that is 4.4 % more than the estimated workload of the project, i.e., 67 KLOCs. Therefore, it is a good assignment based on this estimation.

Now, Bob needs to analyze the reason why the project is delayed. Bob meets Kris, the project manager, to investigate the reason that affects the actual team performance. Kris suggests a vector of parallelism factors and one of coordination factors (Rows 3 and 4 in Table 3).

Table 3. The required positions

Vectors	Position			
	Project manager	Analyst	Programmer	Tester
L	1	2	4	2
U	2	3	8	8
P_r	0.5	0.6	0.8	0.8
C_r	0.8	0.8	0.7	0.7
L'	1	2	8	6

Now, we can analyze this problem with the proposed method in this paper. Because $\mu = 0.5$, $P_r = [0.5\ 0.6\ 0.8\ 0.8]$ and

$C_r = [0.8\ 0.8\ 0.7\ 0.7]$; we get the following performance data:

$\sigma_1 = 6.96$; $\sigma_2 = 5.29$; and $\sigma_3 = 9.89$.

Note that $\sigma_2 = 5.29$ (52.9 KLOCs) tells the reason why the project seems late.

Now, $\sigma_3 \times (1-\mu) + \sigma_2 \times \mu = 7.13$ when the new assignment is shown in Table 5 as **bold** numbers, where the new vector L' is shown in the 4th row in Table 3, i.e., the total KLOCs to be completed will be 71.3 with the new assignment.

Table 4. The candidates, evaluations, and assignments with GRA

Candidates	Positions			
	Project manager	Senior	Programmer	Tester
Adam	0.18	**0.82**	0.29	0.01
Brett	0.35	**0.80**	0.58	0.35
Chris	0.84	0.85	**0.86**	0.36
Doug	0.96	0.51	0.45	**0.64**
Edward	0.22	0.33	**0.68**	0.33
Fred	0.96	0.50	0.10	**0.73**
George	0.25	0.18	0.23	0.39
Harry	0.56	0.35	**0.80**	0.62
Ice	0.49	0.09	0.33	0.58
Joe	0.38	0.54	**0.72**	0.20
Kris	**0.91**	0.31	0.34	0.15
Larry	0.85	0.34	0.43	0.18
Matt	0.44	0.06	0.66	0.37

Table 5. The productivity of candidates on positions

Candidate	Positions			
	Project manager	Analyst	Programmer	Tester
Adam	0.18	**0.82**	0.29	0.01
Brett	0.35	**0.80**	0.58	0.35
Chris	0.84	0.85	**0.86**	0.36
Doug	0.96	0.51	0.45	**0.64**
Edward	0.22	0.33	**0.68**	0.33
Fred	0.96	0.50	0.10	**0.73**
George	0.25	0.18	0.23	**0.39**
Harry	0.56	0.35	**0.80**	0.62
Ice	0.49	0.09	0.33	**0.58**
Joe	0.38	0.54	**0.72**	0.20
Kris	**0.91**	0.31	0.34	0.15
Larry	0.85	0.34	**0.43**	0.18
Matt	0.44	0.06	**0.66**	0.37
Nancy	0.25	0.67	0.56	**0.75**
Peter	0.20	0.78	**0.86**	0.78
Scott	0.46	0.85	0.79	**0.90**
Ted	0.45	0.78	**0.68**	0.64

This result in fact breaks Brooks' law. It means that under some situations such as this case study, the late project can be made to meet the deadline.

Please note, Ann is lucky to perceive that the project might be late before the start of the 6th month. If she notices the same situation in the beginning of the 9th month, there will not be any luck for Ann to accomplish this project on time. If $\mu = 0.8$, then $\sigma_3 \times (1-\mu) + \sigma_2 \times \mu = 6.20$ (62KLOCS) means that the team cannot make the deadline.

The above case study demonstrates a typical software engineering practice that "It helps to manage the project if we could *find the problems as early as possible*."

7   Conclusions

Brooks' law is popular in software development. Even though it is stated that "adding programmers to a late project makes it even later", there are exceptions. It is required to develop a way to know if there is a solution for a late project and if it there is no solution, why is this the case? In this paper, we contribute an innovative quantitative way to check if there is solution to a late project. The proposed approach is verified by simulations, experiments, and a case study.

The simulation presents a clear conclusion: *there is a critical time in a project that follows the Brooks' law, i.e., μ is directly related with the coordination factors of each position of a team. The higher the coordination rate is, the larger the μ is*.

Another contribution of this paper is a practical way for agile teams to compute the latest time to adjust the team staffing for a determined project to guarantee its completion before the deadline.

Following the model and methods proposed in this paper, we may conduct further studies along the following directions.

- A better way may express more pertinently Brooks' law. In this paper, the formalized team performance is still increased when a new agent is added to the team, even though the increase becomes smaller while more agents are added.
- A more efficient way can be found to solve the addressed problem in this paper.
- An integrated software tool can be developed to help administrators control the state of a team.

Acknowledgment. Thanks go to Mike Brewes of Nipissing University for his assistance in proofreading this article.

References

1. Abdel-Hamid, T.K.: The dynamics of software project staffing: a system dynamics based simulation approach. IEEE Trans. Softw. Eng. **15**(2), 109–119 (1989)
2. Amdahl, G.M.: Validity of the single processor approach to achieving large-scale computing capabilities. In: Proceedings of the American Federation of Information Processing Societies (AFIPS), April 1967, pp. 483–485. Spring Joint Computer Conference, New Jersey (1967)

3. Berkun, S.: Exceptions to Brooks' Law. http://scottberkun.com/2006/exceptions-to-brooks-law/. Accessed 10 Jan 2016
4. Blackburn, J., Lapre, M.A., van Wassenhove, L.N.: Brooks' Law Revisited: Improving Software Productivity by Managing Complexity, May 2006. http://ssrn.com/abstract=922768 or http://dx.doi.org/10.2139/ssrn.922768
5. Brooks Jr., F.P.: The Mythical Man-Month, Anniversary Edition: Essays On Software Engineering. Addison-Wesley Longman Co., Crawfordsville (1995)
6. Chen, W.-N., Zhang, J.: Ant colony optimization for software project scheduling and staffing with an event-based scheduler. IEEE Trans. Softw. Eng. **39**(1), 1–17 (2013)
7. Gordon, R.L., Lamb, J.C.: A close look at Brooks' law. In: Datamation, pp. 81–86, June 1977
8. Gustafson, J.L.: Reevaluating Amdahl's law. Commun. ACM **31**(5), 532–533 (1988)
9. Hsia, P., Hsu, C., Kung, D.C.: Brooks' law revisited: a system dynamics approach. In: Proceedings of Twenty-Third Annual International Computer Software and Applications Conference, p. 370 (1999)
10. IBM, ILOG CPLEX Optimization Studio (2013). http://www-01.ibm.com/software/integration/optimization/cplex-optimization-studio/
11. Kuhn, H.W.: The Hungarian method for the assignment problem. Naval Res. Logistic Q. **2**, 83–97 (1955). (Reprinted in **52**(1): 7–21 (2005))
12. McCain, K.W., Salvucci, L.J.: How influential is Brooks' law? A longitudinal citation context analysis of Frederick Brooks' the Mythical Man-Month. J. Inf. Sci. **32**(3), 277–295 (2006)
13. McConnell, S.: Brooks' law repealed. IEEE Softw. **16**(6), 6–8 (1999)
14. Munkres, J.: Algorithms for the assignment and transportation problems. J. Soc. Ind. Appl. Math. **5**(1), 32–38 (1957)
15. Pressman, R., Maxim, B.: Software Engineering: A Practitioner's Approach, 8th edn. McGraw-Hill Education, Columbus (2014)
16. Schweik, C.M., English, R., Kitsing, M., Haire, S.: Brooks' versus Linus' law: an empirical test of open source projects. In: Proceeding of the International Conference on Digital Government Research, Montreal, Canada, pp. 423–424, 18–21, May 2008
17. Weinberg, G.M.: Quality Software Management: volume 1, System Thinking. Dorset House Publishing, New York (1992)
18. Zhu, H.: Avoiding conflicts by group role assignment. IEEE Trans. Syst. Man Cybern.: Syst. **46**(4), 535–547 (2016)
19. Zhu, H., Zhou, M.C.: Efficient role transfer based on Kuhn-Munkres algorithm. IEEE Trans. Syst. Man Cybern. Part A: Syst. Hum. **42**(2), 491–496 (2012)
20. Zhu, H., Zhou, M.: M-M role-transfer problems and their solutions. IEEE Trans. Syst. Man Cybern. Part A: Syst. Hum. **39**(2), 448–459 (2009)
21. Zhu, H., Zhou, M.C.: Role-Based collaboration and its kernel mechanisms. IEEE Trans. Syst. Man Cybern. Part C **36**(4), 578–589 (2006)
22. Zhu, H., Zhou, M.C., Alkins, R.: Group role assignment via a Kuhn-Munkres algorithm-based solution. IEEE Trans. Syst. Man Cybern. Part A **42**(3), 739–750 (2012)

An Experience of Constructing a Service API for Corporate Data Delivery

Itamir de Morais Barroca Filho[1]([✉]), Mario Melo[2], Cicero Alves Silva[2],
Gibeon Soares de Aquino Jr.[2], Vinicius Campos[1], and Viviane Costa[2]

[1] Metropole Digital Institute, Natal, Brazil
itamir.filho@imd.ufrn.br, viniciuscampos120@gmail.com
[2] Department of Informatics and Applied Mathematics,
Federal University of Rio Grande do Norte Natal, Natal, Brazil
mariovmelo@gmail.com, cicerojpm@gmail.com, gibeon@dimap.ufrn.br,
vivianecosta2794@gmail.com
http://www.imd.ufrn.br
http://www.dimap.ufrn.br

Abstract. The Federal University of Rio Grande do Norte (UFRN) has information systems which capture different kinds of data related to the institution itself, as well as its academics and students. Much of this data produces information that is available to users through reports found in the systems. However, there are demands that often come from these users, who wish to obtain data not available in the reports, which requires the need for manual extraction and, thereafter, generates costs to the institution. Furthermore, there are several studies, research projects and applications developed by academics, students and companies that require the use of this information, either to offer more transparency or to develop new ideas and promoting innovation. In this aspect, this paper aims to present an experience of development of a solution for providing corporate data using a service API.

Keywords: Service API · Corporate systems · Corporate data · Innovation

1 Introduction

The Federal University of Rio Grande do Norte (UFRN) has three corporate systems called: Integrated System for Property, Administration and Contracts (SIPAC), Integrated System for Management of Academic Activities (SIGAA) and Integrated System for Management of Human Resources (SIGRH). Figure 1 shows an overview of the relationship between these systems and UFRN, other government systems and other institutional systems.

The SIPAC provides operations for management of the units that are responsible for UFRN's finance, property and contracts, computerizing both requests - equipment, service providers, daily wages, tickets, lodging, books, infrastructure

O. Gervasi et al. (Eds.): ICCSA 2016, Part V, LNCS 9790, pp. 218–227, 2016.
DOI: 10.1007/978-3-319-42092-9_17

maintenance - and control of the institution's budget. In addition, it manages purchases, public tenders, inventory, contracts, conventions, public works and repairs, campus maintenance, invoices, educational grants and educational grants' payments, legal procedures, among other features. As for SIGRH, it provides procedures for management of human resources, such as: vacations, retirements, work force scaling, attendance, among others. The SIGAA computerizes academic procedures by using the following modules: undergraduate, postgraduate (strict and lato sensu), technical school, high school and kindergarten, research, extension and a virtual learning environment called Virtual Class [9].

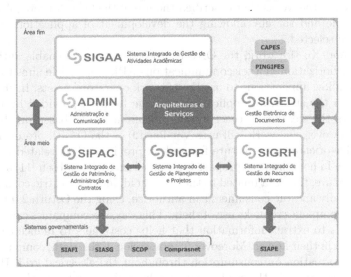

Fig. 1. UFRN's management integrated systems [15].

The corporate data obtained by these systems is organized in five databases that have, altogether, 3,824 tables arranged in 115 schemes. Much of this data produces information that is available to users through reports that can be found in these systems. However, there are demands that often come from these users who wish to obtain data that is not available in the reports, which requires the need for manual extraction and, thereafter, generates costs to the institution. Furthermore, there are several studies, research projects and applications developed by teachers, students and companies that require the use of this information, either to offer more transparency or to develop new ideas, promoting innovation. In this aspect, this a common problem for companies or institutions that manage large amounts of corporate data and have to provide it to its users.

Thus, this paper aims to present an experience of development of a solution for providing corporate data using a service API. The paper is structured as follows: in Sect. 2, related works will be presented; Sect. 3 describes the architecture of the solution developed to make the data available; Sect. 4 presents UFRN's service API and its use cases; and, finally, in Sect. 5 the conclusions and future works are described.

2   Related Works

Due to the companies' needs to provide their data to be used in the development of applications by third parties, many works using the Open Authorization Protocol (OAuth) have been developed in order to ensure users' authenticity, as well as permission for applications that consume the users' data obtained from these APIs [1]. OAuth is an open authorization protocol that aims to standardize the way applications access an API to use the users' information [8].

The related works shown below were taken from the IEEE Digital Library database. Only the works that described the use of the OAuth protocol to make data available and thereby allowing the development of applications by third parties were selected.

In Liu and Xu [10] work, the OAuth protocol is used to enable authentication and authorization in a telecommunications API. The article aims to use the solution to allow applications to use the network operators' assets. In the example shown in the study, an application uses the location provided by the API and displays the weather forecast for the upcoming weeks based on the user's location. However, the work still uses Version 1.0 of the OAuth protocol.

A service-oriented architecture focused on providing data related to energy consumption in households is proposed in Mikkelsen and Jacobsen [11] article. In this architecture, the data related to the household devices' electric consumption is stored locally and the consumers can authorize, using the OAuth 2.0 protocol, the electric companies to make use of them. This way, the companies can use their analysis tools to extract information that helps costumers to reduce electricity consumption in their homes. Moreover, in possession of this data, companies have real-time information on consumption throughout the electrical grid. However, the solution proposed in the study focuses only on data related to the electricity sector, while this article focuses on the provision of corporate data.

The OAuth-based Authorization Service Architecture (IoT-OAS) is described in Cirani et al. [2] paper. It also uses the OAuth protocol and focuses on providing authorization for Internet of Things services, aiming to reduce the developers' efforts when implementing applications. Thus, this work differs from the one described in this article when it comes to the solution's application scenario. In this paper, we seek to provide an API that allows access to corporate data, while IoT-OAS provides authorization for Internet of Things services.

Gouveia et al. [7] describes a solution to access, in one single place, a person's private data and files in multiple cloud storage services. In the authors' approach, the user authenticates using the Portuguese Citizen Card and the authorization process related to the storage services is done through the OAuth 2.0 protocol. The approach outlined in this article was implemented in a web application, which enables the user to connect to Dropbox, Cloudpt and Skydrive. In the next section, the solution's architecture created to make UFRNs corporate data available will be presented and discussed.

3   Solution's Architecture

Most institutions or companies do not have automated ways to make their data available to users, even though this is an obligation for them. Usually, when users finally have access to these information, they are outdated or incomplete. In order to make corporate data available, some institutions have portals where data is made available to users. However, this type of solution limits the use of the information, as these portals usually produce dynamic content with formats like PDF, making it difficult to create applications that use such information.

In UFRN, in order to enable users to access the institution's information, it is necessary to use this type of portal present in its corporate systems. Currently, the three corporate information systems share the same infrastructure, whose architecture is shown in Fig. 2, as the existing infrastructure. This architecture has a load balancer (LB) [13] whose main function is to distribute the requests between the application servers (AS) according to their capabilities. Each application server provides a system (SIPAC, SIGAA and SIGRH), which is connected to a set of database (BD) servers. This architecture serves about 50,000 users, seven days a week, 24 h a day.

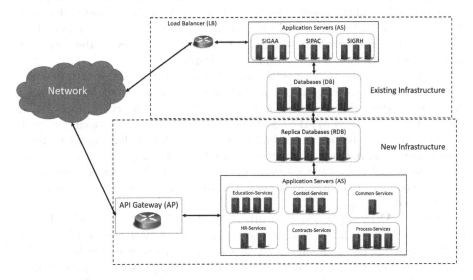

Fig. 2. Architecture of the service API to be used to make UFRN's corporate data available.

In order to solve the issue of making corporate data available in "real time", it is possible to use a very common strategy, which is to build a web services layer in each of the existing systems [4]. However, this approach could bring a serious risk to the systems, because if the number of users using them and the number of requests asking the web services for information is too high, it could overload all of the systems, making them unavailable. Given this problem and aiming to create an ecosystem of applications around such information, it is necessary to design an architecture that meets some basic assumptions, such as: availability, security and scalability. It is also important that the availability of this information does not interfere in the operation of the corporate systems.

Thus, in Fig. 2, the architecture of the Service API to be used to make the corporate data available is shown. This architecture is service-oriented and follows the design principles defined by [5]. In addition, a microservice architecture was used aiming to make the provided services have high availability and be independent from each other. The microservice architecture is intended to replace monolithic systems, providing many services through a distributed system with light, independent services that have a specific purpose [17].

To ensure that the assumptions that were defined and mentioned earlier in this paper are met, the architecture was designed so that its execution occurs in a separate branch of the corporate systems, since the two architectures have a single point of failure: the database servers. If all of the corporate systems and services used the same database servers, there could be a collapse of the entire infrastructure depending on the demand the services were submitted to. To mitigate this risk, we used the strategy of the replication of data from these servers, creating a replicated database environment in which synchronization happens asynchronously and the information is made available in the replicated database in less than 1 s, that is, practically in real time [12]. Thus, it is guaranteed that making service data available will not affect the performance of the corporate systems and, therefore, a higher availability is guaranteed.

For service scalability, we used the standard microservice called Gateway API [14], whose goal is to set a single point of entry and which is the only one that knows the servers that contain the services. This way, the client applications need to know the Gateway API's address, and the Gateway API itself is responsible for carrying out routing, as well as performing load balancing for multiple service instances. Another guarantee provided by this component is security, in which the OAuth 2.0 authorization protocol is used to allow service usage. This protocol enables secure authorization in third-party applications in a simple and standardized way [8].

Once the API's physical architecture is detailed, Fig. 3 shows the API's logical operation and how client applications can use the services provided by it. All of the client applications that wish to use the services available in the API must be registered in the Authorizer. This service is responsible for providing all of the authorization mechanisms to the services provided by the API to client applications. Once registered, the client application may request an authorization token, as shown in step 1. To make this request, it is required that the client application sends its credentials. Once the request is received in the Gateway API, it redirects the request to the Authorizer, as shown in step 2. If all of the information submitted is valid, the Authorizer generates an authorization key called "Authorization Token" and sends it to the client application, as shown in steps 3 and 4.

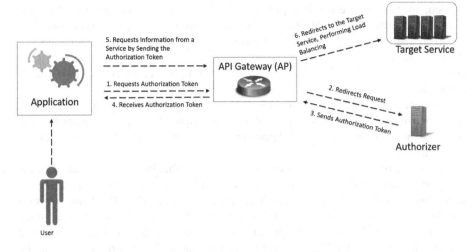

Fig. 3. Services API's logical operation.

In possession of the Token, the client application has permission to access the services that are its credentials allow it to access. Thus, for all of the requests made by the client application to the service resources, the Authorization Token should be informed in the HTTP packet header, as shown in step 5. The request that was sent in the previous step is received in the Gateway API, which validates the sent Token. If the Token is valid and is authorized to access the requested service, the request is redirected to the server that will process it. Step 6 illustrates this behavior. So, after the design of the solution's architecture, the UFRN's service API was developed and will be presented in the next section.

4   UFRN's Service API

UFRN's service API[1] aims to provide an unified, standardized and available way to access the institution's corporate data, enabling the academic community and the society to have access to this information in a simple way. The main contributions to be achieved with this API are:

- Make the largest amount of institutional information available;
- Allow the creation of third-party applications using UFRN's institutional data;
- Encourage the development of solutions that help people's lives when it comes to the university environment.

Once the general goals of the API are defined, the Representational State Transfer (REST) architectural style was chosen to be used for the

---

[1] https://api.ufrn.br.

implementation of the services provided by it. SNAC [16] explains that REST is an architectural principle rather than a standard or a protocol, separating the responsibility of the implementation's details control and focusing on the simplification   of the operations and the use of the HTTP's GET, PUT, POST and DELETE methods.

In addition to this principle, the proposed solution uses the RESTful Web Services concept which, due to the design principles for service-oriented architectures, fits perfectly in the architectural pattern established for the provision of services [5,6].

The data provided by the services is made available in JSON format [3]. This is an open, light and text-based format defined for the exchange of understandable information between humans. This way, we ensure that the traffic of data will be as small as possible and that it is fully independent from the technology used in client applications. The present services make available both public and private data from UFRN's corporate systems' users. To ensure security of the data provided by the API, we use the OAuth in its 2.0 version as the authorization protocol.

4.1 Documentation

The documentation of the services available in UFRN's service API are being made available in the Restful API framework called Swagger [18]. The Swagger was used to provide a language-independent standard, providing information from REST APIs in a clear and simple way. Information such as type of request, returned data, URL and parameters used to access functionalities are presented in the Swagger interface, as shown in Fig. 4 below. A set of services is already developed and open to applications registered for consumption. The services that are currently available are described in at the UFRN's API site.

4.2 Applications

There are currently a few applications that use the data provided by UFRN's service API to perform their functionalities. They are:

- SIGAA Mobile: It is SIGAA's mobile system and has the main functionalities found in the web version, such as: choice of bond that the user has with the institution, preview of the classes associated to the chosen bond and preview of information related to a certain class;
- RUUFRN: It is a mobile application that uses the API to obtain data for the university cafeteria;
- Multiprova: It is a system whose main functionality is the automatic creation of tests. It uses information from classes, students and teachers taken from the service API.
- Inova: It is a enterprise system to manage startups. It consumes data from payments registered in SIPAC.

Ensino-services

Created by suporte@info.ufrn.br

Atividades   Show/Hide   List Operations   Expand Operations

| GET | /consulta/atividade/avaliacao/usuario/{idTurma} | Retorna a lista de avaliações de uma turma. |

Response Class (Status 200)
Model   Model Schema

```
[
  {
    "data": "2015-12-04T17:15:56.134Z",
    "hora": "string",
    "descricao": "string"
  }
]
```

Response Content Type application/json ▾

Parameters

Parameter	Value	Description	Parameter Type	Data Type
idTurma	(required)		path	integer

Try it out!

Fig. 4. Example of documentation of a service that can be found in the API.

Fig. 5. RUUFRN's main screens.

RUUFRN's and SIGAA Mobile main screens are shown in Figs. 5 and 6. The others main screens from this apps can be found at the UFRN's API site. Today, we are receivings lots of request to register new applications use this API, and new services to be created.

Fig. 6. SIGAA mobile main screens.

5   Conclusion and Future Works

This paper presented an experience of development of a solution for providing corporate data using a service API. Today, some corporate data from UFRN's integrated management systems (SIPAC, SIGAA and SIGRH) is available using a service API, that uses the OAuth 2.0 protocol for authentication and authorization of the services provided, which were implemented using the REST architectural style, as shown in Sect. 4.

Nowadays, there are applications that use the data provided by this API, as shown in Sect. 4.2. New developers are also requesting their applications' registration and new services. This API is making it possible to foster innovation by using data from UFRN's integrated management systems. Consequently, UFRN information technology sector will be able to focus on other activities that are not related to demands to make data available to the academic community using spreadsheets.

As future works, we intend to conduct a wide dissemination of this API for teachers, students and companies, as well as offer technical support for the implementation of new applications using the available data. Additionally, it is intended to develop an open data portal using the experience gained from the development, delivery and evaluation of this service API.

Acknowledgment. This work was partially supported by the National Institute of Science and Technology for Software Engineering (INES) funded by CNPq under grant 573964/2008-4.

References

1. Boyd, R.: Getting started with OAuth 2.0. O'Reilly Media Inc., Sebastopol (2012)
2. Cirani, S., Picone, M., Gonizzi, P., Veltri, L., Ferrari, G.: IoT-OAS: an OAuth-based authorization service architecture for secure services in IoT scenarios. Sens. J. IEEE **15**(2), 1224–1234 (2015)
3. Crockford, D.: The application/json media type for JavaScript Object Notation (JSON). Internet RFC 4627, July 2006
4. Erl, T.: Service-Oriented Architecture: A Field Guide to Integrating XML and Web Services. Prentice Hall PTR, Upper Saddle River (2004)
5. Erl, T.: Soa: Principles of Service Design, vol. 1. Prentice Hall, Upper Saddle River (2008)
6. Fielding, R.T.: Architectural styles and the design of network-based software architectures (2000)
7. Gouveia, J., Crocker, P.A., Sousa, S.M., Azevedo, R.: E-Id authentication and uniform access to cloud storage service providers. In: 2013 IEEE 5th International Conference on Cloud Computing Technology and Science (CloudCom), vol. 1, pp. 487–492, December 2013
8. Hardt, D.: The OAuth 2.0 authorization framework (2012)
9. de Aquino Jr., G.S., de Filho, I.M.B., de Neto, M.A.V.M., Barbosa, T.E.: Turma virtual para dispositivos móveis no sistema integrado de gestão de atividades acadêmicas. RENOTE **12**(2)
10. Liu, K., Xu, K.: OAuth based authentication and authorization in open telco API. In: 2012 International Conference on Computer Science and Electronics Engineering (ICCSEE), pp. 176–179, March 2012
11. Mikkelsen, S.A., Jacobsen, R.H.: Consumer-centric and service-oriented architecture for the envisioned energy internet. In: 2015 Euromicro Conference on Digital System Design (DSD), pp. 301–305, August 2015
12. PostgreSQL: high availability, load balancing, and replication (2015). http://www.postgresql.org/docs/current/static/warm-standby.html. Accessed 23 Nov 2015
13. Rahman, M., Iqbal, S., Gao, J.: Load balancer as a service in cloud computing. In: 2014 IEEE 8th International Symposium on Service Oriented System Engineering (SOSE), pp. 204–211, April 2014
14. Richardson, C.: API gateway (2015). http://microservices.io/patterns/apigateway.html. Accessed 23 Nov 2015
15. SINFO: Sistemas institucionais integrados de gesto - sig (2015). https://www.info.ufrn.br/wikisistemas/doku.php/. Accessed 23 Nov 2015
16. SNAC: Guidelines for Implementation of REST. Enterprise Applications Division of the Systems and Network Analysis Center - NSA (2011)
17. Stubbs, J., Moreira, W., Dooley, R.: Distributed systems of microservices using docker and serfnode. In: 2015 7th International Workshop on Science Gateways (IWSG), pp. 34–39, June 2015
18. Swagger: The world's most popular framework for APIs (2015). http://swagger.io/. Accessed 23 Nov 2015

Cat Swarm Optimization with Different Transfer Functions for Solving Set Covering Problems

Broderick Crawford[1,2,3], Ricardo Soto[1,4,5], Natalia Berrios[1(✉)],
Eduardo Olguín[3], and Sanjay Misra[6]

[1] Pontificia Universidad Católica de Valparaíso, 2362807 Valparaíso, Chile
natalia.berrios.p@mail.pucv.cl
[2] Universidad Central de Chile, 8370178 Santiago, Chile
[3] Facultad de Ingeniería y Tecnología, Universidad San Sebastián, Bellavista 7,
8420524 Santiago, Chile
[4] Universidad Autónoma de Chile, 7500138 Santiago, Chile
[5] Universidad Científica Del Sur, Lima 18, Peru
[6] Covenant University, Ogun 110001, Nigeria

Abstract. This work presents a study of a new binary cat swarm optimization. The cat swarm algorithm is a recent swarm metaheuristic technique based on the behaviour of discrete cats. We test the proposed binary cat swarm optimization solving the set covering problem which is a well-known NP-hard discrete optimization problem with many practical applications, such as: political districting, information retrieval, production planning in industry, sector location and fire companies, among others. To tackle the mapping from a continuous search space to a discrete search space we use different transfer functions, S-shaped family and V-shaped family, which are investigated in terms of convergence speed and accuracy of results. The experimental results show the effectiveness of our approach where the binary cat swarm algorithm produce competitive results solving a portfolio of set covering problems from the OR-Library.

Keywords: Binary cat swarm optimization · Set covering problem · Metaheuristic

1 Introduction

The Set Covering Problem (SCP) [9,11,18] is a class of representative combinatorial optimization problem that has been applied to many real world problems, such as: sector location [25,30], fire companies [35], musical composition [28], political districting [17], production planning in industry [32–34] and information retrieval [37], among many others. The SCP is a well-known NP-hard in the strong sense [20].

Exact algorithms are mostly based on branch-and-bound and branch-and-cut [1,3,16]. However, these algorithms are rather time consuming and can only solve instances of very limited size. For this reason, many research efforts have

© Springer International Publishing Switzerland 2016
O. Gervasi et al. (Eds.): ICCSA 2016, Part V, LNCS 9790, pp. 228–240, 2016.
DOI: 10.1007/978-3-319-42092-9_18

been focused on the development of heuristics to find good or near-optimal solutions within a reasonable period of time. The most efficient ones are those proposed in. As top-level general search strategies, metaheuristics such as The Ant Colony Optimization (ACO) [12,14], Artificial Bee Colony (ABC) [9,10, 19], The Shuffled Frog Leaping Algorithm (SFLA) [13,15] and Cuckoo Search Algorithm (CSA) [29,36] have been also successfully applied to solve the SCP.

In this paper we propose a new Binary Cat Swarm Optimization (BCSO) for solving the SCP. Chu and Tsai [6,7] have proposed a new optimization algorithm which imitates the natural behavior of cats. Cats have a strong curiosity towards moving objects and possess good hunting skill. Even though cats spend most of their time in resting, they always remain alert and move very slowly. When the presence of a prey is sensed, they chase it very quickly spending large amount of energy. These two characteristics of resting with slow movement and chasing with high speed are represented by seeking and tracing, respectively. In Cat Swarm Optimization (CSO) [24] these two modes of operations are mathematically modeled for solving complex optimization problems.

In this work we tested eight binarization techniques [22,23,26] with the idea of finding the function that best results are obtained and this replaced by the original Eq. 8. Moreover the discretization technique is changed by the Complement Function, that is explain in the next sections. To perform this test we use 65 problems of the OR-Library of Beasley [3,4]. Finally we choose the best technique through an analysis of the results with the main objective of improve results obtained in previous work [8].

The rest of this paper is organized as follows. In Sect. 2, we give a formal definition of the SCP. The Sect. 3, describes the BCSO. In Sect. 4, describes the binarization techniques that we use. In Sect. 5, we explain how to solve the SCP with BCSO. In Sect. 6, we describe the experimental results obtained when applying the algorithm for solving the 65 instances of SCP contained in the OR-Library. Finally, the conclusions in Sect. 7.

2   Problem Description

The SCP [3,5,21] can be formally defined as follows. Let $A = (a_{ij})$ be an m-row, n-column, zero-one matrix. We say that a column j can cover a row if $a_{ij} = 1$. Each column j is associated with a nonnegative real cost c_j. Let $I=\{1,...,m\}$ and $J=\{1,...,n\}$ be the row set and column set, respectively. The SCP calls for a minimum cost subset $S \subseteq J$, such that each row $i \in I$ is covered by at least one column $j \in S$. A mathematical model for the SCP is

$$v(\text{SCP})= \min \sum_{j \in J} c_j x_j \tag{1}$$

subject to

$$\sum_{j \in J} a_{ij} x_j \geq 1, \quad \forall\, i \in I, \tag{2}$$

$$x_j \in \{0,1\}, \forall\, j \in J \tag{3}$$

The objective is to minimize the sum of the costs of the selected columns, where $x_j = 1$ if column j is in the solution, 0 otherwise. The constraints ensure that each row i is covered by at least one column.

3   Binary Cat Swarm Optimization

Binary Cat Swarm Optimization [27] is an optimization algorithm that imitates the natural behavior of cats [6,31]. Cats have curiosity by objects in motion and have a great hunting ability. It might be thought that cats spend most of the time resting, but in fact they are constantly alert and moving slowly. This behavior corresponds to the seeking mode. Furthermore, when cats detect a prey, they spend lots of energy because of their fast movements. This behavior corresponds to the tracing mode. In BCSO these two behaviors are modeled mathematically to solve complex optimization problems.

In BCSO, the first decision is the number of cats needed for each iteration. Each cat, represented by cat_k, where $k \in [1, C]$, has its own position consisting of M dimensions, which are composed by ones and zeros. Besides, they have speed for each dimension d, a flag for indicating if the cat is on seeking mode or tracing mode and finally a fitness value that is calculated based on the SCP. The BCSO keeps to search the best solution until the end of iterations.

In BCSO the bits of the cat positions are $x_j = 1$ if column j is in the solution, 0 otherwise (Eq. 1). Cat position represents the solution of the SCP and the constraint matrix ensure that each row i is covered by at least one column.

Next is described the BCSO general pseudocode where MR is a percentage that determine the number of cats that undertake the seeking mode.

Algorithm 1. *BCSO*()

1: Create C cats;
2: Initialize the cat positions randomly with values between 1 and 0;
3: Initialize velocities and flag of every cat;
4: Set the cats into seeking mode according to MR, and the others set into tracing mode;
5: Evaluate the cats according to the fitness function;
6: Keep the best cat which has the best fitness value into *bestcat* variable;
7: Move the cats according to their flags, if cat_k is in seeking mode, apply the cat to the seeking mode process, otherwise apply it to the tracing mode process. The process steps are presented above;
8: Re-pick number of cats and set them into tracing mode according to MR, then set the other cats into seeking mode;
9: Check the termination condition, if satisfied, terminate the program, and otherwise repeat since step 5;

3.1 Seeking Mode

This sub-model is used to model the situation of the cat, which is resting, looking around and seeking the next position to move to. Seeking mode has essential factors:

- Probability of Mutation Operation (PMO)
- Counts of Dimensions to Change (CDC), it indicates how many of the dimensions varied
- Seeking Memory Pool (SMP), it is used to define the size of seeking memory for each cat. SMP indicates the points explored by the cat, this parameter can be different for different cats.

The following pseudocode describe cat behavior seeking mode:

Step1: Create SMP copies of cat_k
Step2: Based on CDC update the position of each copy by randomly according to PMO
Step3: Evaluate the fitness of all copies
Step4: Calculate the selecting probability of each copy according to

$$P_i = \left| \frac{FS_i - FS_{max}}{FS_{max} - FS_{min}} \right| \tag{4}$$

Step5: Apply roulette wheel to the candidate points and select one
Step6: Replace the current position with the selected candidate

3.2 Tracing Mode

Tracing mode is the sub-model for modeling the case of the cat in tracing targets. In the tracing mode, cats are moving towards the best target. Once a cat goes into tracing mode, it moves according to its own velocities for each dimension. Every cat has two velocity vector are defined as V_{kd}^1 and V_{kd}^0.

- V_{kd}^0 is the probability that the bits of the cat change to zero
- V_{kd}^1 is the probability that bits of cat change to one.

The velocity vector changes its meaning to the probability of mutation in each dimension of a cat. The tracing mode action is described in the next pseudocode:
Step1: Calculate d_{kd}^1 and d_{kd}^0 where $X_{best,d}$ is the d-th dimension of the best cat, r_1 has a random values in the interval of $[0,1]$ and c_1 is a constant which is defined by the user

$$\begin{aligned} \text{if } X_{best,d} = 1 \text{ then } d_{kd}^1 = r_1 c_1 \text{ and } d_{kd}^0 = -r_1 c_1 \\ \text{if } X_{best,d} = 0 \text{ then } d_{kd}^1 = -r_1 c_1 \text{ and } d_{kd}^0 = r_1 c_1 \end{aligned} \tag{5}$$

Step2: Update process of V_{kd}^1 and V_{kd}^0 are as follows, where w is the inertia weight and M is the column numbers

$$V_{kd}^1 = wV_{kd}^1 + d_{kd}^1$$
$$V_{kd}^0 = wV_{kd}^0 + d_{kd}^0 \qquad d = 1,...,M \qquad (6)$$

Step3: Calculate the velocity of cat_k, V_{kd}', according to

$$V_{kd}' = \begin{cases} V_{kd}^1 \text{ if } X_{kd} = 0 \\ V_{kd}^0 \text{ if } X_{kd} = 1 \end{cases} \qquad (7)$$

Step4: Calculate the probability of mutation in each dimension, this is defined by parameter t_{kd}, t_{kd} takes a value in the inverval of [0,1]

$$t_{kd} = \frac{1}{1 + e^{-V_{kd}'}} \qquad (8)$$

Step5: Based on the value of t_{kd} the new value of each dimension of cat is update as follows where $rand$ is an aleatory variable $\in [0,1]$

$$X_{kd} = \begin{cases} X_{best,d} \text{ if } rand < t_{kd} \\ X_{kd} \text{ if } t_{kd} < rand \end{cases} \qquad d = 1,...,M \qquad (9)$$

4   Binarization Methods

Following are the new binarization techniques, these are made by combining eight transfer functions and Complement discretization technique.

4.1   Complement

In the original BCSO, the transfer function that we use is based on the best cat found, Eq. 9. Now we test tracing mode with Complement technique, this is described in the following model:

$$x_i^d(t+1) = \begin{cases} complement(x_i^k) \text{ if } rand \leq V_i^d(t+1) \\ \\ 0 \qquad\qquad\qquad otherwise \end{cases} \qquad (10)$$

4.2   Transfer Functions

The transfer functions define a probability to change an element of solution from 1 to 0, or vice versa (Table 1). The eight transfer functions that was proposed by Mirjalili and Lewis in [22] are:

Table 1. Transfer functions [22].

S-shaped family		V-shaped family	
Name	Transfer function	Name	Transfer function
S1	$T(V_i^d) = \frac{1}{1+e^{-2V_i^d}}$	V1	$T(V_i^d) = \left\lvert \text{erf}\left(\frac{\sqrt{\pi}}{2}V_i^d\right)\right\rvert$
S2	$T(V_i^d) = \frac{1}{1+e^{-V_i^d}}$	V2	$T(V_i^d) = \lvert tanh(V_i^d)\rvert$
S3	$T(V_i^d) = \frac{1}{1+e^{\frac{-V_i^d}{2}}}$	V3	$T(V_i^d) = \left\lvert \frac{V_i^d}{\sqrt{1+(V_i^d)^2}}\right\rvert$
S4	$T(V_i^d) = \frac{1}{1+e^{\frac{-V_i^d}{3}}}$	V4	$T(V_i^d) = \left\lvert \frac{2}{\pi}arctan\left(\frac{\pi}{2}V_i^d\right)\right\rvert$

5   Solving the Set Covering Problem

The BCSO performance was evaluated experimentally using 65 SCP test instances from the OR-Library of Beasley [2]. For solving the SCP with BCSO we use the following procedure:

Algorithm 2. *Solving SCP()*

1: Initialize parameters in cats;
2: Initialization of cat positions, randomly initialize cat positions with values between 0 and 1;
3: Initialization of all parameter of BCSO;
4: Evaluation of the fitness of the population. In this case the fintess function is equal to the objective function of the SCP;
5: Change of the position of the cat. A cat produces a modification in the position based in one of the behaviors. i.e. seeking mode or tracing mode;
6: If solution is not feasible then repaired. Eeach row i must be covered by at least one columns, to choose the missing columns do: the cost of a column/(number of not covered row that can cover column j);
7: Eliminate the redundant columns. A redundant column is one that if removed, the solution remains feasible;
8: Memorize the best found solution. Increase the number of iterations;
9: Stop the process and show the result if the completion criteria are met. Completion criteria used in this work are the number specified maximum of iterations. Otherwise, go to step 3;

5.1   Repair Operator

Generally, the solutions not satisfied all the constraints of the problems. For guarantee the feasibility of the solution, they must be repaired by a repair process.

Algorithm 3 shows a repair method where all rows not covered are identified and the columns required are added. So in this way all the constraints will be covered. The search of these columns are based in the relationship showed in the Eq. 11.

$$\frac{cost\ of\ one\ column}{amount\ of\ columns\ not\ covered} \tag{11}$$

Once the columns are added and the solution is feasible, a method is applied to remove redundant columns of the solution. A redundant column are those that are removed, the solution remains a feasible solution. The algorithm of this repair method is detailed in the Algorithm 3. Where

- I is the set of all rows
- J is the set of all columns
- J_i is the set of columns that cover the row i, $i \in I$
- I_j is the set of rows covered by the column j, $j \in J$
- S is the set of columns of the solution
- U, is the set of columns not covered
- w_i is the number of columns that cover the row $i, \forall i \in I$ in S

Algorithm 3. *Repair Operator*()

1: $w_i \leftarrow |S \cap J_i| \ \forall i \in I$;
2: $U \leftarrow \{i | w_i = 0\}, \forall i \in I$;
3: **for** $i \in U$ **do**
4:     find the first column j in J_i that minimize $\frac{c_j}{|U \cap I_j|} S \leftarrow S \cap j$;
5:     $w_i \leftarrow w_i + 1, \forall i \in I_j$;
6:     $U \leftarrow U - I_j$;
7: **end for**
8: **for** $j \in S$ **do**
9:     **if** $w_i \geq 2, \forall i \in I_j$ **then**
10:         $S \leftarrow S - j$;
11:         $w_i \leftarrow w_i - 1, \forall i \in I_j$
12:     **end if**
13: **end for**

5.2   Parameter Setting

All the algorithms were configured before performing the experiments. To this end and starting from default values, a parameter of the algorithm is selected to be turned. Then, 30 independent runs are performed for each configuration of the parameter. Next, the configuration which provides the best performance on average is selected. Next, another parameter is selected so long as all of them are fixed. Table 2 shows the range of values considered and the configurations selected. These values were obtained experimentally.

This procedure was performed for each set of instances, Table 2. In all experiments the BCSO was executed with 40000 iterations. Moreover, the results of the eight different transfer functions and complement discretization techniques were considered to select the final parameter.

Table 2. Parameter values

Name	Parameter	Instance set	Selected
Number of cats	C	4, 5 and 6	100
		A and B	50
		C and D	30
		NRE and NRF	25
		NRG and NRH	20
Mixture ratio	MR	4 and 5	0.7
		A and B	0.65
		C, D, NRE, NRF, NRG and NRH	0.5
Seeking memory pool	SMP	4, 5 , A and B	5
		C and D	10
		NRE and NRF	15
		NRG and NRH	20
Probabily of mutation operation	PMO	4, 5 and 6	0,97
		A and B	0,93
		C and D	0,9
		NRE, NRF, NRG and NRH	1
Counts of dimension to change	CDC	4, 5, 6, A and B	0.001
		C and D	0.002
		NRE and NRF	0.002
		NRG and NRH	0.01
Weight	w	All	1
Factor c_1	c_1	All	1

6   Results

The Tables 3 and 4 show the average RPD Eqs. (12) and (13) for 11 instances set from the OR-Library of Beasley [2]. The Table 3 describe the average RPD from the minimum results, of 30 runs, obtained for each instance (we use the Eq. 12). In Table 4 the difference is that the average RPD is of the average obtained from the 30 runs (we use the Eq. 13). The columns of the tables represent the results obtained by each transfer function and are marked the best results.

Analyzing both RPD tables we can see that for the V_3 function in both tables best results were obtained. In RPD_{min} were five the best results and in RPD_{avg} were six the best results. Because of this we generate the Table 5 with the results obtained for each instance to use the technique with complement function V_3. About the solutions obtained we reach 6 optimum but most of the results are very close to the desired optimum. The best results from the all instances set we found in 4, 5 amd 6 instance set with values between 1,1 and 1,8 for the RPD_{min}.

$$RPD_{min} = \left(\frac{Z_{min} - Z_{opt}}{Z_{opt}} \right) * 100 \qquad (12)$$

Table 3. RPD_{min}

	S_1	S_2	S_3	S_4	V_1	V_2	V_3	V_4
4	1,3	2,7	2,9	2,2	1,3	*1,1*	*1,1*	1,6
5	*1,2*	3,2	3,5	3,1	*1,2*	1,5	*1,2*	1,6
6	2,2	3,4	2,9	3,7	2,0	2,1	*1,8*	*1,8*
A	3,3	4,3	4,2	3,7	3,2	*2,9*	3,0	3,4
B	2,7	4,8	5,8	4,0	*2,1*	*2,1*	3,2	3,0
C	*3,2*	5,5	5,5	5,7	3,5	4,4	3,9	3,3
D	*3,0*	3,7	4,2	4,0	4,5	3,6	3,3	3,6
NRE	11,3	11,3	12,1	10,6	10,6	10,6	9,9	*9,1*
NRF	18,7	20,0	18,7	*17,3*	20,1	18,6	*17,3*	20,0
NRG	8,2	8,2	8,3	8,7	8,2	8,3	*7,8*	8,6
NRH	11,5	11,5	*10,8*	11,5	11,5	11,5	11,5	11,5

$$RPD_{avg} = \left(\frac{Z_{avg} - Z_{opt}}{Z_{opt}} \right) * 100 \qquad (13)$$

Table 4. RPD_{avg}

	S_1	S_2	S_3	S_4	V_1	V_2	V_3	V_4
4	*3,4*	7,0	6,9	6,7	3,9	3,8	3,5	4,5
5	*3,7*	6,6	6,9	6,5	3,9	3,8	3,9	3,8
6	4,7	7,1	6,9	6,9	4,7	*4,3*	*4,3*	4,8
A	5,0	6,8	6,6	6,5	4,9	5,1	*4,8*	5,1
B	6,4	8,0	8,8	8,0	6,2	*6,0*	6,3	6,1
C	*6,8*	9,3	9,2	8,9	7,0	7,3	*6,8*	6,9
D	7,6	*7,4*	*7,4*	7,6	7,6	*7,4*	*7,4*	7,7
NRE	13,9	13,9	*12,4*	13,5	13,9	13,9	13,5	14,2
NRF	*21,5*	22,2	*21,5*	*21,5*	*21,5*	*21,5*	*21,5*	*21,5*
NRG	9,6	*9,4*	9,6	9,6	9,6	9,7	9,7	9,7
NRH	*13,1*	*13,1*	*13,1*	*13,1*	*13,1*	*13,1*	*13,1*	*13,1*

Table 5. All results of complement with V_3 technique.

Inst.	Z_{opt}	Z_{min}	Z_{avg}	RPD_{min}	RPD_{avg}
scp41	429	433	444,6	0,9	3,6
scp42	512	519	536,7	1,4	4,8
scp43	516	529	560,6	2,5	8,6
scp44	494	497	513,4	0,6	3,9
scp45	512	514	525,2	0,4	2,6
scp46	560	560	567,9	0,0	1,4
scp47	430	433	439,4	0,7	2,2
scp48	492	494	512,6	0,4	4,2
scp49	641	661	677,2	3,1	5,6
scp410	514	518	524,9	0,8	2,1
scp51	253	253	263,9	0,0	4,3
scp52	302	304	314,5	0,7	4,1
scp53	226	229	233,5	1,3	3,3
scp54	242	242	245,7	0,0	1,5
scp55	211	216	220,4	2,4	4,5
scp56	213	214	225,6	0,5	5,9
scp57	293	297	306,5	1,4	4,6
scp58	288	298	306,6	3,5	6,5
scp59	279	280	281,7	0,4	1,0
scp510	265	271	275,4	2,3	3,9
scp61	138	143	147	3,6	6,5
scp62	146	146	150,3	0,0	2,9
scp63	145	147	152,1	1,4	4,9
scp64	131	133	134,6	1,5	2,7
scp65	161	165	169,2	2,5	5,1
scpa1	253	271	275,9	7,1	9,1
scpa2	252	259	265	2,8	5,2
scpa3	232	238	242,5	2,6	4,5
scpa4	234	239	245,7	2,1	5,0
scpa5	236	237	238,5	0,4	1,1
scpb1	69	72	74,6	4,3	8,1
scpb2	76	82	84,9	7,9	11,7
scpb3	80	80	82,7	0,0	3,4
scpb4	79	81	84	2,5	6,3
scpb5	72	73	73	1,4	1,4
scpc1	227	232	234,9	2,2	3,5
scpc2	219	225	229,5	2,7	4,8
scpc3	243	258	265,7	6,2	9,3
scpc4	219	230	239,2	5,0	9,2
scpc5	215	222	229,7	3,3	6,8
scpd1	60	60	64,4	0,0	7,3
scpd2	66	69	69,9	4,5	5,9
scpd3	72	76	78,6	5,6	9,2
scpd4	62	64	65,7	3,2	6,0
scpd5	61	63	64,9	3,3	6,4
scpnre1	29	30	30	3,4	3,4
scpnre2	30	34	34,5	13,3	15,0
scpnre3	27	30	33,1	11,1	22,6
scpnre4	28	32	33	14,3	17,9
scpnre5	28	30	30	7,1	7,1
scpnrf1	14	17	17	21,4	21,4
scpnrf2	15	17	18	13,3	20,0
scpnrf3	14	16	17	14,3	21,4
scpnrf4	14	16	17,3	14,3	23,6
scpnrf5	13	16	16	23,1	23,1
scpnrg1	176	188	194,1	6,8	10,3
scpnrg2	154	165	167,7	7,1	8,9
scpnrg3	166	182	182,6	9,6	10,0
scpnrg4	168	179	183,3	6,5	9,1
scpnrg5	168	183	184,7	8,9	9,9
scpnrh1	63	70	72,4	11,1	14,9
scpnrh2	63	67	67	6,3	6,3
scpnrh3	59	68	69	15,3	16,9
scpnrh4	58	66	67,1	13,8	15,7
scpnrh5	55	61	61	10,9	10,9

7   Conclusions

Eight new transfer functions divided into two families, s-shaped and v-shaped, are introduced and evaluated. In order to evaluate the performance of all the modified transfer function-based versions. For most cases the results were close to the desired optimum. For future studies, it would be interesting to investigate the impact of the new v-shaped family of transfer function with other discretization techniques like Roulette Wheel and Elitist. Moreover, it could also better solutions using different parameter setting for each set of instances. As can be seen from the results, metaheuristic performs well in all cases observed according to old RPD works [8]. This paper has shown that the BCSO is a valid alternative to solve the SCP. For future works the objective will be make them highly immune to be trapped in local optima and thus less vulnerable to premature convergence problem. Thus, we could propose an algorithm that shows improved results in terms of both computational time and quality of solution.

Acknowledgements. The author Broderick Crawford is supported by Grant CONICYT/FONDECYT/REGULAR/1140897, Ricardo Soto is supported by Grant CONICYT/FONDE- CYT/REGULAR/1160455.

References

1. Balas, E., Carrera, M.C.: A dynamic subgradient-based branch-and-bound procedure for set covering. Oper. Res. **44**(6), 875–890 (1996)
2. Beasley, J.: A lagrangian heuristic for set covering problems. Nav. Res. Logistics **37**, 151–164 (1990)
3. Beasley, J., Jornsten, K.: Enhancing an algorithm for set covering problems. Eur. J. Oper. Res. **58**(2), 293–300 (1992)
4. Beasley, J.E., Chu, P.C.: A genetic algorithm for the set covering problem. Eur. J. Oper. Res. **94**(2), 392–404 (1996)
5. Caprara, A., Fischetti, M., Toth, P.: Algorithms for the set covering problem. Ann. Oper. Res. **98**, 353–371 (2000)
6. Chu, S., Tsai, P.: Computational intelligence based on the behavior of cats. Int. J. Innovative Comput. Inf. Control **3**, 163–173 (2007)
7. Chu, S., Tsai, P., Pan, J.: Cat swarm optimization. In: Yang, Q., Webb, G. (eds.) PRICAI 2006. LNCS (LNAI), vol. 4099, pp. 854–858. Springer, Heidelberg (2006)
8. Crawford, B., Soto, R., Berrios, N., Johnson, F., Paredes, F., Castro, C., Norero, E.: A binary cat swarm optimization algorithm for the non-unicost set covering problem. Math. Probl. Eng. **2015**, 1–8 (2015). (Article ID 578541)
9. Crawford, B., Soto, R., Cuesta, R., Paredes, F.: Application of the artificial bee colony algorithm for solving the set covering problem. Sci. World J. **2014**, 1–8 (2014). (Article ID 189164)
10. Crawford, B., Soto, R., Cuesta, R., Paredes, F.: Using the bee colony optimization method to solve the weighted set covering problem. In: Stephanidis, C. (ed.) HCI 2014, Part I. CCIS, vol. 434, pp. 493–497. Springer, Heidelberg (2014)
11. Crawford, B., Soto, R., Monfroy, E.: Cultural algorithms for the set covering problem. In: Tan, Y., Shi, Y., Mo, H. (eds.) ICSI 2013, Part II. LNCS, vol. 7929, pp. 27–34. Springer, Heidelberg (2013)

12. Crawford, B., Soto, R., Monfroy, E., Paredes, F., Palma, W.: A hybrid ant algorithm for the set covering problem. Int. J. Phys. Sci. **6**(19), 4667–4673 (2011)
13. Crawford, B., Soto, R., Peña, C., Palma, W., Johnson, F., Paredes, F.: Solving the set covering problem with a shuffled frog leaping algorithm. In: Nguyen, N.T., Trawiński, B., Kosala, R. (eds.) ACIIDS 2015. LNCS, vol. 9012, pp. 41–50. Springer, Heidelberg (2015)
14. Dorigo, M., Birattari, M., Stützle, T.: Ant colony optimization. IEEE Comput. Intell. Mag. **1**(4), 28–39 (2006)
15. Eusuff, M., Lansey, K., Pasha, F.: Shuffled frog-leaping algorithm: a memetic metaheuristic for discrete optimization. Eng. Optim. **38**(2), 129–154 (2006)
16. Fisher, M.L., Kedia, P.: Optimal solution of set covering/partitioning problems using dual heuristics. Manag. Sci. **36**(6), 674–688 (1990)
17. Garfinkel, R.S., Nemhauser, G.L.: Optimal political districting by implicit enumeration techniques. Manag. Sci. **16**(8), B495–B508 (1970)
18. Gouwanda, D., Ponnambalam, S.: Evolutionary search techniques to solve set covering problems. World Acad. Sci. Eng. Technol. **39**, 20–25 (2008)
19. Karaboga, D., Basturk, B.: A powerful and efficient algorithm for numerical function optimization: artificial bee colony (abc) algorithm. J. Glob. Optim. **39**(3), 459–471 (2007)
20. Karp, R.M.: Reducibility Among Combinatorial Problems. Springer, Heidelberg (1972)
21. Lessing, L., Dumitrescu, I., Stützle, T.: A comparison between ACO algorithms for the set covering problem. In: Dorigo, M., Birattari, M., Blum, C., Gambardella, L.M., Mondada, F., Stützle, T. (eds.) ANTS 2004. LNCS, vol. 3172, pp. 1–12. Springer, Heidelberg (2004)
22. Mirjalili, S., Lewis, A.: S-shaped versus v-shaped transfer functions for binary particle swarm optimization. Swarm Evol. Comput. **9**, 1–14 (2013)
23. Mirjalili, S., Mohd, S., Taherzadeh, G., Mirjalili, S., Salehi, S.: A study of different transfer functions for binary version of particle swarm optimization. In: Swarm and Evolutionary Computation, pp. 169–174 (2011)
24. Panda, G., Pradhan, P., Majhi, B.: IIR system identification using cat swarm optimization. Expert Syst. Appl. **38**, 12671–12683 (2011)
25. Revelle, C., Marks, D., Liebman, J.C.: An analysis of private and public sector location models. Manag. Sci. **16**(11), 692–707 (1970)
26. Saremi, S., Mirjalili, S., Lewis, A.: How important is a transfer function in discrete heuristic algorithms. Neural Comput. Appl. **26**(3), 625–640 (2015)
27. Sharafi, Y., Khanesar, M., Teshnehlab, M.: Discrete binary cat swarm optimization algorithm. In: Computer, Control and Communication, pp. 1–6 (2013)
28. Simeone, B., Nouno, G., Mezzadri, M., Lari, I.: A boolean theory of signatures for tonal scales. Discrete Appl. Math. **165**, 283–294 (2014)
29. Soto, R., Crawford, B., Barraza, J., Johnson, F., Paredes, F.: Solving pre-processed set covering problems via cuckoo search and lévy flights. In: 2015 10th Iberian Conference on Information Systems and Technologies (CISTI), pp. 1–6 (2015)
30. Toregas, C., Swain, R., ReVelle, C., Bergman, L.: The location of emergency service facilities. Oper. Res. **19**(6), 1363–1373 (1971)
31. Tsai, P., Pan, J., Chen, S., Liao, B.: Enhanced parallel cat swarm optimization based on the taguchi method. Expert Syst. Appl. **39**, 6309–6319 (2012)
32. Vasko, F.J., Wolf, F.E., Stott, K.L.: Optimal selection of ingot sizes via set covering. Oper. Res. **35**(3), 346–353 (1987)
33. Vasko, F.J., Wolf, F.E., Stott, K.L.: A set covering approach to metallurgical grade assignment. Eur. J. Oper. Res. **38**(1), 27–34 (1989)

34. Vasko, F.J., Wolf, F.E., Stott, K.L., Scheirer, J.W.: Selecting optimal ingot sizes for bethlehem steel. Interfaces **19**(1), 68–84 (1989)
35. Walker, W.: Using the set-covering problem to assign fire companies to fire houses. Oper. Res. **22**, 275–277 (1974)
36. Yang, X.-S., Deb, S.: Cuckoo search via lévy flights. In: World Congress on Nature & Biologically Inspired Computing, NaBIC 2009, pp. 210–214 (2009)
37. Zhang, J., Wei, Q., Chen, G.: A heuristic approach for λ-representative information retrieval from large-scale data. Inf. Sci. **277**, 825–841 (2014)

A Data Warehouse Model for Business Processes Data Analytics

Maribel Yasmina Santos[(✉)] and Jorge Oliveira e Sá

ALGORITMI Research Centre, University of Minho, Guimarães, Portugal
{maribel,jos}@dsi.uminho.pt

Abstract. Business Process Management and Business Intelligence initiatives are commonly seen as separated organizational projects, suffering from lack of coordination, leading to a poor alignment between strategic management and operational business processes execution. Researchers and professionals of information systems have recognized that business processes are the key for identifying the user needs for developing the software that supports those needs. In this case, a process-driven approach could be used to obtain a Data Warehouse model for the Business Intelligence supporting software. This paper presents a process-based approach for identifying an analytical data model using as input a set of interrelated business processes, modeled with Business Process Model and Notation version 2.0, and the corresponding operational data model. The proposed approach ensures the identification of an analytical data model for a Data Warehouse repository, integrating dimensions, facts, relationships and measures, providing useful data analytics perspectives of the data under analysis.

Keywords: Analytical data model · Business intelligence · Business process management · Business process model and notation · Data warehousing · Operational data model

1 Introduction

In recent years, more and more organizations are concerned about processes and how to manage them, which means that they started Business Process Management (BPM) initiatives, related with designing, enacting, managing, and analyzing operational business processes [1].

Moreover, organizations need to know and understand how the business is evolving, which led them moving towards Business Intelligence (BI) initiatives. BI has become a strategic initiative and is recognized as essential in driving business effectiveness and innovation [2]. A key component in a BI system is the Data Warehouse (DW) that supports the integration, storage, processing and analysis of the relevant data, reason why this paper is focused in the data model of this repository, in order to make available a physical data model that can support On-Line Analytical Processing (OLAP), a foundation for delivering the information needed for the decision making processes, providing autonomy and flexibility to users in browsing and analyzing data.

© Springer International Publishing Switzerland 2016
O. Gervasi et al. (Eds.): ICCSA 2016, Part V, LNCS 9790, pp. 241–256, 2016.
DOI: 10.1007/978-3-319-42092-9_19

BPM and BI initiatives are commonly viewed as separate organizational information systems projects, suffering from lack of coordination, leading to a poor alignment between strategic management and operational business process execution [3].

Researchers and professionals in information systems have recognized that understanding the business processes is the key for identifying the user needs of the software systems that will support those needs [4]. As a consequence, researchers suggest a process-driven approach to obtain a DW model [5–7] that can support a BI system. Despite these concerns, [8] shows in a survey of techniques for transforming an operational data model into a DW model that existing techniques should be improved, as they suffer from: (i) the amount of needed manual work; (ii) the lack of assistance in discovering facts and the relevant dimensional attributes; (iii) the inappropriate treatment of the process perspective; and, (iv) the possible mismatch between the DW content and the content of the organizational data sources.

Having in mind the integration of BPM and BI, and the advantages that this can bring to the decision-making process, this paper presents a fully automatic approach for identifying a DW model (a set of Star or Constellation Schemas) using as input a set of business processes modeled by Business Process Modeling and Notation version 2.0 (BPMN) [9] and the corresponding operational data model expressed as an entity-relationship model. The proposed approach extends the work of [10], addressing the main problems and pending issues identified by [8]. It is based on the premise that BPMN models can be a foundation for OLAP development, so the proposed approach benefits from the availability of detailed business process models, as those could lead to more suitable DW data models.

In this research, this paper intends to answer the following question: Can a DW model (Star or Constellation Schemas), based on *Dimensions*, *Facts*, *Relationships* and *Measures*, be automatically identified based on a set of business processes models and the corresponding operational model?

This paper is organized in five sections: in the first section the problem is described and a research question is placed; the second section presents the related work by describing the process-based approach; the third section presents the theoretical proposal of this work, based on three definitions and four rules; the forth section describes a demonstration case, showing the derivation from BPMN 2.0 processes to an operational data model, and from the operational data model to the analytical data model of the supporting DW, by using a well-known example of a School Library System with five processes; finally, in the fifth section, the conclusions are presented and some guidelines for future work are addressed.

2 Related Work

Several approaches have been proposed for helping DW designers to obtain the data models of these repositories. These approaches are based on operational systems specification and can be divided into two distinct groups [5]:

1. structure-based approach – the motivation for this approach (also known as a data-driven approach) is that the conceptual and the logical design of a DW depend on the data sources available in the operational systems [11, 12]. There are some techniques that follow this approach [8, 11, 12] and that use mainly the entity-relationship model as the foundation for the proposed solutions. These techniques suffer from some limitations [8] like: (1) the lack of guidance for identifying the DW model, (2) the behavioral aspects of the system are ignored, and (3) several manual decisions and transformations are needed to obtain a DW model; and,

2. process-based approach – the motivation for this approach is to understand that the fundamental role of a DW is to provide business performance measurement and redesign support [6, 13]. A process-based approach also requires understanding the business processes and their relationships in order to identify the relevant Key Performance Indicators (KPIs) and the data sources needed to compute them. This will give to the process performer (who executes a process) the capability to be a decision-maker, incorporating the KPIs into the process activities, having an accurate and timely picture of the business [14]. There are some techniques for identifying the DW models from business process models [5, 7, 8], but those are mainly based on manual steps and do not consider the operational system structure.

In this paper, the process-based approach is followed and the principles suggested by [10, 15] are extended for identifying an analytical data model complying with a set of interrelated business processes models and with the operational data model can be used in the development of the software system that supports those business processes.

Among the several existing modelling languages [16], BPMN 2.0 [9] is here used not only because it is a standard, but mainly because it is actually used in organizations, describing which activities are performed and the dependencies between them [17]. The role of data is changing with the evolution from BPMN 1.2 to BPMN 2.0, increasing data relevance by the introduction of new model constructs such as *data objects* and *data stores*. Although those improvements, BPMN neither provide modeling elements for describing data structures nor a language for data manipulation [5].

BPMN 2.0 basic process models can be grouped into *Private Business Processes* or *Public Processes* [9]. *Public Processes* represent the interactions between a *Private Business Process* and other processes or participants. A *Private Business Process* is an internal organizational process that is represented within a *Pool* with a *Participant*. A *Participant* represents a role played in the process by a person, an organization's department or something involved in the process. The process flow must be in one *Pool* and should never cross the *Pool* boundaries. A *Pool* can be divided into several *Lanes*. A *Lane* represents, for example, the different departments of an organization involved in the process. The interaction between distinct *Private Business Processes* (represented by different *Pools*) can be represented by incoming and outgoing messages.

The main BPMN diagrams' graphical elements to define the behavior of a business process are *Activities*, *Gateways* and *Events* [9]. An *Activity* represents a piece of work.

A *Gateway* is used to control how the process flows and it can diverge (splitting gateways) or converge (merging gateways) the sequence flow. An *Event* is something that happens during the process course and affects the process's flow. *Activities, Gateways* and *Events* are connected by sequence flows, representing the execution order [9].

When a process is executed, resources and/or data are used and produced. The data received, created or used can be represented by *message flows* or *data associations*. A *message flow* connects two *Pools* and represents the content of a communication between two *Participants*. *Data associations* connect activities and *data objects* or *data stores* [9].

Data that flow through a process are represented by *data objects*. Persistent data are represented by *data stores* and are the ones that remain beyond the process life cycle, or after the process execution ends [9].

Data objects and *data stores* are exclusively used in *Private Process Diagrams* [9], reason why the approach presented in [10] is based on *Private Process Diagrams*. In BPMN's most recent version, the number of graphical element has increased, including the *data stores* element, which represents persistent data, allowing business process models to be highly detailed [9]. To define a persistent data model there is the need of identifying the domain entities, their attributes, and the relationships $((1 : n), (m : n)$ or $(1 : 1))$ between entities [17]. The approach presented in [10] defines four assumptions (A_i) and three rules (R_i) for the identification of an operational data model from a set of interrelated business processes. The assumptions are:

- (A_1) A set of private business processes is considered, and, in this scenario, each identified business process is represented in one (main) pool that can be divided into several lanes. The other participants (pools) involved in the business process are considered as "external participants";
- (A_2) When an activity receives information (messages) from an "external participant" and when that information must be kept beyond the process execution, that activity must write the received information into a data store;
- (A_3) A data object represents data (a document, a message, etc.) that an activity sends or receives. When the information contained in that data object must be kept beyond the process life cycle, the activity must write the information in a data store; and,
- (A_4) When a business process reads information from a data store and no other business process writes information in that data store, this means that something is wrong (for example a business process is missing) or that a link with another application exists. The same happens when a business process writes information in a data store that is never used (or no other activity reads information from that data store).

Regarding the rules for identifying an operational data model from a set of interrelated business processes, [10] proposes:

- (R_1) To identify the data model entities;
- (R_2) To identify the relationships between entities; and
- (R_3) To identify the entities attributes.

In [10], the data stores and the roles played by the different participants give origin to entities in the data model. The relationship between the identified entities is deduced from the information exchanged between participants and the activities that manipulate the data store in two ways: directly by the participant that performs the activity and indirectly by the participant that sends or receives information to/from the activity that operates the data store.

Based on this work [10], and on the advantages it brings for the identification of the relevant data that must support the business processes, this paper extends it showing how the operational data model can also be used for identifying an analytical data model, here expressed as a multidimensional model, which can be used for implementing a DW supporting the analytical needs of those business processes. The approach here presented makes use of the operational data model, itself, and also a very strict group of information present in the business process models. To the best of our knowledge, no other work follows a process-based approach for identifying the DW data model in a fully automatic way, being able of identifying facts, dimensions, relationships and measures, either present in the operational data model or specifically proposed for analytical contexts.

3 Analytical Data Models for Business Processes

BPMN 2.0 includes several elements that are used to represent the manipulation of data, as mentioned in the previous section, as the item-aware elements, which are elements that allow the storage and transmission of items during the process flows [9]. These elements are the basic foundations of the entities, attributes and relationships when identifying an operational data model. Besides these elements, and of relevance for the work here presented, are the Lanes objects that represent the participants in the business processes.

In formal terms, the Entity-Relationship Diagram (ERD) represents an operational data model formally described in this work as follows:

Definition 1. An Operational Data Model representing a data reality, $ODM = (E, A, R)$, includes a set of entities $E = \{E^1, E^2, \ldots, E^n\}$, the corresponding attributes for the addressed entities, $A = \{A^1, A^2, \ldots, A^n\}$, where A^1 is the set of attributes for entity E^1, $A^1 = \{A_1^1, A_2^1, \ldots, A_k^1\}$, and a set of relationships among entities $R = \{R_1^1(c_o : c_d), R_2^1(c_o : c_d), \ldots, R_m^l(c_o : c_d)\}$, where R_m^l express a relationship between E^m and E^l of cardinality c_o in the origin entity E^m and c_d in the destination entity E^l. Cardinalities can be of type $1 : n$, $m : n$ or $1 : 1$, with the optional 0 when needed.

The participants, identified in the set of interrelated business processes models, can be defined as follows:

Definition 2. The list of parcipants ia set of interrelated business process models is represented by $P = \{P^1, P^2, \ldots, P^n\}$, being a participant a person or something involved in one of the modeled processes.

Having these elements, *ODM* and P, it is possible to propose the automatic transformation of an *ODM* into an Analytical Data Model that includes a set of fact tables, a set of dimension tables, a set of relationships, and a set of measures that are useful for analyzing the business processes by several perspectives. These measures include direct measures (inherited from the *ODM*) and calculated measures able to provide useful insights of the data under analysis. These calculated measures are automatically proposed by the approach presented in this paper, taking into consideration the implicit relationships existing between the direct measures.

Definition 3. An Analytical Data Model,$ADM = (D, F, R, M)$, includes a set of dimension tables, $D = \{D^1, D^2, \ldots, D^l\}$, a set of fact tables, $F = \{F^1, F^2, \ldots, F^m\}$, a set of relationships, $R = \{R_{F^1}^{D^1}(c_f : c_d), R_{F^1}^{D^2}(c_f : c_d), \ldots, R_{F^m}^{D^l}(c_f : c_d)\}$, which associate facts included in a specific fact table with a set of dimensions, being the cardinality of the relationship of $n : 1$ between the fact table and the dimension table. An *ADM* also includes a set of measures, $M = \{M^1, M^2, \ldots, M^n\}$, used for analyzing the business processes.

Having as starting point the elements expressed by Definitions 1 and 2, it is possible to automatically identify an *ADM*, expressed by Definition 3, following the rules proposed in this paper, for the identification of dimension tables (Rule ADM.1), fact tables (Rule ADM.2), measures (Rule ADM.3) and relationships (Rule ADM.4).

Rule ADM.1: Identification of Dimensions. The identification of the dimensions D of an *ADM* follows a three-step approach:

Rule ADM.1.1: Identification of Dimensions derived from Participants. Each participant identified in the set of interrelated business processes models gives origin to a dimension in the *ADM*. For each dimension, its attributes are constituted by a surrogate key (SK) used as primary key (PK) of the dimension, by a natural key (NK) provided by the *ODM* and by the set of descriptive attributes of the entity present in the *ODM*.

Rule ADM.1.2: Identification of Dimensions derived from the *ODM*. Each entity only receiving cardinalities of 1 ($c_o = 1$ and $c_d = 1$) in the *ODM* corresponds to a dimension in the *ADM*, as this entity is used to complement the description of other entities (entity usually derived from the normal forms). For each dimension, its attributes are constituted by a SK used as PK of the dimension, by a NK provided by the *ODM* and by the set of descriptive attributes of the entity present in the *ODM*.

Rule ADM.1.3: Identification of the Time and Calendar Dimensions. Given the temporal and historical characteristics of an *ADM*, the *Time* and *Calendar* dimensions, D^{Time} and $D^{Calendar}$, are derived from the temporal characteristics expressed in the *ODM*, considering those entities that were not identified as dimension tables by Rule ADM.1.1 or by Rule ADM.1.2. The time and calendar dimensions are set to the lower level of detail allowed by the corresponding attributes, in order to have fact tables able to provide detailed analyses of the business processes. The attributes usually used to express the temporal characteristics are time (hh:mm:ss) and date (dd/mm/yyyy), for which the default dimensions are defined as:

$$D^{Time} = \{TimeID(SK), Time(NK), Second, Minute, Hour\}$$

$$D^{Calendar} = \{DateID(SK), Date(NK), Day, Month, Quarter, Semester, Year\}$$

Rule ADM.2: Identification of Fact Tables. All entities present in the *ODM* and not identified by Rule ADM.1 as dimension tables, represent fact tables (*F*) in the *ADM*. Each fact table includes a set of foreign keys (FK) and a set of measures. The foreign keys are derived from the relationships with the dimensions (Rule ADM.4) and are used to set the PK of the table.

Rule ADM.3: Identification of Measures. The identification of measures *M* in the *ADM* follows a three-step approach:

Rule ADM.3.1: Identification of Event-tracking Measures. To each fact table, an event-tracking measure (*Event Count*) is added, allowing the identification of analytical summaries of the business process with regard to each dimension.

Rule ADM.3.2: Identification of Direct Measures. Direct measures are numeric attributes present in the entities of the *ODM* that gave origin to fact tables by Rule ADM.2. These numeric attributes exclude those that are related with date or time specifications, or any type of keys (NK, PK or FK).

Rule ADM.3.3: Identification of Calculated Measures. Calculated measures are measures that intend to provide a different perspective of the measures identified by Rule ADM.3.1 and Rule ADM.3.2, being able to provide percentages of the incidence of each one of these measures, when analyzed by the different perspectives provided by the dimensions. A calculated measure (M^{i_c}) is derived for each measure (M^i) considering Eq. (1).

$$M^{i_c} = \frac{M^i_j}{\sum_{j=1}^{length(M^i)} M^i_j} \times 100 \ \%, \ 1 \leq j \leq length(M^i) \tag{1}$$

where $length(M^i)$ is the number of rows of the fact table of M^i and M^i_j is a specific value of a row j of M^i.

The measures identified by Rule ADM.3 can be analyzed using different aggregation functions, like SUM (sum), AVG (average), MIN (minimum), MAX (maximum), COUNT (count), among others, available in OLAP systems. For each measure, one or more aggregation functions can be applied.

Rule ADM.4: Identification of Relationships. The identification of the relationships between the dimension tables and the fact tables makes use of the corresponding entities in the *ODM*. The relationships in the multidimensional model are inherited from the *ODM*, considering that between a fact table and a dimensional table, a $n : 1$ relationship exists.

4 Demonstration Case

The work presented in [10] used a set of five interrelated business processes models, specified in BPMN 2.0, for identifying the corresponding *ODM*. The demonstration case is associated to an example of a School Library System and the modeled business processes are: *Register User* (Fig. 1), *Lend a Book* (Fig. 2), *Reserve a Book* (Fig. 3), *Renew a Loan* (Fig. 4), and *Return a Book* (Fig. 5). The participants in these five processes are *Borrower* and *Attendant*.

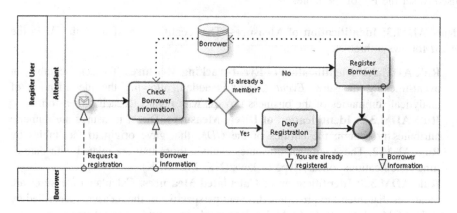

Fig. 1. Register User business process

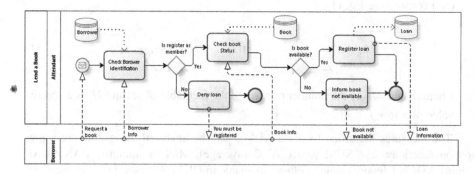

Fig. 2. Lend a Book business process

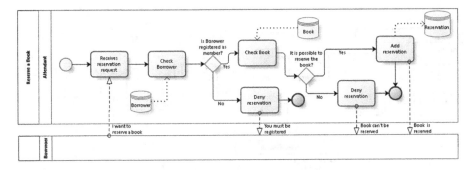

Fig. 3. Reserve a Book business process

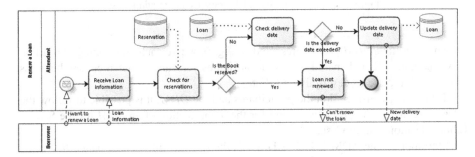

Fig. 4. Renew a Loan business process

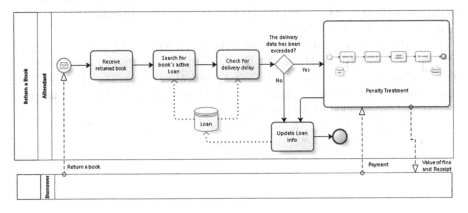

Fig. 5. Return a Book business process

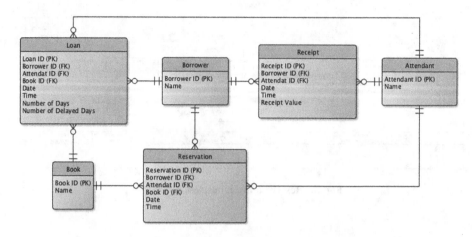

Fig. 6. *ODM* for the School Library Demonstration Case

The identified *ODM* [10] is presented in Fig. 6, where each entity integrates a PK, may integrate FK and includes specific attributes relevant to the modeled business processes (numeric or descriptive attributes).

Taking into consideration the participants and the *ODM*, it is possible to follow the rules presented in the previous section for the identification of an *ADM* for the School Library Demonstration Case.

Starting by Rule ADM.1, Rule ADM.1.1, the two identified participants, *Borrower* and *Attendant*, give origin to the two dimension tables $D^{Borrower} = \{BorrowerKey(PK),$ $BorrowerID(NK), Name\}$ and $D^{Attendant} = \{AttendantKey(PK), AttendantID(NK),$ $Name\}$, respectively. By Rule ADM.1.2 and analyzing the *ODM* previously presented, the *Book* entity gives origin to the $D^{Book} = \{BookKey(PK), BookID(NK), Name\}$ dimension. As Rule ADM.1.1 already addressed *Borrower* and *Attendant*, these entities are not considered in Rule ADM.1.2.

Rule ADM.1.3 allows the identification of the *Calendar* and *Time* dimensions. Considering the data available in the *ODM*, these two dimensions are defined as $D^{Time} = \{TimeKey(PK), TimeID(NK), Time, Second, Minute, Hour\}$ and $D^{Calendar} =$ $\{DateKey(PK), DateID(NK), Date, Day, Month, Quarter, Semestre, Year\}$, assuming a conventional composition of these dimensions.

At this point, all the dimensions were identified, being $D = \{D^{Calendar},$ $D^{Time}, D^{Borrower}, D^{Attendant}, D^{Book}\}$.

Continuing with Rule ADM.2 for the identification of the fact tables, all other tables presented in the *ODM* and not identified as dimensions in the previous rules, constitute fact tables. The fact tables are F^{Loan}, $F^{Reservation}$ and $F^{Receipt}$. Looking to the five business processes that gave origin to the *ODM*, it is possible to see that these fact tables will make possible the analysis of all the loans (either with renews and penalties) and the verification of the evolution in reservations. The set of fact tables is $F = \{F^{Loan}, F^{Reservation}, F^{Receipt}\}$.

Looking now to the measures that will constitute each fact table, by Rule ADM.3.1 an event-tracking measure will be added (*Event Counter*) to each fact table. Following with Rule ADM.3.2 for the direct measures, *Number of Days*, *Number of Delayed Days* and *Receipt Value* are identified. Lastly, by Rule ADM.3.3, calculated measures are created, one for each direct measure: *Event Counter^c*, *Number of Days^c*, *Number of Delayed Days^c* and *Receipt Value^c*, all renamed to % *of*, which leads to the calculated measures % *of Event Counter*, % *of Number of Days*, % *of Number of Delayed Days* and % *of Receipt Value*. Now it is possible to summarize each fact table in terms of its measures:

$$
F^{Loan} = \left\{
\begin{array}{c}
Event\ Counter, Number\ of\ Days, Number\ of\ Delayed\ Days, \\
\%\ of\ Event\ Counter, \%\ ofNumber\ of\ Days, \\
\%\ of\ Number\ of\ Delayed\ Days
\end{array}
\right\}
$$

$$
F^{Reservation} = \{Event\ Counter, \%\ of\ Event\ Counter\}
$$

$$
F^{Receipt} = \left\{
\begin{array}{c}
Event\ Counter, Receipt\ Value, \%\ of\ Event\ Counter, \\
\%\ of\ Receipt\ Value
\end{array}
\right\}
$$

Each fact table also includes a set of foreign keys, one for each dimension to which it is related. All the relationships are inherited from the entities in the *ODM*, emphasizing the links between dimension tables and fact tables (Rule ADM.4).

Having followed the defined rules, Fig. 7 presents the obtained multidimensional model (the *ADM* model), integrating the several dimensions, fact tables, measures, relationships and the corresponding keys.

In order to verify the usefulness of the proposed approach, let us consider one of the available fact tables and verify the analyses that can now be automatically done. Taking the *Loan* fact table as an example, and assuming that the corresponding star schema is storing the rows presented in Figs. 8 and 9 shows that while the direct measures allow the identification of the number of times a borrower loaned a book or the total number of days he/she stayed with the book, the calculated measures provide an overall

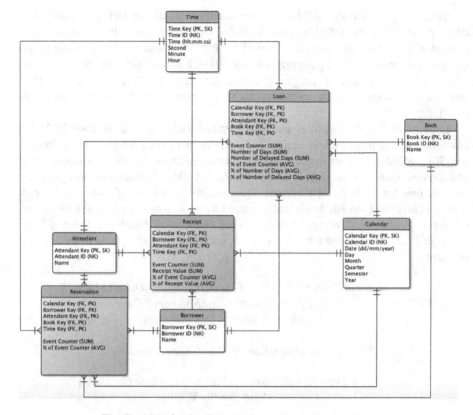

Fig. 7. *ADM* for the School Library Demonstration Case

perspective of the more interesting clients (in these case borrowers) or of the top books asked by those clients. See, for instance, Borrower 1, which loaned 3 times Book 1 (direct measure), representing 30 % of all loans made so far (calculated measured).

One of the main advantages of an *ADM* supported by a DW system is that OLAP can be used in the decision-making process, allowing users to query the data in an *ad-hoc* fashion. This way, specific applications, dashboards or reports provide different perspectives on data, and several graphs or tables can be integrated and combined to enhance the analysis of the business processes that were modelled. In the work presented in this paper, not only business processes are considered but also new measures are proposed with the aim of enhancing the analysis of those business processes. Users can combine direct and calculated measures, complementing operational indicators (Number of Days) with strategic indicators (% of Number of Days). While the first

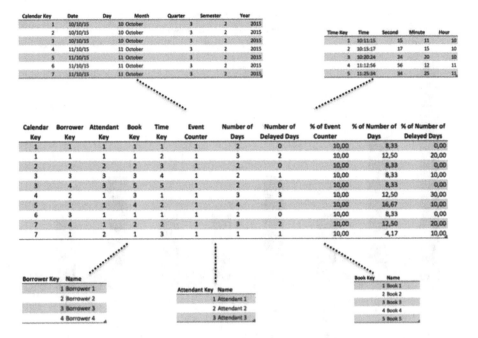

Fig. 8. Example data set for the *Loan* star schema

Name (Book)	Name (Borrower)	Event Counter	Number of Days	Number of Delay Days
Book 1	Borrower 1	3.000	6.000	3.000
	Borrower 3	1.000	2.000	0.000
Book 2	Borrower 2	1.000	2.000	0.000
	Borrower 4	1.000	3.000	2.000
Book 3	Borrower 2	1.000	3.000	3.000
	Borrower 3	1.000	2.000	1.000
Book 4	Borrower 1	1.000	4.000	1.000
Book 5	Borrower 4	1.000	2.000	0.000

Name (Book)	Name (Borrower)	% of Event Counter	% of Number of Days	% of Number of Delay..
Book 1	Borrower 1	30.0	25.0	30.0
	Borrower 3	10.0	8.3	0.0
Book 2	Borrower 2	10.0	8.3	0.0
	Borrower 4	10.0	12.5	20.0
Book 3	Borrower 2	10.0	12.5	30.0
	Borrower 3	10.0	8.3	10.0
Book 4	Borrower 1	10.0	16.7	10.0
Book 5	Borrower 4	10.0	8.3	0.0
Grand Total		100.0	100.0	100.0

Fig. 9. Measures (direct and calculated) of the *Loan* star schema

gives an overall overview of the total (SUM) or average (AVG) number of days by borrower, book or attendant, the second clearly shows the importance of each borrower, attendant or book for the business (Fig. 10).

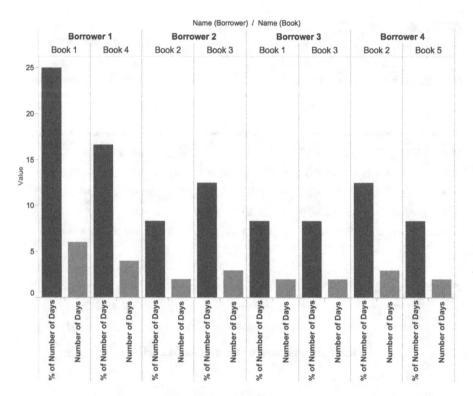

Fig. 10. Integrated analysis of direct and calculated measures

5 Conclusions

In the identification of the data model for a DW, data-driven approaches are commonly used, having as limitations the fact that behavioral aspects of the system are ignored and that several manual decisions and transformations are needed to obtain a DW model. The approach presented in this paper, based on a process-based approach, derives a DW model, an *ADM*, using as input a set of interrelated business processes modeled in BPMN 2.0 and the corresponding *ODM*, in a process where a set of rules automatically produce a multidimensional model.

Dimensions, facts, measures and relationships, used for analyzing the business processes, integrate the *ADM*. A subset of measures, denominated calculated measures, is derived from the direct measures inherited from the *ODM*, providing useful insights of the data under analysis.

This approach enables business processes to be adjusted accordingly to organizational needs, which might directly implicate a new DW model when the business processes change. Thereby, BI and BPM initiatives can now be seen as more integrated projects, benefiting organizations and their corresponding decision-making process.

As future work, it is pointed out the need to validate the proposed approach with a more complex scenario in a real context, including a larger set of interrelated business processes. Moreover, the inclusion of more calculated measures, useful for decision making, is envisaged.

Acknowledgments. This work has been supported by COMPETE: POCI-01-0145-FEDER-007043 and FCT (*Fundação para a Ciência e Tecnologia*) within the Project Scope: UID/CEC/00319/2013, and by Portugal Incentive System for Research and Technological Development, Project in co-promotion n° 002814/2015 (iFACTORY 2015-2018).

References

1. van der Aalst, W.M.: Business process management demystified: a tutorial on models, systems and standards for workflow management. In: Desel, J., Reisig, W., Rozenberg, G. (eds.) Lectures on Concurrency and Petri Nets, Advances in Petri Nets. LNCS, vol. 3098, pp. 1–65. Springer, Heidelberg (2004)
2. Watson, H.J., Wixom, B.H.: The current state of business intelligence. IEEE Comput. **40**(9), 96–99 (2007). doi:10.1109/MC.2007.331
3. Melchert, F., Winter, R., Klesse, M.: Aligning process automation and business intelligence to support corporate performance management. In: Proceedings of the Tenth Americas Conference on Information Systems, pp 4053–4063 (2004)
4. Mili, H., Tremblay, G., Jaoude, G.B., Lefebvre, É., Elabed, L., Boussaidi, G.: Business process modeling languages: sorting through the alphabet soup. ACM Comput. Surv. **43**(1) (2010). Doi:10.1145/1824795.1824799, http://doi.acm.org/10.1145/1824795.1824799
5. Sturm, A.: Enabling off-line business process analysis: a transformation based approach. In: BPMDS (2008)
6. Kaldeich, C., Oliveira e Sá, J.: Data warehouse methodology: a process driven approach. In: Persson, A., Stirna, J. (eds.) CAiSE 2004. LNCS, vol. 3084, pp. 536–549. Springer, Heidelberg (2004)
7. Böehnlein, M., Ulbrich vom Ende A.: Business process oriented development of data warehouse structures. In: Proceedings of Data Warehousing, Physica Verlag (2000)
8. Dori, D., Feldman, R., Sturm, A.: Transforming an operational system model to a data warehouse model: a survey of techniques. In: Proceedings of the IEEE International Conference on Software – Science, Technology and Engineering (SwSTE), pp. 47–56 (2005)
9. OMG: Business process model and notation (BPMN), version 2.0. Technical report, Object Management Group (2011)
10. Cruz, E., Machado, R.J., Santos, M.Y.: Deriving a data model from a set of interrelated business process models. In: ICEIS 2015 - 17th International Conference on Enterprise Information Systems, vol. 1, pp 49–59 (2015)
11. Golfarelli, M., Rizzi, S., Vrdoljak, B.: Data warehouse design from XML sources. In: Proceedings of the 4th DOLAP (2001)
12. Hüsemann, B., Lechtenbörger, J., Vossen, G.: Conceptual data warehouse design. In: Proceedings of the 2nd DMDW, Stockholm, Sweeden (2000)
13. List, B., Schiefer, J., Tjoa, A.M., Quirchmayr, G.: Multidimensional business process analysis with the process warehouse. In: Abramowicz, W., Zurada, J. (eds.) Knowledge Discovery for Business Information Systems, pp. 211–227. Kluwer Academic Publishing, Dordrecht (2000)

14. Rizzi, S., Abelló, A., Lechtenbörger, J., Trujillo, J.: Research in data warehouse modeling and design: dead or alive? In: Proceedings of the 9th ACM International Workshop on Data Warehousing and OLAP (DOLAP 2006), pp. 3–10. ACM (2006)
15. Cruz, E., Machado, R.J., Santos, M.Y.: From business process modeling to data model: a systematic approach. In: 2012 Eighth International Conference on the Quality of Information and Communications Technology, pp. 205–210 (2012). doi:10.1109/QUATIC.2012.31
16. Weske, M.: Business Process Management Concepts, Languages, Architectures. Springer, Heidelberg (2010)
17. Meyer, A.: Data in business process modeling. In: 5th Ph.D. Retreat of the HPI Research School on Service-oriented Systems Engineering (2010)

A Novel Voting Mathematical Rule Classification for Image Recognition

Sadrollah Abbasi[✉], Afshin Shahriari, and Yaser Nemati

Department of Computer Engineering,
Iran Health Insurance Organization, Yasouj, Iran
sadrollahsadeghi@gmail.com,
yasernamati@gmail.com

Abstract. In machine learning, the accuracy of the system depends upon classification result. Classification accuracy plays an imperative role in various domains. Non-parametric classifier like K-Nearest Neighbor (KNN) is the most widely used classifier for pattern analysis. Besides its easiness, simplicity and effectiveness characteristics, the main problem associated with KNN classifier is the selection of number of nearest neighbors i.e. 'k' for computation. At present it is hard to find the optimal value of 'k' using any statistical algorithm, which gives perfect accuracy in terms of low misclassification error rate.

Motivated by prescribed problem, a new sample space reduction weighted voting mathematical rule (AVNM) is proposed for classification in machine learning. Proposed AVNM rule is also non-parametric in nature like KNN. AVNM uses the weighted voting mechanism with sample space reduction to learn and examine the predicted class label for unidentified sample. AVNM is free from any initial selection of predefined variable and neighbor selection as found in KNN algorithm. Proposed classifier also reduces the effect of outliers.

To verify the performance of the proposed AVNM classifier, experiments are made on 10 standard dataset taken from UCI database and one manually created dataset. Experimental result shows that the proposed AVNM rule outperforms KNN classifier and its variants. Experimentation results based on confusion matrix accuracy parameter proves higher accuracy value with AVNM rule.

Proposed AVNM rule is based on sample space reduction mechanism for identification of optimal number of nearest neighbor selections. AVNM results in better classification accuracy and minimum error rate as compared with state-of-art algorithm, KNN and its variants. Proposed rule automates the selection of nearest neighbor selection and improves classification rate for UCI dataset and manual created dataset.

Keywords: KNN · Classification · Lazy learning

© Springer International Publishing Switzerland 2016
O. Gervasi et al. (Eds.): ICCSA 2016, Part V, LNCS 9790, pp. 257–270, 2016.
DOI: 10.1007/978-3-319-42092-9_20

1 Introduction

Machine learning is the branch of engineering discipline that sightsees the description and learns through algorithmic perspective using data. The machine learning algorithms use the input pattern of the sample data to learn and create their knowledge base. Using their pre-stored knowledge base and algorithmic computation, machine learning based systems help to identify the undiscovered sample to its corresponding class. One of the most widely used non-parametric classifier is Nearest Neighbor (NN) rule proposed by the Fix and Hodges [1] in 1951 for pattern analysis. NN rule is simple and most effective among all non-parametric classifiers. Besides its simplicity, it is also considered as lazy classifier, because it doesn't use the training set for any type of generalization. In 1967, K-Nearest Neighbor was proposed by Cover and Hart [2] which used 'k' nearest samples for decision making. The working methodology of KNN classifier for an unidentified class sample is based on approximate number of the nearest neighbors represented by the 'k'. Among training sample the class instance associated with the majority of the 'k', be the class label of unidentified sample in testing dataset.

The performance of the classifier is dependent on the selection of the user defined value of 'k'. This value is range specific and dependent of data set i.e., $k = [1, n]$, where n is the number of data samples in training set. If k is taken as 1 i.e. $k = 1$, then the test sample is assigned a label of its most nearest trained data sample but the effect of noise is noticed on classification result. While the higher value of k, reduces the effect of noise but makes the frontiers of classes less distinct. Thus the proper selection of the value of 'k' is always a key concern area for various applications.

Besides the nearest neighbor selection parameter, another issue with 'k' value was that it must be taken as an odd number [2]. Odd value of 'k' results in good performance for binary classification problems. Using KNN in multiclass problems, the performance of KNN degrades as compared with binary class problems. Another factor which was associated with KNN is the metric used for neighborhood selection. Distance measures using Euclidean distance, Manhattan distance and Minkowski distance are the most widely used distance metrics for neighborhood selection. Each of the distance metric has its own importance in finding the similarity or dissimilarity between the sample points. Thus selection of the distance metric is another issue in machine learning based classification.

Motivated by the above key issues in respect of finding the most appropriate value of nearest neighbor for any application domain, a new mathematical rule is proposed for finding the closest sample to the query vector. Proposed rule is based on the sub space reduction of the training space based on the mathematical formulation which helps in automatically finding the range variation of number of closest sample. Further, the weighting factor is associated with the class to provide the highest weight to most closely linked sample and minimum weight to farthest sample. This helps the model to find the nearest group of the unidentified sample to which it most probably relates to. The weighted mechanism also helps in resolving the issue of tie, i.e. when the nearest neighbors associate with equal number of class label count. In such situation proposed weighted mechanism assign proper biasing to sample space to get nearest class sample label to unidentified data point.

2 Related Work

In machine learning based accuracy dependent systems, the choice of the classifier plays most informative role. The analysis of the performance of the system relates to the outcome of the classifier, i.e. how precisely the classifier assigns the samples to their specific class. Thus the choice of the classifier for a specific problem must have certain characteristics like ease of understanding sample data, simple to understand and properly classify the unidentified sample.

K-Nearest Neighbor (KNN) is one such classifier which came into existence in 1967 by Cover and Hart [2]. It is defined as a linear and non-parametric classifier in pattern recognition. Due to its simplicity and effectiveness it becomes most likely used classifier in research community. KNN rule based classification is based on the concept of finding 'k' nearest neighbors in sample data and based on the majority class label of nearest neighbors the unidentified sample provided its class label.

The main characteristics of KNN rule is defined on the basis of its 'k' value in [3]. The convergence of the probability error of nearest neighbor rule when the samples lie in n-dimensional sample space converges to factor 'R' with probability 1. Among the 'n' known samples, the nearest neighbor rule convergence to 'R' is given as $R^n \leq R \leq 2R^n(1 - R^n)$ as 'n' approaches to ∞ [2]. While the extension of the convergence on the basis of asymptotic performance measure over 'k' nearest neighbor is presented in detail in [4]. The main characteristics discovered over the KNN rule are given as: (a) As the number of samples in the sample space tends towards infinity, then the error rate of the KNN approximates the optimal Bayesian error rate and (b) at $k = 1$, i.e. the minimum closest distance sample to query sample in sample space, the error rate of KNN is constrained higher than twice the Bayesian error rate.

Over the certain advantages of KNN, there are some limitations of KNN which are addressed in [5]. These limitations are (a) Expensive Testing Issue- when the number of attributes is large in number with huge amount of instances in the database. (b) Sensitivity issue- when the dataset are noisy or consists of irrelevant and unbalanced features and (c) Similarity/Dissimilarity metric issue- when the nearest neighbor selection is based over pre initialized value of 'k' with variety of metrics like Euclidean distance, Manhattan distance or Minkowski distance.

One another issue regarding the assigning of the membership to group on basis of nearest neighbor and its solution is proposed in [6, 7]. The concept of fuzzy membership is introduced in KNN to find the membership of the sample to particular class. Three different type of membership assignment technique are introduced in [6], first method is based on the crisp labeling of sample data, second method is based on the predefined procedure presented in [8] and third method is based on the labeled samples according to its nearest neighbor. While in [7], the concept of fuzzy in nearest neighbor selection was proposed for document classification. Besides the advantage of membership value there is still a problem left with selection of the optimal value of 'k' which was unanswered.

In [9], another variant of KNN is proposed named Informative K-Nearest Neighbor (IKNN) classifier for pattern recognition. Here authors focused on the problem associated with selection of the optimal value of 'k' for neighborhood selection. They proposed two novel algorithms based on object informative property. These approaches are local informative KNN and global informative KNN based on the concept of selecting most informative objects in neighborhood. Where the informative property is defined as the object which is close to the query point and provide maximum information while on the other hand be far away from objects with different class labels.

Another modification on KNN was proposed in [10] based on idea of assigning weights to data objects and thus considered as weighted KNN approach. The proposed approach consists of two parts i.e. initially compute the validity factor of complete dataset followed by the weight assignment. Still the validation of the sample points is based on considering 'H' nearest neighbor points. Finally the weight is assigned to nearest sample using the multiplication of validity factor of data point with inverse of Euclidean distance between points with summation of constant value 0.5.

A more versatile weighted approach based on the idea of assigning more weight to most nearest point to query sample and least to distant sample is presented in [11]. The weighted mechanism uses the distance measure of 'k' nearest samples to minimum distance sample in dataset for obtaining weighting factor which lies in the range of 0 to 1. While a uniform weighted mechanism proposed in [12] uses the ratio of $\frac{1}{k}$ to assign the weight to sample data points. This approach was also dependent of the selection of the 'k' nearest neighbors initially and on basis of the ratio the nearest sample got maximum weight of '1' which the 'kth' sample got ratio of $\frac{1}{k}$ as weight.

The main advantage of above discussed uniform weighted approach is to give maximum weight to nearest neighbor and took less computational time as compared with other weighted nearest neighbor algorithms. In all the discussed weighted nearest neighbor based classification algorithms, the selection of the optimal value of 'k' is the main problem, this issue was raised and discussed in detail in [13]. Besides it, the above discussed algorithms still lack in adjusting the distribution of weights based on neighborhood to query object. The solution of the above said problem was introduced using dual weighted nearest neighbor selection algorithm in [14]. Here the sample data object got the dual weighting adjustment based on the uniformity and nearest neighbor. This proposed solution is the extended version of the weighted approaches discussed in [11, 12]. The dual weighting factor is the linear multiplication of both the weighted factors to get dual weight to sample data object in sample space.

To address the various issues over the choice of appropriate number of nearest neighbors with improvement in the classification accuracy in machine learning and pattern recognition domain was presented in [15–17, 21]. While issues related to the size of training sample set and its impact on classification performance using nearest neighbor rule was presented in [18].

3 Proposed Algorithm

As per the literature, in machine learning the most widely used classifier is KNN. Besides its simplicity KNN has certain limitations like selection of the number of nearest neighbor taken into consideration and type of distance metric used. Besides the above discussed issues certain flaws are discovered in respect of associating weight to nearest neighbors, which are discussed in detail in previous section.

In order to resolve the issue of selection of the appropriate number of nearest neighbors for unidentified sample a new mathematical rule is proposed which is named as AVNM algorithm. AVNM is a like a variant of KNN, which automatically identifies the optimal number of nearest neighbors using sample space reduction mechanism. The description of AVNM is discussed in detail below: Indeed the main idea behind the proposed clustering framework is the usage of ensemble learning concept in the field of ACC. The MACCA is very sensitive to the proper initialization of its parameters. If one employs a proper initialization for parameters of the MACCA, the final discovered clusters can be desirable. On the other hand, an ensemble needs diversity to be successful. It can be inferred that by several runs of MACCA with different initializations for parameters, we can reach several partitions that are very diverse. So we use an ensemble approach to overcome the problem of well-tuning of its parameters.

3.1 Working Algorithm of AVNM

The summarized working algorithm of the proposed AVNM is given below.

Input:
     TR: Training vector space.
     $\{Si, c\},$ $i = \{1, 2, 3, ..., N\};$ $Si \in$ Training Samples;
        $c \in$ class label; $N \in$ no. of training samples;
     x' : Query vector/Test sample.

Step 1: Calculate sum of sample and query vector
     for $i = 1\,to\,N$
          $SSum_i = \sum_k f_k$: $f_k \in$ feature in i^{th} sample
         // $SSum_i \in$ Summation of features for i^{th} sample
     end
          $QSum = \sum_k f_k$; // Sum of query vector features

Step 2: Compute distance between all samples to query vector
     for $i = 1\,to\,N$
          $dist_i = |\,SSum_i - QSum\,|$
     end

Step 3: Ranking closely related samples
     $Rank_i = sort\,(dist)$; // sort in increasing order
     //where, $dist = \{\,dist_1,\,dist_2, ...,\,dist_N\}$

Step 4: Feature Distance computation
     $Fmin = Rank_1$; Fetch I^{st} minimum sample
     for $i = 1\,to\,k$ // for each 'k' features
          $SFmean_i = mean\,(\,f_k)$
//where, $SFmean_i$ is the mean value of ith feature for all
// samples
     end

  Step 5: Sample Space Reduction
     // Generating threshold
     While (! conserve or no elimination of samples)
       Do
         $T = (\sum SFmean_i\,) - QSum$
         for $i = 1\,to\,N$
           if ($dist_i > T$)
             eliminate S_i ;
         // eliminate sample having distance > T
         // Where $S_i \in S$; S is a sample set
           update S;
         end if
        end for
       goto: step 4
       end while

Step 6: Fetching number of closest samples to Query
$$k = size\ (S)\ //\ \text{leftover samples in sample set}$$

Step 7: Weight Initialization
$$D_1 = dist\ (Rank_1)$$
$$D_k = dist\ (Rank_k)$$
$$for\ i = 1\ to\ N'$$
$$//\ where\ N'\ is\ no.\ of\ samples\ in\ updates\ S$$
$$W_i = \frac{D_k - D_i}{D_k - D_1}\ //\text{Assigning weight to each sample}$$
$$end$$

Step 8: Weighted neighbor computation
$$for\ i = 1\ to\ N'$$
$$WS = \sum W_{ic}\ ;\ //\text{Where}\ c\ is\ sample\ class\ label$$
$$//\ \text{fetching total weight for respective class}$$
$$end$$

Step 9: Class label assignment
$$if\ (size\ (S) == 1)$$
$$Qc = c(S)\ //\ \text{Query class label} = \text{sample class label}$$
$$else$$
$$Qc = arg\ max\ (\sum WS_k)$$
$$//WS_k\ is\ total\ weighted\ sum\ for\ class\ k$$

3.2 Descriptive AVNM

In machine learning, classification is one such task which help the system to maintain its accuracy. Generally it is called that as better the classification task is, better the system accuracy formed. One such algorithm is KNN, a linear non parametric nearest neighbor rule. KNN uses the distance metric function to select the pre initialized number of nearest neighbors with supervised class label associated with each sample. The majority winning class label is associated to unidentified sample based of nearest neighbors.

Considering all such issues of the KNN and other algorithms, as discussed in literature in Sect. 2, a new mathematical rule is proposed named AVNM. AVNM is free from any initial selection of the parameter like number of nearest neighbor in KNN and its variants proposed in [7–11]. The main strength of proposed rule is iteratively reduction of the samples which are taken into consideration for neighborhood computation. The detail description of the algorithm is given as,

Let 'S' be an initial sample space consisting of 'S' sample points associated with certain class label, 'S' be the query vector or unidentified sample found in sample space as shown in Fig. 1.

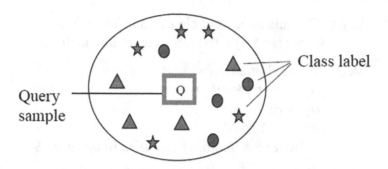

Fig. 1. Initialized sample space '*S*'

In Fig. 1, there are three classes shown in sample space with one query sample whose class to be predicted based on the 'k' closely associated sample set. The goal of the approach is to assign a proper group to query sample based on the property of 'most closely associativity. To find the associativity of the query sample, it is calculated that how far is 'q' from every sample 'S_i' in sample space.

Proposed AVNM algorithm is a nine step procedure which is discussed step by step. In step 1, the sample data and the query vector is initialized. Here each sample consists of several features which are associated with the class label which is given as,

$$S = \{f_1, f_2, f_3, \ldots \ldots, f_n | c_i\} \tag{1}$$

$$q = \{f_1', f_2', f_3', \ldots \ldots, f_n'\}$$

where, f_x is the feature value and c_i be the class label for any sample. Further, the algorithm computes the sum of the feature value for all samples in sample space and query vector, given in Eq. (2). These value are used to compute the distance between all sample point to query vector in sample space as given by distance Eq. (3),

$$SSum_i = \sum_k f_k \text{ and } Q_{sum} = \sum_k f_k \tag{2}$$

$$dist_i = |SSum_i - Qsum| \tag{3}$$

On the basis of the distance the algorithm rank the samples in increasing order as given below,

$$Rank = sort(dist) \tag{4}$$

After computation of the ranking of samples, the main task is to find the threshold using which the sample space is reduced by eliminating samples which are at maximum dissimilarity. At first step the mean of each feature in sample space is computed as given in step 4 of algorithm and after then the absolute distance is calculated between the sum of mean feature vector and query vector as directed in Eq. 5. The computed difference value is used as a threshold to eliminate the samples from sample

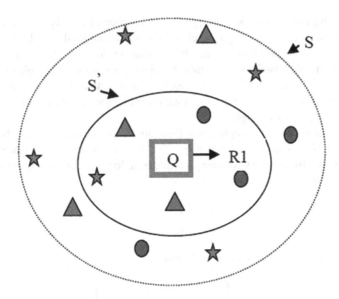

Fig. 2. Reduced sample space from S to S'

space having distance greater than computed threshold, the detailed procedure are shown in Fig. 2.

$$T = \left(\sum SFmean \right) - QSum_j \qquad (5)$$

In the above Fig. 2, it is shown that using the threshold value T, the original sample space S is reduced to S' by eliminating samples having distance greater than T. The process iteratively computes the threshold at each run of the algorithm, as threshold computation is based on the number of samples exist in sample space and sample space reduces on every threshold value till there is no change in sample space seen. The value R1 shown in Fig. 2, is the first minimum value (called ranked 1) sample. This sample is the closest sample to query vector.

At the state of convergence of algorithm, when no change in sample space is observed, the number of sample points exist in the sample space S' be considered as the number of closely related samples to query vector 'q', called 'k'. The formed sub sample space S' helps to get the required number of closely associated samples to query vector for a particular application domain. Thus it helps to remove the pre initialization of the value of 'k' as seen in KNN. The most closely linked sample to query vector is computed as shown in Eq. 3. This closely linked value is the absolute difference between the sum of all feature values of sample data points and query vector. Rather than taking feature by feature difference as in KNN, the sample distance is preferred over feature based distance because a particular sample is formed by associating various features which have their own importance. Thus the overall weighting factor sum of sample is used to compute absolute distance between sample.

After getting the reduced sample space and number of closely linked samples to query vector, next step is to assign the class label to unidentified sample. The solution of the class prediction problem is given by the concept to assign the proper weighted factor to each sample based on its neighborhood to query vector. The rule of assigning weight is in such order that the most closely related sample to query sample in the reduced sample space S' get the maximum weight while the sample which is far enough to query sample get least weight.

Suppose that in reduced sample space there are 'k' points. The distance from query sample to this 'kth' sample is denoted by D_k. The nearest sample distance to query sample is denoted by D_1. Now for every sample in reduced sample space the weighting factor is given as,

$$W_i = \frac{D_k - D_i}{D_k - D_1} \tag{6}$$

It is noticed that the value of W_i is found as,

$$W_i = \left\{ \begin{array}{c} 1; i = 1 \\ 0 < W < 1; 1 < i < k \\ 0; i = k \end{array} \right\} \tag{7}$$

It is noted that the sample having weight 1 is closest sample to query vector while sample having weight 0 is considered as the farthest sample to query and lie on the boundary of reduced sample space. Samples which lie beyond the boundary of sample space are the eliminated samples by the proposed rule. As all samples are restricted the hold position either inside sample space, including boundary region, or outside sample space, and as discussed above that sample lie up to boundary are considered in voting for class prediction while beyond the boundary are eliminated samples which play no role in voting, thus there is no outlier point observed using proposed rule.

4 Experimental Study

To evaluate the performance of the proposed mathematical rule in machine learning system, the analysis of the error rate or classification accuracy is most effective metric used in model building. Some of the performance analysis measures used in machine learning system are given in Table 1. These measures are used for evaluating the classifier accuracy.

Table 1. Performance analysis measure.

TP	FP	Precision = $TP/(TP + FP)$
FN	TN	Negative Predict value = $TN/(FN + TN)$
Sensitivity = $TP/(TP + FN)$	Specificity = $TN/(FP + TN)$	Accuracy = $(TP + TN)/Total$

Where TP, FP, FN, TN are described as True Positive, False Positive, False Negative, True Negative values respectively in Table 1.

In experimental work, the proposed classifier is tested on two phases. In the first phase, the experiments were performed with proposed AVNM rule over 10 standard datasets collected from the UCI machine learning repository [19]. All the selected datasets consists of real, integer values with multivariate data type. The dataset description taken for experimental work is shown in Table 2.

Table 2. Selected dataset from UCI data source [19].

Dataset	Attributes	Instances	Classes
Pen digits	16	10992	10
Optical digits	64	5620	10
Ionosphere	34	351	2
Glass	10	214	7
Libras	90	360	15
Wine red	11	1599	11
Zoo	17	101	7
Vehicle	18	946	4
Wine white	11	4898	11
Image segmendation	19	2310	7

All the datasets have high range of variability between number of attributes, instances and classes. In the second phase of the experimentation, a manually created dataset of T2-Weighted post contrast axial MR images of high grade malignant brain tumors is used, which was collected from Department of Radiology, SMS Medical College Jaipur, Rajasthan, India. The description about the manually created dataset is shown in Table 3.

In experimental work, data is selected using cross validation Leave-One-Out (LOO) methodology [13]. Cross validation separates the whole dataset into two parts named training and testing. LOO method gives equal weightage to each object in database, thus every object in a database act as a query sample once and rest all act like training samples in particular dataset.

4.1 Experimental Results Using UCI Datasets

The experimental results using proposed AVNM rule over 10 standard datasets from UCI data repository is presented in Table 4. In order for comparison purpose with existing methodologies present in literature, the experimental results presented in [13] over similar dataset is used in addition with proposed AVNM rule.

The values presented above in Table 4 in form $x(k)$ where x represents the error percentage and k represents the selected nearest neighbors value at which algorithm produces best result [13].

Table 3. Manual dataset used in experimental work.

Tumor type	#samples
Central Neuro Cytoma (CNC)	133
Glioblastoma Multiforme (GBM)	160
Gliomas (GLI)	155
Intra Ventricular Malignant Mass (IVMM)	152
Metastasis (MTS)	160

Table 4. Experimental result showing error (%) with 'k' value.

Dataset	KNN	WKNN	UWKNN	DWKNN	AVNM
Pen digits	0.58 (3)	0.53 (6)	0.55 (4)	0.55 (22)	0.50 (13)
Optical digits	1.00 (4)	0.96 (7)	1.01 (8)	1.03 (32)	0.96 (7)
Ionosphere	13.39 (1)	13.39 (1)	13.39 (1)	12.54 (14)	9.52 (9)
Glass	26.64 (1)	26.64 (1)	26.64 (1)	25.23 (8)	24.68 (8)
Libras	12.78 (1)	12.78 (1)	12.78 (1)	12.50 (6)	12.62 (9)
Wine red	38.46 (1)	38.15 (6)	38.46 (1)	36.52 (44)	34.54 (17)
Zoo	1.98 (1)	1.98 (1)	1.98 (1)	1.98 (1)	1.98 (4)
Vehicle	33.45 (5)	32.74 (6)	33.45 (7)	33.92 (30)	33.74 (19)
Wine white	38.38 (1)	38.36 (4)	38.38 (1)	38.10 (17)	34.47 (16)
Image segmentation	3.33 (1)	3.12 (6)	3.33 (1)	3.39 (9)	1.98 (5)

4.2 Experimental Results with Manual Dataset

In order to test the proposed AVNM rule in classification of malignant brain tumors in MR images, five types of Grade IV tumors are taken into consideration as shown in Table 3. Before applying the proposed AVNM classification rule to the dataset, the dataset is preprocessed initially. The preprocessing of the dataset includes feature extraction of dataset images, followed by selection of the relevant features among extracted feature set pool. The various feature extraction methods and relevant feature selection mechanism used for selection of features are presented in [20]. The experimental results are shown in Table 5.

Table 5. experimental results for tumor classification.

Dataset	K-NN	WKNN	UWKNN	DWKNN	AVNM
(CNC)	6.01 (9)	5.38 (7)	5.8 (1)	5.38 (7)	4.51 (22)
(GBM)	3.75 (7)	3.36 (5)	3.66 (3)	3.20 (3)	2.50 (11)
(GLI)	2.58 (3)	2.58 (3)	2.33 (11)	2.58 (3)	1.98 (15)
(IVMM)	4.60 (7)	3.58 (5)	4.06 (8)	3.22 (9)	2.63 (14)
(MTS)	3.75 (5)	3.75 (3)	3.7 (7)	3.57 (5)	2.50 (16)

5 Conclusion and Future Works

This paper presents the novel weighted voting based mathematical rule (AVNM) for classification in machine learning based systems. Proposed AVNM rule uses the sample space reduction mechanism by eliminating samples having higher dissimilarity based on threshold T. The idea behind this is to reduce the sample space of nearest neighborhood selection. The sample space keeps on reducing till it gets converged. Once the sample space converged then the number of samples formed inside the final reduced sample space is considered as the maximum number of nearest samples to unidentified query sample. To reduce this sample space initially the mean sample is computed. Using the distance between the mean and query sample, the threshold is computed as discussed in algorithm. Once the threshold is set then, the sample's whose distances are greater than the threshold limit be dropped out thus resulting in sample space reduction. Finally the numbers of samples present in the reduced sample space are considered as the nearest neighbors, reducing the pre initialization problem of 'k' value as present in algorithms like KNN and its variants.

Further, the proposed rule uses the weighting mechanism for giving more weight to closest sample to query point and minimum weight to farthest point to query sample. Later this weighted factor is summed up for respective class samples found in reduced sample space to get desired class label to unidentified query sample using voting mechanism of nearest neighbors.

Proposed AVNM rule is experimented with 10 standard datasets taken from UCI data repository and one manually created dataset of malignant tumors. The experimented results are presented in Tables 4 and 5. The bold indicated values represent the lowest error rate achieved using particular mechanism in dataset classification. It is noticed that the proposed AVNM rule gives satisfactory results for both the experiments i.e. standard dataset and manually created dataset. In Table 4, out of the 10 datasets, 8 times AVNM outperforms the other classification rules present in literature. While in case of manually created dataset AVNM outperforms in all tumor images dataset as compared with KNN algorithm.

References

1. Fix, E., Hodges, J.L.: Discriminatory analysis, non parametric discrimination: consistency properties. Technical report No. 4, U.S. Air Force school of aviation Medicine, Randolf field Texas (1951)
2. Cover, T.M., Hart, P.E.: Nearest neighbor pattern classification. IEEE Trans. Inf. Theor. **13** (1), 21–27 (1967)
3. Wagner, T.: Convergence of the nearest neighbor rule. IEEE Trans. Inf. Theor. **17**(5), 566–571 (1971)
4. Hastie, T., Tibshirani, R., Friedman, T.: The Elements of Statistical Learning, Chap. 2, 2nd edn., pp. 9–38, springer, Heidelberg (2009)
5. Data Mining Algorithms in R. http://en.wikibooks.org/wiki/Data_Mining_Algorithms_In_R/Classification/kNN

6. Keller, J.M., Gray, M.R., Givens, J.A.: A fuzzy K-nearest neighbor algorithm. IEEE Trans. Syst. Man Cybern. **15**(4), 580–585 (1985)
7. Shang, W., Huang, H.-K., Zhu, H., Lin, Y., Wang, Z., Qu, Y.: An improved kNN algorithm – fuzzy kNN. In: Hao, Y., Liu, J., Wang, Y.-P., Cheung, Y.-M., Yin, H., Jiao, L., Ma, J., Jiao, Y.-C. (eds.) CIS 2005. LNCS (LNAI), vol. 3801, pp. 741–746. Springer, Heidelberg (2005)
8. Keller, J., Hunt, D.: Incorporating fuzzy membership functions into the perceptron algorithm. IEEE Trans. Pattern Anal. Mach. Intell. **7**(6), 693–699 (1985)
9. Song, Y., Huang, J., Zhou, D., Zha, H., Giles, C.: IKNN: informative K-Nearest neighbor pattern classification. In: Kok, J.N., Koronacki, J., Lopez de Mantaras, R., Matwin, S., Mladenič, D., Skowron, A. (eds.) PKDD 2007. LNCS (LNAI), vol. 4702, pp. 248–264. Springer, Heidelberg (2007)
10. Parvin, H., Alizadeh, H., Bidgoli, B.M.: MKNN: modified K-nearest neighbor. In: Proceedings of the World Congress on Engineering and Computer Science (WCECS), pp. 22–25 (2008)
11. Dudani, S.A.: The distance weighted K-nearest neighbor rule. IEEE Trans. Syst. Man Cybern. **6**, 325–327 (1976)
12. kang, P., Cho, S.: Locally linear reconstruction for instanced based learning. J. Pattern Recogn. **41**(11), 3507–3518 (2008)
13. Wu, X.D., et al.: Top 10 algorithms in data mining. J. Knowl. Based Inf. Syst. **14**, 1–37 (2008)
14. Gou, J.: A novel weighted voting for K-nearest neighbor rule. J. Comput. **6**(5), 833–840 (2011)
15. Mitani, Y., Hamamoto, Y.A.: Local mean based nonparametric classifier. Pattern Recogn. Lett. **27**, 1151–1159 (2006)
16. Zeng, Y., Yang, Y., Zhao, L.: Pseudo nearest neighbor rule for pattern classification. J. Expert Syst. Appl. **36**, 3587–3595 (2009)
17. Wang, J., Neskovic, P., Cooper, L.N.: Neighborhood size selection in K-nearest neighbor rule using statistical confidence. J. Pattern Recogn. **39**, 417–423 (2006)
18. Fukunaga, K.: Introduction to Statistical Pattern Recognition, 2nd edn. Academic press, Cambridge (1990)
19. Lichman, M.: UCI Machine Learning Repository, Irvine, CA: University of California, School of Information and Computer Science (2013). https://archive.ics.uci.edu/ml/datasets.html
20. Vidyarthi, A., Mittal, N.: Utilization of shape and texture features with statistical feature selection mechanism for classification of malignant brain tumours in MR images. J. Biomed. Technol. Walter de Gruyter **59**, 155–159 (2014)
21. Bhattacharya, G., et al.: Outlier detection using neighborhood rank difference. J. Pattern Recogn. **60**, 24–31 (2015)

Towards a Common Data Framework
for Analytical Environments

Maribel Yasmina Santos[(✉)], Jorge Oliveira e Sá, and Carina Andrade

ALGORITMI Research Centre, University of Minho, Guimarães, Portugal
{maribel,jos,carina.andrade}@dsi.uminho.pt

Abstract. In business contexts, data from several sources are usually collected, integrated, stored and analyzed to support decision-making at the operational, tactical or strategic organizational levels. For these several levels, with different data needs, this paper proposes a Common Data Framework for identifying and modeling the different data repositories that support those data needs. The proposed approach considers all the business needs expressed in business processes models and that were useful in the identification of the operational data model, for on-line transaction processing, the analytical data model, for a data warehousing system, and on-line analytical processing cubes for a traditional Business Intelligence environment. Besides providing an integrated framework in which the data models complement each other, this approach also allows checking the conformity of the data stored in the different repositories, being possible to identify business requirements that were not properly modeled.

Keywords: OLAP cubes · Analytical data model · Business Intelligence · Business Process Management · Business process model and notation · Data warehousing · Operational data model

1 Introduction

For benefiting from processes management and decision-making improvements, organizations should first see these two business facilitators working together, rather than working apart and serving separate purposes. Usually, Business Intelligence (BI) initiatives do not take into consideration business processes, in terms of how they work, and Business Process Management (BPM) initiatives do not specify metrics to measure the business, providing an aggregate and overall perspective of it. Both initiatives deliver value to the business, but each one in its own way.

Organizations are linking BPM and BI [1, 2] and, in this work, it is proposed that BI and BPM initiatives should be integrated by a process-driven approach, which means by modelling business processes and managing the flow of data through an organization and by making explicit decision points at which human interaction is required. Working in an integrated way, BI can place the right data in the hands of the right users exactly when they need it, covering operational, tactical or strategic organizational levels, which processes were characterized and modeled by BPM initiatives.

In this paper, it is proposed an integrated approach supported by a Common Data Framework (CDF) to fulfill all the business needs expressed in business processes

© Springer International Publishing Switzerland 2016
O. Gervasi et al. (Eds.): ICCSA 2016, Part V, LNCS 9790, pp. 271–287, 2016.
DOI: 10.1007/978-3-319-42092-9_21

models, which were useful in the identification of the operational data model [3, 4] which is the basis of on-line transaction processing and the analytical data model [5] that supports a Data Warehouse (DW) system. To complete the overall approach, on–line analytical processing cubes are needed to support analysis and visualization of data in a traditional BI environment. Given this context, the contribution of this paper is the proposal of the CDF and derivation of the data cubes, based on the data available in the business models and other repositories used in the framework.

This paper is organized as follows. Section 2 presents the related work, pointing how analytical data models are usually identified. Section 3 presents and discusses the CDF. Section 4 presents a set of rules for the automatic identification of data cubes, based on other information available in the CDF. Section 5 shows, using a demonstration case, how the overall approach works together. Section 6 concludes with a discussion of the main findings and advantages, and pointing guidelines for future work.

2 Related Work

Based on the principles suggested by [3, 4], in which a set of interrelated business processes models are used to derive an operational data model supporting On-Line Transaction Processing (OLTP), the work of [5] extends this approach and makes use of the same set of interrelated business process models and information available in the operational data model to derive an analytical data model that constitutes the basis for a DW system. As already mentioned, the work here proposed will integrate all these findings in a CDF and will propose the automatic identification of the data cubes for the analytical data model, in order to have an overall and comprehensive overview of the data needs for the several business managers, as these data cubes will support On-Line Analytical Processing (OLAP) tasks.

Looking into the related literature, there are several authors suggesting approaches for obtaining analytical data models in an automatic or nearly automatic way [6–10]. [6] suggested a tool that generates a schema for OLAP from conceptual graphical models and uses graph grammars for specifying and parsing graphical multidimensional schema descriptions. [7] proposed an approach towards the automation of DW relational design through a rule-based mechanism, which automatically generates the DW schema by applying existing DW design knowledge. The proposed rules embed design strategies, which are triggered by conditions on requirements and source databases, and performs the schema generation through the application of predefined DW design oriented transformations. [8] presented a conceptual model (StarER) for DW design based on the user's modeling requirements, which combines the star structure, dominant in DW, with the semantically rich constructs of the entity-relationship (ER) model. [9] described an automatic tool for generating star schemas implemented by a prototype named SAMSTAR, which was used for the automatic generation of star schemas from entity-relationship diagrams (ERD). Finally, [10] proposed a different approach that is able to generate the analytical data model and to populate it with operational data.

All these works use data available in ERD or conceptual graphical models to generate analytical data models, in an approach usually known as data-driven. The approach proposed in this paper is called process-driven. It considers business processes models, expressed in Business Process Model and Notation (BPMN) 2.0, and is able to reach the data cubes that support the analytical needs of the modeled processes. This process-driven approach has as advantages the fact that it takes into account the actual business needs expressed in the business processes models and, also, the construction of an integrated view of the data, expressed in the CDF, from the operational to the strategic data analytics tasks. Moreover, as the several data models are obtained in an iterative way, taking previous information into consideration, it is possible to check the conformity of the data, detecting business requirements that were not properly modeled.

3   Common Data Framework

In a business context, data from several sources are usually collected, integrated, stored and analyzed to support decision-making at several levels of the organization, including operational, tactical or strategic. At these several levels, different data repositories need to be implemented to fulfill different data needs. Data modeling is a key task in this process, as incomplete or inadequate models could lead to the inability to perform specific analytical tasks or to consider specific business indicators.

Data models are defined considering the analytical tasks or queries that need to be answered. If we consider that BPM is concerned about processes and how to manage them, modeling those processes allows the identification of the activities that are performed and the data that flows in those activities. For these, specific repositories are needed at the several managerial levels, from the operational to the strategic.

Previous work showed how business process models could be used in a traditional BI environment, where an operational data model derived from BPMN models [3] can feed an analytical data model [5]. To complete the informational needs in a BI environment, this paper proposes the automatic identification of the data cubes needed for the analytical context, where several analytical perspectives on data are provided. This integrated perspective in the data-modeling task is better depicted in the proposed CDF (Fig. 1), which leverages data modeling capabilities to the BI environment. The advantage of the proposed approach is that it considers all the business needs expressed in the BPMN models and that were useful in the identification of the operational data model that supports OLTP and the analytical data model that supports OLAP in a traditional BI environment.

As Fig. 1 shows, BPM is transversal to all managerial levels, identifying the appropriate business indicators and performance measurements for the business processes activities. In an operational context, where users usually require real-time capabilities, relational database systems are usually adopted, with models optimized for updating the operational databases. This technological environment can reach the tactical managerial level where, usually, multidimensional databases and OLAP for the implementation of analytical data models and the corresponding data cubes are used. This is also the reality of the strategic managerial level, as historic data is usually

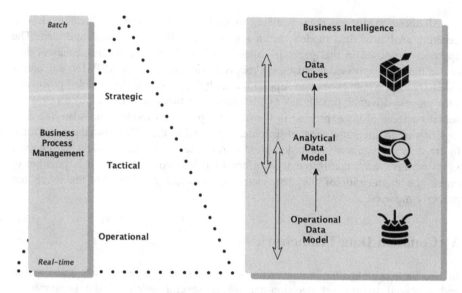

Fig. 1. The common data framework

needed to understand what happened in the past and to verify the tendencies for the future. In this analytical context, a DW allows the storage and analysis of several years of data, being loaded and refreshed from internal (as the operational databases) or external data.

4 Data Cubes for Analytical Data Models

Previous work already undertaken [5] shows how the several elements in BPMN 2.0 can be used to complement the information available in an operational data model, proposing an analytical data model for that context. This analytical data model is expressed in the form of a star or constellation schema, integrating fact tables, dimension tables, relationships between them and measures. In an analytical context, those multidimensional schemas integrate the metadata for implementing the DW repository. Besides the DW, those models are also usually mapped into a multidimensional database where specific data cubes contain pre-aggregations of the data, which are used by end-users to query the DW. Although the importance of the data cubes, the approaches usually followed in the dimensional modeling overlook the necessity of generating the metadata needed for the creation of those cubes [11].

This paper presents an integrated and fully automatic approach to derive the metadata for the OLAP cubes of an analytical data model, which is expressed through a star or a constellation schema. The main advantage of the proposed approach is that it considers the information previously expressed in the modeling of the business processes that support the organizational activities, integrating that information in the data cubes that need to be generated.

Taking into consideration the information present in BPMN models, the more relevant are the *item-aware* elements, which are elements that allow the storage and transmission of items during the process flows [12]. These elements are the basic foundations of the entities, attributes and relationships when identifying an operational data model. Besides these elements, and of relevance for the work here presented, are the *lanes* objects that represent the participants in the business processes.

In formal terms, a multidimensional data model represents an analytical data model formally described in this work as follows:

Definition 1. An Analytical Data Model, $ADM = (D, F, R, M)$, includes a set of dimension tables, $D = \{D^1, D^2, \ldots, D^l\}$, a set of fact tables, $F = \{F^1, F^2, \ldots, F^m\}$ and a set of relationships, $R = \{R_{F^1}^{D^1}(c_f{:}c_d), R_{F^1}^{D^2}(c_f{:}c_d), \ldots, R_{F^m}^{D^l}(c_f{:}c_d)\}$, which associate the facts included in a specific fact table with a set of dimensions, being the cardinality $(c_f{:}c_d)$ of the relationship of $(n{:}1)$ between the fact table and the dimension table. An ADM also includes a set of measures, $M = \{M^1, M^2, \ldots, M^n\}$, used for analyzing the business processes.

The data stores, representing the persistent data, identified in the set of interrelated business processes models, can be defined as follows:

Definition 2. The data stores supporting the business processes are represented by $DS = \{DS^1, DS^2, \ldots, DS^n\}$, being a data store a table or component of a database where persistent data are stored or retrieved by one activity integrated in a business process. As a data store can be shared by different business processes (BP), $BP = \{BP^1, BP^2, \ldots, BP^m\}$, the data stores manipulated by each business process are identified as $BP_{DS}^i = \{DS^1, DS^2, \ldots, DS^n\}$, $1 \leq i \leq m$.

Having these elements, ADM, DS and BP_{DS}, it is possible to propose the automatic transformation of the ADM in the data cubes with the aggregations that will enhance the analysis of the data by the end user. For that, it is relevant to know how the data sources are related with each other, in the context of the business processes under analysis, as this will allow the identification of the relevant data cubes, instead of the definition and implementation of all possible data cubes (lattice data cube). The data cubes can be formally defined as:

Definition 3. An OLAP Data Model ($OLAP_{DM}$), $OLAP_{DM} = (D, C, R, M)$, includes a set of dimension tables, $D = \{D^1, D^2, \ldots, D^l\}$, a set of cubes, $C = \{C^1, C^2, \ldots, C^m\}$ and a set of relationships, $R = \{R_{C^1}^{D^1}(c_c{:}c_d), R_{C^1}^{D^2}(c_c{:}c_d), \ldots, R_{C^m}^{D^l}(c_c{:}c_d)\}$, which associate measures included in a specific cube with a set of dimensions, being the cardinality $(c_c{:}c_d)$ of the relationship of $(n{:}1)$ between the cube and the dimension table. An $OLAP_{DM}$ also includes a set of measures, $M = \{M^1, M^2, \ldots, M^n\}$, used for analyzing the business processes.

Having as starting point the elements expressed by Definitions 1 and 2, it is possible to automatically identify an $OLAP_{DM}$, expressed by Definition 3, following the rules proposed in this paper, for the identification of dimension tables (Rule $OLAP_{DM}.1$), data cubes (Rule $OLAP_{DM}.2$), relationships (Rule $OLAP_{DM}.3$) and measures (Rule $OLAP_{DM}.4$).

Rule OLAP$_{DM}$.1: Identification of Dimensions. Dimensions are tables that add semantics to the cubes and the measures under analysis. Each dimension integrates a key, a set of attributes, and one or more hierarchies. For the identification of the dimensions, associates attributes and hierarchies, the following rules must be followed:

Rule OLAP$_{DM}$.1.1: Dimensions. All dimensions present in the *ADM* represent dimension tables (*D*) in the *OLAP$_{DM}$*.

Rule OLAP$_{DM}$.1.2: Primary Key. The primary key of a dimension in the *OLAP$_{DM}$* is inherited from the corresponding dimension in the *ADM*.

Rule OLAP$_{DM}$.1.3: Attributes. The attributes of a dimension in the *OLAP$_{DM}$* are inherited from the corresponding dimension in the *ADM*.

Rule OLAP$_{DM}$.1.4: Hierarchies. Aggregation hierarchies are defined identifying the relationships between the attributes of a dimension. When a hierarchical relationship is identified between two or more attributes, formally defined as a *n:1* (many to one) relation, an aggregation hierarchy is defined. All hierarchies are functionally dependent of the primary key of the corresponding dimension and are here specified as $H = \{At_1 \prec At_2 \prec \ldots \prec At_l\}$, where *At* are the different attributes that constitute the hierarchy, from the lower level *1* to the upper level *l*, following this aggregation path. Between the several attributes that share *n:1* relationships, different hierarchies can be defined by different combinations of those attributes. For the identification of the different hierarchies, the hierarchies' lattice must be used. Considering a scenario where 3 attributes have *n:1* relationships between them, At_1, At_2 and At_3, Fig. 2 presents the four hierarchies that can be identified, giving also an example considering a temporal relationship between the Day, Month and Year attributes. Considering the high number of hierarchies that can be obtained when combining more than three attributes, in this rule it is recommended the use, for analytical purposes, of those that emerge when combining *l* and *l−1* number of attributes. In a hierarchy of 3, this considers the combinations of 2 attributes and all the 3 attributes.

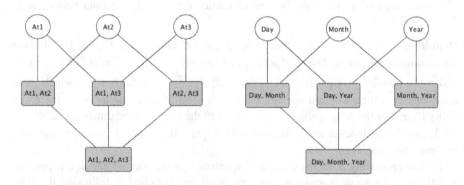

a) The hierarchies' lattice   b) An example of the hierarchies' lattice

Fig. 2. Identification of the several hierarchies from a set of related attributes.

Rule OLAP$_{DM}$.2: Identification of Cubes. Cubes are derived from the fact tables present in the *ADM*. In an analytical context, each one of these tables can give origin to a data cube, although the combination of one or more fact tables can also give origin to data cubes that combine different measures and provide an integrated perspective of the data under analysis. For the identification of the relevant data cubes, this rule proposes the identification of the lattice structure for the fact tables and selection of the cubes that either exist in the *ADM* as fact tables (primary cubes) or that relate different data sources in a business process (derived cubes), as depicted in Definition 2.

Rule OLAP$_{DM}$.2.1: Primary Cubes. Each fact table present in the *ADM* gives origin to a data cube in the *OLAP$_{DM}$*.

Rule OLAP$_{DM}$.2.2: Derived Cubes. Cubes that combine measures from different fact tables in the *ADM* are identified considering the possible combinations that emerge from a lattice of fact tables and considering those combinations that relate two or more of these fact tables in the business processes, information available in the BPMN models and expressed in the business processes data sources (*BP$_{DS}$*). For the sake of clarity, let us consider a business context with four business processes, *BP^1*, *BP^2*, *BP^3* and *BP^4*, using the following data sources: $BP_{DS}^1 = \{DS^1, DS^2, DS^5\}$, $BP_{DS}^2 = \{DS^1, DS^3\}$, $BP_{DS}^3 = \{DS^2, DS^4, DS^5\}$ and $BP_{DS}^4 = \{DS^3, DS^4\}$. Let us also consider that three of these five data sources, *DS^1*, *DS^2* and *DS^3*, gave origin to three fact tables in the *ADM*, *F^1*, *F^2* and *F^3* respectively, representing, by Rule OLAP$_{DM}$.2.1, primary cubes *C^1*, *C^2* and *C^3*. Considering the fact tables' lattice and the business processes data sources, Fig. 3 shows the identified derived cubes, with a derived cube *C^4* integrating measures from *F^1* and *F^2* (as *BP^1* needs data from *DS^1* and *DS^2*), and another derived cube *C^5*, integrating measures from *F^1* and *F^3* (as *BP^2* needs data from *DS^1* and *DS^3*).

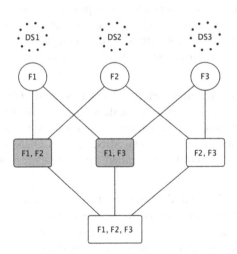

Fig. 3. Identification of derived cubes

Each identified cube, primary or derived, includes a set of foreign keys (FK) used to set the PK of the cube, and a set of measures. The FK are derived from the relationships between the dimensions and the cubes.

Rule OLAP$_{DM}$.3: Identification of Relationships. The identification of the relationships between the dimension tables and the fact tables makes use of the corresponding entities in the *ADM*, following the next rules, considering that exists a *n*:1 relationship between a cube and a dimensional table.

Rule OLAP$_{DM}$.3.1: Relationships for Primary Cubes. The relationships in the *OLAP$_{DM}$* model for the primary cubes are inherited from the *ADM*.

Rule OLAP$_{DM}$.3.2: Relationships for Derived Cubes. The relationships in the *OLAP$_{DM}$* model for the derived cubes are inherited from the *ADM* considering the dimensions that are shared by the fact tables that derived the cube.

Rule OLAP$_{DM}$.4: Identification of Measures. Previous work [5] allowed the identification of three types of measures to include in the fact tables of an *ADM*: event-tracking measures, direct measures and calculated measures. These different types of measures will also be considered in the data cubes, following a two-step approach for the identification of measures for the primary cubes and for the derived cubes:

Rule OLAP$_{DM}$.4.1: Measures for Primary Cubes. The measures for the primary cubes are inherited from the measures present in the fact tables of the *ADM*, as well as the aggregation function defined for those measures (SUM, AVG, MIN, MAX, COUNT), as all the considered measures maintain the same granularity for the considered dimensions.

Rule OLAP$_{DM}$.4.2: Measures for Derived Cubes. Derived cubes integrate two or more fact tables of the *ADM*. In this integration process, the derived cubes assume as dimensions those that are common among the integrated fact tables. As some fact tables may reduce one or more dimensions, the aggregation of records needs to be performed in order to summarize those records. For derived cubes, two different rules need to be applied:

Rule OLAP$_{DM}$.4.2.1: Event-Tracking Measures and Direct Measures. The aggregation of event-tracking measures and direct measures is performed applying the GROUP BY operator along the dimensions considered for the derived cube. As different aggregation functions (SUM, AVG, MIN, MAX, COUNT) may be considered, the aggregations here calculated are inherited from the aggregations specified in the *ADM*.

Rule OLAP$_{DM}$.4.2.2: Calculated Measures. Calculated measures are measures that intend to provide a different perspective of the measures identified by the previous rule, being able to provide percentages of the incidence of each one of these measures, when analyzed by the different perspectives provided by the dimensions. Therefore, as in [5], a calculated measure (M^{i_c}) is derived for each measure (M^i) identified by Rule OLAP$_{DM}$.4.2.1, considering Eq. 1.

$$M^{ic} = \frac{M^i_j}{\sum_{j=1}^{length(M^i)} M^i_j} x100\%, \ 1 \leq j \leq length(M^i) \qquad (1)$$

where $length(M^i)$ is used for identifying the number of rows of the fact table of M^i and M^i_j makes reference to a specific value of a row j of M^i. In the case of the calculated measures, percentages, the aggregation function usually applied is AVG.

5 Demonstration Case

The work presented in [3] used a set of five interrelated business process models specified in BPMN 2.0 to identify the corresponding operational data model and, later, the *ADM* [5]. The demonstration case is associated to an example of a School Library System and the modeled processes are: *Register User, Lend a Book, Reserve a Book, Renew a Loan* and *Return a Book*, see, for example, the Fig. 4 with the *Lend a Book* and Fig. 5 with the *Renew a Loan* business processes, respectively. The participants in these five processes are *Borrower* and *Attendant*.

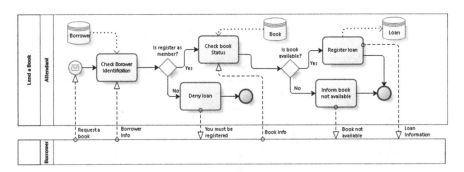

Fig. 4. Lend a book

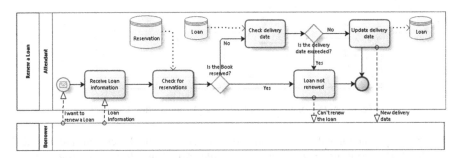

Fig. 5. Renew a loan

As previously mentioned, a key factor in the identification of the fact tables is the identification of the data stores that are manipulated in the same business process, as this is an indication of the data that is usually integrated in a decision making process. In this example, and following Definition 2, the specification of the business processes and the data stores [5] is: *Lend a Book = {Borrower, Book, Loan, Attendant}, Renew a Loan = {Borrower, Reservation, Loan, Attendant}, Reserve a Book = {Borrower, Book, Reservation, Attendant}, Return a Book = {Borrower, Loan, Attendant}* and *Penalty = {Borrower, Loan, Receipt, Attendant}*.

The identified *ADM* [5] is presented in Fig. 6, with three fact tables (*Loan, Reservation* and *Receipt*) and five dimension tables (*Calendar, Time, Borrower, Book* and *Attendant*).

Taking into consideration the definitions and the rules presented in the previous section, it is possible to identify the $OLAP_{DM}$ for the School Library Demonstration Case.

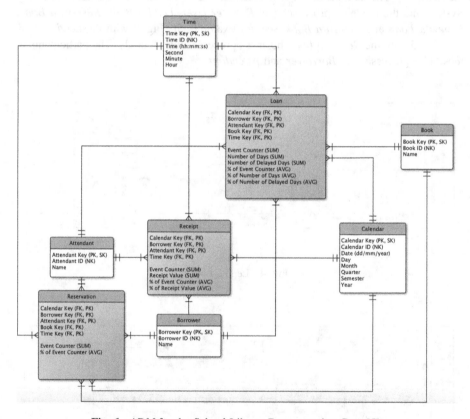

Fig. 6. *ADM* for the School Library Demonstration Case [5]

Starting by Rule $OLAP_{DM}.1$, in Rule $OLAP_{DM}.1.1$, all the dimensions are inherited from the *ADM*, resulting in *Dimensions = {Calendar, Time, Borrower, Book, Attendant}*. Each one of these dimensions inherits the PK and the attributes from the corresponding dimensions, Rules $OLAP_{DM}.1.2$ and $OLAP_{DM}.1.3$, respectively, resulting in *Borrower = {Borrower Key (PK), Borrower ID (NK), Name}, Book = {Book Key (PK), Book ID (NK), Name}, Attendant = {Attendant Key (PK), Attendant ID (NK), Name}, Calendar = {Calendar Key (PK), Calendar ID (NK), Date, Day, Month, Quarter, Semester, Year}* and *Time = {Time Key (PK), Time ID (NK), Time, Second, Minute, Hour}*. By Rule OLAP$_{DM}$.1.4, the identification of the hierarchies to be integrated in these dimensions is needed, as this will allow the computation of pre-aggregates for the cubes. Two dimensions, *Calendar* and *Time*, present attributes with *n:1* relationships among them. In general terms, considering all the attributes with hierarchical relationships, the following hierarchies can be defined:

$$Calendar_{All} = \{Day \prec Month \prec Quarter \prec Semester \prec Year\}$$
$$Time_{All} = \{Second \prec Minute \prec Hour\}$$

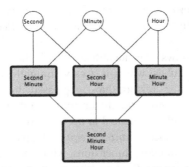

a) Lattice for the time dimension and its hierarchies

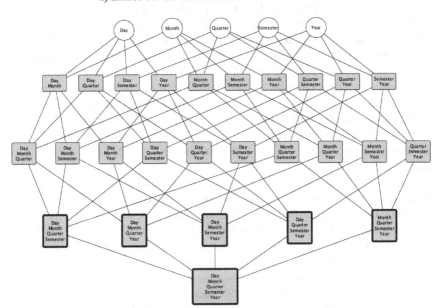

b) Lattice for the calendar dimension and its hierarchies

Fig. 7. Lattices for the time and calendar hierarchies

These overall hierarchies represent the aggregation of all attributes in the hierarchies' lattice and, therefore, the other combinations must be identified computing the corresponding lattices as shown in Fig. 7. In this case, for both dimensions, the hierarchies considering the l level (3 for *Time* and 5 for *Calendar*) and the $l-1$ level (2 for *Time* and 4 for *Calendar*), are suggested attending to the recommendation explicit in Rule OLAP$_{DM}$.1.4.

Following with Rule OLAP$_{DM}$.2 for the identification of the data cubes, the distinction between primary cubes and derived cubes is done. The primary data cubes are derived from the fact tables present in the *ADM* (Rule OLAP$_{DM}$.2.1). In this case, the primary cubes *Loan*, *Reservation* and *Receipt* are identified. Derived cubes (Rule OLAP$_{DM}$.2.2) are identified verifying the combinations of the data sources in the BPMN models, which are fact tables in the *ADM*. Two derived cubes emerge, one integrating the *Loan* and *Receipt* fact tables (from the *Penalty = {Borrower, Loan, Receipt, Attendant}*) and the other integrating the *Loan* and *Reservation* fact tables (*Renew a Loan = {Borrower, Reservation, Loan, Attendant}*). The identification of the data cubes ends with five cubes: *Loan, Reservation, Receipt, Loan&Receipt and Loan&Reservation*.

Rule OLAP$_{DM}$.3 allows the identification of the relationships between the dimensions and the data cubes, which is a very straightforward process. For primary cubes, they inherit all the relationships expressed in the *ADM* (Fig. 6) (Rule OLAP$_{DM}$.3.1). For derived cubes, the dimensions of a derived cube are those that are common among the fact tables integrated in the derived cube (Rule OLAP$_{DM}$.3.2). For the two identified derived cubes, *Loan&Receipt* will have relationships with the *Time*, *Calendar*, *Borrower* and *Attendant* dimensions, while the *Loan&Reservation* will have with *Time*, *Calendar*, *Borrower*, *Book* and *Attendant* dimensions.

The last rule, Rule OLAP$_{DM}$.4, is associated with the measures of each cube. Once again, a distinction is made between primary and derived cubes. For primary cubes (Rule OLAP$_{DM}$.4.1), all the measures are inherited from the fact tables in the *ADM*. For derived cubes, the event-tracking and the direct measures must be aggregated (*agg*) using the aggregation functions defined in the *ADM* and considering the dimensions of the derived cube (Rule OLAP$_{DM}$.4.2.1). Besides this, the calculated measures must be added to the data cube (Rule OLAP$_{DM}$.4.2.2). All measures are presented later in the corresponding data model (Fig. 8). Besides measures, each data cube also includes a set of foreign keys, one for each dimension to which it is related. All the relationships are inherited from the tables in the *ADM*, stressing the links between dimension tables and fact tables.

Having followed the defined rules, Fig. 8 presents the obtained data cubes (the *OLAP$_{DM}$* model), integrating the several dimensions, data cubes, measures, relationships and the corresponding keys. The notation used was ADAPT [13] and the dimensions' description (all the attributes) is depicted in the bottom of the figure.

In order to verify the usefulness of the proposed approach, let us consider two of the available fact tables, *Loan* and *Receipt*, and the corresponding dimension tables (Fig. 9) and verify how the derived cubes can be integrated and how any analyses on the data can be afterwards automatically done.

As previously mentioned, one of the derived cubes will integrate the *Loan* and the *Receipt* fact tables. While *Loan* has five dimensions tables, *Receipt* has four.

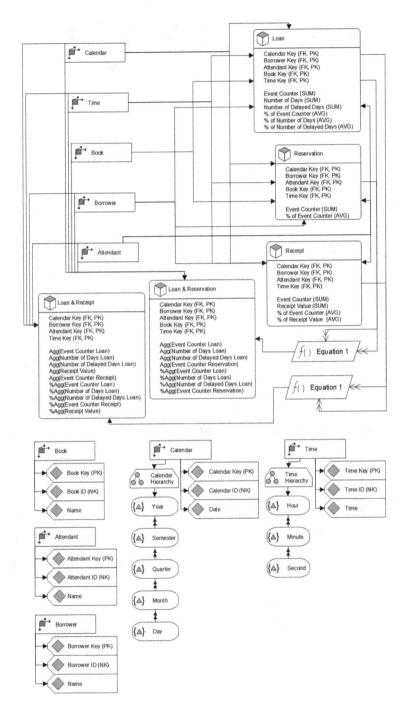

Fig. 8. $OLAP_{DM}$ for the school library demonstration case (in ADAPT Notation [13])

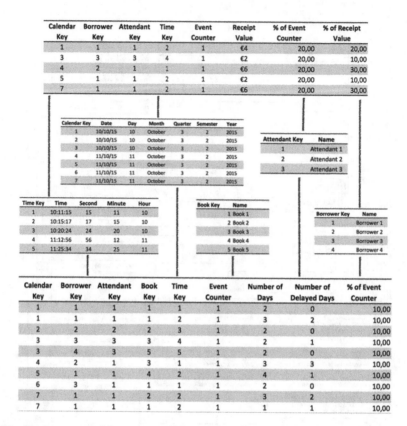

Fig. 9. An extract of the content of the *ADM* (with the *Loan and Receipt* fact tables)

Aggregated Loan Table

Calendar Key	Borrower Key	Attendant Key	Time Key	SUM(Event Counter)	SUM(Number of Days)	SUM(Number of Delayed Days)
1	1	1	1	1	2	0
1	1	1	2	1	3	2
2	2	2	3	1	2	0
3	3	3	4	1	2	1
3	4	3	5	1	2	0
4	2	1	1	1	3	3
5	1	1	2	1	4	1
6	3	1	1	1	2	0
7	1	1	2	2	4	3

Receipt Table

Calendar Key	Borrower Key	Attendant Key	Time Key	Event Counter	Receipt Value
1	1	1	2	1	€4
3	3	3	4	1	€2
4	2	1	1	1	€6
5	1	1	2	1	€2
7	1	1	2	1	€6

Loan&Receipt Cube

Calendar Key	Borrower Key	Attendant Key	Time Key	Event Counter Loan	Number of Days Loan	Number of Delayed Days Loan	% Event Counter Loan	% Number of Days Loan	% Number of Delayed Days Loan	Event Counter Receipt	Receipt Value	% Event Counter Receipt	% Receipt Value
1	1	1	1	1	2	0	10,00	8,33	0,00				
1	1	1	2	1	3	2	10,00	12,50	20,00	1	€4	20,00	20,00
2	2	2	3	1	2	0	10,00	8,33	0,00				
3	3	3	4	1	2	1	10,00	8,33	10,00	1	€2	20,00	10,00
3	4	3	5	1	2	0	10,00	8,33	0,00				
4	2	1	1	1	3	3	10,00	12,50	30,00	1	€6	20,00	30,00
5	1	1	2	1	4	1	10,00	16,67	10,00	1	€2	20,00	10,00
6	3	1	1	1	2	0	10,00	8,33	0,00				
7	1	1	2	2	4	3	20,00	16,67	30,00	1	€6	20,00	30,00

Fig. 10. *Loan&Receipt* data cube

This means that the derived cube will have the four dimensions that are shared. Analyzing the facts in the *Loan* fact table and the reduction of one dimension, *Book*, two records need to be aggregated, as the granularity of this table, regarding the dimensions, has changed. Attending to the aggregation functions specified in the *ADM*, the SUM function will be used for all event-tracking and direct measures. After aggregating the data for *Loan*, the integration with *Receipt* is possible, producing a new cube, *Loan&Recepit*, as depicted in Fig. 10. The figure highlights, in a different color, the calculated measures computed after the integration of the records of the two tables.

After the integration of fact tables in cubes, as expressed in the rules of the previous section, it is possible to use appropriate dashboards to analyze the available data. Besides data cubes that are automatically proposed by the presented approach, specific metrics (the calculated measures) can also be automatically calculated, complementing the primary facts that are usually included in a fact table.

6 Discussion of Results and Conclusions

This paper presented a CDF that intends to provide an integrated view of the operational data schemas needed for managing the daily activities of the business, the analytical repository of data with a historical perspective (the DW) and, also, the OLAP schemas (data cubes) that enhance analytical capabilities by making available data and aggregates of business indicators to support the decision-making process at the several managerial levels: operational, tactical and strategic.

The proposed approach allows checking the business processes, modeled in BPMN 2.0, verifying that the generated models, *ODM*, *ADM* and OLAP cubes are in compliance with each other, as they are iteratively defined in a step-wise approach. This bottom-up approach is then complemented with a top-down verification as missing requirements can be detected, once we arrive to the data cubes. To give an example, in the Demonstration Case described above, we can detect that in the *Loan&Reservation* data cube a tricky situation emerges, related with the integration of the *Loan* and the *Reservation* data cubes. This occurs because the loan renew will check if the book has a reservation and, if it has, the loan renew cannot be made. However, as this information is not considered in the *ODM*, the *ADM* has also no information about these cases. When we reach the data cubes, and considering the rule that allows the identification of derived cubes, we verified that the facts of the two cubes are not related, as no information is stored in the business process after verifying if a loan can be renewed or not. This situation is depicted in Fig. 11.

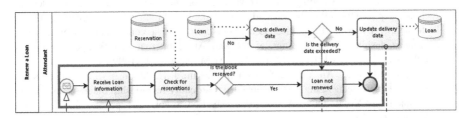

Fig. 11. Renew process missing requirement

Although this important validation of the business requirements and the obtained models, there are other perspectives of the data that were not yet identified and modeled, but that are considered for future work. Because the CFD is based on business processes modeled in BPMN 2.0, it is possible viewing these processes in three perspectives [14]: (1) the process perspective – is concerned with answering the "How?" question and focuses on the control-flow, i.e., the order of the activities and the identification of all possible control-flow paths; (2) the organizational perspective – is concerned with the "Who?" question and focuses on identifying processes' performance; and (3) the case perspective - is concerned with the "What?" question and focuses on properties of cases and their path in the process execution. As the case perspective was already addressed in this work, future work will focus the organizational one.

Acknowledgments. This work has been supported by COMPETE: POCI-01-0145-FEDER-007043 and FCT (*Fundação para a Ciência e Tecnologia*) within the Project Scope: UID/CEC/00319/2013, and by Portugal Incentive System for Research and Technological Development, Project in co-promotion n° 002814/2015 (iFACTORY 2015–2018). Some of the figures in this paper use icons made by Freepik, from www.flaticon.com.

References

1. Cunningham, D.: Aligning business intelligence with business processes. In: TDWI Research, vol. 20 (2005)
2. Marjanovic, O.: Business value creation through business processes management and operational business intelligence integration. In: Proceedings of the 43rd Hawaii International Conference on System Sciences (2010)
3. Cruz, E., Machado, R.J., Santos, M.Y.: Deriving a data model from a set of interrelated business process models. In: ICEIS 2015 - 17th International Conference on Enterprise Information Systems, vol. I, pp. 49–59 (2015)
4. Cruz, E., Machado, R.J., Santos, M.Y.: From business process modeling to data model: a systematic approach. In: 2012 Eighth International Conference on the Quality of Information and Communications Technology, pp. 205–210 (2012)
5. Santos, M.Y., Oliveira e Sá, J.: A data warehouse model for business processes data analytics. In: Proceedings of the 16th International Conference on Computational Science and Its Applications (ICCSA 2016), China, July 2016
6. Hann, K., Sapia, C., Balaschka, M.: Automatically generating OLAP schemata from conceptual graphical models. In: Proceedings of the 3rd ACM International Workshop on Data Warehousing and OLAP (DOLAP), pp. 9–16. ACM, New York (2000)
7. Peralta, V., Marotta, A., Ruggia, R.: Towards the automation of data warehouse design. Technical report TR-03-09, Universidad de la República, Montevideo, Uruguay (2003)
8. Tryfona, N., Busborg, F., Christiansen, J.G.B.: StarER: a conceptual model for data warehouse design. In: Proceedings of the 2nd ACM International Workshop on Data Warehousing and OLAP (DOLAP), pp. 3–8. ACM, New York (1999)
9. Song, I.-Y., Khare, R., An, Y., Lee, S., Kim, S.-P., Kim, J.-H., Moon, Y.-S.: SAMSTAR: an automatic tool for generating star schemas from an entity-relationship diagram. In: Li, Q., Spaccapietra, S., Yu, E., Olivé, A. (eds.) ER 2008. LNCS, vol. 5231, pp. 522–523. Springer, Heidelberg (2008)

10. Usman, M., Asghar, S., Fong, S.: Data mining and automatic OLAP schema generation. In: Fifth International Conference on Digital Information Management, pp. 35–43 (2010)
11. Pardillo, J., Mazón, J.-N., Trujillo, J.: Model-driven metadata for OLAP cubes from the conceptual modelling of data warehouses. In: Song, I.-Y., Eder, J., Nguyen, T.M. (eds.) DaWaK 2008. LNCS, vol. 5182, pp. 13–22. Springer, Heidelberg (2008)
12. OMG: Business process model and notation (BPMN), version 2.0. Technical report, Object Management Group (2011)
13. Bulos, D., Forsman, S.: Getting started with ADAPT™ - OLAP database design, symmetry corporation (2002)
14. Aalst, W.M.P., Medeiros, A.K.A., Weijters, A.J.M.M.: Genetic process mining. In: 26th International Conference Applications and Theory of Petri Nets, ICATPN 2005, Miami, USA, pp. 48–69, June 20–25 2005

Dependency Analysis Between PMI Portfolio Management Processes

Ana Lima[1,2(✉)], Paula Monteiro[1,2], Gabriela Fernandes[2],
and Ricardo J. Machado[1,2]

[1] CCG/ZGDV Institute, Guimarães, Portugal
{ana.lima,paula.monteiro}@ccg.pt, rmac@dsi.uminho.pt
[2] ALGORITMI Research Centre, Universidade do Minho, Guimarães, Portugal
g.fernandes@dps.uminho.pt

Abstract. Software projects are no longer managed in isolation, but as a core business activity increasingly exposed to a high-level of rigor and responsibility. In this context, portfolio management efforts should be adopted in order to meet the organization's strategic goals. The Project Management Institute (PMI) has developed the 'Standard for Portfolio Management', whose objective is to propose a strategy for project portfolio management through processes. An efficient adoption of this PMI standard to manage projects in software development companies benefits from a thoroughly understanding of the existing dependencies among all portfolio management process. This paper presents the dependencies among all portfolio management process from the standard for portfolio management from PMI. The presented dependencies were identified by performing a systematic analysis of the process groups and knowledge areas of the PMI standard.

Keywords: Dependency analysis · Standard portfolio · Process group · Knowledge area

1 Introduction

The portfolio management for software development projects of a company should be as dynamic as the environment in which is inserted. The projects characteristics such as the differentiated complexity by project, the constant evolution of technologies to use and to develop and the capacity of highly qualified resource allocation for projects, are challenges for portfolio management that should be reviewed and prioritized continuously for a short time. Additionally, the management of multiples projects have an independent existence with separate goals and problems, and yet draw at least some resources from a common pool, and they must be integrated into the management control and reporting systems of the resource pool owner. The characteristics of software development projects, justifies the need to capacitate managers of those teams to apply advanced techniques of portfolio management for software development projects.

To manage multiple projects successfully the company needs to maintain control over a varied range of projects, balance often between conflicting requirements, limited resources, and coordinate the project portfolio to ensure that optimum organizational outcome is achieved [1].

O. Gervasi et al. (Eds.): ICCSA 2016, Part V, LNCS 9790, pp. 288–300, 2016.
DOI: 10.1007/978-3-319-42092-9_22

In perspective to optimize their portfolio, companies are focused on accepting projects that are aligned with its strategic objectives, since projects add greater value to both the business and the stakeholders.

A key point in portfolio management is the balancing of portfolios; i.e., investments in the projects should maintain the balance between risk and return, growth and maintenance in the short and long term [2]. While project management, traditionally, is directed to "do the project right", portfolio management is concerned with "doing the right project" [3].

Portfolios are groups of projects which are relatively independent of one another but share and compete for scarce resources [4]. These resources can be money, people or technology, and the performance of projects can be improved through its coordination [5].

The traditional approach to project management considers projects as being independent of each other [6]. Rodney Turner et al. [5] defends that only ten per cent of all projects activity are managed in an isolated way, the majority of projects being part of a portfolio (or program).

The PMI [7] defines portfolio management as the centralized management of one or more portfolios, which includes identifying, prioritizing, authorizing, managing, and controlling projects, programs, and other related work, to achieve specific strategic business objectives.

The portfolio management consists of a dynamic decision process, whereby a business's list of active new product projects is constantly up-dated and revised. In this process, new projects are evaluated, selected and prioritized; existing projects may be accelerated, killed or de-prioritized; and resources are allocated and reallocated to active projects [8]. Rajegopal et al. [9] complete the definition as the process for identifying and selecting the right projects, given the company's ability to accomplish these projects established against the financial and human resources available. This can also be defined as how to optimize the overall investment portfolio, programs and approved projects related to business strategy [9].

The portfolio decision process is characterized by uncertain and changing information, dynamic opportunities, multiple goals and strategic considerations, interdependence among projects, and multiple decision-makers and locations. Far beyond the strategic resource allocation decisions, the portfolio decision process encompasses or overlaps a number of decision-making processes, including periodic reviews of the total portfolio of all projects (looking at all projects holistically, and against each other), making go/kill decisions on individual projects on an on-going basis, and developing a new product strategy for the company [10].

The main goals of portfolio management are the maximization of the financial value of the portfolio, linking the portfolio to the company's strategy, and balancing the projects within the portfolio, taking into account the company's capacities [11].

Several inputs can be used to guide a company in improving portfolio management, by selecting the most appropriate tools and techniques in a given context, including various bodies of knowledge (BoKs). The portfolio management body of knowledge is the sum of knowledge within the profession of portfolio management. The complete portfolio management body of knowledge includes proven traditional practices that are widely applied, as well as innovative practices that are emerging in the profession [4].

Other BoK from PMI referred in this paper is Standard OPM3, maturity model from PMI for project management, program management and portfolio management.

This paper presents the dependencies among all portfolio management processes from the standard for portfolio management from PMI, indicating which processes have greater influence and dependencies between inputs and outputs. The presented dependencies were identified by performing a systematic analysis of the process groups and knowledge areas of the PMI standard using the Dependency Analysis Graph. The Dependency Analysis Graph allows a better and faster overview on the project portfolio management processes, allowing a friendly representation of the processes with greater and lesser influence, as well as the dependencies between them both in terms of process groups and the knowledge areas. This paper wasn't performed based on a systematic literature review or systematic mapping study, but on author's knowledge and experience in the area of project portfolio management.

This paper is organized as follows. Section 2 presents a brief description of the Standard for Portfolio Management from PMI. Section 3 describes the dependencies between Portfolio Management Processes and in Sect. 4 conclusions are presented, as well as some highlights for further research.

2 The Standard for Portfolio Management from PMI

The process defined by PMI [3] for portfolio management assumes that the company has a strategic plan, knows its mission, has established its vision and goals.

An efficient portfolio management depends on the degree of maturity of a company and its processes. Thus, the knowledge of the maturity of a company is critical to determine its abilities and to select the correct methods to evaluate, select, prioritize and balance the projects which will be part of its portfolio, preferring the achievement of its objectives and defined goals in the strategic planning.

The Standard for Portfolio Management from PMI is composed of a set of 16 portfolio processes divided into five knowledge areas and three process groups.

The PMI proposes three process groups for project portfolio management: the *defining* process group, the *aligning* process group, and the *authorizing and controlling* process group.

The objective of the *defining* process group is to establish the strategy and the company's objectives that will be implemented in a portfolio. The objective of the *alignment* process group is to manage and optimize the portfolio. And finally, the objective of the *authorizing and controlling* process group is to determine who authorizes the portfolio, as well as the ongoing oversight of the portfolio.

The knowledge areas identified in the PMI Standard of the portfolio management are: strategic management, governance management, performance management, communication management and risk management [3].

Table 1 presents the PMI 16 portfolio management processes organized by knowledge areas and by process groups. These 16 processes are executed sequentially by each portfolio independently of the application area of the company. A process group includes a set of portfolio management processes, each one demanding inputs and providing outputs, where the outcome of one process becomes the input to another [3].

Table 1. Portfolio management processes organized by groups and knowledge areas [13]

Portfolio Management Knowledge Areas (PMKA)	Portfolio Management Process Groups (PMPG)		
	Defining Process Group	*Aligning Process Group*	*Authorizing and Controlling Process Group*
Portfolio Strategic Management (PSM)	Develop Portfolio Strategic Plan	Manage Strategic Change	
	Develop Portfolio Charter		
	Define Portfolio Roadmap		
Portfolio Governance Management (PGM)	Develop Portfolio Management Plan	Optimize Portfolio	Authorize Portfolio
	Define Portfolio		Provide Portfolio Oversight
Portfolio Performance Management (PPM)	Develop Portfolio Performance Management Plan	Manage Supply and Demand	
		Manage Portfolio Value	
Portfolio Communication Management (PCM)	Develop Portfolio Communication Management Plan	Manage Portfolio Information	
Portfolio Risk Management (PRM)	Develop Portfolio Risk Management Plan	Manage Portfolio Risks	

Table 2 depicts the mapping between each process groups and the knowledge areas and process improvement stage from OPM3 [3, 12]. The last column of Table 2 is the mapping between the portfolio management standard and the maturity model for portfolio management from OPM3. To help the discussion in Sect. 3, in Table 2 includes a column to define an acronym for each portfolio process {PP*n*}, where PP stands for portfolio process, and *n* corresponds to the number of the process.

Table 2. Mapping between process groups, knowledge areas and process improvement stages from OPM3

Portfolio Management Process Groups (PMPG)	Portfolio Management Knowledge Areas (PMKA)	Portfolio Processes (PP)	Acronym	OPM3 Process Improvement Stage (PIS)
Defining Process Group	Portfolio Strategic Management (PSM)	Develop Portfolio Strategic Plan	{PP 1} DPSP	S,M,C,I
	Portfolio Strategic Management (PSM)	Develop Portfolio Charter	{PP 2} DPC	S,M,C,I
	Portfolio Strategic Management (PSM)	Define Portfolio Roadmap	{PP 3} DPR	S,M,C,I
	Portfolio Governance Management (PGM)	Develop Portfolio Management Plan	{PP 4} DPMP	S,M,C,I
	Portfolio Governance Management (PGM)	Define Portfolio	{PP 5} DP	S,M,C,I
	Portfolio Performance Management (PPM)	Develop Portfolio Performance Management Plan	{PP 6} DPPMP	S,M,C,I
	Portfolio Communication Management (PCM)	Develop Portfolio Communication Management Plan	{PP 7} DPCMP	S,M,C,I
	Portfolio Risk Management (PRM)	Develop Portfolio Risk Management Plan	{PP 8} DPRMP	S,M,C,I
Aligning Process Group	Portfolio Strategic Management (PSM)	Manage Strategic Change	{PP 9} MSC	S,M,C,I
	Portfolio Governance Management (PGM)	Optimize Portfolio	{PP 10} OP	S,M,C,I
	Portfolio Performance Management (PPM)	Manage Supply and Demand	{PP 11} MSD	S,M,C,I
	Portfolio Performance Management (PPM)	Manage Portfolio Value	{PP 12} MPV	S,M,C,I
	Portfolio Communication Management (PCM)	Manage Portfolio Information	{PP 13} MPI	S,M,C,I
	Portfolio Risk Management (PRM)	Manage Portfolio Risks	{PP 14} MPR	S,M,C,I
Authorizing and Controlling Process Group	Portfolio Governance Management (PGM)	Authorize Portfolio	{PP 15} AP	S,M,C,I
	Portfolio Governance Management (PGM)	Provide Portfolio Oversight	{PP 16} PPO	S,M,C,I

The OPM3 is organized in three areas related to three elements for application in companies: knowledge, assessment and improvement. Combining these three elements in a continuous cycle of five steps: (1) prepare for assessment; (2) perform assessment; (3) plan improvements; (4) implement improvements; and (5) repeating the process. In OPM3, companies can then be classified into 4 stages of development in each portfolio process [14]: (1) standardize (S) - structured processes are adopted; (2) measure (M) - data is used to evaluate process performance; (3) control (C) - control plan developed for measures; and (4) continuously improve (I) - processes are optimized.

3 Dependencies Between Portfolio Processes from the PMI Standard for Portfolio Management

At a first glance to the PMI Standard for Portfolio Management it is not easy to perceive the existing dependencies. Based on the detailed information about the inputs and the outputs of the processes, our efforts to highlight the existing dependencies are aimed to explicit both the implementation order of the processes and the input-output interrelation they establish.

3.1 Elementary Dependency Analysis

In this section, we describe how we characterize the elementary dependency of a particular portfolio process; what we call the PPn-centric dependency analysis (n is the number of the process portfolio; see Table 2).

As an example, we will analyse the {PP4} DPMP portfolio process depicted in Fig. 1. The {PP4} DPMP "Develop Portfolio Management Plan" process receives information of the {PP1} DPSP "Develop Portfolio Strategic Plan" process and sends information to the {PP14} MPR "Manage Portfolio Risks" process and to the {PP16} PPO "Provide Portfolio Oversight" process. All processes in the depicted graph are positioned in the respective process group lane (as an example, the PP4 {DPMP} is located in the lane of the *Defining* process group).

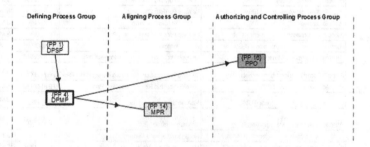

Fig. 1. Elementary dependency analysis graph

Elementary dependencies between processes are perfectly identified in the standard portfolio management from PMI. However, the overview of all portfolio management processes organized by process groups or knowledge areas is not easily perceived. This is why our systematic analysis is applied to highlight all the detailed overall dependencies between the complete set of portfolio processes.

3.2 Portfolio Processes Dependencies

In order to obtain the complete set of all the dependencies between all portfolio processes we start to analyse the processes' inputs and outputs (see Fig. 2).

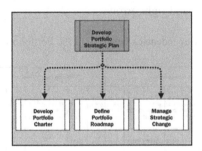

Fig. 2. Example of dependency between portfolio process [3]

For the {PP4} DPMP "Develop Portfolio Management Plan" process, the corresponding PP4-centric dependency analysis is next explained. In the "input and output processes" section of the PMI standard we can read that: the {PP1} DPSP is an input process of the {PP4} DPMP; the {PP3} DPR is an input and output process of the {PP4} DPMP; the {PP7} DPCMP is output process of the {PP4} DPMP. This means that the input processes of {PP4} DPMP are the {PP1} DPMP, {PP2} DPC and {PP3} DPR; the output processes are the {PP3} DPR, {PP5} DP, {PP6} DPPMP, {PP7} DPCMP, {PP10} OP, {PP15} AP, and {PP16} PPO. All these relations are represented in the matrix of Table 3, where an "IN" stands for input process, "OUT" for output process, and "I/O" for input and output process. The matrix contains the information of all the perceived dependencies. Each matrix row represents the portfolio process source under analysis and the columns represent the depended portfolio processes, both in the input and output perspectives.

Table 3. {PP4} DPMP matrix line

depended PP \ PP depends	Defining Process Group								Aligning Process Group						Authorizing and Controlling Process Group		Input Number of dependency	Output Number of dependency
	PSM {PP 1} DPSP S,M,C,I	PSM {PP 2} DPC S,M,C,I	PSM {PP 3} DPR S,M,C,I	PGM {PP 4} DPMP S,M,C,I	PGM {PP 5} DP S,M,C,I	PPM {PP 6} DPPMP S,M,C,I	PCM {PP 7} DPCMP M,C,I	PRM {PP 8} DPRMP S,M,C,I	PSM {PP 9} MSC M,C,I	PGM {PP 10} OP S,M,C,I	PPM {PP 11} MSD S,M,C,I	PPM {PP 12} MPV S,M,C,I	PCM {PP 13} MPI M,C,I	PRM {PP 14} MPR M,C,I	PGM {PP 15} AP S,M,C,I	PGM {PP 16} PPO S,M,C,I		
PGM {PP 4} DPMP S,M,I	IN	IN	I/O		OUT	OUT	OUT			OUT					OUT	OUT	3	7

3.3 Portfolio Processes Centric Dependency Analysis

To create the complete matrix of the portfolio management processes the elementary dependency analysis must be performed for all the portfolio management processes. The resulting matrix of this overall analysis is represented in Table 4. In order to easily understand the effective impact of the dependencies between all the portfolio processes the matrix is sorted by process groups in Table 4 (note the red gradient). In Fig. 3, we depict the corresponding graph representation of the global matrix of Table 4. This global graph (also called *Global Portfolio Process Dependency Analysis Graph*) shows the global view of the dependencies between the portfolio processes. Bi-directional dependencies between the portfolio processes of different process groups are represented with lines with left and right arrows.

Table 4. Dependencies between all the Portfolio Processes

PP \ PP	PSM (PP 1) DPSP S,M,C,I	PSM (PP 2) DPC S,M,C,I	PSM (PP 3) DPR S,M,C,I	PGM (PP 4) DPMP S,M,C,I	PGM (PP 5) DP S,M,C,I	PPM (PP 6) DPPMP S,M,C,I	PCM (PP 7) DPCMP M,C,I	PRM (PP 8) DPRMP S,M,C,I	PSM (PP 9) MSC M,C,I	PGM (PP 10) OP S,M,C,I	PPM (PP 11) MSD S,M,C,I	PPM (PP 12) MPV S,M,C,I	PCM (PP 13) MPI M,C,I	PRM (PP 14) MPR M,C,I	PGM (PP 15) AP S,M,C,I	PGM (PP 16) PPO S,M,C,I	Input Number of dependency	Output Number of dependency
PGM (PP 4) DPMP S,M,C,I	IN	IN	I/O		OUT	OUT	OUT		OUT						OUT	OUT	3	7
PGM (PP 5) DP S,M,C,I	IN	IN	IN	IN					OUT						OUT	OUT	4	3
PPM (PP 6) DPPMP S,M,C,I				IN						OUT	OUT						1	2
PCM (PP 7) DPCMP M,C,I			IN	IN	IN								OUT				3	1
PRM (PP 8) DPRMP S,M,C,I				IN										OUT			1	1
PSM (PP 9) MSC M,C,I	OUT	OUT	OUT														0	3
PGM (PP 10) OP S,M,C,I	IN	IN	IN	IN	IN										OUT	OUT	5	2
PPM (PP 11) MSD S,M,C,I			IN	IN	IN												3	0
PPM (PP 12) MPV S,M,C,I	IN		IN	IN		IN											4	0
PCM (PP 13) MPI M,C,I			IN	IN		I/O											3	1
PRM (PP 14) MPR M,C,I			IN	IN			I/O										3	1
PGM (PP 15) AP S,M,C,I			IN	IN							IN					I/O	4	1
PGM (PP 16) PPO S,M,C,I	IN	IN	IN	IN	IN						IN				I/O		7	1
Input Number of dependency	7	5	6	11	7	2	1	1	0	2	0	0	0	0	1	1		
Output Number of dependency	2	1	2	0	0	0	1	1	0	1	1	1	1	1	3	3		

Process group headers: **Defining Process Group** (PP 1–PP 8), **Aligning Process Group** (PP 9–PP 14), **Authorizing and Controlling Process Group** (PP 15–PP 16).

Legend: **OUT** Output Process   **IN** Input Process   **I/O** Input and Output Process

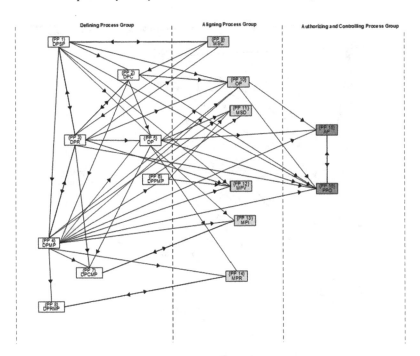

Fig. 3. Global portfolio process dependency analysis graph

3.4 Process Groups Centric Dependency Analysis

To study, discover and analyse in detail the specific dependencies of the portfolio process of one defining process group, based on the information in the global matrix, three additional graphs have been created; we call them *PG-n Centric Dependency Analysis Graph* (where *n* corresponds to the process group under study, 1 – defining, 2 – aligning and 3 – authorizing and controlling). The main idea behind the creation of these PG-n centric graphs is to focus only on the dependencies that are concerned to the process group under study, by eliminating from the global graph a huge number of dependencies that we do not want to take into account when we are studying a particular process group.

Figures 4, 5 and 6 present, respectively, the PG-1, PG-2 and PG-3 Centric Dependency Analysis Graphs. As an example, the construction of the PG-1 uses the information in the first 8 rows of the global matrix that correspond to the *defining* process group.

To better understand the creation of the PG-1 graph, as an example we analyse the {PP4} Develop Portfolio Management Plan. To represent in the graph the dependencies faced by the {PP4} process from the others portfolio processes we must parse the matrix row that corresponds to {PP4} DPMP as shown in Table 5. It is possible to see that the {PP4} presents dependencies from other 9 portfolio processes: {PP1} DPSP, {PP2} DPC, {PP3} DPR, {PP5} DP, {PP6} DPPMP, {PP7} DPCMP, {PP10} OP, {PP15} AP, and {PP16} PPO. The {PP4} DPMP process performs a key role in the PMI standard since it is the process that sends the most information to the other processes, since it presents 7 'OUT'- and 'I/O'-type dependencies.

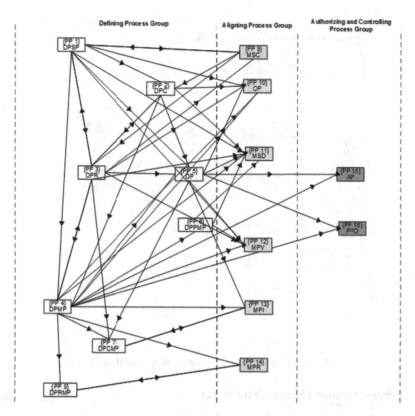

Fig. 4. PG-1 (Defining Process Group) Centric dependency analysis graph

Within the context of the *aligning* process group, PG-2 graph is presented in Fig. 5. The graph emphasizes the fact that the *aligning* process group receives information from the *defining* process group and produces outputs for the *authorizing and controlling* and the *defining* process groups. It is also possible to perceive that some threads of processes of the *aligning* process group conclude their activities inside the group itself; see, for example, {PP11} MSD and {PP12} MPV.

Table 5. PP-1 centric dependency analysis for {PP4} DPMP

	Defining Process Group								Aligning Process Group						Authorizing and Controlling Process Group		Input Number of dependency	Output Number of dependency
depended PP ⟍ PP depends	PSM {PP 1} DPSP S,M,C,I	PSM {PP 2} DPC S,M,C,I	PSM {PP 3} DPR S,M,C,I	PGM {PP 4} DPMP S,M,C,	PGM {PP 5} DP S,M,C,I	PPM {PP 6} DPPMP S,M,C,I	PCM {PP 7} DPCMP M,C,I	PRM {PP 8} DPRMP S,M,C,I	PSM {PP 9} MSC M,C,I	PGM {PP 10} OP S,M,C,I	PPM {PP 11} MSD S,M,C,I	PPM {PP 12} MPV S,M,C,I	PCM {PP 13} MPI M,C,I	PRM {PP 14} MPR M,C,I	PGM {PP 15} AP S,M,C,I	PGM {PP 16} PPO S,M,C,I		
PSM {PP 3} DPR S,M,C,I																		
PGM {PP 4} DPMP S,M,C,I	IN	IN	I/O		OUT	OUT	OUT				OUT				OUT	OUT	3	7
PGM {PP 5} DP S,M,C,I																		

Figure 6 shows the PG-3 centric dependency analysis graph that supports the dependency analysis of the only two existing processes within the *authorizing and controlling* process group: the {PP15} AP and the {PP16} PPO. These two processes are mainly recipients of information from the other two process groups and do not produce information back. By analysing the graph it is possible to perceive that the two processes of the *authorizing and controlling* process group are relevant closing processes of the project portfolio management life-cycle.

3.5 Knowledge Areas and Portfolio Processes Centric Dependency Analysis

The PMI Standard for Portfolio Management classifies each portfolio management processes by one of the following five knowledge areas: portfolio strategic management, portfolio governance management, portfolio performance management, portfolio communication management, and portfolio risk management. To better understand the genuine nature of the existing dependencies between the portfolio management processes, based on the information made available by the PMI standard, we have constructed the graph depicted in Fig. 7. This new graph results from the annotation of the Global Portfolio Process Dependency Analysis Graph presented in Fig. 3 with the reference to the knowledge areas.

The analysis of the graph in Fig. 7 permits to conclude that: (1) the processes under the *portfolio strategic management* knowledge area are the first processes to be executed; (2) the *portfolio governance management* is the only knowledge area that comprises portfolio management processes from all the three process groups; (3) the portfolio management processes classified by the *portfolio governance management* knowledge area are the ones that present the more number of dependencies between within all the portfolio; and (4) the *performance management, risk management*, and *communication management* knowledge areas present a limited number of process dependencies.

4 Conclusions and Future Work

This paper presents the result of the systematic analysis of the dependencies between processes and artefacts with the Standard for Portfolio Management. Briefly, we can conclude that: (1) the {PP4} Develop Portfolio Strategic Plan process plays a key role in the whole PMI standard, since it is the main process producing information to the other processes; (2) the {PP16} Provide Portfolio Oversight process is the one that receives more input from the other processes, denoting the dependency of making good governance decisions in response to portfolio performance, changes, issues and risks with a well set of established portfolio management processes; (3) the aligning process group corresponds to the processes with more quantity of inputs and outputs within the PMI standard, implying the most intensity of processes execution; (4) the authorizing and controlling process group presents a very high dependence from the defining process group, which means that the corresponding processes can only be executed after the full implementation of the former process group.

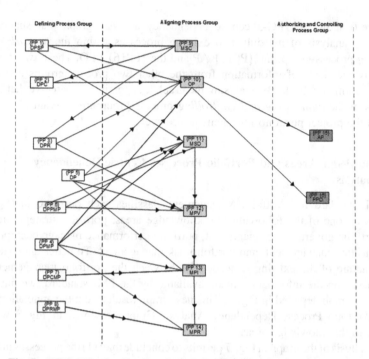

Fig. 5. PG-2 (Aligning Process Group) Centric dependency analysis graph

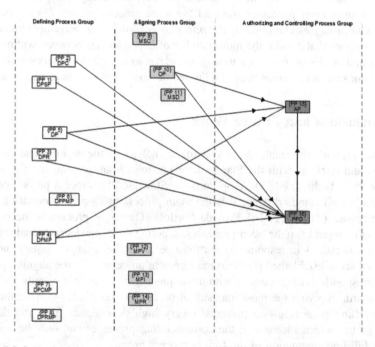

Fig. 6. PG-3 (Authorizing and Controlling Process Group) Centric dependency analysis graph

The theoretical contribution of the research reported in this paper is twofold. Firstly, this research builds knowledge in the area of portfolio management, for which there is limited understanding. Secondly, the paper contributes to a better understanding portfolio management process, by identifying the dependencies between processes for portfolio management. In software development companies, the {PP4} Develop Portfolio Strategic Plan process is particular important in order to establish the project requirements boundaries. Requirements definition in software projects is often very complex namely because of the high number of stakeholders involved and the complexity of the scope definition. The implementation of the {PP16} Provide Portfolio Oversight Process needs to take into account the different approaches to manage software projects, in software development companies it is common to coexist more traditional approaches and more agile approaches to manage different types of software projects, which brings many implications on how to monitor the portfolio to ensure alignment with the company strategy and objectives.

The results presented in this research study are exploratory. As is often the case with exploratory research, the results open up many avenues for future research [15]. One avenue is to apply in a set of software projects to confirm the implications of dependencies processes in these kind of projects, as we known the Standard Portfolio Management from PMI is generic for any portfolio. A second avenue is to deepen what are the inputs and outputs between processes based on artefacts. Namely, by

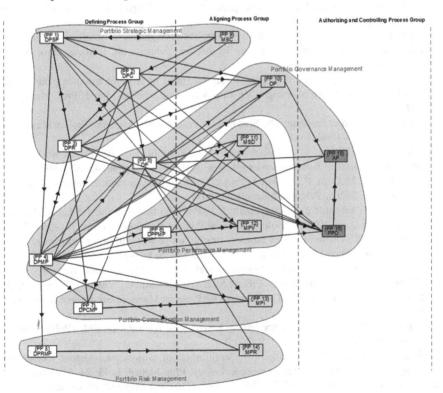

Fig. 7. Global portfolio process dependency analysis graph with annotated knowledge areas

identification the artefacts generated by each of the processes and respective dependencies between them using SPEM 2.0, and a thoroughly discussion of concrete implications for the portfolio management of software development projects. Finally, the third avenue is to compare the dependencies between the processes of the Office of Government Commerce (OGC) standard [16] with the PMI standard [3] in a contrasting basis, expecting different results, since the OGC has an higher commitment to the software development domain.

Acknowledgements. This work has been supported by COMPETE: POCI-01-0145-FEDER-007043 and FCT – Fundação para a Ciência e Tecnologia within the Project Scope: UID/CEC/00319/2013.

References

1. Dooley, L., Lupton, G., O'Sullivan, D.: Multiple project management: a modern competitive necessity. J. Manuf. Technol. Manag. **16**, 466–482 (2005)
2. Schelini, A.L.S., Martens, C.D.P.: Seleção de Projetos orientados para o mercado: Um estudo em uma instituição de desenvolvimento. In: SINGEP 2012 (2012)
3. Project Management Institute: The Standard for Portfolio Management. Project Managment Institute, Newtown Square (2013)
4. Archer, N.P., Ghasemzadeh, F.: An integrated framework for project portfolio selection. Int. J. Proj. Manag. **17**, 207–216 (1999)
5. Rodney Turner, J., Ledwith, A., Kelly, J.: Project management in small to medium-sized enterprises: a comparison between firms by size and industry. Int. J. Manag. Proj. Bus. **2**, 282–296 (2009)
6. Laslo, Z.: Project portfolio management: an integrated method for resource planning and scheduling to minimize planning/scheduling-dependent expenses. Int. J. Proj. Manag. **28**, 609–618 (2010)
7. Project Management Institute: The Standard for Portfolio Management. Project Managment Institute, Estados Unidos (2008)
8. Cooper, R.G., Edgett, S.J., Kleinschmidt, E.J.: Portfolio management in new product development: lessons from the leaders-II. Res. Technol. Manag. **40**, 43–52 (1997)
9. Rajegopal, S., McGuin, P., Waller, J.: Project Portfolio Management: Earning an Execution Premium, pp. 1–268. Palgrave Macmillan (2007)
10. Cooper, R.G., Edgett, S.J., Kleinschmidt, E.J.: Portfolio management: fundamental to new product success. In: PDMA Toolbox for New Product Development, pp. 331–364 (2000)
11. Meskendahl, S.: The influence of business strategy on project portfolio management and its success a conceptual framework. Int. J. Proj. Manag. **28**, 807–817 (2010)
12. Project Management Institute: Organizational Project Management Maturity Model (OPM3): Knowledge Foundation. Project Management Institute, Newtown Square (2013)
13. Project Management Institute: The Standard for Program Management. Project Management Institute, Newtown Square (2013)
14. Pinto, J.A., Williams, N.: Country project management maturity. In: PMI Global Congress Proceedings, Istanbul, Turkey, p. 8 (2013)
15. Besner, C., Hobbs, B.: An empirical identification of project management toolsets and comparasion project outcome: an empirical study. Proj. Manag. J. **43**, 24–43 (2012)
16. Axelos, Management of Portfolios (MOP), TSO (The Stationery Office), London, February 2011. ISBN: 9780113312955

Representation and Reasoning of Topological Relations Between Enclave and Exclave Regions

Wan Neng[✉], Deng Zhongliang, and Yang Guangyong

Beijing University of Posts and Telecommunications, Beijing, China
{wanneng, dengzhl, yangguangyong}@bupt.edu.cn

Abstract. Topological relation is the most important spatial relation. Region Connection Calculus (RCC) is a famous representation of topological relations between regions. But RCC and the other theories improved from RCC cannot represent topological relations between enclave and exclave regions well. An extend RCC theory (ERCC) is proposed with the concept of convex hull that can solve the problem. ERCC can describe eleven kinds of relations between complex regions, which is embedded into an ontology in OWL with rules. A rule based reasoning system is designed to reasoning the topological relations with their semantics. The ontology has more abundant description ability and the reasoning system has more powerful reasoning ability with ERCC theory.

Keywords: Topological relations · RCC · Spatial reasoning · Reasoning system · Knowledge representation · OWL

1 Introduction

The theory of spatial representations and spatial reasoning came from AI (Artificial Intelligence) and now is widely used in GIS (Geographic Information System), image processing, pattern recognition, robot, computer vision, and so on. During the recent years, many research institutes have been set up in this field, and many universities have opened courses of spatial reasoning. Moreover, spatial reasoning is an important topic in the international authoritative academic conferences on AI such as IJCAI and AAAI.

Formal definitions of spatial relations are essential for spatial reasoning, and most of which are about topological relations. The research of spatial topological relations includes point based spatial primitive and region based spatial primitive, both of which have a relatively complete theoretical system. But the latter is much more popular. A well-known case of region based spatial primitive is the Region Connection Calculus (RCC) which was first present in 1992 by Randell [1]. RCC was proposed based on Clarke's space calculus logical axioms [2]. RCC can be used to represent regions with no-empty regular subsets of some topological space, that is region without holes. And then it was further improved and developed by many scientists [3–8]. Cohn extend RCC with an additional 'convex hull' primitive to allow a much finer-grained representation than a purely topological representation allows, which is called RCC-23 with

© Springer International Publishing Switzerland 2016
O. Gervasi et al. (Eds.): ICCSA 2016, Part V, LNCS 9790, pp. 301–311, 2016.
DOI: 10.1007/978-3-319-42092-9_23

23 binary relations forming a JEPD set [3]. If DC and EC aren't distinguished, RCC-23 turns into RCC-15. Dong presented RCC+ theory which added a new axiom to govern the characteristic property of connection relations [4]. Egenhofer and Zhao improved RCC to support Spatial reasoning with a single hole [6, 7].

But these theories cannot describe topological relations between enclave and exclave regions very well because they couldn't represent the topological relations of non-connected regions. According to Wikipedia, an enclave is a territory, or a part of a territory, that is entirely surrounded by the territory of another state, an exclave is a portion of a state geographically separated from the main part by surrounding alien territory [9]. Many exclaves are also enclaves. In reality, a region can have multiple borders. For example, Beijing Capital International Airport is located in Shunyi District of Beijing, but it is governed by Chaoyang District of Beijing, so the airport is an enclave of Shunyi District and an exclave of Chaoyang District. Sanhe city, Xianghe County and Dachang Hui Autonomous County of Hebei Province are entirely surrounded by Beijing and Tianjing, so Sanhe–Xianghe–Dachang is an exclave of Hebei Province, which is the biggest exclave of China. But Sanhe–Xianghe–Dachang isn't an enclave of any region.

The organization of this paper is as follows, Sect. 2 introduces the related work include what is a JEPD set and the theory of RCC. A new theory named Extended RCC extended from RCC is proposed in Sect. 3. Topological relations between enclave and exclave regions can be described by Extended RCC. In Sect. 4 an OWL-based ontology is presented for ERCC theory which can be used to reasoning spatial topological relations. Finally, Sect. 5 concludes.

2   Related Work

JEPD (Jointly Exhaustive and Pair-wise Disjoint) is an important concept for research on modeling for topological relations. The concept of JEPD and the theory of RCC are introduced here.

2.1   The Concept of JEPD

JEPD is a special set of binary relations. The related definitions are as follows.

Definition 1. Jointly Exhaustive Set. Give R is a finite set of binary relations in set S, which means $R = \{r|r \subseteq S \times S\}$. If $\forall x, y \in S \; \exists r \in R, r(x, y)$, then R is called Jointly Exhaustive Set in S.

Definition 2. Pair-wise Disjoint Set. Give R is a finite set of binary relations in set S, which means $R = \{r|r \subseteq S \times S\}$. If $\forall x, y \in S, r_1(x, y)$ and $r_2(x, y)$ is true if and only if $r_1 = r_2$, then R is called Pair-wise Disjoint set in S.

Definition 3. Jointly Exhaustive and Pair-wise Disjoint. Give R is a finite set of binary relations in set S, which means $R = \{r|r \subseteq S \times S\}$. If R is both Jointly Exhaustive and Pair-wise Disjoint, then R is called Jointly Exhaustive and Pair-wise Disjoint in S, abbreviated as JEPD set.

JEPD relations are also named base, basic or atomic relations. Given a set of JEPD relations, the relationship between any two spatial entities of the considered domain must be exactly one of the JEPD relations. So JEPD already became the basic data structure of qualitative reasoning and was adopted by many kinds of temporal and spatial reasoning theories such as Allen's interval algebra and RCC [1, 10, 11].

2.2   The Theory of RCC

RCC is a well-known spatial representation theory of topological relations between regions, which believes connection relation is primitive in the spatial domain. Regions rather than points are taken as primitive units in RCC. The regions may be of any dimension. But dimension of different regions in the same model must be the same. In addition, RCC also claims that the collection of points composing a region must be continuous.

Given two regions x, y, if x connects y, denoted as $C(x, y)$. RCC defined that any two regions must share at least one point. The binary relation $C(x, y)$ is reflexive and symmetric, but not transitive. The RCC theory has two axioms for the connection relation $C(x, y)$.

- Any region x, x connects with itself
- Any regions x, y, if x connects with y, then y connects with x

The two axioms can be formalized as follows.

$$\forall x \in R, C(x, x) \tag{1}$$

$$\forall x, y \in R, C(x, y) \to C(y, x) \tag{2}$$

The relation $C(x, y)$ is not transitive, which can be formalized as follows.

$$\forall x, y, z \in R, C(x, y) \wedge C(y, z) \nrightarrow C(x, z) \tag{3}$$

The other binary relations of RCC are all based on the connection C. Multiple topological relations model such as RCC-5, RCC-8, RCC-15 can be derived from $C(x, y)$, which correspond to regardless of border, considering of border, considering of temporal state respectively. And RCC-8 is the best well-known now, which consists of eight basic relations that are possible between any two regions. The eight relations are as follows.

- DC: disconnected
- EC: externally connected
- EQ: equal
- PO: partially overlapping
- TPP: tangential proper part
- TPPi: tangential proper part inverse
- NTPP: non-tangential proper part
- NTPPi: non-tangential proper part inverse

The eight binary relations form a JEPD set, any two regions must stand in exactly one of the eight relations. RCC-8 can be simplified to RCC-8 if regardless of border. RCC-5 has five relations DR, PO, PP, PPi, and EQ, which form a JEPD set, too. The relation between RCC-8 and RCC-5 is shown in Fig. 1.

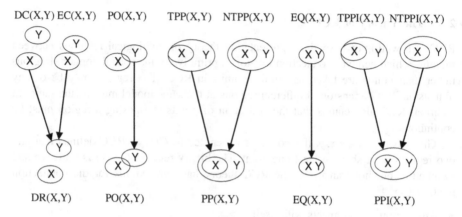

Fig. 1. The relation between RCC-8 and RCC-5

3 A New Theory Extended from RCC

Enclave can be called region with holes, and exclave can be called non-connected region. As they are both objective reality, topology relation model should be able to describe them formally. But RCC-5 only applies to connected region. Although Egenhofer used a form of matrix based on RCC-8 to describing the topological relations between regions with holes [6], but it cannot describe the topological relations of non-connected regions whose topologies differ conceptually. Therefore an extended RCC (ERCC) theory is proposed based on RCC-5, which can support both enclave and exclave regions. In ERCC, the border of region can be ignored, that is, the border's width is zero.

Definition 4. Complex Region: Complex region A is the set of points on a plane formed by n (n ≥ 1)) disjoint closed curves{Ei}, Ei (1 ≤ i ≤ n) is the border of A, A = {x| Starting from x along a direction of any straight line to infinity, it is required to pass through the border of A odd times.}

Definition 5. Convex hull: the smallest convex region which can surround the set of points S on the plane called the convex hull of S, denoted by ChS.

In a convex hull every point on every line joining any two points in the region. According to the above two definitions, ERCC define eleven topological relations for complex region A and B, which form a JEPD set.

Definition 6. Topological relations between complex regions: the topological relation between complex region A and B are as follows:

1. If $Ch_A = Ch_B$,
   (a) If $\forall x((x \in A \rightarrow x \in B) \wedge (x \in B \rightarrow x \in A))$, then **equals**(A,B)
   (b) Else **externallyEquals**(A,B)
2. Else if ChA contains ChB,
   (a) If $\forall x(x \in B \rightarrow x \in A)$, then **contains**(A,B);
   (b) Else if $\forall x(x \in B \rightarrow x \notin A)$, then **disjointlyContains**(A,B);
   (c) Else **externallyContains**(A,B).
3. Else if ChB contains ChA,
   (a) If $\forall x(x \in A \rightarrow x \in B)$, then **containedBy**(A,B).
   (b) Else if $\forall x(x \in A \rightarrow x \notin B)$, then **disjointlyContainedBy**(A,B).
   (c) Else **externallyContainedBy**(A,B).
4. Else if ChA joints with ChB,
   (a) If $\exists x(x \in A \wedge x \in B)$, then **overlap**(A,B).
   (b) Else **disjointlyOverlap**(A,B).
5. Else **disjoint**(A,B), which means ChA disjointing with ChB.

The above eleven kinds of relations are diagramed in Fig. 2, in which equals, contains, containedBy, overlap and disjoint correspond to EQ, PP, PPI, PO, and DR in RCC5.

Theorem 1. The eleven kinds of relations defined in Definition 6 form a JEPD set.

According to Definition 6, the relationship between A and B must be one of the eleven kinds of relationships, and so they form a JEPD set.

Theorem 2. Equals, externallyEquals, overlap, externallyOverlap and disjoint are all symmetrical binary relations. Equals, contains, containedBy are all transitive. The relationship between contains and containedBy, disjointlyContains and disjointlyContainedBy, externallyContains and externallyContainedBy are reciprocal relationship respectively.

Theorem 2 can be deduced directly from Definition 6.

The topological relationship between enclave and exclave regions would be described by the ERCC theory in the next section.

4 Reasoning with ERCC and OWL

Location representation is not a new idea. A map was known as the first kind of location representation. And now ontology become one of the most suitable formats to manage location context. There were a lot of ontology-based location models. In order to represent and reasoning spatial topological relations between enclave and exclave regions by using the semantic web achievements [12–15], an OWL-based ontology is proposed based on ERCC theory. And then, rule-based reasoning system is designed. An instance is introduced to reasoning with this system which has the advantages of both ERCC theory and semantic reasoning theory.

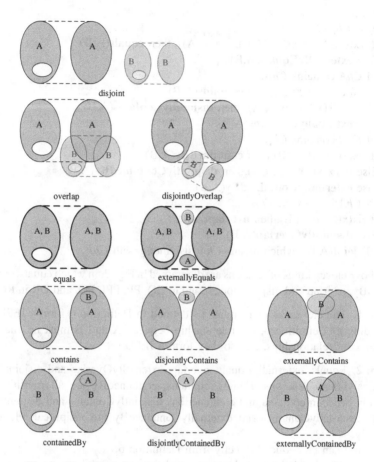

Fig. 2. The 11 relations between complex regions defined in ERCC

4.1   Ontology in OWL Based on ERCC

ERCC calculus includes eleven topological relations between regions on the plane, the semantic of which is embedded into an ontology with rules. The ontology diagram is outlined in Fig. 3 that supports the representation and reasoning topological relations between enclave and exclave regions. The figure shows the class and property in the ontology of location information model. The class and property consist of the TBox part of description logic knowledge base, which is similar to data pattern in database system. The model describes the eleven kinds of relationship in OWL language. One part of the ontology in OWL is cut in Fig. 4.

Compared with other ontologies and models, this ontology has the following characteristics:

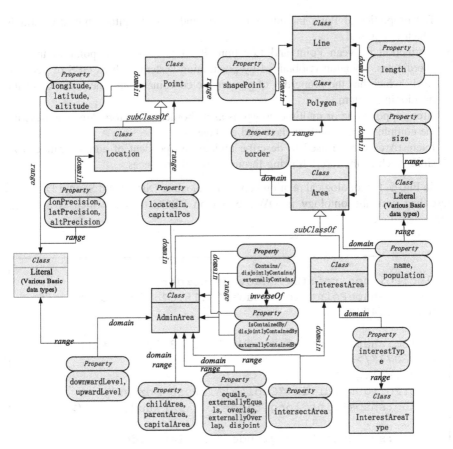

Fig. 3. Ontology diagrams based on ERCC

```
<owl:ObjectProperty rdf:about="&loc;externallyOverlap">
  <rdfs:domain rdf:resource="&loc;Area" />
  <rdfs:range rdf:resource="&loc;Area" />
  <rdf:type rdf:resource="&owl;SymmetricProperty" />
</owl:ObjectProperty>
<owl:ObjectProperty rdf:about="&loc;equals">
  <rdfs:domain rdf:resource="&loc;Area" />
  <rdfs:range rdf:resource="&loc;Area" />
  <rdf:type rdf:resource="&owl;TransitiveProperty" />
  <rdf:type rdf:resource="&owl;SymmetricProperty" />
</owl:ObjectProperty>
```

Fig. 4. Part of ontology in OWL based on ERCC

- The properties of Point include not only latitude and longitude but also altitude which is optional.
- Distinction between Point and Location. Point is an explicit point in the real coordinates. Location is obtained by positioning system such as GPS or cell network, which has the properties of geographic coordinates and the accuracy of the positioning system.
- Area supports one or multiple polygons as its borders, therefore Area can describe the complex region with enclave and exclave.
- InterestArea can be used to describe regions independent of administrative divisions, such as the rivers flowing through several provinces.

In Fig. 5, the topological relation between Chaoyang District and Shunyi District is described with the ontology in OWL based on ERCC theory, which is a typical enclave/exclave.

```
<loc:Polygon rdf:ID="Chaoyang_border1">
  <loc:shapePoint longitude=" 116.345169" latitude="
40.025314"/>
  ... ...
  <loc:shapePoint longitude=" 116.38398" latitude="
40.03078"/>
</loc:Polygon>
<loc:Polygon rdf:ID="CapitalAirPort_border">
  <loc:shapePoint longitude="... " latitude=" ..."/>
  ... ...
  <loc:shapePoint longitude=" ..." latitude=" ..."/>
</loc:Polygon>
<loc:Polygon rdf:ID="Shunyi_border1">
  <loc:shapePoint longitude=" ..." latitude=" ..."/>
  ... ...
  <loc:shapePoint longitude=" ..." latitude="..."/>
</loc:Polygon>
<loc:AdminArea rdfID="ChaoyangDistrict">
  <loc:border rdf:resource="&loc;Chaoyang_border1"/>
  <loc:border rdf:resource="&loc; CapitalAirPort
_border"/>
  <loc:overlap rdf:resource="&loc;ShunyiDistrict"/>
</loc:AdminArea>
<loc:AdminArea rdfID="ShunyiDistrict">
  <loc:border rdf:resource="&loc;Shunyi_border1"/>
  <loc:border rdf:resource="&loc; CapitalAirPort
_border"/>
</loc:AdminArea>
```

Fig. 5. Enclave/exclave described with the ontology based on ERCC

4.2   Rule-Based Reasoning System with ERCC and OWL

OWL and description logic are adapted to express the concept system, but rules are widely used to describe other types of knowledge in semantic web and knowledge of engineering. Therefore it is necessary to unite rule system and ontology based on ERCC to absorb their advantages. There are many variants of rule-based system with different expression ability. Generic/Normal Rule is adopted in this rule system, which is an important rule in Prolog language. It is as follows.

$$H :- B1, B2, \ldots, Bm, \, not\, Bm + 1, \ldots, \, not\, Bn \tag{4}$$

The left part of ':-' is called Head or Conclusion. The right part is called Body or Premise. H and Bi are both atomic formula. 'not'is called Default Negation or Negation As Failure (NAF). If n = m, then formula (4) can be named Definite Rule. If n = m = 0, then named Fact. The formula and rule without variable are ground.

Intuitively, formula (4) means that the conclusion is true if each formula in the premise is true. Default Negation represents 'does not exist', which is different from Classic Negative of FOL. The basic query supported by a rule-based system Π is atomic query, including two kinds:

- Whether a basic atomic can be satisfied
- Given an atomic formula A containing variables, get all binding of these variables, with which the rule-based system Π implies A

There are two kinds of method for resolution, one is backward search from the goal, the other is forward search from the fact. The use of SLD in Prolog language belongs to the former.

A rule which will be used in Sect. 4.3 is as follows.

$$locatesIn(p, a2) :- locatesIn(p, a1), isContainedBy(a1, a2) \tag{5}$$

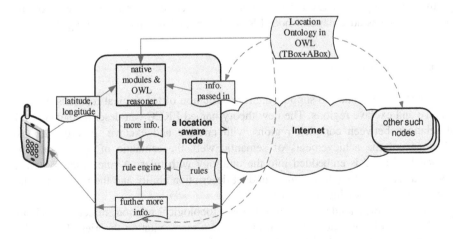

Fig. 6. A scenario for OWL and rule based engine

4.3   An Instance of Reasoning with OWL and Rule System

A typical application scenario of the ontology and rule-based system proposed above is shown in Fig. 6. A location-aware node can get some implicit information by reasoning with the original location information represented by latitude and longitude from the mobile phone, information from other nodes, ontology based knowledge base and OWL reasoner. Furthermore, more implicit information can be inferred by reasoning with the former knowledge and other knowledge in the rule-based system.

For example, given there are three instances of Class AdminArea in an ontology-based knowledge base ABox, the relations of which are as follows.

$$isContainedBy(ChaoyangDistrict, BeijingCity) \tag{6}$$

$$isContainedBy(BeijingCity, China) \tag{7}$$

Then OWL reasoner can infer the following conclusion based on the axioms in TBox.

$$isContainedBy(ChaoyangDistrict, China) \tag{8}$$

Given the location of a mobile phone T was calculated as follows.

$$locatesIn(T, ChaoyangDistrict) \tag{9}$$

Then the rule-based engine can infer the following two conclusions based on formulas (6) and (8) and the rule in formula (5):

$$locatesIn(T, BeijingCity) \tag{10}$$

$$locatesIn(T, China) \tag{11}$$

In the above process, OWL reasoner and rule-based reasoner are both used. And the rule system takes advantage of both ERCC theory and semantic reasoning theory.

5   Conclusion

RCC theory is extended to support the representation of topological relations between enclave and exclave regions. The new theory named ERCC can describe eleven kinds of relations between complex regions with enclave and exclave. In order to take advantage of the achievements of semantic web, the semantic of spatial relations defined by ERCC is embedded into the ontology with rules. A rule based system is designed. The ontology has more abundant description ability and the rule system has more powerful reasoning ability with ERCC.

There are three kinds of spatial relations: topological relation, directional relation and distance relation, each of which has its own representation theories. To translate these theories into a unified format in OWL will be the future work.

Acknowledgments. This work has been supported by the Fundamental Research Funds for the Central Universities, National High Technology Research and Development Program of China (No.2015AA016501).

References

1. Randell, D.A., Cui, Z., Cohn, A.: A spatial logic based on regions and connection. In: Principles of Knowledge Representation and Reasoning, pp. 165–176. Morgan Kaufmann (1992)
2. Clarke, B.L.: A calculus of individuals based on 'connection'. Notre Dame J. Formal Logic **22**(3), 204–218 (1981)
3. Cohn, A.G., Renz, J.: Qualitative spatial representation and reasoning. In: van Harmelen, F., Lifschitz, V., Porter, B. (eds.) Handbook of Knowledge Representation, pp. 551–596. Elsevier, Amsterdam (2007)
4. Dong, T.: A comment on RCC: from RCC to RCC^{++}. J. Philos. Logic **34**(2), 319–352 (2008)
5. Bai, L., Li, Y., Ma, Z.M.: Modeling topological relations between fuzzy spatiotemporal regions over time. In: IEEE International Conference on Fuzzy Systems, pp. 1–8. IEEE Press, Brisbane (2012)
6. Egenhofer, M., Vasardani, M.: Spatial reasoning with a hole. In: Winter, S., Duckham, M., Kulik, L., Kuipers, B. (eds.) COSIT 2007. LNCS, vol. 4736, pp. 303–320. Springer, Heidelberg (2007)
7. Zhao, R.: Research on reasoning and topological relations between spatial region with holes and simple unclosed line. MS thesis, Jilin University, Changchun (2009)
8. Wolter, D., Kreutzmann, A.: Analogical representation of RCC-8 for neighborhood-based qualitative spatial reasoning. In: Hölldobler, S., et al. (eds.) KI 2015. LNCS, vol. 9324, pp. 194–207. Springer, Heidelberg (2015). doi:10.1007/978-3-319-24489-1_15
9. Enclave and exclave. https://en.wikipedia.org/wiki/Enclave_and_exclave
10. Allen, J.F.: Maintaining knowledge about temporal intervals. Commun. ACM **26**(11), 832–843 (1983)
11. Allen, J.F.: Planning as temporal reasoning. In: Proceedings of the Second International Conference on Principles of Knowledge Representation and Reasoning, pp. 3–14. Morgan Kaufmann (1991)
12. Web Ontology Language (OWL). https://www.w3.org/2001/sw/wiki/OWL
13. Batsakis, S., Antoniou, G., Tachmazidis, I.: Representing and reasoning over topological relations in OWL. In: 4th International Conference on Web Intelligence, Mining and Semantics (WIMS14). ACM, New York (2014)
14. Stocker, M., Sirin, E.: Pelletspatial: a hybrid RCC-8 and RDF/OWL reasoning and query engine. In: Proceedings of the 5th International Workshop on OWL: Experiences and Directions. Chantilly (2009)
15. Mainas, N., Petrakis, E.G.M.: CHOROS 2: improving the performance of qualitative spatial reasoning in OWL. In: 26th International Conference on Tools with Artificial Intelligence, pp. 283–290. IEEE Press, Limassol (2014)

Toward a Robust Spell Checker for Arabic Text

Mourad Mars[✉]

Umm Al-Qura University KSA - LIDILEM Laboratory, Grenoble, France
mourad.mars@univ-grenoble-alpes.fr

Abstract. Spell checking is the process of detecting misspelled words in a written text and recommending alternative spellings. The first stage consists of detecting real-word errors and non-word errors in a given text. The second stage consists of error correction. In this paper we propose a novel method for spell checking Arabic text. Our system is a sequential combination of approaches including lexicon based, rule based, and statistical based. The experimental results show that the proposed method achieved good performance in terms of recall rate or precision rate in error detection, and correction comparing to other systems.

Keywords: Spell checking · Arabic language · Error detection · Error correction · Hybrid approach

1 Introduction

Spell checking is the process of detecting misspelled words in a written text and recommending alternative spellings. Spell checkers system can be used in different Natural language processing (NLP) applications such as Optical Character Recognition (OCR) systems where Hidden Markov Model (HMM) based approaches using n-grams have been shown to be quite successful [30], Automatic Speech Recognition (ASR) systems, Intelligent Computer Aided Language Learning (ICALL) platforms, Machine Translation (MT) system, Text-to-Speech (TTS) systems, and others topics.

The spelling correction problem is formally defined [17] as: given an alphabet $\sum$, a dictionary D consisting of strings in $\sum *$, and a spelling error s, where s $\notin$ D and s $\in \sum *$, find the correction c, where c $\in$ D, and c is most likely to have been erroneously typed as s. This is treated as a probabilistic problem formulated as [3]:

$$argmax_c P(s|c)P(c). \tag{1}$$

The history of automatic spelling correction goes back to the 1960s. Even after decades of extensive research and development, the effectiveness of spell checkers remains a challenge today [28].

In this paper, we focus on detecting and correcting errors for Arabic language. The rest of this paper is organized as follows: Sect. 2 gives an overview of the different types of errors. Section 3 presents a list of related work in this area. The architecture of the spell checker, the data used and the algorithm implemented are detailed in Sect. 4. While Sect. 5 presents results of applying our appraoch with different settings (Table 1).

© Springer International Publishing Switzerland 2016
O. Gervasi et al. (Eds.): ICCSA 2016, Part V, LNCS 9790, pp. 312–322, 2016.
DOI: 10.1007/978-3-319-42092-9_24

Table 1. A sample of an original erroneous Arabic text with its manual correction. Third line contains Buckwalter transliteration of both texts. The underlined words are misspelling words. Missing comma at the end of line 2. Refer to Table 2 for details about types of errors and corrections.

Arabic text with misspelling words	Manual correction
تفيد أرقام الأم المتحدة بأن أكثر من عشرين ألف شخص يتج معون في المعبر في حينتفيد تقديرات أخرى بأن أكصر من خمسين ألفا أجبروا على نرك منازلهم منذ الإثنين الماضي خوفا من الغارات.	تفيد أرقام الأم المتحدة بأن أكثر من عشرين ألف شخص يتجمعون في المعبر ، في حين تفيد تقديرات أخرى بأن أكثر من خمسين ألفا أجبروا على ترك منازلهم منذ الإثنين الماضي خوفا من الغارات.
tfyd < rqAm Al<mm AlmtHdp b<n <kvr mn Eryn > lfxS ytj mEwn fy AlmEbr fy Hyntfyd tqdyrAt <xrY b<n <kvr mn xmsyn <lfA <jbrwA ElY nrk mnAzlhm mn* Al<vnyn AlmADy xwfA mn AlgArAt.	tfyd < rqAm Al<mm AlmtHdp b<n <kSr mn Eryn > lfxS ytjmEwn fy AlmEbr , fy Hyn tfyd tqdyrAt <xrY b<n <kvr mn xmsyn <lfA <jbrwA ElY trk mnAzlhm mn* Al<vnyn AlmADy xwfA mn AlgArAt.

2 Types of Spelling Errors

In order to develop a spell checker for any language, we have to identify, analyze and classify all kinds of spelling errors. In the following, we summarize the different types of classification of spelling errors for Arabic language. Common misspelling errors can be classified into the following two main types of errors: non-word errors and real-word errors [28].

2.1 Non-word Errors

Non-word errors, where the word itself is invalid (i.e. words that do not exist in the Arabic language).

Example: الكلب ينبح والقافلة تسير — Alkllb ynbH wAlqAflp tsyr

The word الكلب does not exist in Arabic, and it probably derives from a typo of the word الكلب.

Here is a brief list of different types of errors that may causes a non-word errors.

- Edit (delete, add, move, etc.): Adding, deleting, moving, or replacing, characters in the word.
- Hamza, Alif maqsoura and yaa, taa marboutah and haa errors: writing instead of أ, or ى instead of ي, or ه instead of ة.
- Space bar issues:
  - Run-on errors: two separate words become one
  - Split errors: one word becomes two separate words

- Keyboard proximity: e.g., حبر becomes حلر since ب, ل are next to each other on a typical keyboard
- Physical similarity
  - Similarity of shape: e.g., mistaking two physically similar letters when typing up something handwritten. like ت and ث, غ and خ, etc.
- Punctuation: missing or adding punctuation

In the previous table, some non-word errors occured. The following table idenfity the type of error and the correction action.

Table 2. Different types of errors and correction chosen from previous table.

Error	Correction	Error Type	Correction Action
يتج معون	يتجمعون	Space bar issues split errors	Merge
Missing comma	،	Punctuation	Add Punctuation
حينتفيد	حين تفيد	Space bar issues run-on errors	Split
أكصر	أكثر	Keyboard proximity	Edit
نرك	ترك	Physical similarity	Edit

2.2 Real-Word Errors

Real-word errors, where the word is valid yet inappropriate in the context. قال الكتاب أن مقاله نشر في كل الصحف. — "qAl AlktAb <n mqAlh n$r fy kl AlSHf".

The noun الكتاب — "AlktAb" exists in Arabic and means "the book", but in this context it is most likely a typo of الكاتب — "AlkAtb" and it means "the author". Such errors are the most difficult to detect and correct, because they cannot be revealed just by a dictionary lookup, but can be detected only taking context into account.

3   Related Work

Different approaches to deal with this issue have been presented in the literature [20,21,24,27]. Many studies have been made and quite good results have been obtained for English [9,10]. In general, different approaches have been proposed in the literature to deal with spelling error detection and correction issues [7,8, 11,31].

For Arabic language, there have been few attempts on error detection and correction [1,13,16]. Shalan et al. [28] presented a theoretical rule-based approach to develop a system for Arabic spelling correction. Unfortunately, no experimental and evaluation results were reported. Hassan et al. [14] developed a method based on finite state automata with a language model to select the best correction in a given context. The best accuracy they reported in their experiments was 89 % on a list of 556 misspelled words. Zribi and Ahmed [4] conducted an experiment for detecting and correcting semantic hidden errors in Arabic texts assuming that there is only one error per sentences. Using a training corpus that contains around 23 k words and a testing set with 1.5 k words, the highest rate of detection accuracy was 89.18 %. In fact, corpus size is an essential aspect of the dictionary-based spelling correction techniques, either for the training set or for the testing one.

The following achievements are all presented in the second QALB shared task on automatic text correction for Arabic [12,20,27]. The same data set for training and evaluation to all participants was provided [25,32]. Ibrahim and Ragheb [15] describe the implementation of an Arabic error correction rule-based system. They used the words patterns to improve the error correction, and a syntactic error correction rules. The system achieves an F-score of 0.7287 on the dataset used in the QALB shared task workshop 2015. Bouamor et al. [5] presented a hybrid pipeline that combines rule-based linguistic techniques with statistical methods using language modeling and machine translation, as well as an error-tolerant finite-state automata method. The system was evaluated using the data provided for the QALB shared task workshop 2015. Arib system [2] contains many components that address various types of spelling error and applies a combination of approaches including rule based, statistical based, and lexicon based in a cascade fashion. The overall recall of their system is 0.51, with a precision of 0.67 and an F1 score of 0.58. Mubarak et al. [22,23] system employ a case-specific correction approach that handles specific error types such as dialectal word substitution and word splits and merges with the aid of a language model. Bougares and Bouamor [6] developed a system using a sequential combination of two statistical machine translation systems (SMT) trained on top of the MADAMIRA[1] output [26]. The first is a Character-based one, used to produce a first correction at the character level. Characters are then glued to form the input to the second system working at the Word level. This sequential combination achieves an F1 score of 69.42. This system is ranked first, on the test data provided, among other submissions [27].

4 Methodology and Experimental Settings

4.1 System Architecture

The general architecture of the proposed spell checker is presented in Fig. 1. The input is an Arabic text with misspelled words. The output is an Arabic corrected text. To perform the tasks of errors detection and corrections, the system uses differents resources presented in Sect. 4.2. How the system works and the differents steps to deal with this issues are detailed in Sect. 4.3.

---

[1] MADAMIRA is a system developed for morphological analysis and Disambiguation of Arabic text.

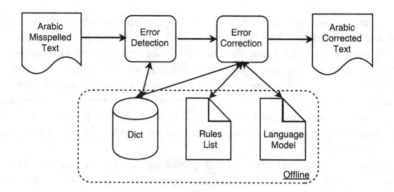

Fig. 1. The architecture of the proposed system.

4.2   Prepared Data

We mean by data preparation and data preprocessing the manipulation of data into a suitable form for further analysis and processing. It is a process that involves many different tasks and resources. In this section we will present all created resources and used tools to detect non-word errors and to suggest corrections.

Dictionary of Arabic words, to create the dictionnary for our system, we downloaded the MultiUN.ar[2] corpus for the years 2000/2005. Of this huge data, a portion for the years 2000 and 2005 were used to create the dictionnary. The whole corpus was used also to create the language model for ranking alternative candidates corrections. The size of the obtained corpora is 1713134 sentences (see Table 3). After processing of this corpus, we obtained a list of around 15.4 Million of words.

Table 3. Prepared data for spell checking (numbers in million).

	Numb. of tokens
Tokens	40820597
Sentences	1713134
Average tokens/sent	23,82

We used also a list of Arabic words created by Attia et al. [3]. This list is automatically generated from the AraComLex[3], an open-source finite-state large-scale morphological transducer, which initially generates about 13 million words. Attia validated the entire list by using Microsoft spell checker (2013) to

---

[2] MultiUN.ar: a corpus available at http://www.euromatrixplus.net/multi-un/.
[3] http://aracomlex.sourceforge.net/.

check the correctness of words and he obtained a list of 9-million words[4]. We also validated this list by our morphological Analyser [18, 19, 33]. All The results are shown in Table 4.

Table 4. Prepared data for spell checking (numbers in million).

	No. of words
MultiUN.ar + WebCorpus	15.4
Valid words in MultiUN.ar + WebCorpus (1)	11.2
AraComLex	13
Valid words in AraComLex (2)	9
Total of valid words in (1) and (2)	**13.6**

Rules for well known errors, we extracted a list of rules from QALB corpus [32] and we also considered Ghalatawi list of rules[5] created for his auto-correct system.

Named entities lexicon, we used our named entities (NE) list[6] to check if the misspelled word is a NE or not. This list contains a total of 65203 Named entities (Person 27480, Organization 17237, and Location 4036).

Language Model, we used MultiUN.ar corpus to create a language model (LM) using SRILM[7] [29]. This resource will be used to rank all alternatives solutions and select the best one for the misspelled word.

4.3 Our Approach

Our automatic spelling detection and corrector system consists of a hybrid pipeline that combines different but complementary approaches to deal with non-word errors:

- **Errors detection**
  - Errors detection (Dic + NE list)
- **Error correction**
  - Error correction proposed by MADAMIRA
  - Rule based + NE errors corrector
  - Space issue
  - Levenshtein distance
  - Ranking alternative candidates
  - Choose the gold correction (ranked first)

---

[4] The list is freely available at: http://sourceforge.net/projects/arabic-wordlist/.
[5] The Ghalatawi autocorrect program is available as an open source program at http://ghalatawi.sourceforge.net.
[6] NE list: Available at www.github.com/mouradmars/Named_Entities_Project.
[7] SRILM: Language Model Toolkit http://www.speech.sri.com/projects/srilm/.

318   M. Mars

Error detection, the process of non-word error detection is done by a simple dictionnary lookup.

MADAMIRA, is a system developed mainly for morphological analysis and Disambiguation of Arabic text. We used the features generated by MADAMIRA to support the spelling error detection and correction. The output of MADAMIRA includes an analysis and correction of the spelling mistakes in the word of character (Alf)(أ) and terminal character (Yaa)(ي).

Spaces insertion/deletion detection, this script is implemented and applied to deal with errors caused by adding spaces in the middle of a real word or by deletion of a space character between two words to form a misspelled word(s).

Punctuation errors algorithm, this script detect punctuation errors (such as missing punctuation or punctuation position).

Levenshtein Algorithm (LD), is an algorithm to measure the similarity between two given strings, which we will refer to as the source string (S) and the target string (T). The distance is the number of deletions, insertions, or substitutions required to transform S into T. For example,

If S is "مكن" and T is "مكن", then LD(s,t) = 0, because no transformations are needed. The strings are already identical.

If S is "مسكن" and T is "مكن", then LD(s,t) = 2, because one substitution (change " س" to " ء ") is sufficient to transform S into T.

Scoring/Ranking using context, We score the different candidates corrections using a LM and consider the gold correction for each misspelled word (Fig. 2).

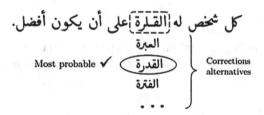

Fig. 2. Select the gold correction using LM.

5   Experimental Results

To evaluate the system, we collected various articles from differents resources to create a test corpus. This final test data contains 53,607 tokens. We extracted 2067 spelling errors, around 24 % of these errors naturally occurring in this corpus and the others was automatically generated (by merging, adding space characters, deleting characters, etc.). We manually prepared the gold correction for each spelling error. Table 5 shows the distribution of errors in the test data set (TDS).

Table 5. Distribution of errors types in % in the TDS.

Data	Edit	Add	Delete	Merge	Split	Punc.	NE. and other
Total of errors	930	351	268	186	165	63	104
% of errors	44,99 %	16,98 %	12,97 %	9,00 %	7,98 %	3,05 %	5,03 %

For evaluation with an annotated gold standard dataset, accuracy are usually reported using measures of precision, recall and F_1 score. We used the standard definitions:

$$Recall = \frac{Number\ of\ annotated\ misspellings\ flagged\ by\ the\ system}{Number\ of\ all\ annotated\ misspellings}. \tag{2}$$

$$Precision = \frac{Number\ of\ annotated\ misspellings\ flagged\ by\ the\ system}{Number\ of\ all\ tokens\ flagged\ by\ the\ system}. \tag{3}$$

$$F_1 = \frac{2 * Precision * Recall}{Precision + Recall}. \tag{4}$$

As a fisrt step of evaluation, we decided to evaluate our system in different TDS settings. We first calculated F_1 of the system on the prepared test data without all punctuations errors (TDS-PUNC). Next, we used TDS-PUNC and we removed also all Alif/Yaa errors (TDS-PUNC-AY). At the end, we took the initial test data set and we removed all named entities errors (TDS-NE) (Table 6).

Table 6. Results on test data set in different settings.

	Precision	Recall	F_1-measure
TDS-PUNC	81,28 %	74,91 %	77,96 %
TDS-PUNC-AY	69,76 %	56,59 %	62,49 %
TDS-NE	68,43 %	63,32 %	65,78 %

The second step in the evaluation task is to compare obtained results with our spell checker to those obtained on similar available systems. So we decided to compare our system with Microsoft Word 2013, and OpenOffice Ayaspell. Results are presented in Table 7.

Table 7. Comparison of our system performance against other available systems.

Systems	Correction Ranked first	Precision	Recall	F_1-measure
Our Spell checker	**95.1%**	72,87 %	65,34 %	**68,90 %**
MS Word 2013	93.74 %	63,65 %	53,92 %	58,38 %
AyaSpell	93.53 %	69,65 %	63,32 %	66,33 %

Basic on these results, our system out performs the two other systems both in the tasks of spell detection and correction and 1st order ranking of the gold correction. In this work, we didn't discussed the performance of our system in term of duration of execution.

6   Conclusion

In this paper, we presented our system for automatic Arabic text correction. Our spelling correction resources (including dictionnary, and rules) and our sequential combination of approaches (Lexicon based, rule-based methods, and LM-based scoring) significantly out performs the other systems Ayaspell, and MS Word especially in first order ranking of candidates.

The experitimental results are promising. However, several points can be investigated in further research mainly combining our system with a SMT system by using available manually corrected corpora dedicated for learners.

References

1. Alkanhal, M.I., Al-Badrashiny, M.A., Alghamdi, M.M., Al-Qabbany, A.O.: Automatic stochastic Arabic spelling correction with emphasis on space insertions and deletions. In: proceeding of IEEE Transactions on Audio, Speech, and Language Processing, vol. 20, no. 7 (2012)
2. AlShenaifi, N., AlNefie, R., Al-Yahya, M., Al-Khalifa, H.: ARIB@QALB-2015 shared task: a hybrid cascade model for Arabic spelling error detection and correction. In: Proceedings of ACL Workshop on Arabic Natural Language Processing, Beijing, China (2015)
3. Attia, M., Al-Badrashiny, M., Diab, M.: GWU-HASP-2015: priming spelling candidates with probability. In: Proceedings of ACL Workshop on Arabic Natural Language Processing, Beijing, China (2015)
4. Zribi, C.B.O., Ahmed, M.B.: Efficient automatic correction of misspelled Arabic words based on contextual information. In: Palade, V., Howlett, R.J., Jain, L. (eds.) KES 2003 Part I. LNAI, vol. 2773, pp. 770–777. Springer, Heidelberg (2003)
5. Bouamor, H., Sajjad, H., Durrani, N., Oflazer, K.: QCMUQ@QALB-2015 shared task: combining character level MT and error-tolerant finite-state recognition for Arabic spelling correction. In: Proceedings of ACL Workshop on Arabic Natural Language Processing, Beijing, China (2015)
6. Bougares, F., Bouamor, H.: UMMU@QALB-2015 shared task: character and word level SMT pipeline for automatic error correction of Arabic text. In: Proceedings of ACL Workshop on Arabic Natural Language Processing, Beijing, China (2015)
7. Brill, E., Moore, R.: An improved error model for noisy channel spelling correction. In: Proceedings of ACL, pp. 286–293 (2000)
8. Church, K., Gale, W.: Probability scoring for spelling correction. Stat. Comput. 1, 93–103 (1991)
9. Dahlmeier, D., Ng, H.T.: Better evaluation for grammatical error correction. In: Proceedings of NAACL (2012)
10. Farra, N., Tomeh, N., Rozovskaya, A., Habash, N.: Generalized character-levelspelling error correction. In: Proceedings of Conference of the Associationfor Computational Linguistics (2014)

11. Fossati, D., Di Eugenio, B.: A Mixed Trigrams Approach for context sensitive spell checking. In: Gelbukh, A. (ed.) CICLing 2007. LNCS, vol. 4394, pp. 623–633. Springer, Heidelberg (2007)

12. Habash, N., Mohit, B., Obeid, O., Oflazer, K., Tomeh, N., Zaghouani, W.: QALB: Qatar Arabic Language bank. In: Proceedings of Qatar Annual Research Conference (2013)

13. Haddad, B., Mustafa, Y.: Detection and correction of non-words in Arabic: a hybrid approach. Int. J. Comput. Process. Orient. Lang. **20**(4), 237–257 (2007)

14. Hassan, Y., Aly, M., Atiya, A.: Arabic spelling correction using supervised learning. In: Proceedings of EMNLP 2014 Workshop on Arabic Natural Language (2014)

15. Ibrahim, M.N., Ragheb, M.M.: CUFE@QALB-2015 shared task: Arabic error correction system. In: Proceedings of ACL Workshop on Arabic Natural Language Processing, Beijing, China (2015)

16. Islam, A., Inkpen, D.: Real-word spelling correction using Google Web IT 3-grams. In: Proceedings of EMPLN 2009, pp. 1241–1249. ACL (2009)

17. Kukich, K.: Techniques for automatically correcting words in text. ACM Comput. Surv. **24**(4), 377–439 (1992)

18. Mars, M., Antoniadis, G., Zrigui, M.: Statistical part of speech tagger for Arabic language. In: ICAI - The 2010 International Conference on Artificial Intelligence (2010)

19. Mars, M., Antoniadis, G., Zrigui, M.: Which algorithm and approach for Arabic part of speech tagging. J. Res. Comput. Sci. (JRCS) (CITII) **50**, 235–245 (2010)

20. Mohit, B., Rozovskaya, A., Habash, N., Zaghouani, W., Obeid, O.: The first QALB shared task on automatic text correction for Arabic. In: Proceedings of EMNLP 2014 Workshop on Arabic Natural Language (2014)

21. Muaidi, H., Al-Tarawneh, R.: Towards Arabic spell-checker based on N-grams scores. Int. J. Comput. Appl. **53**(3), 12–16 (2012)

22. Mubarak, H., Darwish, K.: Automatic correction of Arabic text a cascaded approach. In: Proceedings of EMNLP 2014 Workshop on Arabic Natural Language (2014)

23. Mubarak, H., Darwish, K., Abdelali, A.: QCRI@QALB-2015 shared task: correction of Arabic text for native and non-native speakers errors. In: Proceedings of ACL Workshop on Arabic Natural Language Processing, Beijing, China (2015)

24. Ng, H.T., Wu, S.M., Wu, Y., Hadiwinoto, C., Tetreault, J.: The CoNLL-2013 shared task on grammatical error correction. In: Proceedings of CoNLL-2013 Shared Task (2013)

25. Obeid, O., Zaghouani, W., Mohit, B., Habash, N., Oflazer, K., Tomeh, N.: A web-based annotation framework for large-scale text correction. In: Proceedings of IJCNLP (2013)

26. Pasha, A., Al-Badrashiny, M., Kholy, A.E., Eskander, R., Diab, M., Habash, N., Pooleery, M., Rambow, O., Roth, R.: MADAMIRA: a fast, comprehensive tool for morphological analysis and disambiguation of Arabic. In: Proceedings of LREC (2014)

27. Rozovskaya, A., Habash, N., Eskander, R., Farra, N., Salloum, W.: The Columbia system in the QALB-2014 shared task on Arabic error correction. In: Proceedings of EMNLP 2014 Workshop on Arabic Natural Language (2014)

28. Shaalan, K., Allam, A., Gomah, A.: Towards automatic spell checking for Arabic. In: Proceedings of 4th Conference on Language Engineering, Egyptian Society of Language Engineering (ELSE), pp. 240–247 (2003)

29. Stolcke, A., Zheng, J., Wang, W., Abrash, V.: SRILM at sixteen: update and outlook. In: Proceeding of IEEE Automatic Speech Recognition and Understanding Workshop (2011)
30. Tong, X., Evans, D.A.: A statistical approach to automatic OCR error correction in context. In: 4th Workshop on Very Large Corpora (1996)
31. Wasala, A., Weerasinghe, R., Pushpananda, R., Liyanage, C., Jayalatharachchi, E.: A data-driven approach to checking and correcting spelling errors in Sinhala. Int. J. Adv. ICT Emerg. Reg. **3**, 11–24 (2010)
32. Zaghouani, W., Mohit, B., Habash, N., Obeid, O., Tomeh, N., Rozovskaya, A., Farra, N., Alkuhlani, S., Oflazer, K.: Large scale Arabic error annotation: guidelines and framework, In: Proceedings of 9th International Conference on Language Resources and Evaluation (LREC 2014) (2014)
33. Zrigui, M., Ayadi, R., Maraoui, M., Mars, M.: Arabic text classification framework based on Latent Dirichlet allocation. CIT J. **20**, 125–140 (2012)

Geographical Communities and Virtual Communities: The Web 2.0 in the Creation of Tourist Information

Ilaria Greco[✉] and Angela Cresta

Department of Low, Economic, Management and Quantitative Methods,
University of Sannio, Via delle Puglie, 1, 82100 Benevento, Italy
{ilagreco,cresta}@unisannio.it

Abstract. The diffusion and easy access to geographic and computer equipment has revolutionized not only the models of communication, but also the techniques of production and diffusion of geographic information, combining more and more objective dimension of knowledge with that subjective and perceptive. At the same time, through the Virtual globe (Google Earth, GPS, etc.) or the so-called Volunteered Geographic Information (Wikimapia, Open street map, Google My Maps, etc.) also the spatial dimension of knowledge has found new life in the world of digital communication and Web 2.0.

This paper, therefore, part from the theoretical constructs object of the Cyber-geography [22], of the Geographical information science [6, 29] or of what many call "Neogeography" or "Geography from below" [44], focusing on new dynamics of interaction between technologies, territories and communities in tourism, or on how the tourist information - which is the basis of the image and attractiveness of a tourist place, and therefore of competitiveness - born more and more by the interaction between the territory and the community in the age of digital communications and social networks.

The literature review on issues specifically dealt with, will be accompanied by the presentation of some experiences of Virtual Community particularly significant as a form of representation and self-representation of local tourist systems through a Tourism 2.0.

Keywords: Geographic information · Web 2.0 · VGI · Tourism · Virtual community

1 Geographic Information, Web 2.0 and Tourism[1]

The global tourism is increasingly using the language of the new technologies that have revolutionized the tourism demand (*from off-line to on-line*), the tourist offer (*low cost, last minute, dynamic packaging*), the distribution channels (*Computer Reservation System of the Seventies, the Global Distribution System of the Eighties*), creating a market and a global tourism competition (*global tourism market's*) [12].

---

[1] The paper is the result of a common reflection of the authors; however, the single sections can thus be attributed to: Ilaria Greco paragraphs 1, 2 and Angela Cresta paragraphs 3 and 4.

© Springer International Publishing Switzerland 2016
O. Gervasi et al. (Eds.): ICCSA 2016, Part V, LNCS 9790, pp. 323–336, 2016.
DOI: 10.1007/978-3-319-42092-9_25

The web and online services have become an essential part of the organization travel, is to search for news and information, both for booking and management of a whole range of tourist services. An increasing number of tourists will move, in fact, from *off-line to on-line*, radically changing the ways of buying and consumption of tourism products: attentive, informed and with growing needs, tourists buy a product only after appropriate research information. Tourism demand is, in fact, much more personalized and connected with the experience and knowledge of the places visited; this has led to the emergence of new niche of tourism, related with the most qualifying and typical aspects of a territory [26].

On the other hand, Internet is also an extraordinary promotion and marketing tool for organizations and for tour operators. Internet and the new technologies offer, in fact, the possibility to contact the customer according to a new consumer paradigm - from the *product* to the *customer*, from the *destination* to *tourist experience* -, compacting in this way the tourism chain and offering multiple keys of access to information.

Today is, however, being a further great technological revolution, that of *Web 2.0*[2], which affects not only the supply system and communication, but also, the production of information and tourist knowledge and, therefore, the field of research and geographic and spatial analysis. The dissemination and use of simple instruments and geographic applications through the web or, even, of the operating environments of the VGI (*Volunteered Geographic Information*) encourages, in fact, the production and circulation of "new geographical information" according to dynamic and interactive models that see the direct participation of the users.

The multidimensionality of Web 2.0 pervade, in fact, the different forms of production, representation and transmission of tourist information, from paper to modern geographic information systems, including the network of *social networks* and the *virtual communities*, contributing to the spread of so-called *Tourism 2.0*. Through blogs, sharing platforms of photographic content like Flickr, or those sharing video content like YouTube or MySpace as a media mix, or Social Network such as Facebook, Twitter, etc., an extremely high number of people, also without specific skills and knowledge, exchange information and personal content easily and immediately.

The Web 2.0 does not define a new technology but a new "way to perceive it and use it" by users: it celebrates interactivity and participation of the users themselves in the creation of web content. With the transition from Web 1.0 to Web 2.0 network becomes, in fact, a great collective intelligence [26], in which the contents are generated not by the individual user or web site, but from conversations [28] and by the interaction of users, thus creating of autopoietic social systems that live, grow and evolve regardless of the initial components [30].

Web 2.0 is not only a set of economic and technology trends that collectively form the basis for the next generation of the Internet, but a new social model in which user-generated content can be as valuable as traditional media, where social networks form

---

[2] In the definition of Tim O'Reilly, "Web 2.0 differs from the initial concept web, Web 1.0, because it deviates from the classic static websites, from e-mail, the use of search engines, the linear navigation and proposes a World Wide Web more dynamic and interactive" [36, p. 4].

and grow with speed, where truly global audiences can be reached more easily, and rich media from photos to videos is a part of everyday life online [36].

Through the Web 2.0 consumer/tourist it is no longer a passive user but becomes *reloaded*, or content creator, veicolatore of information and promoter of services and products. There is, in fact, the exponential growth of a new public, no more "consumer" but users active, aware, informed, researching and proposing solutions and who promote new forms of communication and sharing. Audience segmentation that results is more and more refined, so as to become "self-segmentation" [25].

The fundamental principle of these new technologies is to consider the Web as a platform that allows you to "liberate" the creative skills and relational of people. In this way, Internet differs most from the concept of hypertext window to become an application platform that transforms the user, through the User Generated Contents (UGC), from *user* to *creator* of content, which can then be published and shared with other people on the Net.

It is, therefore, a new successful model that fits into the genre of e-WOM (*Electronic Word-of-mouth*), or electronic word of mouth that, through the different in-line communication tools (website, email, blogs, web-community, forums, instant messaging, etc.), using the online interpersonal influence to generate competitive advantages: the power/success of new channels for the dissemination of the information resides not only in the ease and speed of transmitting information, but, as in the traditional word of mouth in the confidence that those who receive the information puts in the transmitter [29] (see Fig. 1).

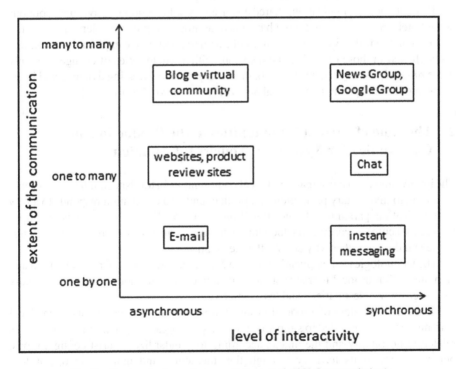

Fig. 1. Electronic channels types (Litvin et al. [28] Our translation)

The real Web 2.0 object is, therefore, *communication* and *sharing*. Communication differs from simple *information* as it aims that transcend the spread its own sake of data and information, a mutual exchange) to establish instead a dialogue (and, as such, between subjects, in order to stimulate a response or, however, create a relationship of exchange. This means moving from *advertising* (which aims to persuade tourists) to *advertainment*, namely a communication aimed to entertain the public, more and more often to impress, to stimulate conversation and trigger mechanisms "viral spread" of information.

Just the concept of *on-line and off-line virality* is the basis of new communication techniques "unconventional" that are emerging on the international scene and Italian [44].

In the field of Tourism Geography, the communication (even the virtual one that makes use of the network) adds "value" to information, avoiding any form of contradiction, if not apparent, between "known places" and "places communicated", between "real spaces" and "virtual spaces", because what is communicated is not a something "different" from reality or imagination, but it is the "perceived space" by the local community and communicated through the Net.

"The communication is important because it transmits emotions, stimulates desire, conveys ideas expression of the cultural identity of a people, represents values that are an expression of a lifestyle, draws an imaginary space in which everyone can recognize [...]; the communication is able to create a continuous relationship between real and virtual, between physical and perceived [...]" [5, p. 9].

Through the communication, shared experiences, told emotions, travel tips, opinions about hotels, restaurants, and more have become more and more the element of greater influence in the formation of preferences of travelers, both in the exploratory phase and research, and in those of choice and decision, with a whole series of changes not only in the *pre and post-experience* but also, in *tourist practices*, creating dynamic and interactive models and phenomena of real *Social Travel and/or Co-travel* [6].

2   The Role of Virtual Communities in the Production and Communication Systems of Tourist Information

The involvement and participation of local communities in tourism communication and, therefore, to involuntary promotion of a product and/or a destination responds to a new communication paradigm "Forget the What, Tell Me Why" (T. Thompson, http://travel2dot0.com), according to which for the traveler is no longer enough to know things to see and do, but rather why choose that destination.

The technologies and applications on the Net open new spaces of representation and communication of local tourist systems; a channel with enormous potential especially for lesser-known destinations and less attractive.

In the new models of production and tourist information communication, local communities are not called to simply be the venue for organizing activities and services related to hospitality and the image, according to standardized tourist configurations, but promoters of its territory through the interaction and multimedia operability, according to a model of dynamic reciprocity between global networks (virtual community) and the local network that they themselves represent.

The network connects two types of communities: the geographical (*local community or territorial*) and the users of the network, the so-called navigators (*virtual community*). Among all the electronic tools, the Internet is definitely one that more than any other has helped to overcome the requirement of geographical contiguity (or physical proximity), to emphasize that instead of virtuality. The communities have evolved, in fact, from groups of subjects linked by the common denominator of shared spaces and community services and common geographical matrices, to groups of individuals who share a system of relations and are interested in starting communication processes exchanging with communities locals.

The virtual community is not opposed to the traditional local community, but is an expression of one or more local communities and, therefore, creator of community feelings and at the same time interlocutor of a potentially infinite network of subjects. Indeed, it is through the interaction between global networks (Internet) and local network (local communities/tourists) on the one hand, and multimedia operability on the other, that virtual communities are called to become promoters of the territory, according a model of dynamic reciprocity between global networks (virtual communities) and the local network that they themselves represent.

In tourism, the local community has an increasingly important role not only in the definition of the tourism product and in welcoming tourists, but in the process of tourism image creation and promotion and communication of the destination. The growing involvement of local communities in tourist communication reflects the very nature of the new forms of tourism and hospitality widespread who in their territorial dimension, rather than sectoral, propose an *offer of such community* through the direct involvement of the local community and interaction of many local actors in the definition and promotion of the local tourist product. It's linked to a new way of understanding travel, not as a moment of consumption, but of contact and cultural and experiential growth, of relationships.

An example is the sustainable rural tourism that discovers, values and diffuses the tourist vocation of a place and allows the local cultural heritage protection, or to all forms of emotional tourism, relaxing, unique, experiential that characterize the so-called minor destinations, permeated by authenticity, uniqueness, identity: from tourism nature to wine tourism, from cultural tourism to tourism villages, etc. [8].

In these new tourist practices, more than in traditional ones, is the local community that, through the Web 2.0, puts itself in the net and it becomes a *virtual community*, becoming a promoter of a new form of tourism communication, defined as "communication from below", understood as an opportunity for educational and cultural exchange and interpersonal enrichment between communities, institutions and travelers. In this case, the experience of life generates "value" which translates into information and tourist consumption.

It is a very different communications from all others: it does not have a commercial purpose focused on the goal of a product/business development (Communicating the offer of a single hotel, structure or more aggregate structures through the network), and is not directed to the prosecution of organizational purposes in view of improvement of quality of services-products offered to users actual and potential, nor has an "institutional" character, or intended to affect the environment in which the subject/promoter is operating, or "trivially" destined to exchange views, experiences, comments about products or services [14]. The objective is to communicate the *destination* in the promotion and enhancement process of local tourist offer, as well as known and perceived by those who live and feels directly responsible for local hospitality as aware of the value of the resources of the area, or simply by those who lived and known as a tourist.

The virtual community has a "cultural mediator" function to unveil values and build, even in marginal areas, tourist images until now not perceived as unitary, attractions for tourists and shared by the inhabitants. The mission is to collect and disseminate (through the web) the enormous information heritage, culture, history and architecture, the elements of identity and traditions held by the local community that from "host community" becomes the protagonist and direct ambassador of the places that lives across the network.

Opening, therefore, new spaces of territorial promotion that comes from below, by local communities, of representation and communication of local tourism systems, a channel with enormous potential especially for lesser-known destinations and less attractive. It is precisely in tourist areas, more than in any other place, it becomes easier to promote a "shared vision" of the local context, where the same is the expression of social cohesion and territorial belonging.

The Tourist Information communicated by the local community is the result of an identification with the places which translates into territorial protagonism by sharing projects and participation in the ordinary management of the territory, considered that "the identity of the places are a product of social actions and how the same people if they give a representation" [35, p. 97].

The localisms and the micro tourist destinations mark Italy and together with the artistic and landscape heritage determining the identity making it an (eco)museum in constant evolution characterized from everyday life, from popular practices, the exclusive landscapes, from the people who inhabit the places and give life to the stories of those places. These forms of representation and self-representation of smaller local tourism systems that can find wide communication through a Tourism 2.0, in a social dimension of tourism.

In recent years many studies on tourism and Web 2.0 (or Travel 2.0) have confirmed this trend: big giants like Google, Tripadvisor, Facebook and Twitter agree that for every destination is essential to be present on social media in order to exercise its influence on the tourist in all phases of travel [15].

The European Digital NTO Benchmark (Tourism Think Tank [43]) which analyzed according to some parameters of the online practices of some tourist destinations, has come to the conclusion that those who "ride the wave of technological change has a good chance to come across a tourist demand increasingly exigent and to maximize the experience of tourists by improving its reputation and increasing brand recognition".

Also the data of Near and Now, the study conducted by Facebook [20] on the behavior of its users about travel and holiday point out that approximately 42 % of the stories shared on individual profiles are travel experiences and inevitably, photo sharing, stories and emotions, but above all the comments between users amplifies the exposure of the community to the holiday theme and feeds the desire to travel.

Even research conducted by Google (The Traveler's Road to Decision 2013), shows that for 84 % of tourists the social media and online reviews influenced and changed radically the way they travel, as well as 76 % say they do not believe the information and communication activities of Tourist Enterprises and to seek feedback with information and advice online and in social channels.

3   Experiences of Virtual Community to Comparison: For a Model of Social Interaction

In recent years we have witnessed a proliferation of regional (Turismointoscana.it, #destinazionemarche, Visit Sicily, …) and business (Bluserena, Bestwestern Italia, …) travel portal more oriented to the *social tourist*, hosting content, comments and photos of tourist matrix (TripAdvisor, Trivago, Virtualtourist, MInube). In these blogs and web communities you can distinguish several forms of Web 2.0 Tourism, along a continuum that ranges from communities of reviews (of users or editor) to those more oriented to user interaction and sharing [13].

Browsing the world of virtual communities, we can come across a user 2.0 (travelers but also enthusiasts, researchers and tourism experts) particularly interested in tourism in the small and micro territories: opportunity to interpersonal enrichment between communities, institutions and travelers [38] but, also, the time for dialogue, discussion, proactive participation in the stories and to experience, in the management of those same territories not always or even smaller tourist destinations.

In light of all this, the objective of this paragraph is to present three experiences of social media web site in tourism for each of them highlighting the contribution that the virtual community through the *Electronic Word-of-mouth* can give visibility, fame and reputation of a destination: three examples different from each other, for model and organizational structure, for radius of action, for users composition that seek and share information, for "other services". They are three examples, however, comparable to the expected mission: to reach all those who for various reasons (tourists/travelers, consultants/experts, tourism enthusiasts, professionals, administrators, etc. …) can contribute to triggering positive effects for the development of a tourist destination.

The website *TopRural* (www.toprural.it) is a search engine of rural houses that, since 2000, promotes rural tourism with the double objective: to offer the traveler the easiest and most effective way to organize their holidays in rural areas and to give, to give owners of cottages and rural houses, the best platform to show their hospitality.

The virtual community, then, is composed by: (1) potential travelers looking for accommodation in rural settings that meet the needs of a unique, authentic and original holiday (landscape features, recreation, proximity to places of tourist interest, value/price, trip type - alone, in pairs - with children - with animals, availability, etc.); (2) owners who have the opportunity to promote, with considerable independence of content and images, their business on a storefront with a wide geographical and linguistic representation (10 European countries, Germany, Andorra, Austria, Belgium, Spain, France,

Italy, Luxembourg, The Netherlands, Portugal and 8 languages); (3) travelers who have visited those facilities, who lived through those places, that on the portal share their holiday experience, with general considerations, more detailed ratings by categories (cleanliness, hospitality, environment, equipment, …) photos, videos, etc.

The web site is one of those portals with a good level of UGC (User Generated Content): essential is, in fact, the freedom and encouragement to the production of content generated by the owners (Owner corner, Register new housing, edit the information of your accommodation, update your availability calendar …) and travelers (Publish your opinions on accommodations, Create your own list of favorite accommodations or destinations …) by users participating in the creation of the site in the same organization which owns the site. This is a crucial aspect for the success of the portal because it can boost the level of visibility, interactivity and community loyalty.

TopRural, in fact, has direct links to Facebook (109.741 subscribers), to Twitter (4.214 followers and 5.097 tweets), shares 4,774 photos on Flickr, more than 200 videos uploaded to Youtube, a recent community on Google+ (with 1225 follower and 1.501.688 views), provides for a monthly service newsletter, but above all, a blog created with the aim of providing users "tips and ideas to organize your rural holidays".

The blog is edited by a team formed by experts (Business, IT, web marketing, communication …) of 10 different nationalities: it offers numerous and high-quality contents to users in order to better guide the traveler in choosing a destination, to support owners and companies in the promotion and enhancement of its accommodation options and activities offered. The information are varied and range from the *most informative content for travelers*, such as, suggested destinations and routes (Rural vacations in …) and events (festivals, tastings, food and wine fairs …), Top10 rankings (top hotel destinations from rural Italian tourists), selections of the farm for thematic (i.e. facilities with wi-fi), *more information-consulting content owners*, such as the Ranking Top3 (3 most viewed rural structures of the month), interviews with experts of rural tourism, segment information, statistical analysis on tourists and on rural tourism (surveys, polls, trend of the tourist sector…). Also there are contest (of opinions, photographic, …), job opportunities to encourage user loyalty, but also the ability to obtain the label "Accommodation published in TopRural" to identify the structures belonging to the most popular portal circuit in Europe for finding farms or other rural houses (Fig. 2).

The portal-depth analysis shows that TopRural has numerous strengths that make it the most useful research tool of rural houses at the service of travelers, but also some weak points, one above all the absence of on-line interaction between the users of the community: in fact there is a good level of interactivity between staff and individual users through the blog, but is totally missing the interaction between owner and actual traveler, and between actual and potential traveler. In addition, the comments of travelers on actual structures is related exclusively to the presence in the TopRural website. Bridging this gap would mean more information and feedback: the owners could find room for improvement of its offer; potential travelers interacting with the community would identify with it, because it consists of people who share the same interests and who have already had the experience that they are planning to live.

Different community that is developing around *costruireturismo.com*, a blog was born a few years ago (2012) and edited by a young scientist, interested in tourism. He says:

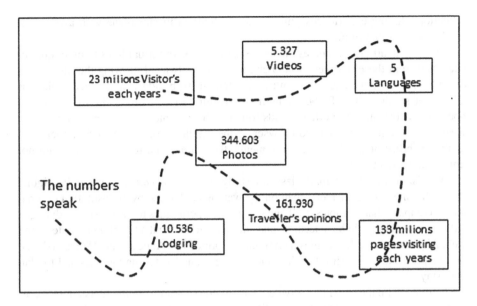

Fig. 2. TopRural (www.toprural.it)

"Tourism is a valuable lever for the economy of the Country and of the countries (…) rich in resources often not valued and difficult to use": for this reason, he is interested in "planning and tourism development with an emphasis on small communities" (with fewer than 5,000 inhabitants), smaller towns with local culinary tradition, with important artistic and cultural resources (Museums, monuments, villages, archaeological sites …), maybe incorporated in landscape valuable contexts, able to satisfy "tourists looking for authentic places where they can live an original experience; increase knowledge and culture; get in touch with the history and traditions; establish human relations with the inhabitants".

At present, the blog presents you with a basic graphics, maybe not too attractive, improvable, but certainly immediate, probably does not have a very high visibility within search engines and probably it is limited to "insiders", but it has good potential for interaction and sharing through the direct connection to Facebook (1136 members) and Twitter (753 followers and 6.592 tweets). The blog provides information on events and activities related to tourism (BIT, BTO, other conventions), on statistical information, and explores tourism on two issues: the destination management of smaller sites and the role of the web and social media. These two issues will inevitably intersect in the content published and provide a new interpretation on the opportunities for development of the smaller tourist destinations.

The creator is also the main and sole editor of the blog, however, open to all forms of collaboration. The unique aspect that caught the attention of the writer is the online community that is being formed and to which reference should be made, through the post, the comments on articles, the re-blog, the virtual place for discussion and debate. Costruireturismo.com is a very interesting universe of travelers, enthusiasts, researchers, consultants, tourism planners, but also professionals, small business owners (managers, animators, artists …) that from the blog broaden the information space, offer insights reflection, sharing experiences, initiatives and tourism master plans with a focus on smaller areas, often rural.

Reading an article stimulates the browser to a series of virtual links to other quality sites, and from these to others, like a chain reaction, always following the idea of the destination management and the contribution that the web and social media can play for takeoff of smaller sites. If the strength to costruireturismo.com is the possibility of opening its community to other worlds (other web sites, blogs, web communities ...), weakness for a less experienced browser, is the smallness of the network, understood as the visibility of the different components, there are a first reading but that are not immediately identifiable.

AngelsForTravellers.com, the last case study, is the *travel community*: highly original and highly innovative character than on the web, is not backed by an institution or a big tourist company that promotes itself and promotes opportunities for discussion or dialogue with users, and it would be inaccurate and simplistic to define it a travel review site on the web. AngelsForTravellers.com is a community formed by "Angels" and "Travelers": the first help the latter to visit a location (city, village, smaller destinations ...) and to plan their stay.

Is frequent on the web meet travelers who express their own judgments on the experience of lived trip, in terms of the quality of accommodation, beauty of places, quality of the products, welcome and hospitality of the communities we have visited: but the originality of this community is the Angel's figure, a person who loves above all their places of origin, is passionate and a great knowledge of tourism resources of the same and that puts this knowledge at the service of travelers.

The added value that the Angel offers to the Traveller is the advice of those that place the daily lives and that through suggestions, opinions, points of view on what to see, where to eat, where to sleep, but also what to avoid, can make the experience travel "less touristy and more real": the feeling of the tourist is to participate in the life of all the days of the city and be part of it at least for a few days, as residents and not as tourists, with no surprises, difficulties and uncertainties; but above all, every Angel can make themselves available to meet the traveler in their city, to guide him in alleys and museums, even to host him in their house. When the trip ends, the angel may at community, read the impressions of the traveler, look at his photos and his movies, and also know the judgment delivered against him.

The *one-to-one interaction* between Angel and Traveler, the set of recommendations and suggestions of the Angel before the trip, the stories, feelings, images that the traveler shares both during and at the end of the experience, is a heritage of unevaluable information that will influence other potential travelers, that organizing a trip to the same destination can make use of the knowledge gained through the community, choose the same Angel or look for another one.

AngelsForTravellers.com, as quite recently (the community was born in Naples in May 2009), is expanding relatively quickly their activity in different cities of Italy and the world and, at present, has more than 250 community territorial: in a few years registered angels reaching nearly 3.000 units. The community also contains *Local partners*, local offices such as targeted promotional and territorial development agencies, socio-cultural associations, tourist information offices (Agora, Herculaneum Centre,

Cilentomania, Irpinia Tourism and Naples City Visible in Campania, Addio Pizzo Travel, NED and Core in Sicily, The Wanderer in Basilicata, Puglia Hub) but also *sedentary angels*, or commercial enterprises managers (bar, restaurant, shop ...) and services to tourism enterprises: Local Partner welcome and they guide tourists even without you was the prior investigation on the platform; it is a network of physical locations that complements and integrates the digital network of Angels based on community and regional offices.

The social network has a good visibility in the major search engines, it promotes high levels of interaction and sharing not only through the very numerous community web but also thanks to the presence of a blog constantly updated on all the initiatives of AngelsForTravellers.com and on the other components of the network: in particular "Welcome to the South", who work tirelessly to promote the community and to do networking with Facebook links (4.353 members) and Twitter (846 followers, 1.983 and 1.133 Following tweet).

The high level of autonomy of the community, the interaction between the Angels and Travelers, but above all the richness and quality of UGC twist the structures of the conventional communication. For several years the tourism communication is not over and it is not only in the hands of tourist enterprises and tourist destinations, but it is in the hands of tourists who speak, comment and converse with each other on the net: for this reason the community enables us to appreciate unusual and singular aspects, peculiarities, authentic resources and apparently not typical tourist resources, in destinations already known exalting the trip with no "perceivable" experiences and feelings through the consultation of a traditional tour guide.

In AngelsForTravellers.com is not uncommon that the Angel, timely and expert of his city, enriching *its content* with travel tips on the main destination but also on smaller places not easily accessible. As a result of the high competence of Angels about the territorial realities of larger scale, but especially the presence of the Angels even in cities and smaller towns (which do not enjoy the prestige of top resorts), it allows you to tell stories, places and communities sometimes ignored or known to a few, but with landscape and environmental excellence, quality typical products, unique ethnographic traditions.

4   What Prospects in Key Geographical?

In all three experiences of virtual community described in the preceding paragraph, editors, actual travelers, potential travelers, bloggers, angels, have helped to increase the visibility and reputation of a place, because through the story of experience that can be derived from affiliation, knowledge, and/or by the visit of the territory, gave voice to individual destinations. The best cases described, therefore, allow us to understand more deeply why for destinations is so important to be present on social media, but mostly why they need the virtual community.

First, the Web 2.0 has encouraged a reinterpretation of the consolidated concept of community: breaking down the essential requirement of geographical contiguity, the meaning of community is extended to persons belonging to territories not physically next, often marked by various official languages and, above all belong under other communities, they are real or virtual.

Secondly, the awareness that today people increasingly feel the need to know each other, to relate, to communicate, to meet virtually, also requires destinations to have to talk, to tell, to share, to listen on the virtual communities: it is essential to know what users think of himself, of its offer (to understand the *sentiment* on the network and monitor their *reputation*), but it is equally essential to understand what other destinations make and how they meet the social tourist; both these elements are essential to organize, to plan, to grow.

Last but not least, the tourist is now on the web and the web is the most immediate way to intercept him and communicate with him: if it's true that not all potential tourists are still using internet, you can not ignore that the phenomenon is growing exponentially, that the traveler is always less traditional tourist and increasingly social tourist who finds stimulation and satisfaction in the social network that identifies and interacts [16].

This scenario requires the territories, small and characterized by a little strong identity or little known, to be *destinations 2.0*: this does not mean being the subject (often involuntary) of comments, travel stories, post on Travel 2.0 sites, blogs or virtual communities, but rather to invest in a project that, starting from the definition of a *concept* around which to develop a tourist industry, through the involvement of local actors, public and private and community, to be able to identify strategies to plan its presence in the social media world.

If we look to the near future, it is likely to greatly increase the use of systems that offer high levels of interactivity, but also that purely demographic factors and the gradual spread of network technologies (along with the increasingly widespread propagation of access to content in mobility systems), will change in a very deep way the types of content, the ways to create them, to transmit them and use them.

References

1. Antonioli Corigliano, M., Baggio, R.: Internet e Turismo 2.0. Egea, Milano (2011)
2. Antonioli Corigliano, M., Baggio, R.: Internet e Turismo in Italia. In: XVIII Rapporto sul Turismo Italiano, Mercury, pp. 193–207 (2013)
3. Baggio, R., Mottironi, C., Antonioli, M.: Turismo e comunicazione istituzionale online in Italia. In: Turistica, vol. XX, pp. 5–20 (2011)
4. Bencardino, F.: Innovare per crescere. In: Turismo informato. Atti del Congresso nazionale sul Geogiornalismo, Il Paradosso, Fondazione Alario per Elea-Velia, n.5/6 (2013)
5. Bencardino, F., Prezioso, M. (eds.): Geografia del turismo. Mc Graw-Hill, Milano (2007)
6. Capineri, C.: Geografia e cambiamenti tecnologici: virtual globes e neogeografia. In: Atti del XXX Congresso Geografico Italiano, Il futuro della geografia: ambiente, culture, economia, Bologna, Patron, vol. II, pp. 93–103 (2008)
7. Corna Pellegrini, G., Paradiso, M. (eds.): Nuove comunicazioni globali e nuove geografie. CUEM, Milano (2009)
8. Cresta, A., Greco, I.: Luoghi e forme del turismo rurale. Evidenze empiriche in Irpinia. Franco Angeli, Milano (2010)
9. Dall'Ara, G., Santinato, M.: Dai turismi alle nicchie. Teamwork srl, Rimini (2004)

10. Dall'Ara, G.: Il Marketing Passaparola nel Turismo. Stimolare e promuovere le tecniche di gestione del passaparola positivo. Agra, Roma (2005)
11. Dall'Ara, G.: Programmare lo sviluppo turistico dei territori. Halley, Matelica (2009)
12. Desinano, P.: Turismo and reti informatiche. L'evoluzione dei mercati turistici elettronici. In: Savelli, A. (ed.) Città, turismo e comunicazione globale, pp. 189–200. F. Angeli, Milano (2004)
13. Di Vittorio, A.: Turismo 2.0: le community on line dei viaggiatori e la condivisone dell'esperienza turistica. In: Mercati e competitività, no. 4, pp. 147–167. Franco Angeli, Milan (2011)
14. Dioguardi, V.: L'immagine della destinazione turistica come costruzione sociale. F. Angeli, Milano (2009)
15. Ejarque, J.: Strategie per hotel: i consigli di Google, Tripadvisor, Facebook e Twitter (2014a). http://www.fourtourismblog.it
16. Ejarque, J.: Il turista social per le destinazioni turistiche. In: Destination & Tourism, no. 17 (2014b)
17. Elgar, E.: The Geography of the internet: cities, regions and internet infrastructure in Europe, Cheltenham (2013)
18. Endrighi, E.: Il turismo community-based per lo sviluppo delle aree rurali. In: Basile, E., Cecchi, C. (eds.) Diritto all'alimentazione, Agricoltura e sviluppo, Atti del XLI Convegno di Studi della Società di Economia Agraria (2006)
19. Enright, M.J., Newton, J.: Tourism destination competitiveness: a quantitative approach. Tour. Manag. 25(6), 777–788 (2004)
20. Facebook: Facebook Travel Near and Now, Allfacebook.de - Facebook Marketing Blog (2013)
21. Franch, M.: Destination Management: Governare il turismo tra locale e globale. Giappichelli Editore, Torino (2002)
22. Giorda, C.: Cybergeografia. Estensione, rappresentazione e percezione dello spazio nell'epoca dell'informazione. Tirrenia Stampatori, Torino (2000)
23. Gnecchi, F.: Comunità virtuali, comunità locali e comunicazione pubblica, vol. 59, pp. 247–264. CUEIM Sinergie, Verona (2002)
24. Google: Today's Traveler: Google's Annual Traveler's Road to Decision Study (2012). http://www.thinkwithgoogle.com
25. Grassi, M.: Nuovi strumenti per il turismo. In: Gatti, F., Romana Pugelli, F. (eds.) Nuove frontiere del turismo, pp. 137–148. Hoepli, Milano (2006)
26. Landini, P., Fuschi, M., Zarrilli, L., Ferrari, F.: Turismo e territorio. L'Italia in competizione. In: Rapporto Annuale 2007, SGI, Roma (2007)
27. Lazzeroni, M.: Geografia della conoscenza e dell'innovazione tecnologica: un'interpretazione dei cambiamenti territoriali. Franco Angeli, Milano (2004)
28. Litvin, SW., et al.: Electronic word-of-mouth in hospitality and tourism management. Tour. Manag. 29, 458–468 (2008)
29. Longley, P.A., Goodchild, M.F., Maguire, D.J., Rhind, D.W.: Geographical Information Systems: Principles, Techniques, Applications and Management. Wiley, New York (1999)
30. Maturana, H., Varela, F.: L'albero della conoscenza. Garzanti, Milano (1987)
31. Meini, M., Spinelli, G.: Ipermappe, sistemi multimediali per l'informazione turistica. In: Bollettino dell'Associazione Italiana di Cartografia, no. 126-127-128, pp. 225-237 (2006)
32. Meini, M., Spinelli, G.: Il territorio nella comunicazione turistica digitale. In: Adamo, F. (ed.) Annali del turismo, vol. 1. Geoprogress Edizioni, Novara (2012)
33. Micelli, S.: Imprese, reti e comunità virtuali. Etas, Milano (2000)

34. Micera, R., Lo Presti, O.: L'immagine delle destinazioni. In: XVIII Rapporto sul Turismo Italiano, Mercury, pp. 331–340 (2013)
35. Jess, P., Massey, D.: Luoghi contestati. In: Massey, D., Jess, P. (eds.) Luoghi, culture e globalizzazione, pp. 97–143. Utet, Torino (2001)
36. O'Reilly, T., Musser, J.: Web 2.0. Principles and Best Practices. O'Reilly RadaR (2006)
37. Paradiso, M.: Geografia e pianificazione territoriale della società dell'informazione. F. Angeli, Milano (2003)
38. Pine, B.J., Gilmore, J.H.: L'economia delle esperienze: oltre il servizio. Etas, Milano (2000)
39. Polizzi, G.: La comunicazione della destinazione turistica al tempo di internet. McGraw-Hill, New York (2011)
40. Rocca, G.: Geografia della comunicazione: metodologie e problematiche dei processi di mobilità territoriale. Pàtron, Bologna (1998)
41. Rocca, L.: Partecipare in rete. Nuove pratiche per lo sviluppo locale e la gestione del territorio, Bologna, Il Mulino (2010)
42. Rossi, C.: In viaggio… verso il digitale. Le imprese della distribuzione turistica di fronte alla sfida del web. In: Congresso Internazionale. le tendenze del mareting, Venezia (2006)
43. Tourism Think Tank: The European Digital NTO Benchmark (2013). http://thinkdigital.travel
44. Turner Andrew, J.: Introduction to Neogeography. O'Reilly Media, Sebastopol (2006)

Analysis Spreading Patterns Generated by Model

Thiago Schons[1], Carolina R. Xavier[1], Alexandre G. Evsukoff[2],
Nelson F.F. Ebecken[2], and Vinícius da F. Vieira[1,2(✉)]

[1] UFSJ - Federal University of São João del Rei, São João del Rei, MG, Brazil
carolrx@gmail.com, carolinaxavier@ufsi.edu.br,vinicius@ufsj.edu.br
[2] COPPE/UFRJ - Federal University of Rio de Janeiro, Rio de Janeiro, Brazil

Abstract. Spreading have been studied in networks from a wide range
of contexts, such as social, biological and technological. Models for
spreading simulation can be applied to real world networks in order
to investigate how spreading phenomena occurs from different perspectives. An usual approach is to analyst a diffusion process by assessing
the number of reached nodes and the depth of a propagation. This work
describes the spreading processes by identifying their patterns, characterized by the canonical name of the propagation trees started by each
seeder. Diffusion was investigated in four real world networks considering
Independent Cascade Model (ICM). The results show that, as observed
in real world scenarios, the occurrence of complex cascades is quite rare
and the majority of propagation trees are very simple.

1 Introduction

A network can be defined as a set of nodes connected by edges able to model elements which show some kind of relationship. The study of phenomena modelled
by networks has attracted growing interest in several areas, such as biology, soci-
ology, physics, economy and urban planning [1,16]. A very intuitive example of a
network frequently studied is the web, where the documents can be represented
by nodes and the hyperlinks can be represented by edges. Another example of
networks broadly explored currently are social networks, in which the nodes and
edges can represent, respectively, persons and the friendship between them.

A very hot topic in the study of social networks is related to the way that
communication occurs among a set of individuals, i.e., how an information is
propagated among them. The spreading of gossips by different sets of people, for
instance, has unique characteristics but it still can reveal some patterns, induced
by different factors which can be investigated. In order to do this, we need to
explore the several factors that allow us to understand how an information is
propagated, like the number of contacts of the individuals and the influence
which each one has on each neighbour to pass this information forward. In this
sense, we are able to study an information cascade when we characterize its
depth, the number of individuals to an information and the enrolment of the
individuals to transmit an information [14].

© Springer International Publishing Switzerland 2016
O. Gervasi et al. (Eds.): ICCSA 2016, Part V, LNCS 9790, pp. 337–349, 2016.
DOI: 10.1007/978-3-319-42092-9_26

Information cascading is one of the most relevant subjects in the study of social networks and shows great interest for marketing professionals, for instance. When an advertising company identifies which are the most influential individuals or, particularly, those more likely to generate enrolment to a campaign, it can perform specific actions on this individuals that allow the diffusion to be optimized and the number of reached people to be maximized.

Several works can be found in the literature aiming on investigating information cascades in networks. Dow *et al.* [4] perform an exploratory study on the features observed during the spreading of two popular memes arose from very distinct Facebook users: one picture posted by the United States first lady, and another one posted by a Norwegian student. In this study, the authors verified that, despite both memes have reached a huge number of people, the way that each meme was propagated was quite different. Another work, performed by Goel *et al.* [6], investigate the formation of cascades in a set of social networks in order to explore patterns which allow the detection of viral information.

Another class of works can be found in the literature proposing and investigating the application of spreading models, which allow the simulation of a diffusion process in networks in different contexts, like epidemiology [15], binary decision processes [9] and occurrence of information cascades [7].

This work explores the application of one of the most adopted models for spreading processes in networks, Independent Cascade Model (ICM), to real networks in order to investigate the formation of cascades in a set of distinct scenarios. The current work analyses the way in which the diffusion of a content occurs, according to a spreading model investigating the formation of patterns in real networks and how an information cascade occurs and discuss which users have more power to influence the diffusion. Thus, more than just assess how many individuals are reached in a diffusion process, this work perform an analysis of the most frequent spreading patterns in networks from different contexts.

The analysis of the way that an information is propagates and which parameters affect the information diffusion in a network is a great contribution for enhancing techniques for the diffusion of products, news and information addressed to a specific subset of consumers.

The remainder of the work is organized as follows. Section 2 shows some concepts and background needed for understanding the work. Section 3 presents some works which are related to the proposed methodology and were considered as base to the current work. Section 4 shows the methodology proposed for this study. In Sect. 5, the performed experiments are presented and discussed. Section 6 shows the conclusions of the present work and some possible directions for future works.

2   Background

2.1   Spreading Processes

The study of spreading processes plays an important role in the understanding of phenomena in several scenarios. In a biological context, epidemiological processes of diseases propagation like HIV, influenza and tuberculosis can be analysed with spreading models [13]. In digital systems, the propagation of computer viruses shows a spreading pattern similar to diseases propagation and can also be studied as an epidemic process.

The analogy of epidemiology with spreading processes is also fundamental in phenomena observed in social networks, frequently modelled as information diffusion processes. The importance of characterizing and modelling spreading in social networks can be highlighted by the large number of studies found in the literature dedicated to them in online environments, like Twitter, Facebook and Stackoverflow. Dow et al. [4] investigate the propagation of memes on Facebook by analysing post shares. The behaviour propagation is studied by Leskovec et al. [11] in a work in which the authors investigate the occurrence of cascades of products recommendation in an online purchase system. Aral and Walker [2] perform an experiment to understand how the influence of a person affects the consumers demand. In order to do this, the authors investigate people who have adopted an specific Facebook app and track the messages sent to their friends to identify the effect caused for the adoption of the app by other users.

Several works can be found in the literature with the purpose of defining spreading models under different perspectives. Among the most studied models, it is important to mention the models based on epidemiology, like suscepti-ble infected (SI), susceptible infected susceptible (SIS) and susceptible infected recovered (SIR). Pastor-Satorras and Vespignani [15] proposed the application of these models to networks and investigated epidemic of computer viruses, bring-ing important insights not only in computer viruses but in communication and social phenomena.

It is also worth to mention the Threshold Model, proposed by Granoveter [9] to model collective behaviour aiming on binary decision problems, like voting. Granovetter used this model to explain, for example, the residential segregation. In this case, the author observed that the threshold is essential for the model and that it can be different for each individual. He also observed that the threshold is deeply influenced by several factors, like socio-economic, educational level, age and others.

Inspired by studies about interactive particle systems [5,12] and by the probability theory, dynamic cascade models were conceived to model diffusion processes. In marketing context, Goldenberg et al. [7,8] studied cascade models in which the dynamics are simulated iteratively: if a node v_i becomes active in a time step t, then it has a single chance to try to influence its neighbours $v_j \in N(v_i)$, inactive so far, in the time step $t + 1$. The probability of success of the attempt of activating v_j is p_{v_i,v_j}. Additionally, if several neighbours of v_j become active in the time step t, their attempts to activate v_j will occur in an arbitrary order. If one of these attempts succeeds in t, then v_j will become active in $t + 1$; however, regardless v_i succeeds or not in activating its neighbours, it will not make any additional attempts in the next steps. Similarly to threshold models, the process stops when no more activation are possible.

The present work considers the Independent Cascade Model, the most simple and widely adopted cascade model found in the literature. This model defines parameters p_{ij}, which represent the probability that a node v_i recently activated to succeed in its attempt to activate its inactive neighbours v_j, regardless the diffusion process history so far. Furthermore, to better define the model, the independence in the order of activation attempts needs to be introduced. Let $\mathbb{S}$ be the set of nodes that have already tried and failed in activating v_j, and $p_{v_j}(v|\mathbb{S})$ the probability of success of v_i in activating v_j. Let $v_1, v_2, ...v_k$, and $v'_1, v'_2, ...v'_k$ be two different permutations of $\mathbb{S}$, and $\mathbb{T}_i = \{v_1, v_2, ...v_k\}$, and $\mathbb{T}'_i = \{v'_1, v'_2, ...v'_k\}$. The independence in the order indicates that the order of attempts performed by each node of $\mathbb{S}$ does not affect the probability of v_j in being activated in the and, which is given by:

$$\prod_{i=1}^{k}(1 - p_{v_j})(v_i|\mathbb{S} \cup \mathbb{T}_i) = \prod_{i=1}^{k}(1 - p_{v_j})(v'_i|\mathbb{S} \cup \mathbb{T}'_i) \tag{1}$$

where $\mathbb{S} \cap \mathbb{T} = \emptyset$.

The products in Eq. 1 indicate that the order in which the neighbours of a node try to activate it does not affect the global probability of a node to be activated.

Other variant of cascade models can incorporate information of success and failure of a node and this memory can be used to affect future attempts of a node to influence its neighbours. An extensive study of cascade models can be found in the book by Wu and Wang [18].

2.2 Canonical Names

After applying the Independent Cascade Model to real networks, the methodology proposed in this work analyses the resulting spreading and, in order to do this, besides assessing the number of nodes reached, it also investigates the way that the diffusion occurs in each studied scenario. For that, it is important to use a method that allows the spreading pattern to be identified.

The diffusion of a certain content in a network can be modelled by a set of trees, from which the roots represent the nodes that started the process, i.e., the seeder nodes. The leaves of the trees represent the nodes in which the diffusion process was terminated. By characterizing the trees observed in the diffusion process, it is possible to assess the frequency of different spreading patterns and enhance the understanding on the topological structure of a network.

In this work, the characterization of a spreading pattern is possible due to the algorithm for assigning canonical names proposed by Smal [17], frequently used to identify isomorph graphs. The algorithm, by assigning the canonical name of a tree, is able to ignore the labels of the nodes and circumvent the issue of visually find spreading patterns.

The algorithm for assigning canonical names, as defined by Smal [17], works as follows. Each leave of the tree receive a label "10". A parent node receive the label "1" concatenated to the ordered labels of the sons concatenated to "0".

Figure 1 show an example of two distinct but isomorph trees that, as a consequence, has the same canonical name. Considering the tree depicted in left side of Fig. 1, node D receives the label "10". As a consequence, node C receives the label "1" + "10" + "0". The label "1 10 1100 0" is assigned to node A. Since node A is the root of the tree, the label "11011000" is assigned to the tree depicted in right side of Fig. 1. Thus, in the context of diffusion, isomorph trees can be considered as result from similar diffusion processes.

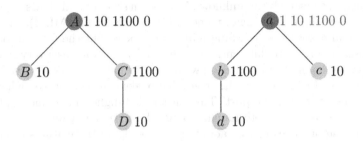

Fig. 1. Canonical names assigned to two isomorph trees.

3   Related Work

Several works found in the literature aim on investigating spreading phenomena in artificial and real world application. This section briefly describes some of them, in order to provide a broader view of the topic and relate the works performed by other authors to the current work.

Other works on information spreading in social networks and network analysis consider the network topology and the characteristic of an specific propagated information, which can "die" prematurely, not being propagated among the nodes or succeed, being propagated somehow. Several factors can be pointed out as causes the potential propagation of an information, like the importance of the seeders.

Dasgupta et al. [3] describe the profiles of a cell phone network by defining an objective function that assigns an influence value and model the accumulated energy for each user in order to investigate churn. In this study, the authors explore the intention to leave the company as a diffusion process and conclude that churn depends not only on the relation between one individual and another one that have left the company, but also on topological properties of the network. They also conclude that the effort to keep a client in the company is considerably lower than the effort to attract new users.

According to Joshi [3], in the last years, the increasing of online social data availability has offered new opportunities to map the diffusion processes network structure. Thus, product adoption and information propagation can be explored in order to investigate if a process can be considered as viral. To identify generic features of online diffusion, Joshi [3] studies some sources of data and observe that several external factors influence information spreading, such as personal contact and media diversity.

Another study aiming on characterizing online diffusion is the one by Dow et al. [4] that analyses the behaviour of two image sharing cascades, started by two quite distinct individuals. One of the images is a picture posted by the fist lady of the United States Michelle Obama to celebrate the election of Barack Obama in 2008, named Obama Victory Picture (OVP) in the original work. The other image was posted by an unknown Norwegian student and shows a picture of a joke between him and a friend, named Million Like Meme (MLM). Both posts resulted in large cascades, reaching millions of nodes, but the features observed by the cascades are quite different. Since the seeder of OVP was a very influential person, a high number of interaction with the original post was observed during the first hours of the process. Other notable peaks can be observed when other influential users shared the post. The authors highlight that demographic and geographical aspects are determinant to an information cascade.

Different approaches can be adopted to investigate the features of an information cascade. Dow et al. [4], for instance, explore two specific cascades and discuss how two distinct seeders can affect the propagation. Bakshy et al. cite13 investigate how a content is adopted and discuss the potential reasons why some people share it. In this study, the authors investigate how the influence of an individual and the number of individuals who adopted a content affect the probability of a person to adopt this content. It is also worth to mention another set of works, like those cited in Sect. 2, investigate the formulation of diffusion models to reproduce real world cascades in artificial scenarios.

4   Methodology

The methodology proposed for the current work comprises several steps, detailed in this section. First of all, Independent Cascade Model (ICM) was chosen to be considered as the model for diffusion simulation, as mentioned in Sect. 2. The implementation and application of ICM demands two parameters to be defined: the criterion for choosing the seeder nodes and the criterion for the probability distribution for activating the activation of the nodes adjacent to an activated node.

In the ICM implemented in this work, the out-degree was used as a criterion for choosing the seeders. In order to do this, for each network considered, a ranking of the nodes with the highest out-degree values was generated and the algorithm prioritizes the selection of the top ranked nodes.

Two approaches were considered to define the criterion of defining the nodes activation, as performed by Jung et al. [10]: Weighted Cascade Model and Trivalency Model.

Weighted Cascade Model assigns a probability of propagation for each edge, given by $P(u, v) = 1/d_v$ where d_v is the in-degree of the node v. This model can be intuitively used to explain the information spreading in social networks, in cases where the listeners capture similar amounts of information, regardless the in-degree.

Trivalency Model randomly selects an activation probability value for each edge from three candidates: 0.001, 0.01 and 0.1. The motivation for this model is to incorporate the diversity of types of personal relationships in the propagation model. In this work, three types of relationship were considered and the probability propagation depends on the type of relation of each link. In the present work, ICM is applied to networks considering only the topological structure of the network, i.e., without any assumptions and ground truth and, for this reason, the choice of the relationship types is performed randomly.

The methodology proposed in the present work applies ICM to real networks. For the experiments presented in Sect. 5, four real networks, freely available on the web[1] and with distinct features were selected. Large scale networks (with more than ten thousand nodes) were selected, in order to keep this study more coherent to real contexts. A brief description of the adopted networks is presented on Table 1.

Table 1. General description of the studied networks.

Network	Nodes	Edges	Brief description
Gnutella	10876	39994	Peer-to-peer in Gnutella network
Epinions	75879	508837	Social network on Epinions.com
Slashdot	77360	905468	Social network on Slashdot.com
Amazon	262111	1234877	Co-purchasing network on Amazon.com

At each evaluated scenario, the spreading induced by each seeder was assessed. To do this, canonical names were assigned to each diffusion tree, as discussed in Sect. 2.2. After applying the canonical names to each diffusion tree, an evaluation of the reached nodes and the frequency of resulting canonical names could be performed. An analysis of the results obtained by this study and a discussion about these results are presented in Sect. 5.

5   Results and Discussion

After applying Independent Cascade Model, as presented in Sect. 4 to the networks described in Table 1 (*Gnutella*, *Slashdot* and *Epinions*), some analysis can be made regarding the number of reached nodes.

In order to explore a broad set of scenarios for the experiments, diffusion was investigated at each network considering ICM varying the number of seeders. In all executions, the seeders were defined as the nodes with the higher number of connections, i.e., the nodes with the greater degree centrality. The number of nodes reached at each execution was assessed considering both strategies adopted for the probability of node to activate its neighbours. Figure 2 shows the variation of reached nodes for each network and each probability criterion: Weighted

---

[1] https://snap.stanford.edu/data/.

Cascade (Figs. 2(a), (c) and (e)) and Trivalency (Figs. 2(b), (d) and (f)). This study involves a high number of executions of ICM algorithm, bringing a high computational cost for evaluating large networks and, for this reason, Amazon network was not considered.

Figure 2 allows some observation to be made. When WC criterion for defining the probability of propagation was applied, as the number of seeders starts to be increased, the number of reached nodes grows more rapidly than when Trivalency criterion is applied. When number of seeders overcomes a certain value, the number of reached nodes follows the number of seeders almost linearly. With a narrow inspection on the networks, it can be noticed that, on Gnutella network, 95 % of the nodes show indegree value lower than 10, which makes that, according to WC criterion, the propagation probability to be greater than 10 %. On Epinions network, a similar degree distribution can be observed and 90 % of the nodes shows indegree value lower than 10, while on Slashdot network the number of nodes with indegree fewer than 10 is 81 %. For the Trivalency criterion, the each of the three defined propagation probabilities has the same chance to occur ate each node, 1/3. Thus, about 33 % of the edges tend to show the higher propagation probability propagation, 10 %, making large cascades quite uncommon.

The canonical name of each spreading tree observed for each execution of Independent Cascade Model was assigned to each corresponding seeder, allowing the investigation of the shapes of cascades in addition of analysing the number of reached nodes. To this end, some labels were defined for the most common canonical names, facilitating manipulation of canonical names and identification of patterns. The labels assigned for each of the most common canonical names are shown in Fig. 3, together with the graphical representation of the canonical tree. For the sake of simplicity, the labels of the canonical names in Fig. 3 will be considered for the remainder analysis when convenient.

Following a similar approach as performed by Goel et al. [6] to assess real spreadings, Independent Cascade Model was applied to the networks described in Table 1 considering a wide range of seeders. Weighted Cascade criterion was considered for the following analysis. The number of distinct canonical trees observed at each execution of the ICM for each network is presented in Fig. 4.

By observing Fig. 4, it can be noticed that the number of canonical trees significantly decreases as the percentage of seeder nodes increases. This phenomenon can be explained by the high connectivity among the seeders, which makes the cascades to be interrupted prematurely due to the fact that, when a propagation tree tries to reach an specific node, there is a high probability that this node has been previously reached by another cascade. As the number of seeders grows, the intersection of propagation trees occurs even earlier, resulting in even simpler trees.

In order to make a deeper investigation in the way that the diffusion process occurs in the different scenarios, Fig. 5 shows the distribution of the canonical trees for the ICM considering different numbers of seeder nodes for the studied networks. Figure 5 only shows the six most frequent canonical trees at

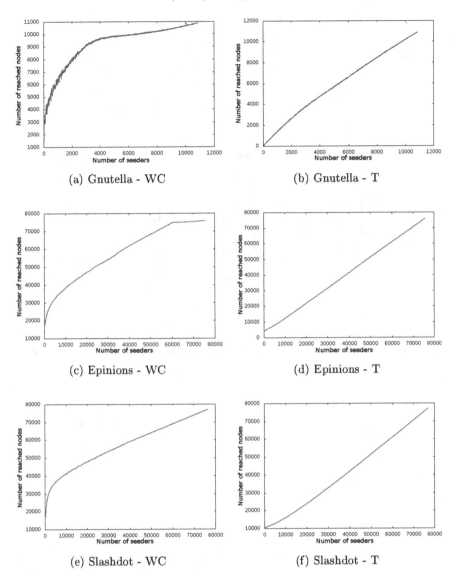

Fig. 2. Independent Cascade Model with Weighted Cascade criterion (WC) applied to the networks (a) Gnutella, (c) Epinions, (e) Slashdot; and with Trivalency criterion (T) applied to the networks (b) Gnutella, (d) Epinions, (f) Slashdot.

each scenarios. The labels annotated in each line of the figure indicate the canonical names of the top six rank for each execution.

Figure 5 presents the rank of most frequent cascades and allows a better understanding of the diffusion process considering ICM in the studied networks. Figure 5(a) shows that, as the number of seeders grows, the complexity

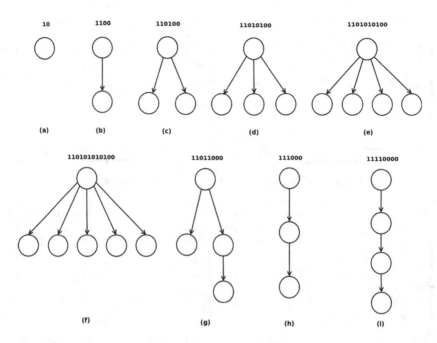

Fig. 3. Most common canonical trees together with the respective names and labels.

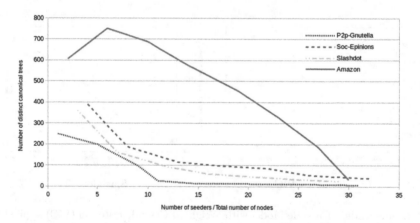

Fig. 4. Number of distinct canonical trees observed for different scenarios of the ICM.

of the cascades decrease and less deeper. With 1000 seeders, the frequency of the cascade labeled as (a), i.e., the cascade that does not reach any node but itself, is ranked in the third position and with 4000 seeders, the same cascade is gets the second position. With 7000 and 10000 seeders, this cascade is top ranked. A study on the neighbourhood of the seeders shows that the 10 % most

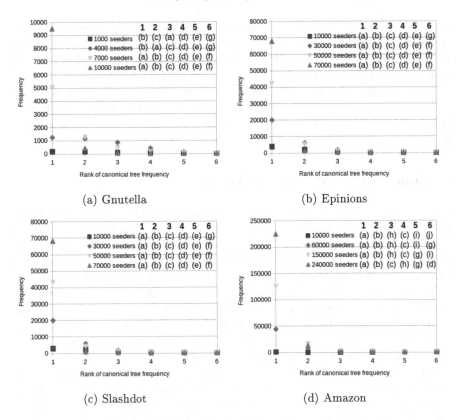

Fig. 5. Frequency of the top ranked canonical trees observed in the networks: (a) Gnutella; (b) Epinions; (c) Slashdot; (d) Amazon.

connected nodes reaches 58 % of the graph nodes in a first level neighbourhood. Thus, the cascade started by these nodes prevents prematurely other cascades to occur. When 4000 seeders are considered (about 35 % of the nodes) the first level neighbourhood comprises 99 % of the nodes and for 7000 seeders, 100 % of the nodes are reached, highly increasing the chance of cascade (a) to be top ranked. From Fig. 5(b), which shows the results for Epinions network, the simplicity of the top ranked cascades can be explained by the fact that 13 % of the most connected nodes are adjacent to 79 % of the nodes of the graph, what is coherent with the low depth of the cascades. A quite similar behaviour can be observed in Fig. 5(c), which presents the same results for Slashdot network. Analogously, simple cascades take the top positions of the rankings. However, with a lower number of seeders, the cascade labeled as (h), with three levels of depth, appears in the rank. This behaviour can also be explained by studying the neighbourhood of the seeders, given that 1000 seeders are adjacent to 86 % of the nodes, 4000 seeders are adjacent to 97 % and 70000 seeders reach 100 %.

The results presented for Amazon network show that simple cascades are still way more frequent. However, deeper cascades (labeled as (g), (h) and (i)) occur in the investigated scenarios. This result can be explained due to the fact that when 60000 seeders are considered, i.e., about 23 % of the nodes, 45 % of the nodes are reached. When 150000 nodes are set as seeders, 80 % of the nodes are reached and even when 24000 seeders are considered (about 91 %), 99.5 % of the nodes are reached in contrast to the other studied networks, in which 100 % of the nodes are reached with a smaller number of seeders.

6   Conclusions and Future Work

This work presents a study on the application of Independent Cascade Model (ICM) to a set of real world networks in order to investigate spreading processes considering different scenarios. The methodology proposed in this work allows us to explore different parameters of ICM and assess, not only the number of nodes reached, but the depth and the complexity of induced cascades.

Two criteria for defining the probability of a node in propagating information were tested: Weighted Cascade and Trivalency. The results obtained by assessing the number of reached nodes suggest that Weighted Cascade is mode coherent to diffusion processes observed in real world phenomena and this criterion was selected to define ICM in other studied scenarios.

The propagation patterns were evaluated by assigning the canonical name of each propagation tree. This allows the investigation of the complexity of the diffusion processes in the proposed scenarios and allows some conclusion to be made. A common behaviour was observed for the studied networks considering the propagation patterns: as the number of seeders is increased, the complexity of the cascades is decreased. This results can be intuitively explained by the high number of nodes that the seeders can reach in a first order.

The results indicate that the proportion of seeders considered and the way that the seeders are selected must be more deeply studied. Other techniques for selecting the seeders must be investigated, in order to better distribute them over the network. Furthermore, the diffusion process must be tested considering each seeder separately, aiming on observing the potential of a seeder to start a cascade without the interference of other cascades.

The analysis of the way that an information is propagates and which para-meters affect the information diffusion in a network is a great contribution for enhancing techniques for the diffusion of products, news and information addressed to a specific subset of consumers.

Acknowledgment. The authors are grateful to the agencies CNPq, CAPES and FAPEMIG for their financial support.

References

1. Anderson, T.K., Laegreid, W.W., Cerutti, F., Osorio, F.A., Nelson, E.A., Christopher-Hennings, J., Goldberg, T.L.: Ranking viruses: measures of positional importance within networks define core viruses for rational polyvalent vaccine development. Bioinformatics **28**(12), 1624–1632 (2012)
2. Aral, S., Walker, D.: Tie strength, embeddedness, and social influence: a large-scale networked experiment. Manag. Sci. **60**(6), 1352–1370 (2014)
3. Dasgupta, K., Singh, R., Viswanathan, B., Chakraborty, D., Mukherjea, S., Nanavati, A.A., Joshi, A.: Social ties, their relevance to churn in mobile telecom networks. In: Proceedings of 11th International Conference on Extending Database Technology: Advances in Database Technology, EDBT 2008, pp. 668–677. ACM, New York (2008)
4. Dow, P.A., Adamic, L.A., Friggeri, A.: The anatomy of large Facebook cascades. In: Kiciman, E., Ellison, N.B., Hogan, B., Resnick, P., Soboroff, I. (eds.) ICWSM. The AAAI Press (2013)
5. Durrett, R.: Lecture Notes on Particle Systems and Percolation. Brooks/Cole Pub. Co., Pacific Grove (1988)
6. Goel, S., Watts, D.J., Goldstein, D.G., The structure of online diffusion networks. In: Proceedings of 13th ACM Conference on Electronic Commerce, EC 2012, pp. 623–638. ACM, New York (2012)
7. Goldenberg, J., Libai, B., Muller, E.: Talk of the network: a complex systems look at the underlying process of word-of-mouth. Mark. Lett. **12**, 211–223 (2001)
8. Goldenberg, J., Libai, B., Muller, E.: Using complex systems analysis to advance marketing theory development: modeling heterogeneity effects on new product growth through stochastic cellular automata. Acad. Mark. Sci. Rev. **9**(3), 1–18 (2001)
9. Granovetter, M.: Threshold models of collective behavior. Am. J. Sociol. **83**(6), 1420–1443 (1978)
10. Jung, K., Heo, W., Chen, W.: Irie: scalable and robust influence maximization in social networks. In: Proceedings of 2012 IEEE 12th International Conference on Data Mining, ICDM 2012, pp. 918–923. IEEE Computer Society, Washington, DC (2012)
11. Leskovec, J., Singh, A., Kleinberg, J.M.: Patterns of influence in a recommendation network. In: Ng, W.-K., Kitsuregawa, M., Li, J., Chang, K. (eds.) PAKDD 2006. LNCS (LNAI), vol. 3918, pp. 380–389. Springer, Heidelberg (2006)
12. Liggett, T.M.: Introduction. In: Liggett, T.M. (ed.) Interacting Particle Systems. Grundlehren der mathematischen Wissenschaften, vol. 276, pp. 1–5. Springer, New York (1985)
13. Moore, C., Newman, M.E.J.: Epidemics and percolation in small-world networks. Phys. Rev. E **61**, 5678–5682 (2000)
14. Newman, M.: Networks: An Introduction. Oxford University Press Inc., New York (2010)
15. Pastor-Satorras, R., Vespignani, A.: Epidemic spreading in scale-free networks. Phys. Rev. Lett. **86**(14), 3200–3203 (2001)
16. Schmith, J., Lemke, N., Mombach, J.C.M., Benelli, P., Barcellos, C.K., Bedin, G.B.: Damage, connectivity and essentiality in protein protein interaction networks. Phys. A: Stat. Mech. Appl. **349**(3), 675–684 (2005)
17. Smal, A.: Explanation for tree isomorphism talk (2008)
18. Wu, J., Wang, Y.: Opportunistic Mobile Social Networks. CRC Press, Boca Raton (2014)

Confederative ERP Systems
for Small-to-Medium Enterprises

Michal Žemlička[1]([⊠]) and Jaroslav Král[2]

[1] The University of Finance and Administration,
Estonská 500, 101 00 Praha 10, Czech Republic
michal.zemlicka@post.cz
[2] Faculty of Informatics, Masaryk University,
Botanická 68a, 602 00 Brno, Czech Republic
kral@fi.muni.cz

Abstract. Small-to-medium enterprises (SME) are frequent. It holds for
SME software users as well as for SME software developers. Both cannot exclusively use products and philosophies of large software vendors.
SME users have not enough resources to apply or implement products
and processes of large vendors. The processes can be based on philosophy not applicable in SME. It follows that SME must collaborate with
SME software vendors and use their solutions. It can happen that even
great users must use solutions of small software vendors solving special
needs. We show that these challenges can be solved if we apply a variant
of service-oriented architecture using document-oriented communication.
The communication is supported by infrastructure services. Our experience shows (see examples) that it can have dramatic effects.

Keywords: Small-to-medium enterprises (SME) · Supporting business
processes in SME · Software adaptation · Document-oriented software
architecture · Software confederations

1 Introduction

Small-to-medium enterprises (SME) business is a frequent phenomenon. SME
produce almost 20 % of GDP in many countries, e.g. in Czech Republic. According to Czech Statistical Office, more than 80 % of IT professionals worked in
2013 as sole traders (craftsmen) or as employees in SME. They are developers,
maintainers or IT operators. Many software firms are SME too.

The software professionals must take into account that business, culture, and
business processes are different from the ones used in large enterprises. The software developed by large vendors for large clients cannot be as a rule seamlessly
used by SME. The systems are inspired by the culture of large organizations
and expect the existence of large resources (tools, professionals, investments,
available data, etc.).

SME must often take part in global business. Their business processes (BP)
must support collaboration with other SME but also with information systems

© Springer International Publishing Switzerland 2016
O. Gervasi et al. (Eds.): ICCSA 2016, Part V, LNCS 9790, pp. 350–362, 2016.
DOI: 10.1007/978-3-319-42092-9_27

of giant firms. Many activities (e.g. bookkeeping) tend to be outsourced and their providers should be changed easily if necessary. Some production sub-processes (parts of BP) can be outsourced too. The business partners should be easily exchangeable. The above properties imply that the systems are very easily maintainable.

We show that the problems can be smoothly solved if the developed software systems have a specific service-oriented architecture (SOA) using document-oriented communication. The communication is supported by infrastructure services. It can (see examples) that it can have dramatic effects.

The paper is structured as follows: Crucial requirements with respect to the needs of small-to-medium business are collected in Sect. 2. Section 3 collects implementation principles of the proposed architecture. Implementation principles are remembered in Sect. 3. Section 4 discusses cooperation of small products developed by companies with large products of large companies. Section 5 remembers systems extending functionality of large systems. Open issues are collected in Sect. 6. Further research and conclusions are in Sects. 7 and 8.

2 Crucial Requirements

Business processes in SME must promptly respond to changing market conditions, changing business partners, or changing customers' needs. They must be agile. SME usually do not have enough time, enough resources, and enough experience. The issues can be reduced if a proper service-oriented architecture is used [21].

The crucial business requirements are:

1. SME should be able to collaborate with its business partners in the way transparent for businessmen. It must enable a smooth implementation of business actions like business agreements conclusion and control, solution of business cases, etc.
2. Activities should be easily outsourced and insourced.
3. The sourcing as well as other business activities should be performable in an agile way.
4. System should enable a smooth integration or use of existing software artifacts, especially existing applications, also if the code of the artifacts is closed and the artifacts have specific interfaces.
5. The system properties should be modifiable without the necessity of extensive coding.
6. It should be possible to reuse existing software artifacts.
7. System should provide log data transparent for business people and easily usable in business operation and management.
8. It is desirable to enable a transparent decentralized organizational structure of the supported enterprise.

3 Implementation Principles

We will show that the above requirements can be met if a specific service-oriented software architecture (SOA, [2,18]) called *software confederation* [5,7,9] is used. Software confederations have the following properties[1]:

- Services forming the SOA are highly autonomous and loosely coupled. They communicate in a peer-to-peer style.
- Structure of the system can and should be hierarchic – services can be again constructed from services (compare with layered approach suggested by OASIS [12] or by OpenGroup [16]).
- The system structure and local interfaces (interfaces between the services) are inspired by real-world counterparts of the services that are to be supported. It is, the communication is easily understandable to the (business/domain expert) users. The interfaces are based on electronic versions of documents typical for given problem domain (business documents).
- Various parts of a confederation can solve problems from various domains. The interfaces of the parts then use language (documents) typical for the corresponding problem domains.
- The users can be highly involved in system design, development, maintenance and use. It is, the system is designed so that the users can understand how it works and can effectively change its behavior if necessary.
- As the interfaces between system parts are based on real-world (preferably business) documents, it is possible to combine manual and computerized processing of the individual tasks. It is advantageous during development (finished parts can be used when available) as well as in emergency events (some parts could be down or unavailable) or in rare situations (some activities of the system are so rare, that it is not advantageous to implement their full computerization; it can also happen that it is required an operation that has not been planned). The services in confederations in fact support information hiding proposed by Parnas in [17]. It is especially the case when business document communication (interface) of services is used.

4 SOA Enabling to Meet the Requirements

Software confederations are a variant of service-oriented architecture (SOA, [2,7]) consisting of a set of (wrapped) applications (WA). Any WA is again wrapped by a wrapper service called *front-end gate* (FEG, [6,8]). WA has only one distinguished link with its FEG. A pair (WA, FEG) is called *service wrapped application* (SWA). FEG can communicate using its non-distinguished links with other SWA's and possibly with the specific wrappers described below.

---

[1] The concept of software confederations can be implemented partly. There is still a chance that the resulting system retains some important properties.

The policy to communicate with the wrapped application only through its front-end gates increases security (individual groups of partners have access only to functions that they really need). The partners therefore see the front-end gates as services (black boxes) providing the desired functionality.

This construction can be usefully generalized. It is reasonable to allow an application to have multiple wrappers. The wrappers can communicate with heads of composite services (HCS). It enables to build a hierarchy of composite services. Front-end gates and heads of composite services can communicate with external services via access point services. The access points are used to implement e.g. necessary security capabilities. This structure can be substantially generalized. The first generalization is that a service can have more than one wrapper.

The services in confederations can play various roles (for details see e.g. [9]). Let us mention some of them:

Front-end gate (FEG) – service modifying application (service) interface to match the needs of a group of the service partners (and to be business document oriented). its main capability is the ability to transform n-tuples of documents in a local format into documents in external formats and send the resulting documents to their addressees as well as accepting documents in external formats from various sources and transform them to n-tuples of documents in the internal format. They are typically used for changing existing service interface to the desired one. Using more front-end gates for one wrapped application customizes its interface for corresponding service partners or partner groups.

Head of composite service – service providing the only access point to a group of services. It allows seeing the group as a single service. This approach allows to create hierarchical structure of the system and to support higher competence of the organizational units for their services.

Access point (AP) – service providing system interface for external partners. It checks the partner's identity. They often use specific protocols for the communication with external partners (services being parts of external information systems). In some cases it is possible to combine Front-end gate and Access point to a single service. It is usually reasonable to keep them separated (see Fig. 1).

Portal – System interfaces intended for human users could be implemented as portal services. It concentrates user interfaces to the system for a specific group of users. There can be multiple portals serving for one system. Typical examples are intranet (in-company users), extranet (users outside the company), and specialized portals (e.g. for process owners).

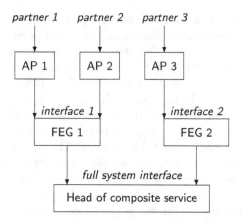

Fig. 1. Separated front-end gate and access points

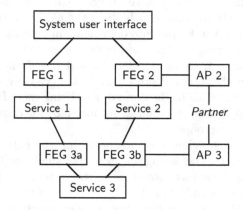

Fig. 2. Connecting a partner to a marketplace-like system

Let us show two approaches to connecting external partners to the system.

The simpler one just creates access points connecting the partner to the required front-end gates of the required services (Fig. 2[2]). When the current service interfaces do not match the partner's needs, new front-end gate(s) could be created. Even if there is an available interface being a superset of the required one, it is good to create a new interface. The reason is that using the exactly matching interface the partner does not have access to the functionality that is not intended for him. It reduces errors and complicates system misuse.

---

[2] When the Service is mentioned, it means the application providing the service/activity/agenda. The application can be a wrapped legacy system, a third-party product, or a newly written application. It is reasonable to preserve its direct interface and, if possible, its code.

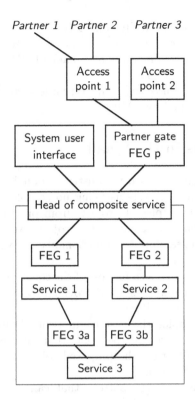

Fig. 3. Connecting partners to a system being a composite service

Decoupling of a partner (e.g. when the cooperation ends) in such case requires finding of all used access points and unconnected them (turn them off).

The more closed approach (Fig. 3) is built on the idea that the system can be viewed as a composite service. It is, the system is accessible from partner systems via a single service being head of the composite service. The head can be then equipped by front-end gates and access points for cooperation with other partners. It could be reasonable to connect the user interfaces (e.g. the extranet one) using the head too.

Decoupling of a partner in this case requires turning off the partner specific access point (there is only one) only.

It is possible to create access points one by partner or one by contract.

Internet of things has in fact the architecture very similar to confederation systems discussed here.

Note that the solution shown in previous figures enables provided that we have a middleware powerful enough the following effects:

- The services can be hierarchically composed.
- The services can be distributed; they can be supported by different hardware and software systems. It is possible to enhance ERP of SME or to integrate various subsystems with them.
- The document orientation is important for large systems as well. It is crucial for the systems developed for SME.

5   Confederations in the Small

Some principles and attitudes used to construct confederation architecture can be used to integrate software systems being no complex multicomponent confederation. Let us give examples of projects from practice of small-to-medium firms in which the authors took part. The projects were used by some Czech small-to-medium firms extending the functionality of existing systems of large users produced by large vendors.

The solution can be viewed as a process of the construction of a confederation consisting of two components – a system provided by a large vendor and a new component developed by a small firm.

In the following examples the first component of the pair (e.g. an ERP) is viewed as a three-tier structure. We show that all the tiers can be used to build a pair solution. Note that the proposed solutions substantially enhance integratability with new systems.

Pair confederation is a design pattern that can have dramatic effects. In OpenCard[3] replacement the development costs have been reduced about 20×. It seems that it can be applied to toll collection system as well (with significant expenses reduction in comparison to the existing implementation).

5.1   Commodity Consumption Control

A small firm was asked by a large hospital to develop new functions and integrate them with its large ERP system delivered by a large vendor as a black box. The

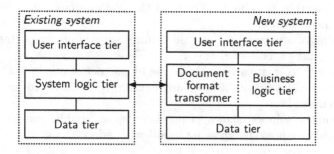

Fig. 4. Integration via business document interface

---

[3] Pre-paid card for Prague public transportation and for accessing other services.

aim was to provide the support for the commodity requests and consumption control and optimization. The system developed by the firm uses autonomous system communicating with ERP system via a simple wrapper enabling to use a document-oriented communication. The documents are business-oriented ones like orders and the responses to them. The interface is provided by the ERP system.

The gate used to implement the communication can be easily adapted, often used without any change, to enable the communication with different ERP systems by other vendors. The systems in fact use a wrapper of (virtual) business logic tier of the ERP system.

The new system is integrated via operation tier (Fig. 4). Existing business document system interface can be used.

5.2 New Banking Service by Autonomous Vendors

A small software firm provides banking services, usually insurance services, that must be integrated (operate) in the framework of several enterprise systems (ES). A very successful solution wrapped the data tier of ES so that that it can be used as a software service having document-oriented interface:

It was a very successful solution. The company using it has grown from a small firm into a middle-sized one. It is successful internationally.

In this case the new systems communicate with data tier using a newly written document-oriented data representation.

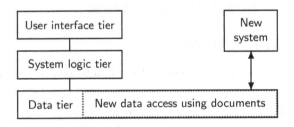

Fig. 5. Integration via document-based data access

5.3 Operation-Oriented User Interface

Usability is a crucial aspect of modern information systems [14]. The quality of the user interfaces (UI) plays the central role in usability. The requirements of business and management users are changing. They are used to use spreadsheet systems and are often masters in using it. This is the reason of the growing popularity of spreadsheet-oriented UI.

It appears that it is possible to (partly) encapsulate the UI into a spreadsheet. Such systems use documents being encapsulated spreadsheet tables in XML. It follows that there should be a document management system enabling and

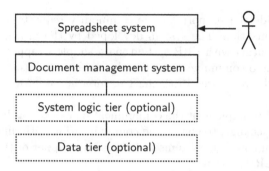

Fig. 6. Interface based on a spreadsheet system

integrating the access to the documents being spreadsheets in XML formats (Fig. 6).

This solution becomes popular as it has, besides the usability, many advantages for end users. Many operations, managing ones inclusive, on input data can be performed immediately. The spreadsheet-oriented attitudes become very popular. A more advanced processes will probably need something like the gateways in confederations (e.g. transformation of n documents into m documents). It is possible to (hiddenly) use capabilities of document management systems here. We applied this attitude in a project for a small Prague firm.

The spreadsheet attitude is becoming so popular that even database access is sometimes designed so that it is "spreadsheet-like". Confederations enable an easy combination of different user interface formats.

5.4   Confederations in Action

Let us discuss remarkable examples of the applications of confederated SOA philosophy.

Application forms – collecting supply requirements within a hospital is in confederative environment simple. It has been simply applied. It is possible to combine paper and electronic requests. It can be combined with various hospital ERP's. The requirements of physicians and nurses are mapped into request forms. The forms must pass several steps of request acceptance process. They are finally transformed into standard order form to be sent to hospital ERP. The order form and the request forms are used to control stock out. It could seem simple but specific conditions in the hospital caused that the process cannot be supported by the ERP. The user interface was provided by a spreadsheet system in the architecture shown in Fig. 6. The system uses documents being encapsulated spreadsheet tables.

Simplity firm – they are able to combine various ERP and banking systems. It used some elements of document orientation. Note that their solutions had great success. The firm have been rewarded as the best company in Czech Republic in 2015.

Chip card – There is a remake of a special chip card holding information necessary for Prague city library and for Prague city transport company. The remake is based on an implementation of some aspects of confederations. The implementation has been many times cheaper than the first one. Moreover, the new implementation is applicable also in other cities.

Highway toll collection – Original supplier as well as other possible suppliers require several years for implementation and hundreds of millions of Czech crowns. A proper application of confederative approach seems to implement it within a year or two and with at least 30 times lower price.

Document-oriented interfaces simplify decomposition of systems into autonomous components. It is a very important aspect of system quality. In the case of OpenCard and highway toll collection it simplifies the use and orchestration of different data collection systems as well as payment tools.

6 Open Issues

Although the concept of confederation is quite old and mature [4], there are still some open practical issues:

- The documents produced and used by the system should be logged and preserved for later use. In practice from the data preservation point of view there are different document groups:
  - permanent documents – documents that should be preserved for a very long time (for ever) like business documents applicable for a long time – e.g. agreements or log books.
  - temporal documents – e.g. documents used for synchronization of the cooperating parts of the company
  - documents that can be potentially used for optimization or at court – i.e. documents with some limited applicability. The issue is that the set of documents that can be required by court and their validity can be changed by law during the lifecycle of the system. It is, such documents must have individually as well as group-wise modifiable preservation time.
- Trustworthiness of the documents. For simple data exchange (temporal documents) electronic signing could be enough. For some documents there is required longer trustworthiness than is the validity of the digital sign. It is, there must be a mechanism allowing protection of the trustworthiness of the documents for the desired period. Currently such systems are quite expensive – both for creation and use.

7 Further Research

Confederative systems using the principles described above have many desirable properties. They are well understood by users. The users can be involved in system requirements specification, development, agile operation, and maintenance. They, as a rule, welcome the principles and attitudes described above. We observed it in all our recent real-world projects.

The principles discussed above are not easily acceptable for most developers. They are trained to work in object-oriented world; i.e. they apply the object-oriented paradigm. It includes also modeling using Unified Modeling Language (UML, [1]) as it is based on object orientation and does not properly model the structure of confederative systems. It leads to relation of sole entity with dozens of other entities what frequently leads to omissions in requirements as well as in system design. It appears that some turns known from structured analysis and design [19,20] suit better in this case as the methodology reflects the recommendations from [13]. The document and service orientations are technologies of a new paradigm. Let us call this paradigm *document service oriented paradigm* (DSO paradigm).

It is known (see e.g. physics) that paradigms are very difficult to change. The change often requires new generation of people. Although it is in IT typically easier to change paradigm than it is in physics, the time needed to adapt to DSO in development and operation of information systems is several years.

There are problems in education of IT experts. The advantages and patterns of DSO are very difficult to teach during standard university software courses. It is very difficult to design good projects as DSO are useful rather for large systems. Coding practices and overall philosophy of IT studies are object oriented. Some hints how to overcome this problem can be found in [10].

It is open whether broad application of complex standards like Universal Business Language (UBL, [3,15]) is not premature and whether some ad hoc solutions like the use of spreadsheet encoding or using document management systems are not preferable now.

8 Conclusions

We have shown how to enhance the business and business processes in small-to-medium business (SMB) using some of the attitudes and power of software confederations. Confederations are a variant of coarse-grained business-document- and service-oriented architecture. Their structure and processes are transparent for users. It is not common for developers. The construction of confederations implies for them that their software development paradigm must be changed. It is very difficult especially for the people trained to apply object-oriented philosophy (compare e.g. [10]). We believe that the use of confederative architecture is a precondition for the smooth enhancement of business management and improvement. The document-oriented services and SOA enable a crucial enhancement of software system quality aspects. The most important quality aspects are: autonomy of system parts and coarse-grained user-oriented communication messages. Details can be found in [11,22].

Acknowledgement. The paper has supported by the Institutional support for long-term strategic development of the research organization University of Finance and Administration.

References

1. Unified Modeling Language (UML), v2.4.1, August 2011. http://www.omg.org/spec/UML/2.4.1/
2. Erl, T.: Service-Oriented Architecture: Concepts, Technology, and Design. Prentice Hall PTR, Upper Saddle River (2005)
3. International Organization for Standardization, International Electrotechnical Commission: ISO/IEC FDIS 19845 information technology - universal business language version 2.1 (UBL v2.1), final draft of international standard (2015)
4. Král, J., Demner, J.: Towards reliable real time software. In: Proceedings of IFIP Conference Construction of Quality Software, North Holland, pp. 1–12 (1979)
5. Král, J., Žemlička, M.: Electronic government and software confederations. In: Tjoa, A.M., Wagner, R.R. (eds.) Twelfth International Workshop on Database and Experts System Application, pp. 383–387. IEEE Computer Society, Los Alamitos (2001)
6. Král, J., Žemlička, M.: Component types in software conferations. In: Hamza, M.H. (ed.) Applied Informatics, pp. 125–130. ACTA Press, Anaheim (2002)
7. Král, J., Žemlička, M.: Software confederations - an architecture for global systems and global management. In: Kamel, S. (ed.) Managing Globally with Information Technology, pp. 57–81. Idea Group Publishing, Hershey (2003)
8. Král, J., Žemlička, M.: Software architecture for evolving environment. In: Kontogiannis, K., Zou, Y., Penta, M.D. (eds.) 13th IEEE International Workshop on Software Technology and Engineering Practice, pp. 49–58. IEEE Computer Society, Los Alamitos (2006). http://doi.ieeecomputersociety.org/10.1109/STEP.2005.25
9. Král, J., Žemlička, M.: Implementation of business processes in service-oriented systems. Int. J. Bus. Process Integr. Manag. **3**(3), 208–219 (2008)
10. Král, J., Žemlička, M.: Experience with real-life students' projects. In: Ganzha, M., Maciaszek, L., Paprzycki, M. (eds.) Proceedings of the 2014 Federated Conference on Computer Science and Information Systems. Annals of Computer Science and Information Systems, vol. 2, pp. 827–833. IEEE (2014). http://dx.doi.org/10.15439/2014F257
11. Král, J., Žemlička, M.: Simplifying maintenance by application of architectural services. In: Murgante, B., Misra, S., Rocha, A.M.A.C., Torre, C., Rocha, J.G., Falcão, M.I., Taniar, D., Apduhan, B.O., Gervasi, O. (eds.) ICCSA 2014, Part V. LNCS, vol. 8583, pp. 476–491. Springer, Heidelberg (2014). http://link.springer.com/chapter/10.1007/978-3-319-09156-3_34
12. MacKenzie, C.M., Laskey, K., McCabe, F., Brown, P.F., Metz, R.: Reference model for service-oriented architecture 1.0, OASIS standard, 12 October 2006. http://docs.oasis-open.org/soa-rm/v1.0/
13. Miller, G.A.: The magical number seven, plus or minus two: some limits on our capacity for processing information. Psychol. Rev. **63**, 81–97 (1956)
14. Nielsen, J.: Usability Engineering. Academic Press, New York (1993)
15. OASIS: Universal business language version 2.1, OASIS standard, November 2013. http://docs.oasis-open.org/ubl/os-UBL-2.1/UBL-2.1.html

16. Open Group: Draft technical standard SOA reference architecture, April 2009. http://www.openinnovations.us/projects/soa-ref-arch/uploads/40/19713/soa-ra-public-050609.pdf
17. Parnas, D.L.: Designing software for ease of extension and contraction. IEEE Trans. Softw. Eng. 5(2), 128–138 (1979)
18. Stojanovic, Z., Dahanayake, A. (eds.): Service-Oriented Software System Engineering: Challenges and Practices. Idea Group Publishing, Hershey (2005)
19. Weinberg, V.: Structured Analysis. Prentice-Hall software series. Prentice-Hall, Englewood Cliffs (1980)
20. Yourdon, E.: Modern Structured Analysis, 2nd edn. Prentice-Hall, Englewood Cliffs (1988)
21. Žemlička, M., Král, J.: Flexible business-oriented service interfaces in information systems. In: Filipe, J., Maciaszek, L. (eds.) Proceedings of ENASE 2014–9th International Conference on Evaluation of Novel Approaches to Software Engineering, pp. 164–171. SCITEPRESS - Science and Technology Publications (2014)
22. Žemlička, M., Král, J.: Software architecture and software quality. In: Gervasi, O., et al. (eds.) ICCSA 2016, Part V. LNCS, vol. 9790, pp. 139–155. Springer, Heidelberg (2016). doi:10.1007/978-3-319-42092-9_12

Longitudinal Analysis of Modularity and Modifications of OSS

Gabriel Rolim[(⊠)], Everaldo Andrade, Danielle Silva,
and Eudisley Anjos

Center of Informatics, Federal University of Paraíba,
S/N 58, João Pessoa, Paraíba 058-600, Brazil
gabrielsrolim@gmail.com,
everaldo.andrade.j@gmail.com,
danielle.rousy@gmail.com, eudisley@gmail.com

Abstract. The open source software systems are always evolving with the additions of new features, bug fixes and collaboration of many developers often around the world. The modularity of the system metrics help to better understand the characteristics of the system and guarantee the quality of software. In this article, we will compare the evolution of some software metrics, in particular complexity and coupling, with the evolution of the number of bug fixes, additions and features contributions from developers over software versions. Showing that the bug fixes, adding features and contribution of developers exerts a strong influence on the increase of the metrics.

Keywords: Software metrics · Bug · Features · Evolution software · Mining software repositories

1 Introduction

Systems of open source software (OSS) are always evolving with the additions of new features, bug fixes and collaboration of many developers often these are scattered around the world. The modularity of the system metrics help to better understand the characteristics of the systems and guarantee a software quality [1]. The higher value of complexity metrics, greater the difficulty to maintain the system. In addition, the effort for adding new features and fixing bugs become increasingly costly.

Increasingly growing interest from developers with great technical ability [2] to contribute to these systems. In addition, it is important to monitor the development of metrics for modularity of the system, because from them we assure the quality of software.

With mining repositories can get information for each phase of the system. For example, number of developers per version, amount of bugs fixed, amount of added, metric features of modularity for each commit system and so on.

In this work we study the evolution of modularity metrics on 3 projects Open Source Apache Jmeter, POI and Tomcat in order to verify the correlation between metrics of modularity and data taken from the mining repositories: number of developers per version, amount of bugs fixed, amount of features added over the project.

© Springer International Publishing Switzerland 2016
O. Gervasi et al. (Eds.): ICCSA 2016, Part V, LNCS 9790, pp. 363–374, 2016.
DOI: 10.1007/978-3-319-42092-9_28

2 Related Work

There is a focus in the literature for the creation of models for prevention appearances of bugs in the systems by using of analysis technics of the Mining Software Repository (MSR).

Some work of literature sought to see the relationship of the density of bugs before and after the release of a version of the system [3] was tempted verify the correlation between software metrics and the appearance of bugs. In a second analysis [4] was made the first analysis of the emergence of post-release with proprietary Microsoft software, for both cases bugs' repositories such as Bugzilla bug were mined.

There is also analysis of data mining in repositories that sought - making bug detection [5] by analyzing metrics like lines of code to create laws to identify the presence of bugs in the system. The detection of failures was also analyzed more deeply in [6] analyzing all existing metrics in the literature to verify the appearance of bugs in the system.

Another project [7] a statistical model was created by analyzing the data analysis method SOBER not identify a model showing differences in the evaluation of predicates between correct and incorrect executions.

In [8] show a brief history of the MSR field, presents definitions about using MSR techniques to support software research and practice. Where discuss several opportunities and challenges in the field.

In all cases, analysis were performed on a few versions designs, often only one project and one version has been analyzed in some work done closed-source and proprietors codes. In this work, we analyze the evolution of software Tomcat (Java Servlet), POI (Java Archives Handler) Jmeter (Program to Test) for a minimum period of 2 years. Using SonarQube to store and remove the key metrics of system modularity having as main objective to verify how the metrics are affected, the amount of fixed bugs, adding new features and the amount of developers so we can see how the human factor acts the modularity metrics that was presented in [3].

3 Analysis of Projects

The projects were selected by their success, being Open Source, by having a significant number of contributors for having GIT repositories, by having a standard referencing bug fix in messages of commits and possible compilation by using Script Apache Ant or Apache Maven both resolution programs dependencies and compiling software. One of the problems was that despite the sonar data to generate almost all metric, some cohesion and coupling metrics was not possible to save the historical evolution of the system. The metrics were analyzed:

1. Response for Class (RFC): Calculates the cardinality of classes of calls [9]. The calls of classes consists of all local methods and all methods called by local methods. It is logical that the higher the RFC complex is class. Intuitively, the larger the RFC, the harder it is to keep the class taking into account that a larger number of methods;

2. Package: Represents the size project. It show the amount of packages in the project. Usually Object Oriented Projects have many packges;
3. Lines: Reports the number of lines of the project;
4. Classes: Reports the number of classes in the project;
5. Complexity: It's cyclomatic complexity, often called metrics McCabe. It used to indicate the complexity of a program by using the control flow graph of the program. The more the number of instructions the more complex the project code [10];
6. Complexity of Function: Informs the average complexity of McCabe by function;
7. Complexity Class: Reports the average complexity per class McCabe;
8. Number of files: Reports the average complexity of McCabe;
9. Cyclic dependency Package (Package Cycles): Minimum number of cycles detected directories to be able to identify all the unwanted dependencies;
10. Number of lines not Discussed (NCLOC): Number of lines without comments;
11. Accesses: Number of functions Getter and Setter.

The sonar is structured into two separate programs, a server is the first one that has a web interface for monitoring and visualization of the results of the analysis of metrics and a second program that runs all algorithms metrics. To do this simply configures a properties file informing the ways of the source code and binaries of system build.

Using a script written in python was possible to travel by commits from GIT repositories and run the sonar for each commit. For a more comprehensive version, jumps 6 commits were made. The steps for obtaining data follow the following steps for each commit (Fig. 1).

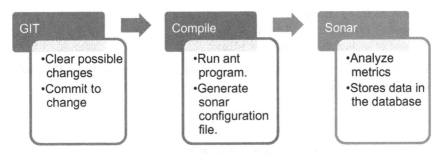

Fig. 1. Step program to obtain data on each commit

Git is a free system and open source version control for large and small projects. Workflow with Git developer at the end of each change or end of a group changes the developer makes a commit or a confirmation of the changes. This will commit your information and a mandatory message explaining what has changed in that commit.

To retrieve social data we use the GitStats. This software is written in python to pull information like:

1. General statistics: total files, lines, commits and author.
2. Activities developers: Commits by hour of day, day of week, month of the year, year and month.
3. Percentage of the amount of commits by author.

The GitStats only generates data in HTML format and graphics is difficult to collect the information from the developers. Because it is written in python program is easy import it as a package, and so with some changes in the code GitStats is possible to collect historical collaboration of developers throughout the project which is very important for our data because we analyze the human factor in the evolution of modularity metrics.

For a better visualization of data generated a centralized database that all information collected. This database history information of maintainability metrics, the program commits each sorting through the commit messages when the commit was to fix a bug, registered in Bugzilla, or if it was an improvement or adjustment or documentation has been saved new functionality (feature), which in this work is considered the same thing.

In each commit the developer puts a message that the changes made to the code and to identify when it was a bug fix or feature was an increase of just look at the message the occurrence of the word "bug" and its variations. This is a standard commit adopted in 3 projects analyzed. Therefore all that was not found commit the presence of the word "bug" in the commit was considered feature.

After all the processing of data was generated 117 graphics with the evolution of all data. For a better understanding a study of the Pearson correlation was made between the following variables:

1. Number of Bug Fixes x Metrics;
2. Number of Developers x Metrics;
3. Amount of additions of Features x Metrics.

Pearson correlation coefficient is a number ranging between −1 and 1. Their values are described below (Table 1):

Table 1. Pearson coefficient.

Pearson coefficient	Correlation
r = 1	Perfectly Positive
$0,8 \leq r < 1$	Strong positive
$0,5 \leq r < 0,8$	Moderate positive
$0,1 \leq r < 0,5$	Weakly positive
$0 < r < 0,1$	Tiny positive
0	Null
$-0,1 < r < 0$	Tiny negative
$-0,5 < r \leq -0,1$	Weak negative
$-0,8 < r \leq -0,8$	Strong negative
r = −1	Perfectly Negative

When the value is positive indicates that the correlation is the growth of data values when the value is negative indicates that the correlation is decreasing values of the data and the closer to 1 or −1 is the most perfect correlation between data.

Were analyzed and compared data from the following periods (Table 2):

Table 2. Periods of analysis of projects.

Project	Period
Jmeter	2010/10/07–2012/10/12
POI	2008/05/08–2014/02/09
Tomcat	2010/11/21–2014/02/09

These periods were possible to analyze the following quantities version (Table 3):

Table 3. Number of versions.

Project	Number of version
Jmeter	22
POI	41
Tomcat	18

4 Results

Looking at the table above, we found that 48.71 % of the correlations with bug metrics compared to three projects are strongly positive and 33.33 % are moderate positive. We can else note that the Cyclomatic Complexity, Class and Line metrics maintains the highest values.

Table 4. Pearson correlation Evolution Metrics with Volume Bug.

Metrics	Jmeter	POI	Tomcat
RFC	0,906	0,647	0,587
package	0,688	0,895	0,901
Line	0,961	0,939	0,615
Complexity Function	0,687	0,021	0,901
Files	0,982	0,599	0,826
Cyclomatic Complexity	0,953	0,981	0,97
Classes	0,953	0,981	0,97
Cycle Packages	0,977	0,139	−0,978
NCLOC	0,973	0,799	0,657
Function	0,991	0,919	0,221
Complexity File	0,526	0,789	−0,34
Complexity Classes	−0,444	0,789	−0,324
Accessors	0,774	0,56	0,922

Table 5. Pearson correlation of the development of metrics to count Developers.

Metrics	Jmeter	POI	Tomcat
RFC	0,91	0,716	0,557
package	0,791	0,795	0,84
Line	0,931	0,939	0,534
Complexity Function	0,694	0,084	0,84
Files	0,95	0,682	0,834
Cyclomatic Complexity	0,962	0,928	0,853
Classes	0,971	0,881	0,855
Cycle Packages	0,949	0,421	−0,997
NCLOC	0,942	0,945	0,594
Function	0,946	0,918	−0,015
Complexity File	0,832	0,656	−0,463
Complexity Classes	−0,259	0,677	−0,298
Accessors	0,899	0,755	0,849

In Table 5, we observed that 53.84 % of the correlations with Developer metrics are strongly positive. Between Developer metrics we deduce that Cyclomatic Complexity, Classes and Accessors metrics have the highest values metrics. While the Complexity Classes, Cycle Packages and Complexity File metrics have the lowest average correlation between projects.

Table 6. Pearson correlation of the evolution of metrics with Discounts Features.

Metrics	Jmeter	POI	Tomcat
RFC	0,976	0,763	0,57
package	0,766	0,883	0,954
Line	0,969	0,944	0,507
Complexity Function	0,719	−0,011	0,954
Files	0,973	0,615	0,846
Cyclomatic Complexity	0,967	0,977	0,987
Classes	0,988	0,892	0,887
Cycle Packages	0,985	0,094	−0,982
NCLOC	0,94	0,962	0,546
Function	0,964	0,974	0,223
Complexity File	0,712	0,76	−0,286
Complexity Classes	−0,378	0,767	−0,205
Accessors	0,845	0,76	0,86

Analyzing the table above, we concluded that 53.84 % of the correlations with Feature metrics are strongly positive. What leads us to affirm that the human factor and the addition of new features has a significant impact on the maintainability of systems, because the more developers more metrics will rise and consequently more features will be added.

Analyzing the Tables 4, 5 and 6, we see how the correlation between metrics and the number of developers, features and bug are significant for all three projects. The complexity metrics of class Jmeter, complexity of function and Cycle package POI, function, complexity and complexity File Class Tomcat did not get a higher correlation than moderate positive/negative thus showing how important it is these variables.

The tables show that 52.99 % of the correlations are strong positive. This way it demonstrates a strong correlation of the three variables with the increase of the metrics.

The coefficient of packet cycle for Tomcat is relevant information which is strongly negative. To improve understanding of this data we will check the following graphics:

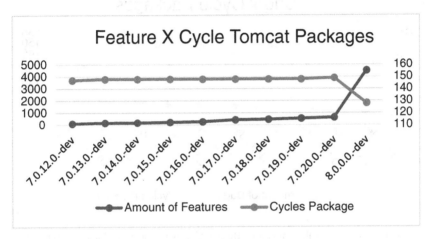

Fig. 2. Evolution of the corrected amount of features x Cycles package. (Color figure online)

The Fig. 2 represents the moment that the number of features increases while Cycles Package metric decrease in version 8.0.0.0-dev resulting in strongly negative correlation.

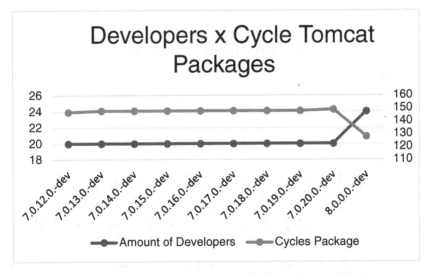

Fig. 3. Evolution of the amount of Developers x Cycles package. (Color figure online)

We verified that on increase number of developers in Tomcat project in version 8.0.0.0.-dev resulted in strong decrease of Cycles Package metric decrease, as shown in Fig. 3. The reducing Cycle Package metric due to increased number of developers shows that, in version 8.0.0.0.-dev, the developers were concerned adjust all project. As the exchange major version resulting in breaks compatibility from an earlier version [11], the version 8.0.0 demonstrate through Cycle Package metric that Tomcat project had considerable management change in last version.

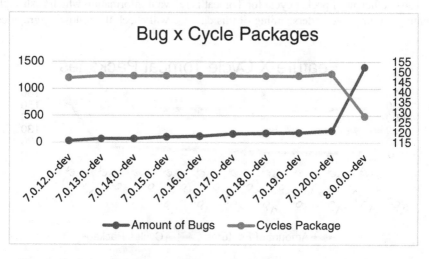

Fig. 4. Evolution of the amount of Bugs x Cycles package. (Color figure online)

The figure above, demonstrates that with increase of Amount of Bugs resulted in reduction of Cycle Package metric. In this way, resulting in a strong negative correlation.

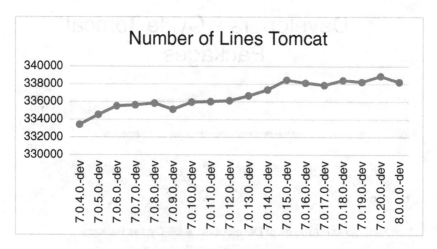

Fig. 5. Evolution of number of lines Tomcat project.

The evolution of quantity lines of Tomcat project, shows in Fig. 5, and despite increasing number of bugs and features described in Figs. 4 and 2, respectively. The number lines of Tomcat project obtained reduction the number of lines between 7.0.20.0.-dev and 8.0.0.0.-dev versions.

We are also seen, in Fig. 5, as the cyclomatic complexity oscillates in passages versions.

We can see in the graphics, where and with metrics behaved for the Pearson correlation was negative. It is observed that at the turn of version 7.x to 8.x there was a big refactoring of the Tomcat code, where you see a big increase in the number of developers, a large increase in bug fixing and adding new features. This shows us how projects tend to change completely when viewed version. One interesting thing is that despite the increase in developers, increasing numbers of bugs correction, increasing the amount of the amount of added features design lines followed by decreased cyclic dependency package, which demonstrate that the coupling of the project decreased thus making it easy to maintain.

Despite this complexity kept growing:

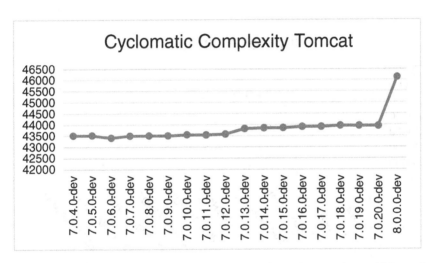

Fig. 6. Evolution of the number of lines in the x-axis versions of Tomcat have withdrawn from the Git repository and the y-axis number of lines.

As it be seen on Fig. 6, there has been a significant increase Cyclomatic Complexity metric in version 8.0.0.0.dev. That explains increase in organization of project.

In the other project also noticed this pattern of increasing complexity metrics as seen in the charts below.

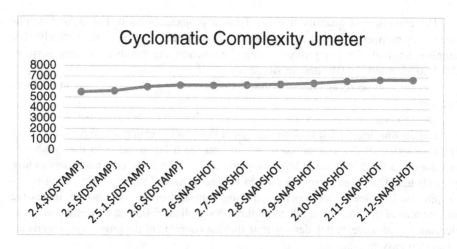

Fig. 7. Evolution of Cyclomatic Complexity Jmeter.

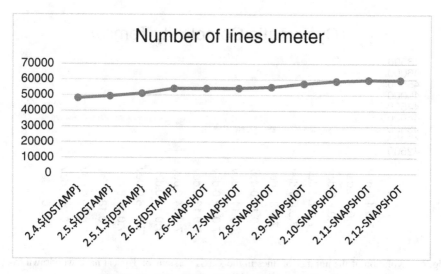

Fig. 8. Evolution of the number of lines of Jmeter.

In both the Cyclomatic Complexity metric and the number of lines metric in Jmeter project evolve in a linear way, as can be seen in Figs. 7 and 8. This shows us that when the number of lines increased in Jmeter, your Cyclomatic Complexity metric has a chance to increased too.

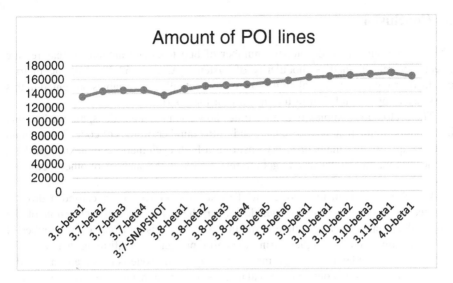

Fig. 9. Evolution of the number of lines of POI.

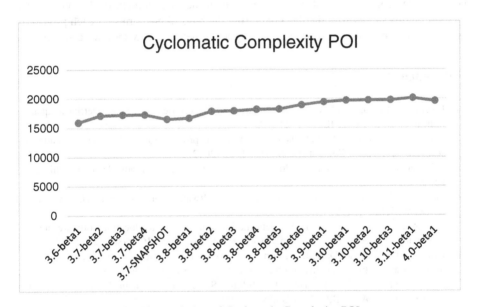

Fig. 10. Evolution of Cyclomatic Complexity POI.

As well as Jmeter project, there is a correlation between the number of lines and Cyclomatic Complexity metrics in POI, as can been in Figs. 9 and 10. This means that the Cyclomatic Complexity increased when number of lines increased too.

5 Conclusion

The variable amount of developers, number of bug fixes and amount of features are directly correlated metrics modularity of systems. As seen 52.99 % of the metrics obtained strong positive correlation showing that actually the variables presented in this work has a high correlation in the increase of the metrics.

The cyclomatic complexity demonstrates how complex the code is, but we find that its evolution in the plan has the same trend as the number of lines of code. To maintain software quality and maintainability ensuring a high cyclomatic complexity can be a metric without much relevance to getting responsibly to the analysis of other metrics of modularity.

We observe that in turn project to Tomcat version 7.x to 8.x caused a drop in package dependencies showing that developers have improved the cohesion of the project thus making it less coupled. We find also that this happens to the number of developers involved was higher during the entire project life reaffirming that codes for lines of high complexity and high maintainability of the code is drastically affected.

This work could not obtain historical metrics for cohesion and coupling metrics over a possible future work would be to check that the same projects adding these metrics and performing the correlation with the same variables, other work would be to analyze all commits the projects, since it was not possible at the first moment of the survey. An increase in this survey would collect data from the bugs' repositories for each project in order to verify how the bugs were appearing and when and how they were solved.

References

1. Basili, V., Briand, L., Melo, W.: A validarion of object-oriented deseign metrics as quality indicators. Softw. Eng. IEEE, 1–24 (1996)
2. Lee, D., Kim, B.C.: Motivations for open source project participation and decisions of software developers. Comput. Econ. **41**, 31–57 (2012)
3. Schröter, A., Zimmermann, T., Premraj, R., Zeller, A.: If your bug database could talk, pp. 2–4
4. Lehman, M., Ramil, J.: Metrics and laws of software evolution-the nineties view. In: Software Metrics (1997)
5. Turhan, B., Kocak, G., Bener, A.: Data mining source code for locating software bugs: a case study in telecommunication industry. Expert Syst. Appl. **36**, 9986–9990 (2009)
6. Briand, L.C., Wüst, J., Daly, J.W., Porter, D.V.: Exploring the relationships between design measures and software quality in object-oriented systems, pp. 1–36 (1999)
7. Liu, C., Yan, X., Fei, L., Han, J., Midkiff, S.: SOBER: statistical model-based bug localization. In: ACM SIGSOFT Software (2005)
8. Hassan, A.E.: The road ahead for mining software repositories. In: Frontiers of Software Maintenance, 2008, pp. 48–57. IEEE (2008)
9. Nagappan, N., Ball, T., Zeller, A.: Object-oriented metrics which predict maintainability. In: Proceedings of the 28th International Conference on Software Engineering (ICSE 2006) (2006)
10. McCabe, T.J.: A complexity measure. IEEE Trans. Softw. Eng. **SE-2** (1976)
11. Preston-Werner, T.: Semantic Versioning 2.0.0. http://www.semver.org

State of Art Survey On:
Large Scale Image Location Recognition

Nuno Amorim$^{(\boxtimes)}$ and Jorge Gustavo Rocha

Algoritmi Research Centre, University of Minho, 4710-057 Braga, Portugal
ntma90@gmail.com, jgr@di.uminho.pt

Abstract. Image location recognition is a well known process of retrieving the precise location from the contents of the photographs. New photographs are compared to a large geocoded database and the result is both the position and orientation.

This is an inexpensive location system, since it only needs a camera, present in all modern smart phones. It has some great advantages over GPS. It give us the orientation and works under occluded environments, making this system highly attractive to a wide variety of applications.

But at a large scale, this process is easily hindered by heavy weighted database representations, expensive computational operations and visually similar environments. As a consequence, low geocoding rates, inaccurate localization and slow queries are obtained.

In the past years, a variety of solutions have been proposed to address these challenges but we are yet to adopt one of them as the image location recognition solution.

In this paper we review and compare recent state of art advances on image geocoding algorithms focusing on the scalability of such solutions.

1 Introduction

Geocoding is the process by which we transform geographic information such street addresses, monuments names and other geographic data into precise coordinates. Image geocoding is a similar process, however, photographs are used as the source to describe environments that we want to locate. At the basis of image geocoding are *image features*. An *image feature* is piece of 2D information which encode visual data into a *feature descriptor* in a 2D pixel position. Both are used to find correspondences between photographs and environments.

Environments can be represented in two ways: with a database of photographs or 3D structures. The first is built by covering environments with several geotagged photographs. Image to image comparison (image matching) is applied to retrieve a set of most similar images to the query. The pose estimated may be approximate or precise if we perform two-view geometry.

The 3D database representation is built by modeling environments as point clouds. The *Structure from Motion* (SfM) is a well known algorithm used to create such point clouds. SfM is an incremental algorithm which builds a 3D sparse

© Springer International Publishing Switzerland 2016
O. Gervasi et al. (Eds.): ICCSA 2016, Part V, LNCS 9790, pp. 375–385, 2016.
DOI: 10.1007/978-3-319-42092-9_29

point cloud by computing 2D correspondences between photographs. Positive correspondences are triangulated into the 3D coordinate system. Overlap between photographs is required since the third coordinate (depth) is computed by simulating bifocal vision with photograph pairs. Geotagged photographs can be used to geocode the resultant point cloud. With the 3D structure present, correspondences are found between query photographs and features associated to the 3D structure. Positive matches are used along with pose estimation methods (i.e. based on RANSAC) to estimate a projection matrix which relates 2D images features to the 3D point cloud. This matrix can be decomposed into a position and orientation.

Image geocoding can be used as a complement to GPS. Location recognition through GPS requires the signal of three or more satellites. However, the signal is often weak or non existent in occluded areas. Taking photographs is a task available at any time. Modern smart phones, now omnipresent in our society, contain the required components for image geocoding: a camera to capture photographs, and a GSM or WiFi communication link to submit these photographs to a location service, if required. The flexibility offered by image geocoding makes this system highly attractive to the guidance services of pedestrians, robots, cars, among others.

However, at a large scale, image geocoding has a few drawbacks. First, both the variants to represent the database of this system are heavyweight. Image features and 3D points can easily scale to millions when representing a single environment. Secondly, the comparison of *feature descriptors* usually requires computationally expensive operations, which greatly slows image geocoding queries. But the worst case scenario is when image geocoding is confused by visually repetitive environments. The same photograph may be positively matched to similar locations and the problem is how to automatically decide which location is the correct one.

In the past years, many researchers have been struggling to solve or at least attenuate these challenges at large scale. Despite this effort, an unique and generic solution is yet to be found. With many algorithms proposed, we believe that compiling the state of art achievements on image geocoding will help this community to focus on the current issues of image-based location recognition.

However, image location recognition is a vast area of research. Therefore, this survey will only focus on algorithms which allow the calculus of the GPS pose (position and orientation) of new images using large scale image datasets.

1.1   Survey Organization

This survey is divided into four sections. First we compile available large scale datasets used for image geocoding in Sect. 2. We describe the process by which each dataset was built and provide relevant statistics. In Sect. 3 we review the state of art on large scale image geocoding algorithms. We start with early approaches based on image content retrieval and *bag-of-words* to more recent research which combine SfM point clouds with intelligent database compression techniques. Afterwards, we cross the two previous sections into Sect. 4, providing performance metrics achieved on large scale datasets using state of art image geocoding algorithms. To conclude this paper, in Sect. 5 we summarize this survey and offer a brief discussion and thoughts over the state of art achievements on large scale image geocoding.

2 State of Art Large Scale Datasets

A dataset of geocoded environments is at the core of every image-based location recognition system. Building such dataset requires a large amount of visually detailed photographs to represent environments. Every aspect counts in respect to quality: the aspect ratio, resolution, the quality of camera lens (lens distortion), the point of view in each photographs were taken, among others. Ideally we want a balance between these factors: a too much perfect scenario will not reflect the reality but a polluted dataset will not contain relevant visual data to represent environments. Additionally, we also want the dataset to be accurately geotagged (or a portion of it). Geocoding errors outputted by an image geocoding algorithm will clearly reflect low accuracy geotags when building the dataset. In short, as one may expect building large scale image geocoding datasets requires some effort. Thankfully, many researchers have focused on building such datasets and made them available to the image geocoding community.

In total, there are eight well known large scale datasets as listed in Table 1. We have tagged each dataset with DsN where N is the identification number that we associated to each dataset. This identifier will be used in the following sections whenever we are referring to a specific dataset.

Table 1. Large scale image datasets used in state of art research.

ID	Dataset	Availability	#Photos	Scope	#Cited
DS1[a]	Aachen	on request	4.4K	city	3
DS2[b]	Dubrovnik	public	6K	city	12
DS3[c]	Landmarks	public	200K	landmarks	2
DS4	Pittsburgh	work	254K	city	2
DS5[d]	Quad	public	6.5K	landmarks	1
DS6[e]	Rome	public	16K	landmarks	9
DS7[f]	San-Francisco	work	1.700K	city	4
DS8	Vienna	on request	1.3K	landmarks	5

[a]https://www.graphics.rwth-aachen.de/software/image-localization
[b]http://www.cs.cornell.edu/projects/p2f/
[c]https://www.landmark3d.codeplex.com/
[d]http://www.vision.soic.indiana.edu/projects/disco/
[e]http://www.cs.cornell.edu/projects/p2f/
[f]https://www.purl.stanford.edu/vn158kj2087

Two different categories of datasets are represented here: datasets built upon a controlled or uncontrolled collection of photographs. For the controlled collections, a specific set of equipment was used to take photographs and measure the GPS coordinates. In contrast, uncontrolled collections of photographs came from public imagery websites such as Flickr[1] and Panoramio[2]. Many different cameras are used and the GPS coordinates may not be accurate.

Related to controlled collections we have datasets DS1, DS3, DS4, DS5, DS7 and DS8. Dataset DS3 was built with a mobile vehicle equipped with several devices such as 360°lidar sensor, high definition cameras, GPS among others. Accurate panorama photographs were taken using this equipment and later cropped into perspective images. In DS5 and DS8 a single calibrated camera was used. On the other hand, DS1 was built from photographs taken with different cameras on over two years. Datasets DS4 and DS7 were built through Google Street View[3] panoramas. Panoramas were then cropped into perspective photographs with ground truth GPS from the associated metadata.

For uncontrolled datasets we have DS2 and DS6. Using the API of image sharing web sites, several photographs of varying resolution where downloaded. For ground truth comparison, the GPS tag associated to some photographs is used to geocode the database and to compute the GPS coordinates of query photographs.

3   Literature Review on Large Scale Image Geocoding

Image location recognition first started with image content retrieval techniques [1,2]. Images were queried to a database of photographs, and the response is a set of photographs similar to the query. However, this method required matches between the entire database and the query. To avoid this, [3–5] and many other research surveyed in [6] proposed a workflow based on bag-of-words [7] to accelerate the matching process. Image features are quantized into visual words and stored into a vocabulary tree. New images are queried to this vocabulary structure which outputs a similarity score between the query and database images.

If the 3D structure of environments is available, photographs can be matched directly to environments structure [8,9]. The geocoded database is composed by two layers of data: the 2D data extracted from database photographs and the respective projected 3D points. To efficiently compare new photographs to the database, data from the 2D layer is propagated into a vocabulary tree. The output is a set of most similar 2D descriptors and associated 3D points. The position and orientation are obtained by computing a projection matrix which validates the 2D-3D correspondences.

However, as stated in [10, p. 5], with the increase of database data, 2D-3D matching often delivers incorrect correspondences due to similar visual features

---

[1] https://www.flickr.com/.

[2] http://www.panoramio.com/.

[3] https://www.google.com/maps/streetview/.

from different parts of the world. In that regard, [10–12] proposed a complementary 3D-2D matching which matches sets of local 3D points to query images. The 3D points are ranked by their *visibility* (a confidence degree given by the number of images in which the point was successfully detected). In [10], point clouds are pre-processed into two unique clouds: a seed and compressed cloud. The first cloud contains a small set of most visible points, and thus, of high level confidence to be found in new images. The second cloud contains a larger set of points than the seed, but with lower visibility. When querying new photographs, they perform *approximate nearest neighbors* on the seed cloud (2D-3D) to find a small set of high level confidence correspondences. If insufficient correspondences are found, the initial set of correspondences is enriched by 3D-2D matching from the compressed cloud to query images. Similarly, [12,13] also use this 2D-3D-2D combination, however, their 2D-3D matching in search of seed correspondences is processed with a vocabulary tree. Despite the improvement on geocoding rates, both research report the need of high memory requirements.

The memory consumption of database representations is a huge concern when considering the large scale image geocoding. Image descriptors are usually heavyweight and structure from motion models can easily scale to millions of points for small environments. To control the growth of the database, [8] used Mean Shift Clustering [14] to reduce the number of *image descriptors* associated to each 3D point. Additionally, they use a greedy algorithm to select a minimal set of views to represent SfM points clouds. Several synthetically generated views are spawned around models to cover points of view not covered by the original photographs. Then, a minimal set of views is selected based on how much point coverage they offer towards the SfM point clouds.

Also based in view compression, [15] reduces the number of database images by computing the *minimum connected dominant set* of a visibility graph (a graph containing SfM connected views). The result is a group image subsets. For each subset, they built individual SfM point clouds and use approximate nearest neighbors to compute correspondences between new images and the subsets.

In contrast to compression based on views, research such [10,12] formulate this problem as a *K-cover* problem. The goal is to find the minimal set of 3D points, prioritized by their visibility, to represent SfM models. With a similar approach, [16] used mixed quadratic programing to reduce the point cloud into a subset of points. Later research [17,18] endorse that point *distinctiveness* should be used along with point visibility to refine the compressed representations.

Considering that modifying the 3D structure and compressing the amount of image descriptors may hinder the geocoding capability of SfM point clouds, [19] use classifiers to find correspondences between query images and the 3D structure. Their approach requires an approximately constant memory profile since they do not need image descriptors. However, they require an initial GPS estimation to avoid confusion in large datasets.

Supporting the same theory about the removal of structural data as a mean to compression, [20] introduce the concept of hyperpoints into point clouds. In this research, image matching is completely replaced by queries on a fine vocabulary tree [21]. This upgraded version requires an higher storage for the vocabulary but delivers more accuracy when querying new photographs on databases of billion of images [22,23]. Database descriptors are not stored (as they are only used for image matching) reducing a large amount of memory required to support the database.

Recent image geocoding methods are based on a database of SfM models. In [24] shows that a database of images (without a 3D structure) together with a vocabulary tree is also effective for large scale image localization. Assuming that vocabulary tree queries deliver coarse matches, Cao et.al. developed a pose estimation workflow based on the connectivity of image graphs to reinforce the confidence of vocabulary queries. Two relevant characteristics of their workflow is that they do not require a large amount of stored image descriptors and do not store the 3D structure built on database images.

4   Discussing Performance Metrics

Summing up the literature review, the current challenge of image geocoding research is how to achieve optimal geocoding rates with compact database representations. However, other performance metrics such as geocoding speed and accuracy are also relevant to evaluate the quality and how generic an image geocoding algorithm is. Unfortunately, some metrics are missing in state of art experimental results.

To begin with, Table 2 provides the amount of queried images and geocoding rate obtained when applying state of art algorithms to large scale datasets from Sect. 2. To simplify this section, we only provide the best performance rates achieved in each paper.

Summarizing Table 2, the number of query images per dataset used varies differently. The minimum amount used was 266 for DS8 and the maximum of 24000 for DS4. Geocoding rates vary from 68.40 % to 100.00 % for all datasets. DS2 and DS6 are the most preferable benchmark datasets with the majority of citation and best geocoding rates. An interesting fact is that the geocoding rate is not progressively positive throughout consecutive research for the same datasets. In [17] the point cloud respective to DS2 is greatly reduced into 2.63 % of its original size, but the geocoding rate is 6.62 % lower than the one achieved in [10]. The same phenomenon is seen in [16,18] for DS2 and DS6. Their compression is based on the selection of minimal subsets of points to refine the database content, and this may be an indicator that this database representation may not be ideal for image geocoding purposes. In contrast, the fine vocabulary representation of [20] marks the best geocoding rates for DS3 and DS7 while reducing the database size.

As we stated before, the majority of experimental results does not register the achieved geocoding accuracy as in Table 3. Naturally, this metric would deliver ambiguous results on datasets DS2 and DS6 which were built under uncontrolled environments. However, researchers chose to emphasize the geocoding rate over other performance indicators for the remaining datasets. The measured accuracy

Table 2. Geocoding rate achieved by state of art research on large scale datasets, sorted by publishing year. Bellow each dataset ID is the number of queries used.

Paper	DS1 369	DS2 800	DS3 10000	DS4 24000	DS5 348	DS6 1000	DS7 803	DS8 266
[8]	-	-	-	-	-	-	-	62.03 %
[11]	-	94.13 %	-	-	-	92.10 %	-	76.69 %
[9]	-	97.75 %	-	-	-	97.46 %	-	77.78 %
[10]	-	100.00 %	-	-	68.40 %	99.70 %	50.20 %	-
[13]	-	98.50 %	-	-	-	97.70 %	-	82.33 %
[12]	-	99.44 %	-	-	-	99.15 %	-	82.71 %
[15]	-	91.75 %	-	-	-	-	-	-
[16]	-	91.15 %	-	-	-	80.60 %	-	-
[25]	-	-	-	78.00 %	-	-	78.00 %	-
[26]	-	-	-	91.00 %	-	-	89.00 %	-
[17]	73.98 %	93.38 %	81.45 %	-	-	-	-	-
[19]	-	98.97 %	-	-	-	97.64 %	-	-
[27]	-	99.75 %	-	-	-	-	-	-
[24]	89.16 %	99.50 %	-	-	-	99.70 %	-	-
[20]	-	-	91.10 %	-	-	-	95.00 %	-
[18]	72.09 %	94.00 %	-	-	-	92.40 %	-	-

Table 3. Geocoding accuracy rate achieved by state of art research on large scale datasets. Percentages correspond to errors lower than 18 meters.

Paper	DS1 369	DS2 800	DS3 10000	DS4 24000	DS5 348	DS6 1000	DS7 803	DS8 266
[11]	-	87.00 %	-	-	-	-	-	-
[9]	-	87.50 %	-	-	-	-	-	-
[10]	-	-	-	-	90.00 %	-	-	-
[12]	-	88.00 %	-	-	-	-	-	-
[13]	-	75.00 %	-	-	-	-	-	-
[27]	-	96.37 %	-	-	-	-	-	-
[20]	-	-	96.90 %	-	-	-	61.90 %	-

in state of art is usually represented as a percentage over the successful geocoded photographs. Other representations are used such as mean average precision (mAP) [24], but we only report accuracy as a percentage to keep the table visually simple. All accuracy rates correspond to errors below 18 m.

Similar to accuracy, few are the research which report the achieved geocoding speed as shown in Table 4. In [8] reported real time queries if a new image match the first of the ten occurrences retrieved from the vocabulary tree query.

Table 4. State of art geocoding speed per query achieved in large scale datasets. Values are in seconds.

Paper	DS1 369	DS2 800	DS3 10000	DS4 24000	DS5 348	DS6 1000	DS7 803	DS8 266
[8]	-	-	-	-	-	-	-	[0.077, 0.302]
[11]	-	-	-	-	-	[0.910, 2.930]	-	[0.550, 1.960]
[9]	-	[0.280, 1.700]	-	-	-	[0.250, 1.660]	-	[0.460, 2.430]
[12]	-	[0.250, 0.560]	-	-	-	[0.280, 2.140]	-	[0.270, 0.520]
[13]	-	[0.250, 0.510]	-	-	-	[0.270, 0.610]	-	[0.400, 0.490]
[20]	-	-	≈0.792	-	-	-	-	-

However, their geocoding rate is low 62.03 %). In [10] claim to require few seconds per query for DS2 and DS3. Other research either report the average time per query [20] or an interval representing the time required for a successful pose or rejection.

With all performance metrics presented, in Table 5 we provide the best geocoding rate for each large scale dataset and associated accuracy rate and geocoding speed. This last table aims to mark the current state of art progress on performance achieved for comparison future research.

5   Conclusions

In this paper the state of art image location recognition algorithms focused on large scale was reviewed. We started by providing the datasets used in most of state of art research and how they were built. Relevant statistics such as the number of photographs contained in each dataset, the number of times cited and the process by which they were built are provided.

Table 5. Best geocoding rate achieved in each state of art dataset and associated accuracy rate, query speed and research. Colored in green and red we have the highest and lowest geocoding rate achieved in all datasets.

Dataset	Geocoding Rate	Accuracy Rate	Query Speed	Paper
DS1	88.08 %	—	—	[17]
DS2	100.00 %	—	—	[10]
DS3	91.10 %	96.90 %	≈0.792	[20]
DS4	91.00 %	—	—	[26]
DS5	68.40 %	90.00 %	—	[10]
DS6	99.70 %	—	—	[10, 24]
DS7	95.00 %	61.90 %	—	[20]
DS8	82.71 %	—	[0.270, 0.520]	[12]

Following the datasets review, an overview on large scale image geocoding pipelines was provided. We started with early algorithms based on content retrieval and bag-of-words to accelerate geocoding queries. Then direct (2D-3D) and indirect (3D-2D) matching was reviewed. Since different database representations require large memory overheads, an overview of a few techniques used to compress the geocoded database was included. Techniques based on a minimal selection of views and 3D points were mentioned and also other representations which avoid removing structural data from the database.

The performance metrics achieved in the reviewed state of art using large scale datasets was presented and discussed. We addressed both the geocoding rate, accuracy and speed, although the majority of researchers are focusing only the geocoding rate rather than other relevant performance metrics such as geocoding speed and accuracy.

This survey reviews the more significant research studies that can provide solutions to the image location problem. The results are interesting and promising, but none of these solutions is able to scale to deal with the imagery of an entire city, for example.

All performance metrics should be considered in the literature rather than just focusing on achieving the optimal balance between geocoding rate and database compression. If image geocoding ever comes to be a wide available location recognition service, users will not tolerate slow queries and inaccurate positioning.

Furthermore, as we previously said, we only registered the best geocoding rates for each algorithm applied to each dataset. The majority of these algorithms are parametrized, and these parameters vary from dataset to dataset resulting in optimal geocoding rates. Therefore it would be interesting to perform a consistent comparison between image geocoding pipelines. However, such comparison requires a deeper study and analysis on state of art image location recognition at large scale.

Researchers should evaluate how generic their image geocoding solutions are. We noticed that the same algorithm applied to different datasets offer different experimental results. Therefore, can we achieve the same performance on other unused datasets? Are we required to tune algorithms to a target dataset? And how do we adapt these algorithms to the increase size and diversity of the database? These are questions yet to be answered.

References

1. Robertson, D., Cipolla, R.: An image-based system for urban navigation. In: Proceedings of British Machine Vision Conference (BMVC 2004), vol. 1, pp. 819–828 (2004)
2. Zhang, W., Kosecka, J.: Image based localization in urban environments. In: Third International Symposium on 3D Data Processing, Visualization, and Transmission, pp. 33–40 (2006)
3. Schindler, G., Brown, M., Szeliski, R.: City-scale location recognition. In: IEEE conference on computer vision and pattern recognition (CVPR 2007), pp. 1–7 (2007)

4. Li, Y., Crandall, D.J., Huttenlocher, D.P.: Landmark classification in large-scale image collections. In: Proceedings of the IEEE International Conference on Computer Vision, pp. 1957–1964 (2009)
5. Knopp, J., Sivic, J., Pajdla, T.: Avoiding confusing features in place recognition. In: Daniilidis, K., Maragos, P., Paragios, N. (eds.) ECCV 2010, Part I. LNCS, vol. 6311, pp. 748–761. Springer, Heidelberg (2010). doi:10.1007/978-3-642-15549-9_54
6. Bhattacharya, P., Gavrilova, M.: A survey of landmark recognition using the bag-of-words framework. In: Plemenos, D., Miaoulis, G. (eds.) Intelligent Computer Graphics, pp. 243–263. Springer, Berlin Heidelberg (2012)
7. Nistér, D., Stew, H.: Scalable recognition with a vocabulary tree. In: IEEE Computer Society Conference on Computer Vision and Pattern Recognition, pp. 2161–2168 (2006)
8. Irschara, A., Zach, C., Frahm, J.-M., Bischof, H.: From structure-from-motion point clouds to fast location recognition. In: IEEE Conference on Computer Vision and Pattern Recognition, pp. 2599–2606 (2009)
9. Sattler, T., Leibe, B., Kobbelt, L.: Fast image-based localization using direct 2D-to-3D matching. In: Proceedings of the IEEE International Conference on Computer Vision, pp. 667–674 (2011)
10. Li, Y., Snavely, N., Huttenlocher, D., Fua, P.: Worldwide pose estimation using 3D point clouds. In: Fitzgibbon, A., Lazebnik, S., Perona, P., Sato, Y., Schmid, C. (eds.) ECCV 2012, Part I. LNCS, vol. 7572, pp. 15–29. Springer, Heidelberg (2012). doi:10.1007/978-3-642-33718-5_2
11. Li, Y., Snavely, N., Huttenlocher, D.P.: Location recognition using prioritized feature matching. In: Daniilidis, K., Maragos, P., Paragios, N. (eds.) ECCV 2010, Part II. LNCS, vol. 6312, pp. 791–804. Springer, Heidelberg (2010). doi:10.1007/978-3-642-15552-9_57
12. Sattler, T., Leibe, B., Kobbelt, L.: Improving image-based localization by active correspondence search. In: Fitzgibbon, A., Lazebnik, S., Perona, P., Sato, Y., Schmid, C. (eds.) ECCV 2012, Part I. LNCS, vol. 7572, pp. 752–765. Springer, Heidelberg (2012). doi:10.1007/978-3-642-33718-5_54
13. Choudhary, S., Narayanan, P.J.: Visibility probability structure from SfM datasets and applications. In: Fitzgibbon, A., Lazebnik, S., Perona, P., Sato, Y., Schmid, C. (eds.) ECCV 2012, Part V. LNCS, vol. 7576, pp. 130–143. Springer, Heidelberg (2012)
14. Comaniciu, D., Meer, P., Member, S.: Mean shift: a robust approach toward feature space analysis. IEEE Trans. Pattern Anal. Mach. Intell. 24(5), 603–619 (2002)
15. Havlena, M., Hartmann, W., Schindler, K.: Optimal reduction of large image databases for location recognition. In: Proceedings of the IEEE International Conference on Computer Vision, pp. 676–683 (2013)
16. Park, H.S., Wang, Y., Nurvitadhi, E., Hoe, J.C., Sheikh, Y., Chen, M.: 3D point cloud reduction using mixed-integer quadratic programming. In: IEEE Conference on Computer Vision and Pattern Recognition Workshops, pp. 229–236 (2013)
17. Cao, S., Snavely, N.: Minimal scene descriptions from structure from motion models. In: 2014 IEEE Conference on Computer Vision and Pattern Recognition, pp. 461–468 (2014)
18. Cheng, W., Lin, W., Sun, M.-T.: 3D point cloud simplification for image-based localization. In: IEEE International Conference on Multimedia & Expo Workshops (ICMEW), pp. 1–6 (2015)
19. Donoser, M., Schmalstieg, D.: Discriminative feature-to-point matching in image-based localization. In: 2014 IEEE Conference on Computer Vision and Pattern Recognition (CVPR), pp. 516–523 (2014) doi:10.1109/CVPR.2014.73

20. Sattler, T., Havlena, M., Radenovic, F., Schindler, K., Pollefeys, M.: Hyperpoints and fine vocabularies for large-scale location recognition. In: The IEEE International Conference on Computer Vision (ICCV), December 2015

21. Mikulik, A., Perdoch, M., Chum, O., Matas, J.: Learning vocabularies over a fine quantization. Int. J. Comput. Vis. **103**(1), 163–175 (2012). doi:10.1007/s11263-012-0600-1

22. Stewénius, H., Gunderson, S.H., Pilet, J.: Size matters: exhaustive geometric verification for image retrieval accepted for ECCV 2012. In: Fitzgibbon, A., Lazebnik, S., Perona, P., Sato, Y., Schmid, C. (eds.) ECCV 2012, Part II. LNCS, vol. 7573, pp. 674–687. Springer, Heidelberg (2012). doi:10.1007/978-3-642-33709-3_48

23. Havlena, M., Schindler, K.: VocMatch: efficient multiview correspondence for structure from motion. In: Fleet, D., Pajdla, T., Schiele, B., Tuytelaars, T. (eds.) ECCV 2014, Part III. LNCS, vol. 8691, pp. 46–60. Springer, Heidelberg (2014)

24. Cao, S., Snavely, N.: Graph-based discriminative learning for location recognition. In: Proceedings of the IEEE Computer Society Conference on Computer Vision and Pattern Recognition, pp. 700–707 (2015)

25. Torii, A., Sivic, J., Pajdla, T., Okutomi, M.: Visual place recognition with repetitive structures. In: Proceedings of the IEEE Computer Society Conference on Computer Vision and Pattern Recognition, pp. 883–890 (2013). doi:10.1109/CVPR.2013.119

26. Arandjelović, R., Zisserman, A.: DisLocation: scalable descriptor distinctiveness for location recognition. In: Cremers, D., Reid, I., Saito, H., Yang, M.-H. (eds.) ACCV 2014. LNCS, vol. 9006, pp. 188–204. Springer, Heidelberg (2015)

27. Svarm, L., Enqvist, O.: Accurate localization and pose estimation for large 3D models. In: 2014 IEEE Conference on Computer Vision and Pattern Recognition (CVPR), pp. 532–539 (2014)

A Felder and Silverman Learning Styles Model Based Personalization Approach to Recommend Learning Objects

Birol Ciloglugil[1](✉) and Mustafa Murat Inceoglu[2]

[1] Department of Computer Engineering,
Ege University, 35100 Bornova, Izmir, Turkey
`birol.ciloglugil@ege.edu.tr`
[2] Department of Computer Education and Instructional Technology,
Ege University, 35100 Bornova, Izmir, Turkey
`mustafa.inceoglu@ege.edu.tr`

Abstract. In this paper, a new algorithmic personalization approach based on Felder and Silverman learning styles model is presented. The proposed approach uses learning objects modeled with the IEEE LOM metadata standard, which serves as the main standard for representation of learning objects' metadata. Personalization is provided with two steps in the proposed approach. At the first step, each learning object is evaluated by taking into account how values of IEEE LOM metadata elements match each dimension of Felder and Silverman learning styles model. The second step involves recommending appropriate learning objects to learners. Four weight values are calculated for each learning object, describing how related the learner and the learning object in question is at each dimension of Felder and Silverman learning styles model. Then, weight values for each dimension is combined by using Manhattan distance metric to provide a single weight value as a fitness function representing the general relatedness of the learner and the learning object. Results of the personalization approach can be used to recommend learning objects ordered according to their weight values to the learners. An example scenario illustrating the proposed approach is provided, as well as a discussion of current limitations and future work directions.

Keywords: E-Learning · Personalization · Felder and silverman learning styles model · Learning objects

1 Introduction

Personalization in e-learning systems is generally based on trying to match learners and learning materials. In order to achieve that, both learners and learning materials should be modeled by standards. The main standards for learning materials are SCORM packaging standard [1] for creating reusable learning objects and IEEE LOM standard [2] for annotating learning objects with appropriate metadata about their content. These standards make it possible for e-learning systems to process learning resources automatically for personalization

© Springer International Publishing Switzerland 2016
O. Gervasi et al. (Eds.): ICCSA 2016, Part V, LNCS 9790, pp. 386–397, 2016.
DOI: 10.1007/978-3-319-42092-9_30

purposes. On the other hand, learners also must be modeled by some standards that provide information about the way they learn better [3], so that personalization can be applied. Learning style models provide a good basis for modeling individual differences of the way the learners learn. There are 71 learning style models in the literature to model individual differences of learners [4]. Felder and Silverman learning styles model [5], Kolb's Experiential Learning Theory [6] and Honey and Mumfold learning styles model [7] are the most commonly used ones. According to [8], among these models, Felder and Silverman learning styles model is the most suitable one to be used in adaptive e-learning systems to provide learning styles based adaptation. Another reason contributing to selection of Felder and Silverman learning styles model to be used in the proposed approach is that it is designed especially for engineering students [5].

Felder and Silverman learning styles model (FSLSM) is a four-dimensional model where each dimension is based on the way learners perceive, receive, process and understand information and can be expressed as a linguistic variable. These dimensions are sensing/intuitive, visual/verbal, active/reflective and sequential/global. In order to have their FSLSM information, learners should take the Index of Learning Styles (ILS) scale [9] which contains 44 questions in total, with 11 questions for each dimension. Each question provides two options to learners and the answers are graded as -1 or $+1$. Therefore, a learner's learning style for each dimension can be expressed as an odd number between -11 and $+11$, expressing how strong the learner's tendency for that dimension is. FSLSM can also be defined as the Cartesian product of the linguistic variables representing its dimension. For example, a learner can be intuitive, visual, reflective and global, while another learner can be sensing, verbal, reflective and sequential. This representation provides a polarized way to classify learners into one of the two categories for each FSLSM dimension. Learners' FSLSM information have been modeled this way for the personalization approach presented in this paper.

In this study, the standards mentioned above have been used as the base of the proposed personalization approach. In Sect. 2, a review of personalization resources in e-learning systems is given with the focus being on the usage of IEEE LOM metadata standard to provide personalization. Section 3 features details of the proposed personalization approach with an example scenario. Section 4 concludes the paper with discussion of the proposed approach and future work directions.

2   How to Provide Personalization in E-Learning Systems with IEEE LOM Metadata Standard

The relationship between learning styles models and IEEE LOM v1.0 metadata elements by taking into account their set of values have been analyzed in detail by [10, 11]. [10] uses learning resource type metadata element of IEEE LOM v1.0 standard in their personalization approach to match learners with appropriate learning objects. 76 relations between IEEE LOM metadata elements and different personalization parameters, that can be used to provide personalization, have been identified by [11]. 28 of these relations are with FSLSM. 20 of the 28 relations are associated with active/reflective dimension of FSLSM, while 6 are related to visual/verbal dimension and 2 are affiliated with sequential/global dimension. No relations has been detected related to sensing/intuitive dimension of FSLSM. Relations detected by [11] can be summarized as given in Table 1. Table 2 features full set of values of the metadata elements that are related to FSLSM dimensions [2]. These relations can be used as a resource for offering different personalization approaches.

"1.7 Structure" metadata element can have five different values; atomic, collection, networked, linear and hierarchical; as shown in Table 2, but only two of these values can be used for personalization purposes in only one FSLSM

Table 1. IEEE LOM v1.0 Matadata Elements and their set of values related to FSLSM Dimensions [11].

IEEE LOM v1.0 Metadata Element	Set of values related to FSLSM Dimensions	Related FSLSM Dimension
1.7 Structure	{hierarchical}	Global
1.7 Structure	{linear}	Sequential
4.1 Format	{video/mpeg, video/...}	Visual
5.1 Interactivity Type	{active}	Active
5.1 Interactivity Type	{expositive}	Reflective
5.2 Learning Resource Type	{diagram, figure, graph}	Visual
5.2 Learning Resource Type	{narrative text, lecture}	Verbal
5.3 Interactivity Level	{very low, low, medium, high, very high}	Active
5.3 Interactivity Level	{very low, low, medium, high, very high}	Reflective

Table 2. IEEE LOM v1.0 Matadata Elements and their full set of values [2]

IEEE LOM v1.0 Metadata Element	Full set of values
1.7 Structure	{atomic, collection, networked, linear, hierarchical}
4.1 Format	{video/mpeg, text/html, application/x-toolbook, ...}
5.1 Interactivity Type	{active, expositive, mixed}
5.2 Learning Resource Type	{exercise, simulation, questionnaire, diagram, figure, graph, index, slide, table, narrative text, exam, experiment, problem statement, self assessment, lecture}
5.3 Interactivity Level	{very low, low, medium, high, very high}

dimension [11]. If "1.7 Structure" metadata of a learning object has the value "hierarchical", it is associated with global learners. If "1.7 Structure" is "linear", the learning object is better suited to sequential learners.

"4.1 Format" metadata element can have various values, but only "video" types have been detected to be associated with visual learners in [11].

"5.1 Interactivity Type" can be active, expositive or mixed. Learning objects with "active" interactivity type data are targeted to active learners, while "expositive" ones are recommended to reflective learners. "Mixed" interactivity type data determines that the learning object follows a mixed approach and can be used for both active and reflective learners, thus it does not provide meaningful information for personalization purposes.

"5.3 Interactivity Level" determines the interactivity level for learning objects' interactivity type. It provides a five scale degree and can have the values; very low, low, medium, high or very high. It is used for both active and reflective learners defining how "active" or "expositive" the learning object is.

As given in Table 1, five IEEE LOM metadata elements are related to FSLSM in [11]. Only one of them, learning resource type, is used by [10] for personalization. However, an effective approach to model and match learners and learning objects (LOs) has been provided by [10].

Five IEEE LOM metadata elements related to FSLSM dimensions have been examined to determine their usability for the proposed personalization approach. "1.7 Structure" element of IEEE LOM standard is involved with "1.8 Aggregation Level" metadata element. Together they describe how the learning resources such as assets and shareable content objects (SCOs) in LOs are organized. An LO may contain a single asset/SCO or various assets/SCOs for a lesson or a whole course in a hierarchical order depending on the aggregation level value. The proposed approach supports LOs containing assets and SCOs only, because course level LOs contain a lot of assets and SCOs and therefore, it is not ideal to recommend an LO containing a whole course to the learner at once. However, [12] suggests working with course level LOs, since their personalization approach is integrated with a learning management system (LMS). It has been decided not to use course level LOs and the "1.7 Structure" metadata element for not limiting the proposed approach's usage with LMSs.

"1.4 Format" metadata element's usage has been extended to include learning resources with "text" format, because "text" format is generally associated with verbal learners. Since a polarized approach is used for learners' FSLSM data modeling, using interactivity type for personalization is sufficient. Interactivity level is used for modeling the degree of interactivity type, and is not relevant in a polarized approach. Therefore, format, interactivity type and learning resource type have been detected as the three metadata elements to be used in the proposed personalization approach.

As stated in Table 2, IEEE LOM v1.0 metadata element "5.2 Learning Resource Type" value set includes 15 elements. Relationship of each element in the value set with FSLSM dimensions are given in Table 3. As a continuation of the polarized approach used for FSLSM modeling, the matching of IEEE LOM metadata tags with FSLSM is annotated the same way. Thus, the relationship value is represented with 0 if the learning resource type in hand is better suited to active, sensing, visual and sequential learners; and represented with 1 if the learning resource type in hand is better suited to reflective, intuitive, verbal and global learners. The relationship values in Table 3 are determined with the help of various resources in the literature [2,10,12,13]. Table 3 can be analyzed both row-wise and column-wise. Column-wise analysis is based on FSLSM dimensions' relationship with learning resource type values. For example, for active/reflective dimension of FSLSM, active learners like learning by doing things actively [13] and therefore, can be associated with exercise, simulation, questionnaire, exam, experiment, problem statement and self assessment which are represented with 0's in Table 3. On the other hand, reflective learners like thinking and observing more than being active. Thus, reflective learners are more related to learning objects with learning resource type values diagram, figure, graph, index, slide, table, narrative text and lecture, which are represented with 1's in Table 3.

Table 3. IEEE LOM v1.0 Metadata Element "Learning Resource Type" value set and each value's relationship with FSLSM dimensions in the proposed personalization approach.

Learning Resource Type value	Active/Reflective dimension	Sensing/Intuitive dimension	Visual/Verbal dimension	Sequential/Global dimension
Exercise	0	0	0	0
Simulation	0	0	0	0
Questionnaire	0	1	0	0
Diagram	1	0	0	0
Figure	1	0	0	0
Graph	1	0	0	0
Index	1	0	0	1
Slide	1	1	0	0
Table	1	0	0	0
Narrative text	1	1	1	0
Exam	0	1	0	0
Experiment	0	0	0	0
Problem statement	0	1	0	0
Self assessment	0	1	0	0
Lecture	1	1	1	0

Row-wise analysis of Table 3 shows how each learning resource type value is related to each FSLSM dimension. As a row-wise example, the first two rows of Table 3 include "exercise" and "simulation" values. These values are related to active, sensing, visual and sequential dimensions of FSLSM [10] that are represented with 0's. "Questionnaire" and "problem statement" labels are related to active, intuitive, visual and sequential learners [2] that are represented with 0,

1, 0 and 0 for each dimension. The values for other learning resource type labels have been assigned in accordance with [2,10,12,13], as well.

3 The Proposed Personalization Approach Based on IEEE LOM Metadata Standard and Felder and Silverman Learning Styles Model

The proposed personalization approach tries to match learners and learning objects by using Felder and Silverman learning styles model and IEEE LOM v1.0 metadata standard, respectively. The use of IEEE LOM metadata elements in personalized e-learning environments for personalization purposes has been investigated and learning resource type, interactivity type and format has been detected as the most suitable IEEE LOM metadata elements to be used in the proposed approach. Learning resource type is the most commonly used element used in [10–12]. However, it should be noted that [12] uses CLEO Lab application profile [14] that extends the value set of learning resource type. Using application profiles is ideal for developing standalone applications offering specialized solutions [15], but in order to offer a generalized solution that works with more LOs, IEEE LOM metadata elements' default value sets have been used in the proposed approach without any extensions.

The proposed personalization approach consists of two steps. At the first step, each learning object is graded by taking into account how values of the mentioned IEEE LOM metadata elements fit each dimension of Felder and Silverman learning styles model. Active/reflective dimension is associated with two IEEE LOM metadata elements, namely learning resource type and interactivity type. Visual/verbal dimension is also related with two elements; learning resource type and format. Sensing/intuitive and sequential/global dimensions work with learning resource type as it is the only metadata element identified for these dimensions. Two steps of the proposed personalization approach is examined in detail at the following subsections.

3.1 Learning Object Evaluation with IEEE LOM Metadata Standard

Table 4 includes properties of a learner $l1$ and a learning object $lo1$. An example scenario involving the learner $l1$ and the learning object $lo1$ will be used to demonstrate the proposed personalization approach.

A learner l's FSLSM information can be modeled as given in Eq. (1) [10].

$$FSLSM_l = << T_1, e_1 >< T_2, e_2 >< T_3, e_3 >< T_4, e_4 >> \tag{1}$$

The "T_1", "T_2", "T_3" and "T_4" values represent FSLSM dimensions; "AR" active/reflective, "SI" sensing/intuitive, "VV" visual/verbal, and "SG" sequential/global, respectively. The "e_i" variables can be 0 or 1 to represent learner l's learning style for each FSLSM dimension. A polarized approach has been

Table 4. An example learner *l1* and learning object *lo1* with their properties used for personalization.

Example	Properties used for personalization
Learner *l1*	$FSLSM_{l1}$: "AR": -1, "SI": 3, "VV": -7, "SG": 9
Learning object *lo1*	learningResourceType: exercise, interactivityType: active, interactivityLevel: high, format: video

used to evaluate learners, thus, if the FSLSM dimension value is negative, the learner belongs to the first FSLSM dimension, namely active, sensing, visual and sequential, represented with the value 0; while the value is 1 for positive FSLSM dimension data. Therefore, learner *l1* in Table 4 is active, intuitive, visual and global that can be represented as given below with Eq. (1):

$$FSLSM_{l1} =<< \text{``AR''},0 >< \text{``SI''},1 >< \text{``VV''},0 >< \text{``SG''},1 >>$$

Equation (2) is used to represent a learning object *lo*'s score (*"$LOScore_{lo}$"*) according to its metadata elements' values with respect to each FSLSM dimension [10].

$$LOScore_{lo} =<< T_1, f_1 >< T_2, f_2 >< T_3, f_3 >< T_4, f_4 >> \qquad (2)$$

Learning resource type is the only metadata element used in [10], but three metadata elements identified by [11] have been used in the proposed approach to provide better recommendation of learning objects to learners. Therefore, Eq. (2) is the total "$LOScore_{lo}$" value of a learning object *lo*, which is calculated by using metadata elements learning resource type (*"LRT_{lo}"*), interactivity type (*"IT_{lo}"*) and format (*"F_{lo}"*) that are represented in Eqs. (3), (4) and (5), respectively.

$$LRT_{lo} =<< T_1, g_1 >< T_2, g_2 >< T_3, g_3 >< T_4, g_4 >> \qquad (3)$$

$$IT_{lo} =< T_1, h_1 > \qquad (4)$$

$$F_{lo} =< T_3, i_3 > \qquad (5)$$

Equation (3) represents the learning resource type element (*"LRT_{lo}"*) with a value for each FSLSM dimension. Equation (4) uses interactivity type to calculate the "h_1" value which is related to only active/reflective dimension of FSLSM. Equation (5) is associated with the format metadata element and format element only affects visual/verbal dimension of FSLSM.

As given in Table 4, the learning object *lo1*'s learning resource type value is "exercise", therefore it is associated with active, sensing, visual and sequential learners [10] and can be represented as given below with the corresponding values in Table 3:

$$LRT_{lo1} =<< \text{``}AR\text{''},0 >< \text{``}SI\text{''},0 >< \text{``}VV\text{''},0 >< \text{``}SG\text{''},0 >>$$

The "h_1" value in Eq. (4) is calculated as given in Eq. (6). Interactivity type ("$InteractivityType_{lo}$") can have the values "active", "expositive" and "mixed". If the value is "mixed", this learning object can be used for both active and reflective learners. Thus, it cannot be used to provide personalization and "h_1" is set as "-1" to determine the invalid input value. According to Eq. (6), since "$InteractivityType_{lo}$" value of the learning object $lo1$ is active, "h_1" of "IT_{lo}" is set as 0.

$$h_1 = \begin{cases} 0, & InteractivityType_{lo} = \text{``active''} \\ 1, & InteractivityType_{lo} = \text{``expositive''} \\ -1, & InteractivityType_{lo} = \text{``mixed''} \end{cases} \quad (6)$$

The "i_3" value in Eq. (5) is calculated as given in Eq. (7). Format value can have many different values, only "video" and "text" can be used to determine visual and verbal learners respectively. If format element doesn't have one of these two values, "i_3" is set to "-1" to determine that format value of this learning object cannot be used for personalization. "i_3" value of "F_{lo1}" for the learning object $lo1$ in Table 4 is 0, because videos are associated with visual learners.

$$i_3 = \begin{cases} 0, & Format_{lo} = \text{``video''} \\ 1, & Format_{lo} = \text{``text''} \\ -1, & Format_{lo} \neq \text{``video''} \wedge Format_{lo} \neq \text{``text''} \end{cases} \quad (7)$$

Since Eqs. (3)–(5) have been calculated for the learning object $lo1$ in Table 4, they can be used to form the total "$LOScore_{lo}$" value given in Eq. (2). "f_1" value is calculated with the help of Eqs. (3) and (4) as given in Eq. (8). Among the two metadata elements related to active/reflective dimension of FSLSM, interactivity type has a priority over learning resource type when they have different values. If "h_1" value, that is associated with interactivity type, is not "-1", "h_1" is assigned to "f_1". Otherwise, "g_1" is assigned to "f_1". Since both "g_1" and "h_1" are equal to 0 for the learning object lo_1 given in Table 4, "f_1" is also 0.

$$f_1 = \begin{cases} g_1, & h_1 = -1 \\ h_1, & h_1 \neq -1 \end{cases} \quad (8)$$

The "f_3" value is calculated with the help of Eqs. (3) and (5) as given in Eq. (9). Among the two metadata elements related to visual/verbal dimension of FSLSM, format has a priority over learning resource type when they have different values. If "i_3" value, that is associated with format, is not "-1", "i_3" is assigned to "f_3". Otherwise, "g_3" is assigned to "f_3". For the learning object $lo1$ in Table 4, "g_3" and "i_3" are both equal to 0. Thus, "f_3" is also 0.

$$f_3 = \begin{cases} g_3, & i_3 = -1 \\ i_3, & i_3 \neq -1 \end{cases} \quad (9)$$

When Eqs. (3)–(5) are used together to calculate the "f_i" values of Eq. (2), the total "$LOScore_{lo1}$" value of the learning object $lo1$ in Table 4 can be represented as given below.

$$LOScore_{lo1} = << \text{"}AR\text{"}, 0 >< \text{"}SI\text{"}, 0 >< \text{"}VV\text{"}, 0 >< \text{"}SG\text{"}, 0 >>$$

The first step of the proposed personalization approach has been completed and can be expressed in an algorithmic way with the "CalculateLOScore()" method given in Algorithm 1. The "CalculateLOScore()" method takes a learning object lo as its input and calculates "$LOScore_{lo}$" as its output. "$LOScore_{lo}$" will be used in the second part of the personalization process to compare how a learner l and a learning object lo match each others.

Relationship of the equations at the first step with Algorithm 1 can be explained as follows: Eq. (3) corresponds to the lines 04–17; Eq. (4) is calculated with the help of Eq. (6) and together they correspond to the lines 18–22; Eq. (5) is calculated with the help of Eq. (7) and together they correspond to the lines 23–27; Eq. (8) corresponds to the lines 28–30; and Eq. (9) corresponds to the lines 32–34 of Algorithm 1. Since output of Algorithm 1 is $LOScore_{lo}$, which corresponds to Eq. (2), Algorithm 1 depicts how Eq. (2) is calculated with the help of Eqs. (3)–(9).

3.2   Matching Learning Objects and Learners

The second step of the proposed approach involves recommending appropriate learning objects to learners. Thus, Eqs. (1) and (2) should be used together to match learners and learning objects. Four weight values, namely "f_1", "f_2", "f_3" and "f_4" in Eq. (2), are calculated for each learning object at the first step. Since Eq. (1) provides FSLSM information of the learner l, the next step is to determine how related the learner l and the learning objects are for each dimension of Felder and Silverman learning styles model. Equation (10) is used for the matching operation, where the "e_i" and "f_i" weight values for each FSLSM dimension is combined by using the Manhattan distance metric to provide a single weight value as a fitness function representing how related the learner and the learning objects are.

$$W_{lo} = \sum_{i=1}^{4} |e_i - f_i| \tag{10}$$

For the example scenario being followed, when Eq. (10) is applied for learner $l1$ and learning object $lo1$ given in Table 4, the "e_i" and "f_i" values match in two of the four FSLSM dimensions. The value pairs "e_1" - "f_1" and "e_3" - "f_3" match as both are 0, but the value pairs "e_2" - "f_2" and "e_4" - "f_4" doesn't match as "e_2" and "e_4" are 1, while "f_2" and "f_4" are 0. Thus, the calculated "W_{lo1}" value is 2, which means an average match. "W_{lo}" values can range from 0 to 4, as the results closer to 0 means a closer match, with 0 meaning an exact match and 4 meaning a mismatch in both FSLSM dimensions.

Finally, depending on the recommendation approach of the personalization strategy being followed, learning objects can be ordered according to their weight

---

Algorithm 1. The algorithm for "CalculateLOScore()" method, which calculates the total LO score "$LOScore_{lo}$" of a learning object lo.

Input: lo
Output: $LOScore_{lo}$
CalculateLOScore():

01: $LRT_{lo} \leftarrow$ "LRT" metadata element of learning object lo
02: $InteractivityType_{lo} \leftarrow$ "InteractivityType" metadata element of learning object lo
03: $Format_{lo} \leftarrow$ "Format" metadata element of learning object lo
04: **select** (LRT_{lo})
05:     **case** "$exercise$", "$simulation$", "$experiment$":
06:         $g_1 \leftarrow 0, g_2 \leftarrow 0, g_3 \leftarrow 0, g_4 \leftarrow 0$
07:     **case** "$questionnaire$", "$exam$", "$problem\,statement$", "$self\,assessment$":
08:         $g_1 \leftarrow 0, g_2 \leftarrow 1, g_3 \leftarrow 0, g_4 \leftarrow 0$
09:     **case** "$diagram$", "$figure$", "$graph$", "$table$":
10:         $g_1 \leftarrow 1, g_2 \leftarrow 0, g_3 \leftarrow 0, g_4 \leftarrow 0$
11:     **case** "$index$":
12:         $g_1 \leftarrow 1, g_2 \leftarrow 0, g_3 \leftarrow 0, g_4 \leftarrow 1$
13:     **case** "$slide$":
14:         $g_1 \leftarrow 1, g_2 \leftarrow 1, g_3 \leftarrow 0, g_4 \leftarrow 0$
15:     **case** "$narrative\,text$", "$lecture$":
16:         $g_1 \leftarrow 1, g_2 \leftarrow 1, g_3 \leftarrow 1, g_4 \leftarrow 0$
17: **end select**
18: **if** $InteractivityType_{lo} =$ "$active$" **then** $h_1 \leftarrow 0$
19:     **else if** $InteractivityType_{lo} =$ "$expositive$" **then** $h_1 \leftarrow 1$
20:         **else** $h_1 \leftarrow -1$
21:     **end if**
22: **end if**
23: **if** $Format_{lo} =$ "$video$" **then** $i_3 \leftarrow 0$
24:     **else if** $Format_{lo} =$ "$text$" **then** $i_3 \leftarrow 1$
25:         **else** $i_3 \leftarrow -1$
26:     **end if**
27: **end if**
28: **if** $h_1 = -1$ **then** $f_1 \leftarrow g_1$
29:     **else** $f_1 \leftarrow h_1$
30: **end if**
31: $f_2 \leftarrow g_2$
32: **if** $i_3 = -1$ **then** $f_3 \leftarrow g_3$
33:     **else** $f_3 \leftarrow i_3$
34: **end if**
35: $f_4 \leftarrow g_4$
36: **return** $LOScore_{lo}$

---

values and recommended to the learner in that order. In this strategy, the learner has the freedom to choose the learning object she wants to work with from the ordered list. As an alternative approach, only the best matching learning object can be recommended to the learner. Therefore, the output of the algorithmic personalization approach proposed in this paper can be used in a flexible way in e-learning systems.

4 Discussion and Conclusion

In this paper, a new personalization approach to recommend learning objects to learners based on their learning styles has been presented. Felder and Silverman learning styles model is used for modeling the learners' learning styles, while IEEE LOM v1.0 standard is used for modeling the learning objects. The proposed approach expands the work provided by [10] by using more metadata elements and Manhattan distance metric to combine dimensions of FSLSM.

Two dimensions of FSLSM (sensing/intuitive and sequential/global) are associated with only one IEEE LOM metadata element, while the other two (active/reflective and visual/verbal dimensions) are related to two metadata elements. This can effect results of the proposed approach in a negative way, because all dimensions have the same input weight in Eq. (10) regardless of the number of IEEE LOM metadata elements they are associated with. However, there are limited IEEE LOM metadata elements identified in the literature to be exploited for personalization purposes. More research is needed to detect more relations between FSLSM and IEEE LOM metadata elements. Until then, eliminating the FSLSM dimensions with only one metadata element as a personalization source can be considered as a solution. Another solution may be redefining the effect of each dimension in Eq. (10) according to the number of IEEE LOM metadata elements associated with that dimension.

Another limitation of the proposed approach is the polarized modeling of learners' FSLSM dimension data and learning objects' IEEE LOM metadata. Providing a wider spectrum to model the input data in future studies may help achieving better results.

The main aim of this study is to provide an initial base for different personalization approaches to recommend learner objects. Felder and Silverman learning styles model is used as the only personalization resource for modeling learners. The algorithmic approach proposed in the paper can be extended with other personalization resources such as learner experience, knowledge level, learning goals and motivation level.

Learning styles modeling applied in this study is based on static learner modeling, where learners' learning styles are considered to be the same over time. However, studies show that learners' learning styles may change with time [13]. Dynamic learning modeling is another future work direction.

The proposed personalization approach can be integrated in e-learning systems and learning management systems to let the learners interact with learning objects and provide feedback about the quality of the recommended learning objects. Thus, working with actual learners to evaluate the proposed approach is considered as future work.

References

1. SCORM 2004, 4th edn. (2009). http://scorm.com/scorm-explained/technical-scorm/content-packaging/metadata-structure/
2. IEEE-LOM, IEEE LOM 1484.12.1 v1 Standard for Learning Object Metadata-(2002). http://grouper.ieee.org/groups/ltsc/wg12/20020612-Final-LOM-Draft.html
3. Ciloglugil, B., Inceoglu, M.M.: User modeling for adaptive e-learning systems. In: Murgante, B., Gervasi, O., Misra, S., Nedjah, N., Rocha, A.M.A.C., Taniar, D., Apduhan, B.O. (eds.) ICCSA 2012, Part III. LNCS, vol. 7335, pp. 550–561. Springer, Heidelberg (2012)
4. Coffield, F., Moseley, D., Hall, E., Ecclestone, K.: Should We Be Using Learning Styles? What Research Has to Say to Practice. Learning and Skills Research Centre/University of Newcastle upon Tyne, London (2004)
5. Felder, R.M., Silverman, L.K.: Learning and teaching styles in engineering education. Eng. Educ. **78**(7), 674–681 (1988)
6. Kolb, D.A.: Experiential Learning: Experience as the Source of Learning and Development. Prentice-Hall, Englewood Cliffs (1984)
7. Honey, P., Mumford, A.: The Manual of Learning Styles. Peter Honey, Maidenhead (1982)
8. Kuljis, J., Liu, F.: A comparison of learning style theories on the suitability for elearning. In: Hamza, M.H. (ed.) Proceedings of the IASTED Conference on Web Technologies, Applications, and Services, pp. 191–197. ACTA Press, Calgary (2005)
9. Felder, R.M., Soloman, B.A.: Index of Learning Styles questionnaire (1997). http://www.engr.ncsu.edu/learningstyles/ilsweb.html
10. Sangineto, E., Capuano, N., Gaeta, M., Micarelli, A.: Adaptive course generation through learning styles representation. Univ. Access Inf. Soc. **7**(1–2), 1–23 (2008)
11. Essalmi, F., Ayed, L.J.B., Jemni, M., Kinshuk, Graf, S.: Selection of appropriate e-learning personalization strategies from ontological perspectives. Interact. Des. Archit. J. - IxD&A **9**(10), 65–84 (2010)
12. Savic, G., Konjovic, Z.: Learning style based personalization of SCORM e-learning courses. In: 7th International Symposium on Intelligent Systems and Informatics, SISY 2009, pp. 349–353 (2009)
13. Felder, R.M., Spurlin, J.: Applications, reliability and validity of the index of learning styles. Int. J. Eng. Educ. **21**(1), 103–112 (2005)
14. CLEO-Lab, CLEO Extensions to the IEEE Learning Object Metadata, Customized Learning Experience Online (CLEO) Lab (2003)
15. Castro-Garca, L., Lpez-Morteo, G.: An international analysis of the extensions to the IEEE LOM v1.0 metadata standard. Comput. Stand. Interfaces **35**(6), 567–581 (2013)

Modeling Software Security Requirements Through Functionality Rank Diagrams

Rajat Goel$^{(\boxtimes)}$, M.C. Govil, and Girdhari Singh

Department of Computer Science and Engineering,
Malaviya National Institute of Technology, Jaipur, India
rajatgoel85@yahoo.co.in

Abstract. Though UML or Unified Modeling Language is a popular language for modeling software requirements, it is mostly useful for functional requirements only and provides limited support for non-functional requirements, like security. In the present scenario when the use of internet-based and cloud-based applications is increasing, such requirements are far more relevant. In this paper Functionality Rank Diagrams are proposed that follow a well-structured requirement elicitation and ranking mechanism, and model these non-functional requirements leading to a better system design.

Keywords: Requirements elicitation · Non-functional requirements · UML · Use cases

1 Introduction

The prominence of non-functional software requirements cannot be undermined. These are important for building a successful application [1]. Security is perhaps the most important of these requirements. To obtain a secure software, researchers [2,3] have emphasized the need of imbibing security in Software Development Life Cycle (SDLC) itself. This is far more essential in the present scenario of internet-based or cloud-based [4] software. Research community finds requirements and design to be the most suitable development phases for applying security [5]. According to Futcher et al. [6] requirement specification is instrumental to obtain a secure software. To achieve this, the best option is to elicit correct requirements initially [7].

An effective modeling technique too plays a critical role when stakeholders have different goals. It can be interpreted unambiguously unlike text. UML [8] is a popular modeling language. However, Woods [9] believes that it mainly presents associations among objects. Researchers [10,11] consider UML semantics as inconsistent, informal and problematic. Moreover, UML is incompetent to model non-functional requirements. Use case diagram, one of the constructs in UML, is not fit for specifying requirements completely and correctly [12].

Functionality Rank Diagrams (FRD), proposed in this paper, model some of the non-functional requirements including security and cloud - adoption

© Springer International Publishing Switzerland 2016
O. Gervasi et al. (Eds.): ICCSA 2016, Part V, LNCS 9790, pp. 398–409, 2016.
DOI: 10.1007/978-3-319-42092-9_31

decision [13]. This decision concerns with whether to adopt cloud for software (or some functionality) or to remain on-premise. An FRD is, in many ways, an enhancement of use case diagram. Both are distinguished in Sect. 3.3. FRD is a result of a formal requirement elicitation and assessment mechanism. An example of library management system of a college is considered throughout this paper to understand the mechanism. Due to space constraints only a limited view of the case study is presented here. The library provides issue, return and search services to faculty members and students. The users can also edit their personal details like username, password etc.

2   Related Works

Controlled Requirement (CORE) Specification Method [14] considers views of all stakeholders and is supported by diagrams but it is more related with flow of information, does not give enough importance to non-functional requirements and proper elicitation of requirements is not easy. Feature Oriented Domain Analysis (FODA) [15] is an iterative process where relevant information is elicited about the systems domain to develop a generic model. Here domain experts and users are involved in the development process but application of this method is constricted to well-understood domains only. Issue-based Information System (IBIS) [16] is a formal elicitation method but lacks in iteration and graphics. Security-aware Software Development Life Cycle (SaSDLC) [20] acknowledges the fact that security ought to be treated as an important functional requirement within the development process. Assets are ranked in this methodology but by three types of stakeholders for all kinds of projects. UMLsec [17] is an extension of UML for Secure Systems Development that uses stereotypes and tags but it doesnt follow any requirement assessment procedure or elaborate parameters.

Some other important works in order to improve modeling and particularly UML are Proliferation of UML 2.0 dialects [18], ADORA [19], formalization of UML through A context free grammar [20] and extension of UML to add context view [9]. An Activity Diagram [21] is proposed which is heavily text-based and in a tabular fashion. Misuse-Cases [22] are an improvement over use cases to represent security threats to a system graphically. Multi-perspective Requirements Engineering or PREview [1] is an iterative approach designed for industrial use.

3   Proposed Methodology

The proposed methodology overcomes the shortcomings of the prior works. It begins with a formal requirement elicitation process which involves ranking of functionalities provided by the system on seven parameters viz. Confidentiality, Authentication, Authorization, Non-repudiation, Integrity, Scalability and Configurability. All types of stakeholders participate in the ranking process. Kamata et al. [27] considers such involvement beneficial for software.

The first five parameters are the security services mentioned by Forouzan [23]. The remaining two are specific to cloud-based software [13], added for making decision for adopting cloud. Each functionality is ranked by only those stakeholders [24] which are relevant to it. The idea of relevance is a novelty of this work. The method is functionality centric. The diagram is supplemented by a textual template to avoid any ambiguities. This is inspired from some of the other works [22,25,26]. The broad steps in building an FRD are Identification, Mapping, Ranking, Analysis and Design.

4   Identification

The Identification phase consists of identification of entities i.e. stakehold-ers and functionalities by the preliminary stakeholders i.e. client, developers and domain experts. The identified stakeholders iteratively identify more stakeholders and functionalities until they are identified exhaustively.

4.1   Obtainment of Stakeholder and Functionality Sets

The two lists of stakeholders and functionalities are checked for any redun-dancy like redundant stakeholders Faculty Member and Teacher. In such a case, a single name must be assigned to the entity. Sometimes, functionality or a stakeholder needs to be decomposed into its constituents. If decomposition takes place, the entity is replaced by its constituents. For example, as per requirement, stakeholders like Students can be decomposed into Undergraduate and Post graduate. Alternatively, the constituents may need to be aggregated. Developer and the experts may add more entities for operational reasons. At the end of this phase final sets S and F for stakeholders and functionalities respectively, are obtained.

P is the set of parameters.
$S = \{S_1, S_2, ...S_n\}$, $F=\{F_1,F_2,... F_m\}$, where, $n \in W$, $m \in W$
P = {Authentication, Authorization, Confidentiality, Integrity, Non-repudiation, calability, Configurability}
For the library management system, the sets S and F are as follows.
F = {Issue, Return, Search, Edit Personal Details}
S = {Student, Faculty Member, Librarian, Administrator, Client, Developer, Domain Expert}

4.2   Consultation Group

One stakeholder of each type enlisted in S is included in the Consultation Group. This group advises the developers all through the development process and takes part in the ranking process later.

4.3   Story and Story Diagram

The requirements will generally be narrated in natural language, referred to as Story in this work. A story as and when elicited is represented by a Story Diagram. There are two kinds of stories as described below.

Simple Story. The generic form of the story is:

S_i, S_{i+1}, S_{i+2}, ... can [do/perform] F_k, F_{k+1}, F_{k+2}, ...   where, $i \geq 1$, $k \geq 1$

For example, Student can do Issue. Here, Student is the stakeholder and Issue is the functionality. The major task of a developer is to correctly identify the entities from these stories. Figure 1 is the corresponding Story Diagram. It is almost like a use case but, an elongated oval is used against the oval. This difference highlights the fact that this diagram represents a functionality and not a scenario or behavior of the system. A story diagram models the properties of the systems functionalities, as will be seen later. A story may contain more than one stakeholder like Student and Faculty Member can do Search. Figure 2 is its story diagram.

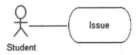

Fig. 1. Student can do Issue

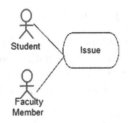

Fig. 2. Student and Faculty Member can do Issue

Story with Requestor and Facilitator. This story is in the form of:

Both S_i and S_{i+1} are stakeholders but one of them requests the service or functionality while the other provides it. Thus, the former is the Requestor while the latter is the Facilitator. For a given functionality $F_k \in F$, both the Requestors set RQ_k and Facilitator set FL_k are sub sets of set S i.e. $RQ_k \subseteq S$, $FL_k \subseteq S$. An example story of this kind is A student can do Issue through librarian. Figure 3 is the diagram depicting that student is the requestor and librarian is the facilitator in the Issue functionality. The requestor(s) are placed on the left side of the functionality and facilitator(s) on the right. There may be more than one requestors and facilitators.

Fig. 3. Functionality with Requestor and Facilitator

The Requestor and Facilitator sets are functionality specific. For example, in Return functionality, these sets are RQ_{Return} and FL_{Return} respectively where, $FL_{Return} = \{Librarian, Administrator\}$

The Requestor and Facilitator sets are functionality specific. For example, in Return functionality, these sets are RQ_{Return} and FL_{Return} respectively where,

A Story Diagram evolves with the information contained in the story and finally develops into an FRD after rank determination. Only a limited number of stories and their diagrams are mentioned in this paper.

5   Mapping

The related functionalities and stakeholders are mapped. This serves as rules for ranking. It is conducted by the developers and the domain experts based on the information solicited in the previous phases. Two relevance tables (Tables 1 and 2), are developed. The shaded cells denote relevance and un-shaded cells denote irrelevance.

5.1   Functionality-Stakeholder Relevance

Table 1 illustrates which stakeholders are involved in which functionality. It represents F R S or relation R between Functionalities and Stakeholders. All entities are relevant to the client, developer and expert. They are not shown in the table although they will participate in the ranking procedure.

Table 1. Functionality stakeholder relevance

Stakeholders	Functionalities			
	Issue	Return	Search	Edit Personal Details
Student	▓	▓	▓	▓
Faculty Member	▓	▓	▓	▓
Librarian	▓	▓	▓	
Administrator		▓	▓	▓

5.2   Stakeholder-Parameter Relevance

Stakeholders are allowed to rank assets on relevant parameters only. Some parameters will be left for ranking by technical stakeholders like developers. Table 2 denotes S R P i.e. relation R between stakeholders and parameters.

Table 2. Stakeholder-parameter relevance

Stakeholders	Authentication	Confidentiality	Integrity	Authorization	Non-repudiation	Configurability	Scalability
Student		■		■			
Faculty Member		■		■			
Librarian		■	■	■	■	■	■
Administrator		■	■	■	■	■	■
Client		■	■	■	■	■	■
Developer	■	■	■	■	■	■	■
Domain Expert	■	■	■	■	■	■	■

5.3 Rank Matrix

Each consultation group member has one version of the rank matrix for himself in accordance to Tables 1 and 2 i.e. he ranks the relevant functionalities on the parameters applicable, which are confidentiality and authorization in this case. Table 3 is a typical rank matrix for Student. It is derived from Tables 1 and 2 i.e. $FRS, SRP \Rightarrow FRP$.

Table 3. Rank matrix for student

Functionalities	Authentication	Confidentiality	Integrity	Authorization	Non-repudiation	Configurability	Scalability
Issue		<Rank>		<Rank>			
Return		<Rank>		<Rank>			
Search		<Rank>		<Rank>			
Edit Personal Details		<Rank>		<Rank>			

6 Ranking

As per the mapping, participants rank functionalities into three classes as Low, Medium and High. Later, these ranks are converted to numeric values for analysis.

6.1 Parameter Values and Ranges

The values alloted to low, medium and high ranks are 2, 4 and 6 respectively. These values, the ranges and corresponding raks are shown in Table 4.

Table 4. Values assigned to ranks

Rank	Value	Range
Low	2	
Medium	4	3 to <5
High	6	5 and more

6.2 Calculation of Parameter Averages

The consolidated average of ranks filled by all stakeholders for any j^{th} parameter P^j for k^{th} functionality F_k is denoted by $Average_p$. It is calculated using (1). If number of relevant stakeholders to a functionality is n then there will be same number of rank matrices. For determining average for security as a whole $Average_{sec}$, the mean of averages of five parameters viz. Authentication, Authorization, Confidentiality, Integrity and Non-repudiation is calculated. To make cloud adoption decision, mean of averages of scalability and configurability is calculated and denoted by $Average_{ca}$. Not Applicable or irrelevant cells are not considered in calculation.

$$Average_p = \sum_{i=1}^{n} values(S_i F_k P_j)/n \qquad (1)$$

7 Analysis

Based on the average values High, Medium and Low ranks are assigned to every functionality. Prioritization of time, cost and effort can be done accordingly. The decision for cloud adoption is binary. So, a threshold value 4 is set, exceeding which cloud should be adopted.

7.1 Parameter Ranks

The consolidated averages and ranks of all parameters, with respect to every functionality, is illustrated through a Parameter Rank Table. Such a table for Return functionality is shown in Table 5.

The corresponding values of $Average_{sec}$ and $Average_{ca}$ are denoted by (2) and (3) respectively. It is seen that since value of $Average_{sec}$ is greater than 5, so functionality rank is High. Similarly, as value of $Average_{ca}$ is less than 4, cloud is not adopted.

$$Average_{sec} = (5.89 + 4.67 + 5.12 + 5.09 + 4.88)/5 = 5.13 \qquad (2)$$

$$Average_{ca} = (2.21 + 1.18)/2 = 1.69 \qquad (3)$$

Table 5. Parameter template

Functionality: Return		
Parameter	Average	Rank
Authentication	5.89	High
Confidentiality	4.67	Medium
Integrity	5.12	High
Authorization	5.09	High
Non-repudiation	4.88	Medium
Scalability	2.21	Low
Configurability	1.18	Low

7.2 Functionality Analysis

A Combined Functionality Template (Table 6) is created, which summarizes all the functionalities of the software. In the example taken, students and faculty members are the requestors in every functionality. Personal Details can be edited by the intervention of Administrator. The security and cloud adoption averages for other functionalities are obtained in the same way as for the Return functionality. However, only decision is mentioned in the template.

Table 6. Combined functionality template

Library management system				
Functionality	Rank	Requestors	Facilitators	Cloud adoption
Issue	Medium	Student, Faculty Member	Librarian, Administrator	No
Return	High	Student, Faculty Member	Librarian, Administrator	No
Search	High	Student, Faculty Member	Librarian	Yes
Edit Personal Details	Low	Student, Faculty Member	Administrator	Yes

7.3 Implications of Ranks

Once the ranks are identified for security, adequate measures can be taken. Appropriate techniques for security may be chosen. A simple example can be biometric system for higher rank, PIN and card for medium rank, and password for low rank. Similarly, complexity of cryptographic algorithms may vary with the ranks to secure data items and messages. The choice of these measures will also be governed by the cost to be incurred.

8 Design

This section describes the gradual evolution of FRDs and their culmination in the Combined Functionality Rank Diagram (CFRD).

8.1   Functionality Rank Diagram (FRD)

The FRD evolves with the rank of the functionalities. The security rank is denoted by the number of concentric ovals around it i.e. a single oval for low rank, double for medium and triple for high. A cloud may be drawn additionally. Figures 4 and 5 show medium and high ranks respectively for search and return functionality. Figures 6 and 7 show cloud adoption for Search (with low rank) and Edit Personal Details (with high rank) respectively.

8.2   Combined Functionality Rank Diagram (CFRD)

A CFRD (Fig. 8), built on the basis of Combined Functionality Template (Table 6), summarizes the complete system. It depicts every functionality with its rank, requestors and facilitators. Optionally, the relevant stakeholders can be displayed with the FRDs.

Fig. 4. Functionality with medium security rank

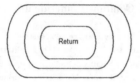

Fig. 5. Functionality with high security rank

Fig. 6. Low security rank functionality on cloud

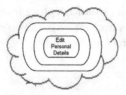

Fig. 7. High security rank functionality on cloud

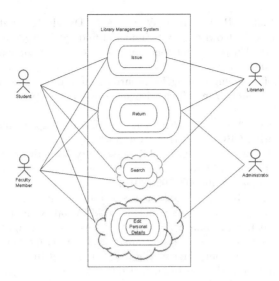

Fig. 8. Combined Functionality Rank Diagram

9   Conclusion and Future Work

The methodology is quite flexible to cater software of different domains. The formation of a comprehensive consultation group, is unique to this me-thodology which results in better customer satisfaction. The two phases of requirements and design are tightly coupled. The FRDs are developed in accordance with a well structured requirement elicitation and assessment procedure. The FRD illustrates the security rank, stakeholders involved, and cloud adoption all in one. The designs are also based on empirical analysis. Ranking is done by only the relevant stakeholders. Use of template avoids any ambiguity. The use of less but orthogonal structures makes the diagrams simple yet efficient. More practically, FRD considers threats in types (of different parameters) and not in numbers unlike [29].

In future, finer granularity in establishing the implications of ranks remains to be done. The methodology ought to be applied on more real-life case studies of different domains to test its mettle. To make the mechanism more domain specific weights can be assigned to parameters and/or stakeholders. There is a scope of improving the methodology by adding more parameters and consequently, newer diagrams will be proposed to depict them.

References

1. Sommerville, I., Sawyer, P., Viller, S.: Viewpoints for requirements elicitation a practical approach. In: IEEE International Symposium on Requirements Engineering, pp. 74–81 (1998)

2. Lindvall, M., Basili, V.R., Boehm, B.W., Costa, P., Dangle, K., Shull, F., Tesoriero, R., Williams, L.A., Zelkowitz, M.V.: Empirical findings in Agile methods. In: XP Universe and Agile Universe Conference on Extreme Programming and Agile Methods, 197–207 (2002)

3. Shreyas, D.: Software engineering for security - towards architecting secure software. In: ICS 221 Seminar in Software Engineering, University of California, Irvine, pp. 1–12 (2001)

4. Rittinghouse, J.W., Ransome, J.F.: Cloud Computing: Implementation, Management, and Security. CRC Press, Boca Raton (2010)

5. Goel, R., Govil, M.C., Singh, G.: Imbibing security in software development life cycle: a review paper. In: Afro - Asian International Conference on Science, Engineering and Technology, pp. 593–599 (2015)

6. Futcher, L., Solms, R.V.: SecSDM: a usable tool to support IT undergraduate students in secure software development. In: HAISA, pp. 86–96 (2012)

7. Sabahat, N., Iqbal, F., Azam, F., Javed, M.Y.: An iterative approach for global requirements elicitation: a case study analysis. In: International Conference on Electronics and Information Engineering, pp. 361–366 (2010)

8. Booch, G., Rumbaugh, J., Jacobson, I.: Unified Modeling Language User Guide. Pearson Education India, Bengaluru (2005)

9. Woods, E.: Harnessing UML for architectural description the context view. IEEE Softw. 31, 30–33 (2014)

10. Choppy, C., Reggio, G.: Requirements capture and specification for enterprise applications: a UML based attempt. In: Australian Software Engineering Conference, pp. 19–28 (2006)

11. Konrad, S., Goldsby, H., Lopez, K., Cheng, B.H.C.: Visualizing requirements in UML models. In: International Workshop on Visualization Requirements Engineering, p. 1 (2007)

12. Chua, B.B., Bernardo, D.V., Verner, J.: Understanding the use of elicitation approaches for effective requirements gathering. In: International Conference on Software Engineering Advances, pp. 325–330 (2010)

13. Ezzat, E.M., Zanfaly, D.S., Mostafa, M.M.: Fly over clouds or drive through the crowd: a cloud adoption framework. In: International Conference and Workshop on the Current Trends in Information Technology, pp. 6–11 (2011)

14. Mullery, G.P.: CORE-A method for controlled requirement specification. In: 4th International Conference on Software Engineering, pp. 126–135 (1979)

15. Kang, K.C., Cohen, S.G., Hess, J.A., Novak, W.E., Peterson, A.S.: Feature-oriented domain analysis (FODA) feasibility study, Carnegie-Mellon University (1990)

16. Douglas, N., Rittel, H.W.J.: Issue based information system for design. In: Association for Computer Aided Design in Architecture, University of Michigan, pp. 275–286 (1988)

17. Talukder, A.K., Maurya, V.K., Santhosh, B.G., Jangam, E., Muni, S.V., Jevitha, K.P., Saurabh, S., Pais, A.R.: Security-aware Software Development Life Cycle (SaSDLC)-Processes and tools. In: IFIP International Conferecne on Wireless and Optical Communication Networks, pp. 1–5 (2009)

18. Jürjens, J.: UMLsec: extending UML for secure systems development. In: Jézéquel, J.-M., Hussmann, H., Cook, S. (eds.) UML 2002. LNCS, vol. 2460, pp. 412–425. Springer, Heidelberg (2002)

19. Kobryn, C.: Experts voice UML 3.0 and the future of modeling. Softw. Syst. Model. 3, 4–8 (2004)

20. Glinz, M.: Problems and deficiencies of UML as a requirements specification language. In: International Workshop on Software Specification and Design, pp. 11–22 (2000)
21. Chanda, J., Kanjilal, A., Sengupta, S., Bhattacharya, S.: Traceability of requirements and consistency verification of UML use case, activity and Class diagram: a formal approach. In: International Conference on Methods and Models in Computer Science, pp. 1–4 (2009)
22. Samuel, B.M., Watkins, L.A., Ehle, A., Khatri, V.: Customizing the representation capabilities of process models: understanding the effects of perceived modeling impediments. IEEE Trans. Softw. Eng. **41**, 19–39 (2015)
23. Sindre, G., Opdahl, A.L.: Eliciting security requirements with misuse cases. Requir. Eng. **10**, 34–44 (2005)
24. Forouzan, B.A.: Data Communications and Networking. McGraw-Hill, New York City (2007)
25. Pressman, R.S.: Software Engineering a Practitioners Approach. McGraw-Hill, New York City (2001)
26. Hernndez, U.I., Rodrguez, F.J., Martin, M.V.: Use Processes - Modeling requirements based on elements of BPMN and UML Use Case Diagrams. In: International Conference on Software Technology Engineering, vol. 2, pp. 36–40 (2010)
27. Santhosh Babu, G., Maurya, V.K., Jangam, E., Muni Sekhar, V., Talukder, A.K., Pais, A.R.: Suraksha: a security designers workbench. In: Hack.in, pp. 59–65 (2009)
28. Kamata, M.I., Tamai, T.: How does requirements quality relate to project success or failure? In: Requirements Engineering Conference, pp. 69–78 (2007)
29. Pauli, J.J., Xu, D.: Misuse case-based design and analysis of secure software architecture. In: International Conference on Information Technology Coding and Computer, vol. 2, pp. 2005–2010 (2005)

Finding Divergent Executions in Asynchronous Programs

Mohamed A. El-Zawawy[1,2(✉)]

[1] College of Computer and Information Sciences,
Al Imam Mohammad Ibn Saud Islamic University (IMSIU),
Riyadh, Kingdom of Saudi Arabia
[2] Department of Mathematics, Faculty of Science,
Cairo University, Giza 12613, Egypt
maelzawawy@cu.edu.eg

Abstract. Developing interactive and distributive softwares is a complicated task for many reasons including the nature of this technique of programming; asynchronous programming. In this style of programming, there is no synchronization between running processes concerning message-posting communications. Therefore the presence of nonterminating executions for asynchronous programs, as a result of simple programming errors, is common.

This paper presents a transformation technique for asynchronous programs to discover their nonterminating (divergent) executions. The proposed technique converts a given asynchronous program P into another form P' such that the nontermination problem of P is equivalent to the state reachability problem of P'. The paper presents the syntax and semantics of source and target languages of the transformation method. Then the transformation is presented using the formal syntax and semantics. The paper also formalizes the relationship between asynchronous programs and their transformations.

Keywords: Divergent executions · Asynchronous programs · Method posting · Message-passing programs · Distributed systems · Reactive asynchronous systems

1 Introduction

A common framework to design efficient and interacting software systems [4,5,28] is asynchronous message-passing programming. The applications of these frameworks ranges from mobile applications to low-level methods. Asynchronous programming is mainly concurrent. Therefore they suffer from involved and complicated programming errors. These errors can be automatically revealed using model checking techniques [18,30].

The absence of harmony among concurrently running unlimited delayed communications (that are message-based) and methods complicates coding reactive and parallel asynchronous systems. In asynchronous programs, errors may cause

© Springer International Publishing Switzerland 2016
O. Gervasi et al. (Eds.): ICCSA 2016, Part V, LNCS 9790, pp. 410–421, 2016.
DOI: 10.1007/978-3-319-42092-9_32

nonterminating behaviors. Therefore an important challenge of asynchronous programs is that finite number of client-generated requests create a finite number of messages [19, 26].

Most exiting techniques of model checking in Asynchronous programming use formal techniques ignoring events synchronization and assuming sequential execution of concurrent tasks. Therefore it is common for these techniques to suffer from false positives (while discovering infeasible errors) and from false negatives, as well, because of mistakes due to concurrent tasks communications [10, 18].

This paper present a new technique to discover divergent behaviors of asynchronous programs. The technique uses input numbers to divide the execution sequence of a given program into two stages. Then using simple data structures, the original program is transformed into another asynchronous program where the nontermination in the original program is equivalent to the state reachability in the program resulted from the transformation. Compared to similar techniques [19], the proposed technique is much simpler yet powerful enough.

Motivation. The paper is motivated by the need of a simple technique for the involved problem of discovering divergent behaviors in asynchronous programs.

Contributions. Contributions of the paper are the following:

1. A precise operational semantics to asynchronous programming and one of their transformations.
2. A transformation technique to reduce the problem of detecting divergent behaviors in asynchronous programs to the problem of state reachability.

Paper Outline. Section 2 presents the main contribution of the paper; it presents a formal operational semantics to asynchronous programs, the syntax of their transformation, a precise operational semantics to the transformation. Section 2 also formalizes the benefits of the transformation. Section 3 reviews the work most related to the technique presented in this paper. Section 4 presents an interesting direction for future work and Sect. 5 concludes the paper.

2   Program Transformation Semantics

This section presents two models of asynchrony programming [6, 27]; $\mathcal{A}$syMp and $\mathcal{F}$AsyMp. For each model the section presents precise syntax and operational semantics. The section also presents a transformation from the programs of the first model $\mathcal{A}$syMp to programs of the second model $\mathcal{F}$AsyMp. The advantage of this transformation is to find the nontermination behaviors of $\mathcal{A}$syMp programs using the their transformation forms in $\mathcal{F}$AsyMp. This simplifies the problem of revealing the nontermination behaviors as it is shown to be equivalent to the reachability problem in programs of $\mathcal{F}$AsyMp.

2.1 $\mathcal{A}$syMp Syntax

Figure 1 presents our adapted model of asynchronous programming, $\mathcal{A}$syMp. This model is inspired by the model used in [19]. Every program in the model is a variable and a sequence of methods. Each method has a local variable and a statement. The set of available statements include a statement to call a method (*call m(e)*) where the execution of the method amounts to executing the call statement. Another available statement is that of posting a method (*post m(e)*) which amounts to announcing that the method $m(e)$ is required to be executed at any later program-point and before termination of program execution.

$$x ::= g \mid l_m.$$
$$s \in Stmts ::= s_1; s_2 \mid x := e \mid if\ e\ then\ s_t\ else\ s_f \mid while\ e\ do\ s \mid$$
$$x := call\ m(e) \mid return\ (e) \mid post\ m(e) \mid skip.$$
$$m \in Mthds ::= (method\ m : Int\ l_m,\ s_m).$$
$$P \in Progs ::= (Int\ g,\ m^*).$$

Fig. 1. Grammar of $\mathcal{A}$syMp; asynchronous message-passing programming langauge.

2.2 $\mathcal{A}$syMp Semantics

This section presents an operational semantics to the constructs of the langauge $\mathcal{A}$syMp. The semantics is necessary to link the asynchronous programs of our model to their transformations presented later in this section towards revealing divergent behaviors. The semantic states are presented in Definition 1.

Definition 1.

- $\mathcal{L} = \{g, l_m, r_m\}$.
- $f_v : \mathcal{L} \rightharpoonup Integers$.
- $f_m : \mathcal{M} \rightharpoonup \mathcal{P}(Integers)$.
- A state σ is a quadrable (n, s, f_v, f_m). The set of all states is denoted by $\mathcal{S}_A$.
- A state σ is a picking if $\sigma(s) = return$.
- A state σ is final if $\sigma(s) = return$ and $\sigma(f_m) = \emptyset$.
- An execution is a sequence of states.
- An execution is "infinitely picking" if it includes an infinite number of picking states.
- For two states $\sigma = (n, s, f_v, f_m)$ and $\sigma' = (n', s', f_v', f_m')$,

$$\sigma \sqsubseteq_1 \sigma' \iff n \leq n' \wedge$$
$$skip = s = s'$$
$$\wedge f_v(g) = f_v'(g) \wedge f_m \leq f_m'.$$

– *An execution is "possibly infinite" if it has two states σ and σ' such that $\sigma < \sigma'$.*

The set of all variables is denoted by $\mathcal{L}$. The contents of the variables during program executions are captured by the partial maps f_v. The variable r_m has the value resulted from running the latest executed method. The sets of posted methods during program executions are captured by the partial maps f_m. Hence a semantic state (denoted by σ) is a quadrable of a number n represents the state number in the sequence of the states representing the program execution, the statement s being executed, and the two maps f_v and f_m. When the statement of a state is *return*, this means that the execution of a method has finished and the control is to pick the following posted method if any. In this case, the state is descried as *picking*. If for a picking state there is no more posted methods, then the state is final. An order relationship $\sqsubseteq_1$ is defined over the set of states $\mathcal{S}_A$. Definition 1 formalizes that $(\sigma, \sigma') \in \sqsubseteq_1$ if σ and σ' are picking and having same valuations for the global variable g, and the set of methods posted at σ is a subset of that posted at σ'.

Lemma 1. *If the set of variable valuations are finite, then the relationship $\sqsubseteq_1$ is a well-quasi-ordering.*

Figure 2 presents the inference rules of the operational semantics of $\mathcal{A}$syMp. Rule 2 presents the semantics of the assignment statement where the value of the expression e is calculated using the valuations of the global variable g and of the local variable of the current running method $l_{f_v(\chi)}$. Upon finishing executing a method, Rule 10 picks one of the posted method and embarks on executing it. The choice of the method is random ($q = random(dom(f_m))$).

Lemma 2. *If (n, s, f_v, f_m) is a well-formed semantic state and $(n, s, f_v, f_m) \rightsquigarrow (n', s', f'_v, f'_m)$, then (n', s', f'_v, f'_m) is a well-formed semantic state.*

2.3 $\mathcal{F}$AsyMp Syntax

The syntax of the language $\mathcal{F}$AsyMp is presented in Fig. 3. This langauge is used to provide new equivalent representations of programs built using the language $\mathcal{A}$syMp.

In $f\mathcal{A}$syMp, a program has a set of global variables and a sequence of methods. The variable *stage* is used to divide the program execution into two stages. The arrays $Int\ post_0[\mu][]$ and $Int\ post_1[\mu][]$, where the number of methods in the program is μ, are used to record method posting in stages 0 and 1, respectively. The variable χ is used to store the index of the currently running method. A method has a local variable and a statement (possibly composed). Among the available statements is $y.add(e)$ used to add a posted method to the appropriate location of the appropriate array of $Int\ post_0[\mu][]$ and $Int\ post_1[\mu][]$.

$$(n, skip; s, f_v, f_m) \rightsquigarrow (n + 1, s, f_v, f_m) \tag{1}$$

$$\frac{v = [e]f_v]\{g, l_{f_v(x)}\}}{(n, x := e; s, f_v, f_m) \rightsquigarrow (n + 1, s, f_v[x \mapsto v], f_m)} \tag{2}$$

$$\frac{(n, s_1, f_v, f_m) \rightsquigarrow (n', skip, f_v', f_m')}{(n, s_1; s_2, f_v, f_m) \rightsquigarrow (n', s_2, f_v', f_m')} \tag{3}$$

$$\frac{(n, s_m, f_v[l_m \mapsto [e], f_m) \rightsquigarrow (n', skip, f_v', f_m')}{(n, x := call\ m(e); s, f_v, f_m) \rightsquigarrow (n', s, f_v'[x \mapsto f_v'(m_r)], f_m')} \tag{4}$$

$$(n, return\ e, f_v, f_m) \rightsquigarrow (n + 1, skip, f_v[m_r \mapsto [e], f_m) \tag{5}$$

$$(n,\ post\ m(e), f_v, f_m) \rightsquigarrow (n + 1, skip, f_v, f_m[m \mapsto f_m(m) \cup \{[e]\}]) \tag{6}$$

$$\frac{[e]f_v]\{g, l_{f_v(x)}\} = 1}{(n, if\ e\ then\ s_f\ else\ s_t, f_v, f_m) \rightsquigarrow (n + 1, s_t, f_v, f_m)} \tag{7}$$

$$\frac{[e]f_v]\{g, l_{f_v(x)}\} = 0}{(n, if\ e\ then\ s_t\ else\ s_f, f_v, f_m) \rightsquigarrow (n + 1, s_f, f_v, f_m)} \tag{8}$$

$$\frac{[e]f_v]\{g, l_{f_v(x)}\} = 1}{(n, while\ e\ do\ s_t, f_v, f_m) \rightsquigarrow (n + 1, s_t; while\ e\ do\ s_t, f_v, f_m)} \tag{9}$$

$$\frac{[e]f_v]\{g, l_{f_v(x)}\} = 0}{(n, while\ e\ do\ s_t, f_v, f_m) \rightsquigarrow (n + 1, skip, f_v, f_m)} \tag{10}$$

$$\frac{\begin{array}{l} q = random(dom(f_m)) \\ v \in f_m(q) \qquad f_m' = f_m[q \mapsto f_v(q) \setminus \{v\}] \\ f_v' = f_v[l_q \mapsto v, \chi \mapsto q] \end{array}}{(n, skip, f_v, f_m) \rightsquigarrow (n + 1, s_q, f_v', f_m')} \tag{11}$$

$$\frac{\emptyset = dom(f_m)}{(n, skip, f_v, f_m) \rightsquigarrow (n + 1, skip, f_v, \emptyset)} \tag{12}$$

Fig. 2. Operational semantics for $\mathcal{A}$syMp.

2.4 $\mathcal{F}$AsyMp Semantics

This section presents an operational semantics to the language $\mathcal{F}$AsyMp. The semantic states are presented in Definition 2.

Definition 2.

- $\mathcal{L} = \{g_0, g_1, stage, \chi, l_m, r_m\}$.
- $f_v : \mathcal{L} \rightharpoonup Integers$.
- $f_m : \mathcal{M} \times \{0, 1\} \rightharpoonup \mathcal{P}(Integers)$.
- A state δ is a quadrable (n, s, f_v, f_m). The set of all states is denoted by $\mathcal{S}_F$.

- *For two states $\delta = (n, s, f_v, f_m)$ and $\delta' = (n', s', f'_v, f'_m)$,*

$$\delta \sqsubseteq_2 \delta' \iff n \leq n' \wedge$$
$$skip = s = s' \wedge$$
$$f_v(stage) = 0 \wedge$$
$$f'_v(stage) = 1 \wedge$$
$$f_v(g_0) = f'_v(g_1) \wedge f_m \leq f'_m.$$

$x ::= g_0 \mid g_1 \mid l_m \mid stage \mid \chi.$

$y ::= Post_0[\mu][] \mid Post_1[\mu][].$

$s \in Stmts ::= s_1; s_2 \mid x := e \mid if\ e\ then\ s_t\ else\ s_f \mid while\ e\ do\ s \mid x := call\ m(e) \mid$
$\qquad\qquad return\ (e) \mid y.add(e) \mid skip.$

$m \in Mthds ::= (method\ m : Int\ l_m, s_m).$

$P \in Progs ::= (Int\ g_0,$
$\qquad\qquad Int\ g_1,$
$\qquad\qquad Int\ post_0[\mu][],$
$\qquad\qquad Int\ post_1[\mu][],$
$\qquad\qquad Boolean\ stage,$
$\qquad\qquad Int\ \chi,$
$\qquad\qquad m^*).$

Fig. 3. Grammar of $f\mathcal{A}syMp$; flat asynchronous message-passing programming langauge.

The set of the variables $\mathcal{L}$ of Definition 2 has two versions of the global variable: g_0 to capture the value of the global variable during the first stage and g_1 to capture the value of the global variable during the second stage. The domain of partial map f_m has the form $\mathcal{M} \times \{0, 1\}$ to record the method posting and to recognize the stage where posting happens. The semantic states have the same components as that of the langauge $\mathcal{A}syMp$. However the order relationship on the set of all states $\mathcal{S}_F$ is more specific than that of the langauge $\mathcal{A}syMp$. This is so as according to Definition 2 $(\delta, \delta') \in \sqsubseteq_2$ if:

1. δ belongs to the first stage and δ' belongs to the second stage,
2. δ and δ' are picking states,
3. g_0 of state δ has same valuation as g_1 of δ', and
4. the set of methods posted at δ is a subset of that posted at δ'.

Lemma 3. *If the set of variable valuations are finite, then the relationship $\sqsubseteq_2$ is a well-quasi-ordering.*

Figure 4 presents the inference rules of the operational semantics of $\mathcal{F}$AsyMp. The figure only shows the rules that are not the same as corresponding ones of Fig. 2. Rules 13 and 14 make sure to use the convenient global variable according to the current stage of running the program. While picking a posted method for execution, Rule 16 ensures picking a method that has been posted in the current stage.

$$\frac{stage = 0 \qquad v = [e]f_v]\{g_0, l_{f_v(x)}\}}{(n, x := e; s, f_v, f_m) \leadsto_2 (n + 1, s, f_v[x \mapsto v], f_m)} \tag{13}$$

$$\frac{stage = 1 \qquad v = [e]f_v]\{g_1, l_{f_v(x)}\}}{(n, x := e; s, f_v, f_m) \leadsto_2 (n + 1, s, f_v[x \mapsto v], f_m)} \tag{14}$$

$$\frac{}{(n, post\ m(e), f_v, f_m) \leadsto_2 (n + 1, skip, f_v, f_m[(m, f_v(stage)) \mapsto f_m(m) \cup \{[e]\}])} \tag{15}$$

$$\frac{\begin{array}{l} f_v(stage) = w \\ (q, w) = random(dom(f_m)) \\ v \in f_m((q, w)) \qquad f'_m = f_m[(q, w) \mapsto f_v((q, w)) \setminus \{v\}] \\ f'_v = f_v[l_q \mapsto v, \chi \mapsto q] \end{array}}{(n, skip, f_v, f_m) \leadsto_2 (n + 1, s_q, f'_v, f'_m)} \tag{16}$$

Fig. 4. Operational semantics for $\mathcal{F}$AsyMp.

Lemma 4. *If (n, s, f_v, f_m) is a well-formed semantic state and $(n, s, f_v, f_m) \leadsto_2 (n', s', f'_v, f'_m)$, then (n', s', f'_v, f'_m) is a well-formed semantic state.*

2.5 Transformation from $\mathcal{A}$syMp to $\mathcal{F}$AsyMp syntax

This section presents a simple transformation of programs written in $\mathcal{A}$syMp to programs written in $\mathcal{F}$AsyMp. The objective of the transformation is to easily discover if the original program is nonterminating due to involved posting of methods. Simple data structures are used to achieve the transformation. For an asynchronous program P of the language model $\mathcal{A}$syMp (Fig. 1), we let P_f denotes the program of the model $\mathcal{F}$AsyMp (Fig. 3) resulting from transforming P.

Table 1 presents the proposed transformation. We note that the global variable g is replaced with a set of global variables:

- two versions for the global variable (g); g_0 and g_1,
- a variable μ to store the number of methods in the program,
- a Boolean variable to determine the stage of execution,
- two arrays to record method-posting in the two stages of executions; *Int* $post_0[\mu][]$ and *Int* $post_1[\mu][]$,
- a variable χ to store the index of the running method.

The post statement is replaced with a composed conditional statement that implements the posting in the first stage with no conditions. In the second stage posting is only carried out for methods already posted in the first stage; otherwise posting statements are replaced with the *skip* statement. The transformation assumes given numbers i and j where states with number $n \leq i$ are in first stage and have *stage* $= 0$ and states with number $i < n \leq j$ are in second stage and have *stage* $= 1$. If the f_m component of the state i is contained in the f_m component of the state j then the original program has nonterminating behaviors due involved method posting. This is true provided that the program data (variables valuations) are finite.

Table 1. Program Transformation from $\mathcal{A}$syMp to $\mathcal{F}$AsyMp.

Int g	*Int* g_0
	Int g_1,
	Int μ,
	Int $post_0[\mu][]$,
	Int $post_1[\mu][]$,
	Boolean peroid,
	Int χ,
postm(e)	*if* $(stage == 0)$
	then
	$\quad post_0[m].add([e])$
	else
	$\quad$ *if* $(stage == 1 \wedge [e] \in post_0[m])$
	$\quad$ *then*
	$\qquad post_1[m].add([e])$
	$\quad$ *else*
	$\qquad$ *if* $(stage == 1 \wedge [e] \notin post_0[m])$
	$\qquad$ *then*
	$\qquad\qquad$ *skip*
	$\qquad$ *else*
	$\qquad\qquad$ *skip*
g := e	*if* $(stage == 0)$
	then
	$\quad g_0 := e$
	else
	$\quad g_1 := e$

The following theorem is proved using structure induction on the program structures.

Theorem 1. *A finite-data program P in AsyMp has an execution with an infinite number of picking states if and only if its transformation P_f in $\mathcal{F}AsyMp$ reaches a final state for some given numbers i and j. This is true provided that executions of P and P_f are initialized from equivalent states.*

3   Literature Review

Asynchrony programs have been studied by many researchers [3,11,15–17,21–23]. This section reviews the work most-related to the technique presented in this paper.

In [19], a several-steps conversion method for revealing nonterminating executions in asynchronous programs was presented. The method starts with a code-to-code conversion transforming nontermination of asynchronous programs to the problem of state-reachability. The method then do another code-to-code conversion to approximate the state-reachability problem of asynchronous programs to that of sequential programs. The technique presented in the current paper is simpler than that presented in [19].

In [18], a formal framework for event-driven programming that is asynchronous is presented. This framework has the advantage of supporting runtime establishments of concurrent threads, events, tasks, and task buffers. Hence the framework to some extent simulates the cooperation among all theses constructs. The decidability of program analyses built using this framework was discussed in [18]. Also the presented framework was proved more expressive compared to frameworks using Petri nets; more precisely using "Data Nets" class of advanced Petri nets. Data Nets utilize ordered sequences of names carried by the net tokens [18].

In [9], and for SCOOP, a semantics framework equipped with a set of semi-automatic techniques for evaluating asynchronous programs against several models of executions was presented. The framework was used to reveal a deadlock-related differences among main models of languages executions. The framework relies on graph-semantics conversions and is implemented using the GROOVE tool. In this work the graph conversions are used to atomically represent languages concepts [9].

A high-reliability langauge, P#, for asynchronous programming was presented in [13]. The language design was associated with a static analysis for data race and automatic parallelism checking framework. Comments on coding experience using P# to code parallel techniques and Microsoft systems was presented in the paper [13].

In [14], the problem of asserting local conditions for systems that are asynchronous message-passing is studied. A reduction method to reveal semantic states whose message stacks have little number of messages is introduced in [14]. These states are called almost-synchronous invariants. This work shows that these invariants exist in most asynchronous programs and this can be used to built proofs for their correctness. This method was shown to be efficient compared to related methods [14].

Multi-pushdown systems modeling shared memory concurrent programs are studied in [2] where the focus is on characteristics that are linear-time and omega-regular. The studied problem is the undecidable model checking problem for Boolean programs including recursive methods and finite count of threads. An execution path is scope-bounded if a bounded number of context-switches of every thread isolate every couple of method call and return statement included in the thread. The effect of the concept of scope-boundedness on the complexity and the decidability of this problem was studied in this paper. Moreover, as the definitions suggest, while scope-bounding is convenient for studying liveness characteristics needed to verify unbounded computations, context-bounding is convenient for studying safety characteristics [2].

Using ordered message queues, a method for analysis of asynchronous programs that are message-passing was presented in [8]. Although the method parameters are bounded, it is not the case for the number context switches and pending messages. However the number of communication cycles is limited. In these cycles any number of processes can send or receive any number of messages using any number of context switches.

Communicating Finite State Machines (CFSMs) was used in [12] to represent parallel systems of processes. This work also designed a framework of Communicating Minimal Prefix Machines (CMPMs) representing the CFSMs. This is not the case in the classical product automaton designed using particular CFSMs with large state-spaces [12].

4 Future Work

An interesting direction for future work is to provide a mathematical semantics for asynchronous programs using concepts from algebra and topology theories [7,25,29]. Then the semantics can be used to prove whether the given asynchronous program has a divergent behavior. Initially results of ongoing research by the author of the paper has led to the believe that bilattices [1,20,24] may be suitable algebraic structures to capture semantics of asynchronous programs.

5 Conclusion

This paper presented a new technique for transformation of asynchronous programs for the sake of revealing divergent behaviors. The idea of the transformation is to convert the source asynchronous program P into a new program P'. Then the divergent problem of P amounts to the problem of state reachability of P'. Precise syntax and semantics for source and target languages of the transformation technique were presented in the paper. The syntax and semantics were then used to formally present the main transformation of the paper.

References

1. Arieli, O., Avron, A.: Reasoning with logical bilattices. J. Log. Lang. Inf. **5**(1), 25–63 (1996)
2. Atig, M.F., Bouajjani, A., Narayan Kumar, K., Saivasan, P.: Linear-time model-checking for multithreaded programs under scope-bounding. In: Chakraborty, S., Mukund, M. (eds.) ATVA 2012. LNCS, vol. 7561, pp. 152–166. Springer, Heidelberg (2012)
3. Atig, M.F., Bouajjani, A., Touili, T.: Analyzing asynchronous programs with preemption. In: LIPIcs-Leibniz International Proceedings in Informatics, vol. 2. Schloss Dagstuhl-Leibniz-Zentrum für Informatik (2008)
4. Backes, M., Pfitzmann, B., Waidner, M.: The reactive simulatability (RSIM) framework for asynchronous systems. Inf. Comput. **205**(12), 1685–1720 (2007)
5. Benveniste, A., Berry, G.: The synchronous approach to reactive and real-time systems. Proc. IEEE **79**(9), 1270–1282 (1991)
6. Berry, G., Gonthier, G.: The esterel synchronous programming language: design, semantics, implementation. Sci. Comput. Program. **19**(2), 87–152 (1992)
7. Birkhoff, G., Birkhoff, G., Birkhoff, G., Birkhoff, G.: Lattice theory. Am. Math. Soc. **25** (1948). New York
8. Bouajjani, A., Emmi, M.: Bounded phase analysis of message-passing programs. Int. J. Softw. Tools Technol. Transf. **16**(2), 127–146 (2014)
9. Strüber, D., Rubin, J., Arendt, T., Chechik, M., Taentzer, G., Plöger, J.: Rule-Merger: automatic construction of variability-based model transformation rules. In: Stevens, P., et al. (eds.) FASE 2016. LNCS, vol. 9633, pp. 122–140. Springer, Heidelberg (2016). doi:10.1007/978-3-662-49665-7_8
10. Cristian, F.: Reaching agreement on processor-group membrship in synchronous distributed systems. Distrib. Comput. **4**(4), 175–187 (1991)
11. Czaplicki, E., Chong, S.: Asynchronous functional reactive programming for GUIs. In: ACM SIGPLAN Notices, vol. 48, pp. 411–422. ACM (2013)
12. Dakshinamurthy, S., Narayanan, V.K.: A component-based approach to verification of formal software models to check safety properties of distributed systems. Lect. Notes Softw. Eng. **1**(2), 186 (2013)
13. Deligiannis, P., Donaldson, A.F., Ketema, J., Lal, A., Thomson, P.: Asynchronous programming, analysis and testing with state machines. In: Proceedings of the 36th ACM SIGPLAN Conference on Programming Language Design and Implementation, pp. 154–164. ACM (2015)
14. Desai, A., Garg, P., Madhusudan, P.: Natural proofs for asynchronous programs using almost-synchronous reductions. In: ACM SIGPLAN Notices, vol. 49, pp. 709–725. ACM (2014)
15. El-Zawawy, M.A.: An efficient layer-aware technique for developing asynchronous context-oriented software (ACOS). In: 2015 15th International Conference on Computational Science and Its Applications (ICCSA), pp. 14–20. IEEE (2015)
16. El-Zawawy, M.A.: A robust framework for asynchronous operations on a functional object-oriented model. In: 2015 International Conference on Cloud Computing (ICCC), pp. 1–6. IEEE (2015)
17. El-Zawawy, M.A., Alanazi, M.N.: An efficient binary technique for trace simplifications of concurrent programs. In: 2014 IEEE 6th International Conference on Adaptive Science & Technology (ICAST), pp. 1–8. IEEE (2014)
18. Emmi, M., Ganty, P., Majumdar, R., Rosa-Velardo, F.: Analysis of asynchronous programs with event-based synchronization. In: Vitek, J. (ed.) ESOP 2015. LNCS, vol. 9032, pp. 535–559. Springer, Heidelberg (2015)

19. Emmi, M., Lal, A.: Finding non-terminating executions in distributed asynchronous programs. In: Miné, A., Schmidt, D. (eds.) SAS 2012. LNCS, vol. 7460, pp. 439–455. Springer, Heidelberg (2012)

20. Fitting, M.: Bilattices and the semantics of logic programming. J. Log. Program. **11**(2), 91–116 (1991)

21. Ganty, P., Majumdar, R.: Algorithmic verification of asynchronous programs. ACM Trans. Program. Lang. Syst. (TOPLAS) **34**(1), 6 (2012)

22. Ganty, P., Majumdar, R., Rybalchenko, A.: Verifying liveness for asynchronous programs. In: ACM SIGPLAN Notices, vol. 44, pp. 102–113. ACM (2009)

23. Jhala, R., Majumdar, R.: Interprocedural analysis of asynchronous programs. In: ACM SIGPLAN Notices, vol. 42, pp. 339–350. ACM (2007)

24. Jung, A., Rivieccio, U.: Kripke semantics for modal bilattice logic. In: 2013 28th Annual IEEE/ACM Symposium on Logic in Computer Science (LICS), pp. 438–447. IEEE (2013)

25. Kelley, J.L.: General Topology. Springer Science & Business Media, New York (1975)

26. Kopetz, H.: Real-Time Systems: Design Principles for Distributed Embedded Applications. Springer Science & Business Media, New York (2011)

27. Liskov, B., Shrira, L.: Promises: linguistic support for efficient asynchronous procedure calls in distributed systems. In: ACM Proceedings, vol. 23 (1988)

28. Pfitzmann, B., Waidner, M.: A model for asynchronous reactive systems and its application to secure message transmission. In: Proceedings of 2001 IEEE Symposium on Security and Privacy S&P 2001, pp. 184–200. IEEE (2001)

29. Stone, M.H.: Applications of the theory of Boolean rings to general topology. Trans. Am. Math. Soc. **41**(3), 375–481 (1937)

30. Wilson, R.P., Lam, M.S.: Efficient context-sensitive pointer analysis for C programs. In: ACM Proceedings, vol. 30 (1995)

Is Scrum Useful to Mitigate Project's Risks in Real Business Contexts?

Joana Oliveira[1(✉)], Margarida Vinhas[1], Filipe da Costa[1], Marcelo Nogueira[1,2], Pedro Ribeiro[1], and Ricardo J. Machado[1]

[1] ALGORITMI Research Centre, University of Minho, Guimarães, Portugal
jmaoliveira.92@gmail.com
[2] Software Engineering Research Group, University Paulista, UNIP,
Campus of Tatuapé, São Paulo, Brazil
marcelo@noginfo.com.br

Abstract. This work aims to determine the impact of the agile methodologies usage in software development, in particular, the usefulness and impact of the adoption of Scrum in a real business scenario. The aim is thus whether the adoption and implementation of this methodology, has contributed, and to what extent, for the mitigation of the risk management of software project and for the quality of the software. In order to be possible this study was carried out an investigation, which was distributed internationally, via discussion groups of professionals, followers and supporters of this methodology. Some of the issues explored were: the impact on elicitation of requirements; communication between team's members and the result of the development - the product. The aim is to also check if the function, role or the culture of the element is an influential factor in your opinion and attitude towards this methodology.

Keywords: Scrum · Utility · Project monitoring · Risk · Project management · Software

1 Introduction

With a continuing increase of the existence of critical services computerized it is fundamental to improve the implementation of the information systems used. The organizations that produce software products seek to adapt its processes, using agile development methodologies in order to promote the quality of its products. The main concern of this project is the risk mitigation in a set of agile methodologies adoption, Scrum in particular, by detecting as soon as possible any risk situation when managing the project development. As possible risk situations, among others, there may be flaws in the communication between those involved, different understanding of the same message, or even disinterest or abandonment of any of the parties involved, so that, given the possible impact in the project, it is very important for a project manager to have a tool that can help him to detect these situations. Scrum is a major help in the involvement of all parties given its nature.

There are already studies, found in literature, conducted around the Scrum impact. Among these studies already carried out, it is possible to highlight three studies similar

© Springer International Publishing Switzerland 2016
O. Gervasi et al. (Eds.): ICCSA 2016, Part V, LNCS 9790, pp. 422–437, 2016.
DOI: 10.1007/978-3-319-42092-9_33

to that undertaken throughout this document, but that differ in scope. Green [5] has determined the Scrum impact in the development of a product in a defined company, in a quantitative manner only, and without any relation to the correspondent's origin. In turn, Cartaxo et al. [2], has attempted to measure the Scrum impact in the customer satisfaction. This was also an empirical study without taking into account, as the first study mentioned, the culture impact of the parties involved in the process. Last of all, Mann and Maurer [7] have conducted a field-work focused only on product quality.

This document aims to present the results of an survey which allows to, when applied to an organization, infer about the impact of the use of Scrum in the process of software development, the attitude impact and the perception that the different team members have of the entire process, as well as all the stakeholders in the project. This survey could be a monitoring base of continuous improvement processes or their optimization, through the detection of possible risk situations, as soon as possible. The obtained data in this survey were analyzed by mainly using the statistical software SPSS, however, the Excel was also used in order to obtain some graphics. If the organization so wishes, it can use this survey as a base of information for performance indicators to be used in the Balanced Scorecard (BSCs) [6]. On the other hand if applied generically to several respondents of different organizations, different areas and even to different regions and cultures, it will allow organizations to verify which trends are being followed as well as the alterations that are being suggested to the process. It is then a tool that can help the project managers to realize how they are and compare to others, in order to be able to share experiences and avoid problems that were already solved by others. It may also be a Benchmarking tool to identify the best existing practices [1].

2   The Scrum

The software process can be executed in different ways and according to different approaches, organized in a specific way (a model). The difficulties faced by the software development teams and their managers lies in defining processes that promotes management mechanisms that keep the project under control [4]. The model is an important decision in order to mitigate the risk.

Unlike the development process in sequential cascade in which the development is seen as a constant ow forward (like a waterfall) through the stages of requirements analysis, project, implementation, tests (validation), integration and software maintenance [11]. Scrum is an iterative and incremental development process to the project management and agile software development [3, 9, 10]. Scrum should have the following characteristics [10, 11].

Many software projects fail for several reasons including the lack of risk management. Nogueira [8] (p. 172) has identified a list of risks to these type of projects having established a ranking for them. Among the risks identified and classified stand out: requirements poorly defined, incomplete or misunderstood; continuous changing requirements; omission of important information during the project; lack of motivation of the team among others. From a total of 39 risks analyzed, 21 have been validated.

Based on the intrinsic characteristics of Scrum, especially by the type of event, artifacts and engaging in the process, this seems to be useful in mitigating the risk of software projects. This paper intends, through a survey presented below, inferences about the Scrum utility and impact in the risk mitigation in software development projects.

3   Research Approach

3.1   Research Method

The problem under consideration refers to the Scrum usefulness in the risk mitigation in a business context of a Software House.

As result of the problem set from this study, it is intended to answer to the following question of investigation: Which use can Scrum have in the risk mitigation, in a business context of a Software House?

Considering the problem under study and the issue of investigation to which this problem origins, the method of investigation chosen is the survey method which will be brief presented in the following section. The survey is divided into two main parts, some open-ended questions and one closed-ended question. The conclusions drawn on the respondents' opinion and their personal suggestions, transmitted in these open-ended questions, are presented in the section correspondent to the conclusions of this study. Since this is an academic study about the Scrums impact, this survey presents a positive perspective of this framework, as it is primarily directed to those who practice it, so it is assumed that this framework has already been accepted by the respondents.

On the one hand, so that it is possible to the respondents to call into question this framework and suggest possible alternatives to Scrum, some close-ended questions were included, as mentioned above, where some respondents demonstrate their opinion on a particular statement, according to a Likert scale of 1 to 5, where 1 represents a total disagreement with the statement and 5 represents a full agreement with the statement proposed. On the other hand, to make it possible to obtain an analysis of the respondents opinion, it has been included an open-ended question in order to contextualize their answers in general and open a space to present their ideas.

3.2   The Survey

The survey used on this research work is organized into 11 groups, each one with the following purpose:

1. Personal and professional data – the information obtained by the answers from group 1 will allow to segment the qualitative answers of the survey.
2. Details of the last participation in project (with Scrum) – the questions from group 2 intend to obtain information on the last participation of the respondent in Scrum.
3. Benefits/constraint of using Scrum – the questions from group 3 intend to infer qualitatively the respondent's perception regarding the benefits and constraints of Scrums utility.

4. Existence and relevance of the Scrums role: Product Owner – the questions from group 4 are focused on the existence and importance of Scrums role: product owner.
5. Existence and relevance of the Scrums role: Scrum Master – the questions from group5 are focused on the existence and importance of Scrums role: Scrum Master.
6. Existence and relevance of Scrums artifact: Scrum Board – the questions from group6 are focused on the existence and importance of Scrums artifact: Scrum board.
7. Product Backlog definition – in the question 7 is intended to verify the product backlog quality, that is, the definition of the software requirements to develop.
8. Events taken into account – with the question in group 8 is intended to determine which events have been taken into account during the process, having as reference the last participation.
9. What would you improve in the implementation of Scrum? – Although it's an open-ended question, and of difficult or even impossible statistical analysis within the same organization, question 9 can provide information or important suggestions to improve the processes, but it may also indicate that the communications between people are not made easier, because if they were, answers are not expected, they would have appeared in the course of the project.
10. The best process for software development – question 10 is very important to understand if the Scrum is the best development of methodology to the stakeholders to follow and, if not, which other methods should be used in substitution (if responding to 10c) or combined with others (10b).
11. About the study – questions 11 and 12 will allow the respondents to be more directly involved, if they so wish. In addition, the respondents will be able to give suggestions to the study anonymously as well as leave their information in order to have access to the study result. The latter situation would be more relevant if the survey is used generically in different organizations, types of market or even in different geographical areas.

The survey evaluation was conducted in a real business environment and in a controlled manner by a well-defined team of software development. Within this software development team, the survey were validated by professionals with different roles in the team and different Scrum roles, to ensure that the questions were interpreted in a consensual way. In this way, the mitigation of the vocabulary ambiguity was ensured and used in the placement of the questions.

3.3 Case Study

The purpose of this project is not just to present a case study about the impact arising from the use of Scrum, as has happened with the work field made by Mann and Maurer [7]. In this project was made one case study in order to test which survey to use in this research. The results of the survey validation among the professionals of software development were interpreted without the use of any technique of statistical analysis, having been possible to get some of the following conclusions:

1. In the studied company is unanimous that the communications between the different team members have improved without any notorious bureaucratic increase in the process.
2. Regarding the different types of events and existent roles, they all agree, which suggest that the stakeholders know the process well.
3. There are disagreements on the size of the sprint (which in the opinion of some developers should be extended from 1 to 2 weeks).
4. Seek to improve the specification and clarification of all the requirements previous to the start of a sprint, so as to avoid the inclusion/modification of the same during the sprint.
5. The Scrum board is unanimously considered (all answers are maximum rating of 5 on a scale of 0 to 5) as an added value. However it is clear that the Scrum board (initially a physical picture on the wall) is currently in an online platform, but this format presents more complexity to the process and hamper its analysis during the meetings, so that it is suggested to return to the physical framework (with post-it on the wall).
6. Overall, the attitude towards the use of Scrum is very positive for which the process should be maintained. This approach is considered useful, at least for the team, since it is possible to have more sense of what is being done by the team instead of each member being working on his own task. The analysis of the process works better as a team than as individual work.
7. Based on the answers of the respondents (questions 9, 10 and 11) is suggested that the Scrum should be complemented with other methodologies including: IBM Rational Unified Process; Adaptive Software Development (ASD); Feature Driven Development (FDD) and Lean Software Development.

Therefore, it will be interesting review the software development process, in order to include, the best practices of these suggested methods.

3.4 Assigning Weights to the Different Roles in the Team and to the Different Scrum Roles

As presented by Nogueira [8] is possible to assign weights that reflect the relevance of the obtained answers according to certain criteria that differentiate the respondents. In this study we used the role that the respondents play in the project team as well as their Scrum role. To reflect the different relevance of each of the answers, were set different weights according to the analysis of each question, 3.1 to 3.17, presented in the previous section and that will be used in the testing of the formulated hypotheses. Weights were assigned, that is, numerical values between 0 and 1, so that the sum of this weight distribution by function/Scrum role is equal to one. For each of the possible segmentation were used four possible profiles.

The weights are likely of different interpretations due to the fact that different people may assign different weights to the answers of these respondents depending on the degree of importance that these functions have to each one of them, that is, the assignment of these weights is subjective. This was the chosen scale to do this paper given the authors experience, with the intention of studying the assumptions made taking into account the different relevance of the given answers.

4 Results and Discussion

4.1 About the Sample

This section is intended to present the results obtained by the analysis of the answers from the online survey, using a statistical analysis.

The people who answered the survey had 36 years old and only two people with 48 years old answered the survey. Note that also answered the survey ten people with 28 years old and eleven with 32 years old, The average age of the individuals that answered the online survey is of approximately 34 years old and that the youngest individual, at the time of the survey, had 24 years old and that the oldest individual, at the time of the survey, had 49 years old. It is still possible to identify 85 individuals who answered the survey, this value constitutes the sample size.

Relating the professional experience with or without the use of Scrum, it can be observed that in this question most of the individuals only have one years of experience, but the second largest number of individuals in study has ten years of professional experience. It is also important to note that there are still some respondents with 25 years of professional experience, which shows that this survey was able to cover a large scale of years in terms professional experience.

Relating the professional experience that the individuals have with the use of Scrum shows that the largest number of respondents and more significant, since it verifies a big difference between these and the remaining number of respondents distributed by the years of experience with Scrum, only have one year of experience in using this agile method. The second and third highest number of individuals with more experience in Scrum have 2 and 5 years of experience, respectively. It was still possible to get answers from two individuals with 7 and 8 years of experience in Scrum. The difference that exists between the total number of years of experience and the number of years of experience using Scrum can be explained by the fact that the use of this agile method is relatively recent in the world of project management.

Related to the country where each of the answers arise verifies that about 47 % of these answers are from Portugal. Remain 53 % were obtained through answers from other parts of the world such as China, USA, Australia, Germany, UK, Switzerland, Spain, India and Brazil.

Relating to the area of intervention of the company it can be verified that about 78 % of the answers match to individuals that conduct their tasks in Software House companies. Relating to the role that each individual who answered to the survey plays in their team, it verifies that 39 % play the role of developer, 29 % are project managers and that 14 % are testers. However it should be noted that this survey was able to cover all the tasks that can be performed in a team that intervenes in the completion of a software project.

4.2   Benefits/Constraints of Using Scrum

It is recalled that the scale of 1 to 5 that is displayed on the x-axis of each graphic is the scale provided in the survey so that the respondents could express their opinion, so 1 corresponds to a total disagreement with the statement presented and 5 corresponds to a total agreement with the statement presented.

The Fig. 1(a) corresponding to the communication between the elements of the team shows that there is a higher percentage of respondents (80 %) who answered that they agree or totally agree with the fact that the use of Scrum have eased the communication between the team members.

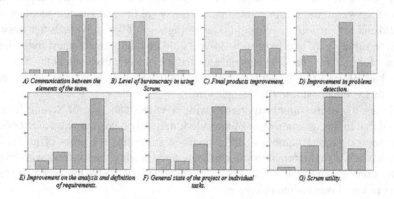

Fig. 1. Benefits/constraints of using Scrum.

There is still 15 % of respondents who have no opinion on this subject and there is also an even smaller percentage of individuals who share the opinion that the use of Scrum in their projects did not translate into an improvement of communication between the team members.

The Fig. 1(b) related to bureaucracy intends to show whether the use of Scrum has given more bureaucracy to their project. It is noted by its observation that the highest percentage of respondents (about 59 %) is in disagree or total disagreement with the fact that the adoption of this agile method translates into more bureaucracy for the project. However, there is a significant percentage of respondents (about 25 %) that have no opinion on this. The remaining percentage of respondents (around 16 %) share the opinion that the use of Scrum resulted in an increased bureaucracy in the project.

The Fig. 1(c) regards the respondents perception given the improvement that was verified in the products final project when using Scrum. By its observation it was concluded that about 72 % of respondents agree or totally agree with the fact that there has been verified an improvement in the final product by using this agile method. Still, about 21 % have no opinion on this subject and about 7 % of the total number of respondents disagree or total disagree with this fact.

The Fig. 1(d) refers to a better problem detection, or not, by those involved in a project with the use of Scrum. By observing this graphic it verifies that about 54 % of respondents agree or totally agree that the use of this agile method has eased the problems detection which will be more difficult to detect with the use of traditional methods. About 31 % of respondents have not opinion on this subject. It is also observed that there are no respondents in disagreement with this ease of problems detection attached to the use of Scrum, but even so 15 % of the individuals disagrees with this fact. However, it is important to point out, that there are no individuals in total disagreement with the use of Scrum in a better detection of problems.

The Fig. 1(e) refers to the fact that the analysis and definition of requirements to the system, may or may not be simplified. It is then verified that the highest percentage of individuals, about 61 %, agree or total agree with the fact that the analysis and definition of requirements to the system may be simplified. About 25 % of respondents have no opinion on this. In addition, about 14 % of the respondents disagree or totally disagree with this fact.

The Fig. 1(f) refers to the idea that at some stage of the project, its general condition or each of the tasks may be simplified. By its observation it is verified that about 69 % of the respondents agree or total agree with this statement. About 18 % have no opinion on this subject. However, there are still about 13 % of respondents who disagree or totally disagree with this statement.

In the case of this Fig. 1(g), is intended to analyze if those involved in team projects perceive Scrum as a useful method. By observing it, it is verified that 78 % of respondents agree or totally agree that Scrum is a useful method. About 20 % have no opinion on this. It is important to point out that there is no respondent that disagree or totally disagree with the statement that Scrum is a useful agile method, yet, about 2 % of respondents do not agree with this statement. It is also important to note that there are no individuals in total disagreement.

4.3   Existence and Relevance of Scrum Roles: Product Owner and Scrum Master and the Artifact Scrum Board

The Fig. 2 indicates to which levels were the Scrum roles, the Scrum Master and the Owner and the artifact Scrum Board were taken into account in the last participation of Scrum. In the graphic it is then clear that the Scrum Master exists in all implementations, as desirable, exclusively in about 95 % of the cases and in the remaining 5 % but not completely, possibly the Scrum Master accumulates other functions in the project. To what Scrum Master concerns, the scenario is slightly different, it does not exists in almost 12 % but exists in a not complete way in almost 25 %. Scrum Board is used in almost 80 % of the cases and in the respondents opinion this role comes as a real added value to the project. The figure also indicates to what extent the Scrum roles of Scrum Master and Owner and the Scrum Board artifact were taken into account in the last Scrum participation. It is then notorious that the Scrum Master exists in all implementations, as desirable, in an exclusive way in about 95 % of the cases and in the remaining 10 % it exists but not completely, possibly the Scrum Master accumulates other functions in the project. With regard to the Product Owner, the picture is slightly different, this does not exist in almost 12 % but exists in a not complete way in almost 25 %.

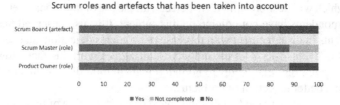

Fig. 2. Scrum roles and artifacts taken into account. (Color figure online)

Scrum Board is used in almost 85 % of the cases and, in the respondent's opinion, this comes as a real added value to the project.

4.4 Product Backlog Definition

Regarding the definition of Product Backlog, that is, the definition of all the features to be included in the product, some questions were made in order to identify at which stage of the project was carried out the requirements elicitation and who were the team members responsible for this process.

Most of the respondents have identified that the product requirements have not been all defined at the beginning of the project, wherefore the Product Backlog was being redefined through out the same, namely during the Sprint Planning meetings.

Regarding the definition of product requirements, in an almost unanimous way the respondents have identified that the Scrum Master (as process manager) and Product Owner (as customer representative in the project) have a fundamental role in its definition, analysis and communication to the other team members.

4.5 Events that Were Taken into Account

In Fig. 3 above, is possible to see which events have been taken into account considering the last Scrum participation in which the respondents have participated. On the graphic we can see that the use of Sprints practically unanimous, while the end meetings and the team meetings to the project closure are not significantly implemented. Note that

Fig. 3. Events that were taken into account in the last Scrum participation (Color figure online).

such meetings are not Scrum general rule. We can see that the more "standard" methodology events are taken into account in most of the Scrum implementation.

4.6   The Best Software Development Process

The previous Fig. 4 shows the respondents opinion about Scrum as the best method. We can see that about 40 % of the respondents argues that the Scrum when used exclusively is the best method while about 25 % support Scrum but when combined with other methodologies. In the following section, are presented the alternatives suggested by these respondents.

Fig. 4.   Scrum as the best method (Color figure online).

4.7   Methodologies Suggested for Combination with Scrum

In the list of suggested methodologies as possible options (Fig. 5) to combine with Scrum notably arises Lean software development followed by Extreme Programming and Agile and then immediately comes Kanban (Development), Feature Driven Development (FDD) among others such as IBM Rational Unified Process.

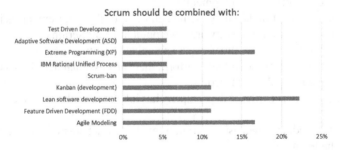

Fig. 5.   Methodologies suggested for combination with Scrum.

In opposition are about 40 % of the respondents, who do not consider the Scrum as the best methodology to follow, presenting in the following section some alternatives.

4.8   Suggested Methodologies for the Replacement of Scrum

The alternatives that are more suggested (Fig. 6), by the respondents are Scrum-ban and Kanban. The first arises already as an alternative to Scrum in combination with Kanban, which also appears as an alternative. Thus it is concluded that the kanban Scrum embodiment, Scrum-ban, appears as an option to follow.

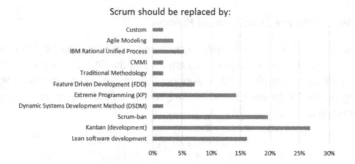

Fig. 6. Suggested methodologies for the replacement of Scrum.

4.9 Hypothesis Analysis

In order to get outcomes more complete and diversified which allows to draw more solid conclusions of the study, a set of hypothesis was analyzed.

To evaluate the hypothesis H1, H2, H3, H4, H5, H6, H7, H8, H9, H10, H11, H12 e H13 was used the Spearman correlation test.

It has been made a descriptive analysis of the averages obtained facing the respondents answers where:

H1: The improvement of communication between elements is influenced by the years of experience in software development.

H2: The improvement in the analysis and definition of the requirements is influenced by the years of experience in software development using Scrum.

H3: The age of individuals influences the perceived reality of Scrum.

H4: The country where individuals perform their professional activity has influence on the perceived usefulness of Scrum.

H5: The area of intervention of the company has influence on the perceived usefulness of Scrum.

H6: The final product, when using Scrum, is evaluated as better depending on the role played in the team.

H7: The role played in the project team has influence on the improvement of the evaluation of the project status or of individual tasks.

H8: The years of experience in software development, using Scrum or not, has influence on the individual's perception facing the bureaucracy increase.

H9: The years of experience in software development, using Scrum or not, have on the improvement of the evaluation of the project status or of individual tasks.

H10: The years of experience in software development using Scrum have influence on the improvement of communication between the elements of team.

H11: The improvement of communication between the elements has influence on the improvement of the analysis and definition of the requirements.

H12: The problems detection has influence on the improvement of the final product quality.

H13: The improvement of communication between the elements has influence on the improvement of the evaluation of the project status or of individual tasks.

Taking into consideration the acceptance criterion of the null hypothesis[1], in the case of the hypothesis tested, it has verified that among the variables tested in H2, H3, H4, H5, H6, H7, H9 and H12 there is no association between the variables. The outputs that support these conclusions are attached in this document.

In the event of the hypotheses in which H0 was rejected, that is, where there's an association between the variables, this association can be positive or negative.

The hypothesis 8 presents an association low inverse, which means that there is a tendency for individuals with higher number of years of experience in software development using Scrum or not, perceive a decrease in the level of bureaucracy of the project.

In the case of H10 and H13 it has been verified a low direct association between the variables in test. It means, for H10, that there is a tendency for the individuals with highest number of years of experience in software development using Scrum, to perceive an improvement in the communication between the team members. For H13, it means that there is a tendency for the improvements in the communication between the team members to translate into improvements in the evaluation of the general state of the project or of individual tasks.

The hypothesis 11 presents a moderate direct association between the variables in test, that is, there is a tendency for improvements in communication between the team members to translate into improvements in the analysis and definition of project requirements.

It refers, after analyzing each test, that although there are variables associated with direct or indirect association, none of these results translate into significant associations since they all are of low to moderate level.

4.10 Analysis of the Hypotheses with Weight Attribution to Variables

From the set of the 13 hypotheses defined initially, it has been selected a subset from those hypotheses, taking into consideration its relevance and possible interest in the study and, in a more concrete way, the different roles and functions of the respondents. The selected hypotheses were: H4, H5, H6, H11 e H12. In each of the hypotheses test were used the average value of each of the variables taking into account their weights.

Analysis of the hypotheses with weights for the performed function in project team.

Having in consideration the acceptance criterion of the null hypothesis, presented previously, for the tested hypotheses matter, it has been verified that among the variables tested in H4 and H5, there is no association between the variables.

---

[1] To conduct a Spearman correlation test is intended to test if: H0: The variables are not associated; H1: The variables are associated. The null hypothesis (H0) is accepted whenever the level of the test significance performed is higher than 0.05. Concluding that the test variables are not associated and that's why there is no influence between them.

In the case of the hypotheses in which H0 was rejected, that is, where there is an association between the variables, this association can be positive or negative. Hypothesis 11 (H11) and the hypothesis 12 (H12) present a low reverse association. In the case of (H11) it means that there is a tendency for improvements in communication between the project team members to translate in a reduction of improvements in the analysis and requirements definition.

In the case of (H12) it means that there is a tendency for improvements in problems detection can result in a reduction of the improvements in the final product quality. Against the odds H11 and H12, it is necessary to note that the results seems to be the unexpected ones. The results are being in contradiction with the same study without the weights according to the function on the team. These contradictions may be related with the selection of weights carried out, it makes sense to define different weights for future work. It is important to point out that when the test was performed without the assignment of weights to the role that the individuals perform in H11 team, in contrast to what was verified now, it presented a direct association on the variables. To (H12) the null hypothesis was accepted so it was not verified any association between the variables.

The hypothesis 6 presents a low direct association which means that there is tendency for the role played in the project team to translate in an improvement of the final product evaluation. Previously, when weights were not considered for the variables, the null hypothesis was rejected, that is, it didn't verifies any association between the variables.

It refers, after analyzing each test that although there are variables with direct or inverse association, none of these results translate into significant associations since they all are of level low to moderate.

Analysis of the hypotheses with weights to Scrum role Taking into consideration the acceptance criterion of the null hypothesis, presented previously, for the tested hypotheses matter, it has been verified that among the variables tested in H4, H5 and H11 that there is no association between the variables. The outputs that support these conclusions are attached in the document.

In the case of the hypotheses in which H0 was rejected, that is, where there is an association between the variables, this association can be positive or negative.

The hypothesis 6 (H6) and the hypothesis 12 (H12) present a low reverse association. According the result for H6 it is possible to conclude that the opinion about the product quality (as a result) is influenced by the Scrum role. In the case of (H12) it means that there is a tendency so that the improvements in detecting problems can be translated in a reduction of improvements in the final product quality. Regarding H12 hypothesis, it should be emphasized that the results for the study of this hypothesis seems to be unexpected, i.e., does not seem reasonable that the improvement in detection problems lead to a reduction in product quality. They are being in contradiction with the same study without the weights according to Scrum role. This contradiction may be related with the selection of weights carried out, it makes sense to define different weights for future work.

When the test was performed without assigning weights to Srcum role (H6), in contrast to what was verified now, it didn't present any relation between the variables in test. When the test was performed with the assignment of weights to the role that the individuals perform in the team (H6) it presented a direct association, although low.

To (H12) the null hypothesis was accepted so it was not verified any association between the variables. When performing the test with the assignment of weight to the role that the individuals play in the team (H12) it presented, similar to the test with weight assignment to Srum role, an inverse low association.

After analyzing each test, it refers that even though there are variables with inverse association, none of the results translates into significant associations since they all are of low level.

5   Conclusion and Future Work

5.1   Conclusion

In addition to the findings described in the analysis of the graphics presented above, it is possible to draw some conclusions of this research in general terms.

During this research it wasn't detected relevant correlations between the studied variables, namely the geographical origin of its stakeholders, or the area of intervention of the respondents organizations, which proves the wide acceptance and use of Scrum. However, it was presented some limitations and some points to improve in this methodology when used in a more purist way.

The biggest limitation pointed out by some respondents was the fact that the Scrum can not handle dispersed geography teams. This methodology foresees periodic meetings such as the Daily Scrum meeting that should take place every day and preferably with all the team standing next to Scrum Board, which should be a physical framework in order to make quick meetings. However, when the elements of a project team are geographically dispersed this meeting is then described in Scrum as not efficient, so it can be a true value to make a Scrum Board in an online platform in a way to be remotely accessible to all the team and thereby, allowing meetings to also be performed remotely, for instance, through videoconference. This way it can be viable the definition of a new Scrum artifact: chat room. In any scenario, the existence of a framework where the state of the project is always visible and of easy access to all the team Scrum Board is considered an added value of this methodology.

The definition of Sprints, namely its fixed duration in time was also pointed as a possible problem of this methodology because at the time of the definition and the assignment of the tasks to each element in a Sprint, the estimated time for each task can be underestimated or estimated in an optimistic way without regarding possible problems, which may overload the team and, consequently, the quality decrease of its results. In this sense it may be more viable the definition of story points or the adoption of the Kanban or its Scrum variant, the Scrum-ban, more exible in the definition of Sprints and even the due dates of intermediary releases.

Since this is an agile methodology that seeks to involve as possible all stakeholders in the project from the development team to the customer representative, their planning meetings, for instance, the Sprint Planning meetings should, whenever possible, function as a workshop presentation, analysis and discussion of requirements, problems and team tasks and not just a simple process of task attribution.

Despite the Scrum does not represent more bureaucracy to the management process a project in teams of small dimension should be adjusted case to case, namely in the Scrum Roles given to the team members, otherwise it can become counter-productive.

From the perspective of some of the respondents this methodology allows a quick reaction to changes but it is not capable, for itself, to measure properly its impacts on the project. Delays to the project plan are hardly perceptive.

Apart from this limitation were presented weaknesses or even challenges such as:

- How to do the reporting to other structures in the organization;
- Which successful metrics can be included in the process;
- Being the Scrum oriented to the team, it can be difficult the coordination between teams when there are more than a team working in the same project;
- How to avoid the tendency to make waterfall throughout the Sprint (making the linear process of development within each sprint, instead of following the agile philosophy).

So this methodology should, in the opinion of its followers, be adapted to combine it with other methods (such as: Lean software development, Kanban (development) or Scrum-ban) and be accompanied of metrics and indicators that facilitates the tracking of the project status when compared to the plan. Migrating the Scrum Board to a computerized platform may contribute to the maintenance of the project metrics.

Despite the limitations that this methodology may have, Scrum, based on the answers of opinion that we have obtained, is unanimously accepted and a framework that functions when well applied. Not being exclusive for software development, not being exclusive for software development, Scrum can be used as a methodology in any development project or in the research of new products.

5.2 Future Work

As suggestions for future project the following are highlighted:

- Engaging more respondents by direct contact, for example, contacting more companies directly in order to obtain answers from the greatest possible number of elements of their project teams in order to get opinions from different elements and functions.
- Internationally, the study presented here is based on the answers given to the inquiry that were released in groups of professionals, participants Scrum via LinkedIn enthusiasts. A possible alternative would be to explore the creation of partnerships with companies and organizations, such as Scrum Alliance, that provides consulting services in Agile Methodologies and in particular in Scrum, in order to get more answers.
- Finally, test the usefulness of this survey, not in the general usefulness of Scrum but on the risk tracking of a project in particular, that is, get answers to this survey in different moments in the same project.

Acknowledgment. This research is sponsored by the Portugal Incentive System for Research and Technological Development PEst-UID/CEC/00319/2013.

References

1. Bogan, C.E., English, M.J.: Benchmarking for Best Practices: Winning Through Innovative Adaptation. McGraw-Hill, New York (2004)
2. Cartaxo, B., Araujo, A., Sa Barreto, A., Soares, S.: The impact of scrum on customer satisfaction: an empirical study. In: 2013 27th Brazilian Symposium on Software Engineering (SBES), pp. 129–136. IEEE, October 2013
3. Confluence. Tanning Documentation (2016). https://confluence.atlassian.com/display/AGILE/Scrum+Board. Accessed 05 May 2016
4. Fernandes, J.M., Machado, R.J.: Requirements in Engineering Projects. Springer, Heidelberg (2016)
5. Green, P.: Measuring the impact of scrum on product development at adobe systems. In: 2011 44th Hawaii International Conference on System Sciences (HICSS), pp. 1–10. IEEE, January 2011
6. Kaplan, R.S., Norton, D.P.: The Balanced Scorecard: Translating Strategy into Action. Harvard Business Press, Boston (1996)
7. Mann, C., Maurer, F.: A case study on the impact of scrum on overtime and customer satisfaction. In: Agile (2005)
8. Nogueira, M.: Engenharia de software: um framework para a gestão de riscos em projetos de software. Ciência Moderna (2009)
9. Schwaber, K.: SCRUM development process. In: Sutherland, J., Casanave, C., Miller, J., Patel, P., Hollowell, G. (eds.) Business Object Design and Implementation, pp. 117–134. Springer, London (1997)
10. Scrum Alliance (2016). Tanning Documentation. https://www.Scrumalliance.org/. Accessed 12 May 2016
11. Scrum Reference Card (2016). http://scrumreferencecard.com/scrum{reference-card/. Accessed 9 May 2016

A Semantic Comparison of Clustering Algorithms for the Evaluation of Web-Based Similarity Measures

Valentina Franzoni[1,3]([✉]) and Alfredo Milani[2,3]

[1] Department of Computer, Control, and Management Engineering,
La Sapienza University of Rome, Rome, Italy
[2] Department of Computer Science, Hong Kong Baptist University,
Kowloon Tong, Hong Kong
[3] Department of Mathematics and Computer Science, University of Perugia,
Perugia, Italy
{valentina.franzoni,milani}@dmi.unipg.it

Abstract. The Internet explosion and the massive diffusion of mobile devices lead to the creation of a worldwide collaborative system, daily used by millions of users through search engines and application interfaces. New paradigms permit to calculate the similarity of terms using only the statistical information returned by a query, or from additional features; also old algorithms and measures have been applied to new domains and scopes, to efficiently find words clusters from the Web. The problem of evaluating such techniques and algorithms in new domains emerges, and highlights a still open field of experimentation.

In this paper, preliminary tests have been held on different semantic proximity measures (average confidence, NGD, PMI, χ^2, PMING Distance), and different clustering algorithms among the most used in literature have been compared (e.g. k-means, Expectation-Maximization, spectral clustering) for evaluating such measures. The suitability of the considered measures and methods to calculate the semantic proximity was verified at the state-of-art, and problems were identified, comparing the results of measurements to a ground truth provided by models of contextualized knowledge, clustering and human perception of semantic relations, which data are already studied in literature.

Keywords: Data mining · Clustering · Semantic evaluation · Semantic similarity · Information retrieval

1 Introduction

One of the main problems that emerge in the classic approach to semantics is the difficulty in acquisition and maintenance of ontologies and semantic annotations [1,2,32]. On the other side, the flow of Web documents is continuously fuelled by the collaborative contribution [4,5] of millions of users. The existing semantic models are expressive enough; on the other hand their basic limitation lies on

© Springer International Publishing Switzerland 2016
O. Gervasi et al. (Eds.): ICCSA 2016, Part V, LNCS 9790, pp. 438–452, 2016.
DOI: 10.1007/978-3-319-42092-9_34

the inability of managing the evolution of ontological models and content annotations, which are not taken into account in the model itself. The lack of automation capabilities and evolutionary maintenance [6] is highly relevant, especially for the generation of context-based semantic annotations or focusing on specific social networks or repositories. Search engines, continually exploring the Web, are a natural source of semantic information on which to base a modern approach to semantic annotation [3,29,36]. A promising idea is that it is possible to generalize the semantic similarity, under the assumption that semantically similar terms behave similarly [7], and define collaborative proximity measures [8,9] based on the indexing information returned by search engines. The general idea is to use search engines as a black box to which submit queries and extract useful statistics, to measure proximity semantics [10,33] about the occurrence of a term or a set of terms from web or offline text entities (e.g. documents or pages), just counting the number of results indexed in the search engine.

One of the main models of evaluation for the performance of semantic proximity measures is clustering. Clustering is generally dependent on the parameters used in the specific algorithms, such as thresholds, initial data and number of desired clusters, as well as on the considered similarity measure [11,12,34]. This study aims to compare the most used clustering algorithms, as an evaluation base for the results of web-based proximity measures, on terms that can be efficiently used for clustering concepts or contexts. In a first phase the features of the main proximity measures used in literature were investigated, to best suit to the use of the statistic information provided by search engines, as a basis to extract semantic content from the Web. In a second experimental phase, the adequacy of the considered measures on the considered search engines was evaluated on clustering, testing different algorithms. In a third experimental phase, the adequacy of the evaluation with clustering was tested, comparing it with other two models of evaluation: use in contexts of knowledge and human perception of semantic associations.

Among the significant contributions of this work, besides the main goal, which is to provide an approach to compare web proximity, there are: the systematic experimental comparison of different proximity measures on the main search engines, both generalist (Google [13], Bing [14], Yahoo Search [15]) and specialized (Flickr [17] for photos and Youtube [16] for videos); the evaluation of the PMING Distance [3], a proximity measure recently introduced, which in the experimental tests carried out the best performances, obtained separately from the measures considered in literature; the use for the experiments of two data sets from literature, both for the data and for the ground truth, thus providing a direct comparison to well-established studies, instead of providing new human evaluations as a proof of validity.

2 The Evaluated Proximity Measures

2.1 Average Confidence (CM)

Given a rule $X \rightarrow Y$, confidence [18] is a statistical measure that, given the number of transaction which contain X, indicates the percentage of transactions which contain also Y. Confidence is a symmetric function. Indicating with $f(x)$, $f(y)$, $f(x,y)$ respectively the cardinalities of the results of a query of term x, y and x AND y in a search engine and N the number of documents which are indexed by the search engine:

$$confidence(x \rightarrow y) = (f(x,y)/N)/(f(x)/N) = f(x,y)/f(x) \qquad (1)$$

From a probabilistic point of view, confidence is an approximation of conditional probability:

$$confidence(x \rightarrow y) = P(x \wedge y)/P(x) = P(y \vee x) \qquad (2)$$

Average Confidence (CM) [4,8,18,35] of two terms x and y can be defined as

$$[confidence(x \rightarrow y) + confidence(y \rightarrow x)]/2 \qquad (3)$$

2.2   Pointwise Mutual Information (PMI)

Pointwise Mutual Information (PMI) [19] is a point-to-point measure of association used in statistics and information theory. Mutual information between two particular events W_1 and W_2, in this case the occurrence of particular words in Web-based text pages, is defined as:

$$PMI(w_1, w_2) = log_2 \frac{P(w_1, w_2)}{P(w_1)P(w_2)} \qquad (4)$$

This type of mutual information is an approximate measure of how much a word gives information on the other word of the pair, in particular the quantity of information provided by the occurrence of the event W_2 about the occurrence of the event W_1, that is the conditional probability $W_1|W_2$. PMI is a good measure of independence, since values near zero indicate frequency, but at the same time is a bad measure of dependence [3], since the dependency score is related to the frequency of individual words. In addition, pairs of terms with low frequency will receive a greater score than pairs of terms with high frequency, so PMI could not always be suitable when the aim is to compare information on different pairs of words [20].

2.3   Chi-Square Coefficient (χ^2)

$\chi 2$ (Chi-squared or Chi-square)[3] makes possible to assess the significance of a relation between two categorical variables, checking if the values, observed by measuring frequency, differ significantly from the frequencies obtained by the theoretical distribution. In common parlance two events are associated where you can define a relationship between them, but in statistics two events are associated only when they are more related than by pure chance. The question to which χ^2 can answer is "how much the observed data deviate from those that would be expected if they were random?". For each value, consider the quantity:

$$[(observed\ value - expected\ value)^2]/expected\ value \qquad (5)$$

where the numerator is squared to always get a positive number, even when the expected value is greater than the observed value. It is evident that the χ^2 increases when the difference between the compared data increases i.e., when the data can be considered significantly different from randomness. The sum of that amount on two values enables to calculate the relative significance of their co-occurrence: the higher the $\chi 2$, the greater the likelihood that the relation is not random, and therefore significant. Given two events W_1 and W_2, in this case the occurrence of particular words in Web-based text pages, define:

$a = W_1 \wedge W_2$ (number of documents where W_1 and W_2 occur);
$b = W_1 \wedge (\neg W_2)$ (number of documents where $W_1 1$ occurs, but not W_2);
$c = W_2 \wedge (\neg W_1)$ (number of documents where W_2 occurs, but not W_1);
$d = (\neg W_1) \wedge (\neg W_2)$ (number of documents where neither W_1 nor W_2 occur);
$n = N = a + b + c + d$.

An algebraically simplified formula to calculate χ^2 is the following [31]:

$$\chi^2 = \frac{(ad - bc)^2 n}{(a+b)(a+c)(b+d)(c+d)} \tag{6}$$

where the coefficient of association can be directly calculated from the observed data, without having to calculate the related expected values. The $\chi 2$ coefficient was also used in clustering of concepts with Newman algorithm [21].

2.4 Normalized Google Distance (NGD)

Rudi Cilibrasi presented in 2006 the Normalized Google Distance (NGD) [7] as a measure of semantic relation based on the assumption that similar concepts occur together in a large number of documents in the Web, i.e. that the frequency of documents returned by a query on Google or any other search engine approximates the distance between related semantic concepts. Notice that the NGD was originally defined for Google, but it is a measure that can be applied to any search engine, so "NGD" is not the same of "distance on Google". The NGD between two terms x and y is formally defined as follows:

$$NGD(x,y) = \frac{\max\{\log f(x), \log f(y) - \log f(x,y)\}}{\log M - \min\{\log f(x), \log f(y)\}} \tag{7}$$

where $f(x), f(y)$ and $f(x,y)$ are the cardinalities of results returned by Google for the query on $x, y, x \; AND \; y$ respectively, and M is the number of pages indexed by Google or a value which is reasonably greater than $f(x)$ and $f(y)$. If the two terms x and y do not ever occur in the same document, but occur separately (x and y not in relation), their NGD should be 1; otherwise if x and y always co-occur (x and y identical), their NGD should be 0. Although it is not a metric, the NGD is a measure of proximity which turns in a variety of experimental applications [3,7].

2.5 PMING Distance

The PMING Distance [3,6] is a hybrid proximity measure introduced in 2012 as a convex weighted linear combination of PMI [19] and NGD [7], which are the proximity measures that obtained the best performance in clustering, in preliminary experimental tests, preserving the characteristics of NGD and PMI relies on the intuition that a suitable combination of the two distances would introduce some contribution of PMI while giving prevalence to the NGD, because the latter has in general a better performance, when considering the tasks of contextualizing knowledge [4] (i.e. ranking of a set of terms with respect to a base term in a context), human perception of semantic relations, and clustering. In PMING, NGD and PMI are locally normalized[1] with the highest value W in the context of evaluation, obtaining values in range $[0,1]$; then complementary values[2] of PMI were calculated and the PMING distance was defined as a convex linear combination of the two locally normalized distances. More formally, the PMING distance of two terms x e y in a context W is defined, for $f(x) \geq f(y)$, as a function $PMING : WXW \rightarrow [0,1]$:

$$PMING(x,y) = \rho \left[1 - \left(\log \frac{f(x,y)M}{f(x)f(y)} \right) \frac{1}{\mu_1} \right] + (1-\rho) \left(\frac{\log f(x) - \log f(x,y)}{(\log M - \log f(y))\mu_2} \right) \tag{8}$$

where:

- ρ is a parameter to balance the weight of components ($\rho = 0.3$ in our tests);
- μ_1 e μ_2 are constant values which depend on the context of evaluation, and are defined as:

$\mu_1 = \max PMI(x,y)$, with $x,y \in W$
$\mu_2 = \max NGD(x,y)$, with $x,y \in W$

Concept clusters by NGD			Concept clusters by FD		
Group 1	Group 2	Group 3	Group 1	Group 2	Group 3
bears	**bowling**	baseball	moon	bears	baseball
horses	dolphin	basketball	saturn	dolphin	basketball
moon	donkey	football	space	donkey	football
space	**saturn**	golf	venus	**golf**	**snake**
-	sharks	soccer	-	horses	soccer
-	snake	tennis	-	sharks	bowling
-	**softball**	volleyball	-	spiders	softball
-	spiders	-	-	**tennis**	volleyball
-	turtle	-	-	turtle	-
-	**venus**	-	-	whale	-
-	whale	-	-	wolf	-
-	wolf	-	-	-	-

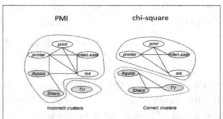

Fig. 1. Results of clustering in [17] *(left)* and in [22] *(right)*

---

[1] The local normalization of a measure η in a evaluation context W consists in evaluating $\max \eta$ of the values $\eta(w_i)$ for $w_i \in W$ and substituting these values with $\eta(w_i)/\eta$.

[2] The complementation to 1 of the locally normalized PMI consists in substituting each locally normalized value with $(1 - \eta(w_i)/\eta)$ and eventually forcing null values in the diagonal of the adjacency matrix.

2.6 Flickr Distance (FD)

Wu et al. in [30] (Microsoft Research Asia) introduced the Flickr Distance (FD) on visual domains on the Web, justifying it as better than other measures, such as the Normalized Google Distance, on the hypothesis that the latter - referring only to textual information - fails to capture some semantic relations that are present in images, and especially those contained in Flickr repository, which is built collaboratively by thousands of independent users. They believe in fact that, aiming to represent the semantics of natural language, is generally better not to refer to particular contexts. For each term in a list data, the FD is built choosing 1000 images from Flickr, obtained querying pairs of words. This large number of images is then given to a concept modeling algorithm (Visual Language Modeling, VLM), which divides each one of them into sectors and analyzes the features of each area, comparing changes with respect to the neighbors. The Flickr distance is finally calculated on the square root of the distance of Jensen-Shannon on the concepts, which correspond to visual models obtained via the VLM. To evaluate the goodness of FD respect to NGD, Wu et al. in [17] created two networks of terms for the measures, building for each one a graph where edges are present only between pairs of terms whose distance is below a given threshold. They carried out on these data two different tests: subjective tests on humans, where 12 people assign a score from 1 to 5 to each of 53 pairs of network concepts, to evaluate the adequacy to human perception; objective tests, comparing the networks of concepts with a third network built from the Word-Net [1] as ground truth. A further comparison is implemented on the networks, reduced to 23 words, with spectral clustering, counting how many concepts are assigned to different clusters and comparing them to the expected clusters. The results obtained from the study by the Microsoft Research team show that FD behaves in all three cases better than NGD, on the lists of concepts and pairs of concepts given in input. Actually the results of the present study partially contradict the argumentation that claims FD to have better performance than NGD, and image analysis to provide significant additional information compared to the related textual semantic information. The result obtained by Microsoft Research Asia for spectral clustering on Google distance and Flickr distance is showed in Fig. 2, where we can see in bold the errors with respect to an ideal and intuitive subjective clustering.

3 Clustering Algorithms

Clustering can be defined as a collection of techniques of multivariate analysis of data, consisting in evaluating a set of points, which are described by a set of attributes and a measure of similarity between them. The output will be partitions (clusters) of the input, where points belonging to the same cluster are homogeneous or similar. The measure of similarity (or rather, dissimilarity) used is usually the Euclidean distance in a multidimensional space, but any other proximity measure or association rule can be used. Among the main clustering techniques we can highlight categories as partitive and hierarchical, divisive and agglomerative clustering [22].

Agglomerative clustering allows to use measures of similarity or dissimilarity; while in divisive clustering the main function will use the density or the sparsity of the clusters.

3.1   Expectation-Maximization

The EM (Expectation-Maximization) is an iterative algorithm for finding maximum likelihood [23] in estimating parameters, in statistical methods where the model depends on latent variables, e.g. from equations which cannot be resolved directly, or from data which weren't observed but the existence of which can be assumed true. EM iteration rotates an expectation step (E), which calculates the expected likelihood on the current estimate of the parameters, and a maximization step (M), which estimates which parameters maximize the expected likelihood, calculated in the E step. The parameters obtained from step M are used in the next E step. Starting from an initial assumption on the model's parameters, the two E and M steps are applied iteratively until convergence, when updating the parameters doesn't increase anymore the likelihood. Although an EM iteration increases the observed parameters and therefore their marginal distribution, there is no guarantee that the sequence converges to a maximum likelihood estimator. For multimodal distributions, where we have continuous distributions with different modes, it's possible to obtain the convergence to a local maximum, depending on the initial data, as we can see in the following Fig. 1.

To avoid local maxima, it is possible to apply random reboot methods from different initial estimates, or the simulated annealing. Consider the function:

$$F(q, \theta) = E_q[\log L(\theta; x; Z)] + H8q) = -D_{KL}(q||p_{z|x}(\cdot|x; \theta)) + \log L(\theta; x) \quad (9)$$

where q is an arbitrary probability distribution of not observed data $z, pZ|X(\cdot|x; \theta)$ is the conditional distribution of the not observed data given

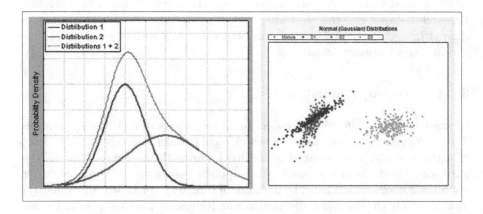

Fig. 2. Probability distributions in EM

the observed data x, H is the entropy and D_{KL} is the Kullback-Leibler divergence [26][3]. The two steps of expectation and maximization can then be seen as follows:

1. choose q to maximize f:
$q^{(t)} = \arg\max_q F(q, \theta^{(t)})$;
2. choose θ to maximize f:
$\theta^{(t+1)} = \arg\max_\theta F(q^{(t)}, \theta)$.

The EM-based clustering is used in machine learning and computer vision, for example in applications of diagnostic imaging, rebuilt in particular from PET (Positron Emission Tomography). In psychometric, EM is essential to estimate the parameters in the IRT (Item Response Theory) models, which make use of questionnaires and tests scores to measure latent abilities and other skills which are not directly observable. The ability of this algorithm to handle missing data and not observed variables makes it useful also for the management of risk in economy. EM can be seen as a generalization of the k-means method.

3.2   K-Means

K-means [25]is a clustering method to partition n observations into k clusters with the closest average (mean). The problem is computationally difficult (NP-hard), but there exist algorithms which make use of heuristics to converge quickly in a local optimum, similar to EM for the mixed Gaussians which form the distribution through finishing steps. Both approaches use central representative values of clusters to model the data, but the k-means tries to find clusters which are comparable basing on their spatial extension, while the EM is able to detect clusters of different shapes. Given a set of n observations $(x_1, x_2, ..., x_n)$, where each observation is a d-dimensional vector of reals, clustering with k-means partitions the observations into k sets $S = S_1, S_2, ..., S_k$ (with $k <= n$), to maximize the sum of squares of the distances between clusters:

$$\arg\min_D \sum_{i=1}^{k} \sum_{x_j \in S_i} \|x_j - \mu_i\|^2 \tag{10}$$

where μ_i is the average of the points in S_i. The steps of the algorithm are the following: 1. initialize k "media" (i.e. k centroids representing desired clusters), random from data sets, generating k clusters; 2. for each point in the data set, calculate the distance from centroid k, by associating points with the less distant average and subsequently assign the point to an existing cluster; 3. recalculate the average or centroid of each cluster, on the basis of the newly made assignments. These steps are repeated until centroids change. The main problem of

---

[3] The Kullback-Leibler divergence, also known as information gain or information divergence [24], measures the difference between two probability distribution P and Q, considering the expected value of the extra bits requested to encode examples from P when a code based on Q is used (i.e. represents the "true" distribution of the observations, or a theoretical distribution which correctly approximates P).

k-means clustering is the initialization of the centroids. In fact, repeating several times the algorithm on different centroids produces different results. Also the used proximity measure, as with any clustering algorithm, affects the quality of the results. Metrics in general lead to the best results.

3.3 Spectral Clustering

Given a set of data points A, the similarity matrix can be defined as the matrix S where S_{ij} represents a similarity measure between points i and j in A. Spectral clustering techniques [27] make use of the spectrum of the similarity matrix to make a size reduction, clustering on less dimensions. Spectral clustering algorithms are used in particular for image segmentation, by partitioning the points in two data sets on the cut of the second smallest eigenvalue of the Laplacian matrix[4]. Partitioning can be done in several ways, for example by taking the mean m of components associated with the eigenvalue v and placing all points with $v > m$ in the first cluster and the others in the second. In this way, the algorithm can be used for hierarchical clustering by repeatedly partitioning the obtained subsets. Other spectral clustering algorithms take into account the eigenvectors corresponding to the k largest eigenvalues, then invoke another clustering algorithm (e.g. k-means) to cluster points in the relative k components of the eigenvector. As a compromise between approaches based on graph theory and those based on optimization, spectral clustering methods are becoming very popular for machine learning and computer vision. An important application which approaches on pairs of points is the content-based image retrieval. Spectral clustering is in fact also used in [17].

4   Experimental Tests

Experimental tests have been run on data sets and ground truth provided by high impact papers in state of the art of web-based similarity [35] and clustering.

4.1   Data Sets Used for Experiments

Dataset1=printer, print, InterLaser, ink, TV, Aquos, Sharp
(7 terms used in [22] for Newman clustering)
Ground truth: clustering in the following clusters:

- printer, print, InterLaser, ink
- TV, Aquos, Sharp

---

[4] The Laplacian matrix, or Kirchhoff matrix [28], is used to calculate the spanning tree number, i.e. the number of the trees which can be constructed starting from an indirect connected graph, formed by every vertex of some of (or all) the edges; in other words, a selection of the edges of a graph which expand on each vertex such as every vertex is inside the tree, without cycles. In spectral graphs, the Laplacian approximates the most sparse cut of the graph, through the second eigenvalue.

Dataset2=soccer-basketball, dolphin-sharks, smallville-superman, dolphin-Eminem, ...
(53 couples of 43 terms used in [30] for evaluation of human perception). Ground truth: score from 1 a 5 by 12 subjects, given to a pair of concepts. Dataset3=moon, saturn, space, venus, bears, dolphin, donkey, golf, horses, sharks, spiders, tennis, turtle, whale, wolf, baseball, basketball, football, snake, soccer, bowling, softball, volleyball (23 terms used in [17] for evaluation on clustering and contextualized knowledge) Ground truth: clustering in the following clusters:

- moon, saturn, space, venus
- bears, dolphin, donkey, horses, sharks, snake, spiders, turtle, whale, wolf
- baseball, basketball, bowling, football, golf, soccer, softball, volleyball, tennis

4.2   Data Collection

To collect data from the Web, two approaches have been attempted, with and without the use of search engines' APIs, disabling the personalized behaviour of the browser and cookies. First, we have implemented two PHP scripts that, given a term or a group of terms, query them in the considered search engines, both through their API, both not, to get the number of web documents returned, that is the frequency of occurrence (in case of single-word queries) or co-occurrence (in case of multiple words queries). Analyzing the results, we observed that Google's results, with and without API, greatly differ, while in the other search engines the difference is much smaller. This issue is explained by Google support as the lack of some additive services with the use of the API, where the lacking services are the ones which Google provides with the use of personal information about browsing, through cookies and/or accounting. Because of this huge difference on Google, we decided to create a script to submit the queries and read the results through page scraping, instead of using the API, to automate the process and at the same time to keep the same results that would be obtained with a manual submission. Page scraping was implemented simply putting the HTML content of the page with query results in a string variable and extracting the data about the number of results matching regular expressions. Note that the cardinality of the results of a query in a search engine can vary and give different results in different times, so manually submitting a bunch of data may not return the best results, because of the time gap between pairs of terms or sets of terms which results have to be compared. Both with and without API is therefore suggested to submit queries with an automated program, to shorten the time lapse.

4.3   Experimental Tests and Clustering

In the experimental phase of the present study (see Fig. 3), the suitability of the considered measures and methods to calculate the semantic proximity was verified at the state-of-art, and problems were identified, comparing the results of measurements with the data sets as ground truth, with respect to the following models: 1. contextualized knowledge; 2. clustering; 3. human perception of semantic relations; to assess the adequacy of the considered measures, implemented on data sets present in literature, both for the terms to query, both as a reference for the ground truth.

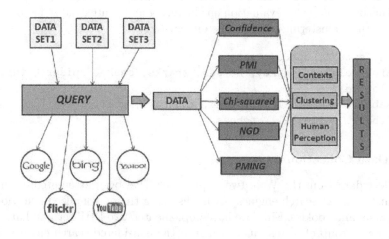

Fig. 3. The architecture of the experimental test B

The evaluation of the considered clustering algorithms has been done comparing the clustering results for the considered measures and engines. A confirmation for the final evaluation trend was found on the other two models of evaluation (e.g. human perception and contextualized knowledge). The expected clustering for dataset3, used as ground truth and obtained from a spectral clustering, as provided in [30] for comparison of NGD and FD, divides the term set into three clusters related to *sports, space* and *animals*. In the comparison of NGD and FD performed in [30] on spectral clustering, the research group by Microsoft always fails to get the exact clustering of ground truth, obtaining 3 errors with the FD, 6 errors with the NGD, where we calculated the number of errors with the swap edit distance, i.e. the minimal number of operation of swap that are necessary to obtain the ground truth, where swaps can be made even on not adjacent elements. This halving of errors is one of the main reasons why Wu et al. justify not only the goodness but also the usefulness of the Flickr Distance. From our experiments, it was observed that both the simple k-means algorithm and spectral clustering are able to identify the sport cluster on more than one of our data sets, which vary on measurements and on engines: this was done with a variable number of errors, but small enough to recognize without ambiguity the context of this group of terms. An even more remarkable result was achieved with the EM: on all engines except Google, at least a measure of proximity was found where this algorithm produce the correct data set, without any error. With the PMI, the EM algorithm produces the optimal clustering on the data obtained from all engines, except Google. The coloured table in Fig. 4 resumes the number of errors with EM for each measure and engine, for the expected clusters. In green, is shown the lowest error, for the PMING distance.

measure/engine	bing	flickr	google	yahoo	youtube	Average
CM	1	11	6	0	1	3,6
NGD	0	0	9	2	0	2,2
PMI	0	0	6	0	0	1,2
χ^2	4	0	5	7	6	4,4
PMING	0	0	7	0	0	1,4

Fig. 4. Number of errors encountered on EM clustering

This result is particularly significant for at least two reasons: the first is the low performance of all measures on Google, including NGD, which is meant to capture the semantics of Google, although it is also applicable to other engines; the second is that, if it is possible to get a good clustering by simple measures such as NGD, χ^2, PMI, then other complex and expensive algorithms such as the one for the calculation of the Flickr Distance, become useless. Because of this complexity and of the need for the FD algorithm to segment images (methodology that goes beyond our study, and therefore hasn't been applied), the FD results on our data set were not calculated and submitted to the EM, and the comparison was not performed on them. Anyway, since the tests of this study obtained the optimal clustering with other simpler measures, even an optimal clustering on the FD measures may not improve outcomes. In addition, the technique of FD is not directly applicable to the other search engines, which index textual documents and videos. As far as we can judge, the two surprising conclusions that could be drawn on FD and NGD may not be so independent. The Microsoft research group has defined the NGD as a calculated measure of co-occurrence in Google, even though the NGD measure is generally applicable to all search engines. Is not entirely clear if Wu et al. [30] has calculated the NGD on Google or if the experiments were extracted from queries on Flickr tags, but if the first case is the real one - as it seems to be - is not surprising that Microsoft Research have found a low benefit from the NGD. In fact, since queries on Google often produce less consistent results, the application domain could have distorted the goodness of the cited study on the NGD. From our experimental tests, in fact, the PMI is the measure with the best performance compared to the average number of errors on all the considered search engines for the pre-existent measures, and the second best fit is just the NGD. Regarding PMING, the experiments on clustering are especially satisfactory, because the results of NGD are improved by PMING as required, both for Yahoo, where clustering on PMING is excellent, both for Google where, despite the problems of reliability of this engine, PMING gets 7 errors against 9 of NGD and 6 of PMI. The trend is confirmed by the other two considered evaluation models: see

Fig. 5, showing the correctly classified pairs with respect to human perception of dataset2. On most of tests it has been further observed that on dataset3, the cluster which contains in the ground truth the terms about *sports* among the ones with terms about animals and space, was clustered with a perfect precision and recall.

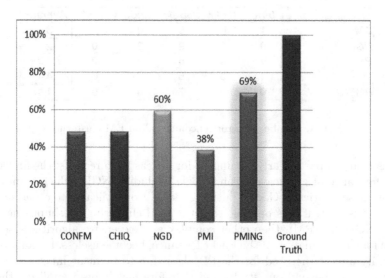

Fig. 5. Correctly classified pairs with respect to the human perception evaluation model (dataset2)

5   Conclusions

Generalist search engines, continuously exploring the Web (e.g., Google, Bing, Yahoo Search) reflect on their indexes the common semantics on which millions of users indirectly agree. The new paradigms which had been defined in last years, which permit to calculate the similarity of terms using only the statistical information returned by a query (e.g. Normalized Google Distance, NGD), or using information from additional features (e.g. Flickr Distance), need to be compared to evaluate the proximity measures. From the experimental tests emerges that the Expectation-Maximization algorithm outperforms the spectral clustering, while the Newman Fast algorithm is not as good as expected for finding clusters using web search engines. The results of testing and benchmarking both in clustering and in the other two models on popular collaborative and generalist engines suggested to focus on Pointwise Mutual Information (PMI) and NGD, as two proximity measures having characteristics of complementary optimality, to fulfil the objectives of adequacy in all the different areas of evaluation. In particular, the PMI measure offers excellent performance in clustering, while the NGD has noticeably better performance in human perception and contexts, but fails in clustering based on Yahoo and Google, getting the worst result. Being the PMING Distance defined as a weighted combination of PMI and NGD, it incorporates the advantages of both. PMING distance offered good performances on all the areas of evaluation, regardless of the search engine.

References

1. Miller, G.A., Beckwith, R., Fellbaum, C., Gross, D., Miller, K.: Introduction to WordNet: An On-line Lexical Database (1993)
2. Budanitsky, A., Hirst, G.: Semantic distance in wordnet: an experimental, application-oriented evaluation of five measures. In: Proceedings of Workshop on WordNet and Other Lexical Resources, p. 641. North American Chapter of the Association for Computational Linguistics, Pittsburgh, PA, USA (2001)
3. Franzoni, V., Milani, A.: PMING distance: a collaborative semantic proximity measure. WI-IAT **2**, 442–449 (2012). IEEE/WIC/ACM
4. Franzoni, V., Milani, A.: Heuristic semantic walk for concept chaining in collaborative networks. Int. J. Web Inf. Syst. **10**(1), 85–103 (2014). doi:10.1108/IJWIS-11-2013-0031
5. Franzoni, V., Milani, A.: Heuristic semantic walk. In: Murgante, B., Misra, S., Carlini, M., Torre, C.M., Nguyen, H.-Q., Taniar, D., Apduhan, B.O., Gervasi, O. (eds.) ICCSA 2013, Part IV. LNCS, vol. 7974, pp. 643–656. Springer, Heidelberg (2013)
6. Leung, C.H.C., Li, Y., Milani, A., Franzoni, V.: Collective evolutionary concept distance based query expansion for effective web document retrieval. In: Murgante, B., Misra, S., Carlini, M., Torre, C.M., Nguyen, H.-Q., Taniar, D., Apduhan, B.O., Gervasi, O. (eds.) ICCSA 2013, Part IV. LNCS, vol. 7974, pp. 657–672. Springer, Heidelberg (2013)
7. Cilibrasi, R., Vitanyi, P.: The google similarity distance. ArXiv.org (2004)
8. Franzoni, V., Milani, A.: Semantic Context extraction from collaborative networks. In: IEEE International Conference on Computer Supported Cooperative Work in Design CSCWD, 2015, Italy (2015)
9. Franzoni, V., Milani, A.: Context extraction by multi-path traces in semantic networks, CEUR-WS. In: Proceedings of RR2015 Doctoral Consortium, Berlin (2015)
10. Franzoni, V., Leung, C.H.C., Li, Y., Mengoni, P., Milani, A.: Set similarity measures for images based on collective knowledge. In: Gervasi, O., Murgante, B., Misra, S., Gavrilova, M.L., Rocha, A.M.A.C., Torre, C., Taniar, D., Apduhan, B.O. (eds.) ICCSA 2015. LNCS, vol. 9155, pp. 408–417. Springer, Heidelberg (2015)
11. Chiancone, A., Niyogi, R., et al.: Improving link ranking quality by quasi-common neighbourhood. In: 2015 IEEE CPS International Conference on Computational Science and Its Applications (2015)
12. Chiancone, A., Madotto, A., et al.: Multistrain bacterial model for link prediction. In: 2015 Proceedings of 11th International Conference on Natural Computation IEEE ICNC. CFP15CNC-CDR, Observation of strains. Infect Dis Ther. **3**(1), 35–43 (2011). ISBN: 978-1-4673-7678-5
13. http://www.google.com
14. http://www.bing.com
15. http://yahoo.com
16. http://www.youtube.com
17. https://www.flickr.com
18. Franzoni, V., Milani, A., Pallottelli, S.: Multi-path traces in semantic graphs for latent knowledge elicitation. In: 2015 IEEE ICNC Proceedings of 11th International Conference on Natural Computation, CFP15CNC-CDR (2015). ISBN: 978-1-4673-7678-5
19. Church, K.W., Hanks, P.: Word association norms, mutual information and lexicography. In: Proceedings of ACL, vol. 27 (1989)

20. Turney, P.D.: Mining the web for synonyms: PMI-IR versus LSA on TOEFL. In: Flach, P.A., Raedt, L. (eds.) ECML 2001. LNCS (LNAI), vol. 2167, p. 491. Springer, Heidelberg (2001)

21. Newman, M.E.J.: Fast algorithm for detecting community structure in networks. University of Michigan, MI (2003)

22. Matsuo, Y., Sakaki, T., Uchiyama, K., Ishizuka, M.: Graph-based word clustering using a web search engine. University of Tokio (2006)

23. Dellaert, F.: The Expectation-Maximization Algorithm. Elsevier, New York (2002)

24. Lin, J.: Divergence measures based on the Shannon entropy. IEEE Trans. Inf. Theory **37**(1), 145–151 (1991)

25. Hartigan, J.A., Wong, M.A.: Algorithm AS 136: A k-means clustering algorithm. J. Roy. Stat. Soc. **28**, 100–108 (1979)

26. Joyce, J.M.: Kullback-Leibler Divergence: International Encyclopedia of Statistical Science. Springer, Heidelberg (2011)

27. Dhillon, I.S., Guan, Y., Kulis, B.: Kernel k-means: spectral clustering and normalized cuts. In: Proceedings of the Tenth ACM SIGKDD (2004)

28. Pozrikidis, C.: Node degree distribution in spanning trees. J. Phys. A: Math. Theoret. **49**(12) (2016)

29. Bastianelli, E., Croce, D., Basili, R., Nardi, D.: Using semantic models for robust natural language human robot interaction. In: Gavanelli, M., et al. (eds.) AI*IA 2015. LNCS, vol. 9336, pp. 343–356. Springer, Heidelberg (2015). doi:10.1007/978-3-319-24309-2_26. http://dx.doi.org/10.1007/978-3-319-24309-2

30. Wu, L., Hua, X.S., Yu, N., Ma, W.Y., Li, S.: Flickr Distance. In: Proceedings of Microsoft Research Asia (2008)

31. Manning, D., Schutze, H.: Foundations of Statistical Natural Language Processing. The MIT Press, London (2002)

32. Franzoni, V., Poggioni, V., Zollo, F.: Automated book classification according to the emotional tags of the social network Zazie. In: CEUR-WS on ESSEM, AI*IA, vol. 1096, pp. 83–94 (2013)

33. Franzoni, V., Leung, C.H.C., Li, Y., Milani, A., Pallottelli, S.: Context-based image semantic similarity. In: 12th International Conference on Fuzzy Systems and Knowledge Discovery (FSKD), Zhangjiajie, pp. 1280–1284 (2015). doi:10.1109/FSKD.2015.7382127

34. Chiancone, A., Franzoni, V., Li, Y., Markov, K., Milani, A.: Leveraging zero tail in neighbourhood based link prediction. In: 2015 IEEE/WIC/ACM International Conference on Web Intelligence and Intelligent Agent Technology (WI-IAT), vol. 3, pp. 135–139 (2015). doi:10.1109/WI-IAT.2015.129

35. Franzoni, V., Milani, A.: A Pheromone-like model for semantic context extraction from collaborative networks. In: 2015 IEEE/WIC/ACM International Conference on Web Intelligence and Intelligent Agent Technology (WI-IAT), Singapore, pp. 540–547 (2015). doi:10.1109/WI-IAT.2015.21

36. Di Iorio, A., Schaerf, M.: Identification semantics for an organization establishing a digital library system. In: Risse, T., Predoiu, L., Nürnberger, A., Ross, S. (eds.) Proceedings of the 4th International Workshop on Semantic Digital Archives (SDA 2014) Co-located with the International Digital Libraries Conference (DL 2014), vol. 1306, London, UK, September 12, 2014, pp. 16–27. CEUR-WS.org (2014). http://ceur-ws.org/Vol-1306

Rating Prediction Based Job Recommendation Service for College Students

Rui Liu[1,2], Yuanxin Ouyang[1,2(✉)], Wenge Rong[1,2], Xin Song[1,2], Cui Tang[1,2], and Zhang Xiong[1,2]

[1] Engineering Research Center of Advanced Computer Application Technology, Ministry of Education, Beihang University, Beijing, China
{liurui,oyyx,w.rong,songxin,tangcui.c,xiongz}@buaa.edu.cn
[2] School of Computer Science and Engineering, Beihang University, Beijing, China

Abstract. When college students enter the job market, one of the main difficulties is that they do not have much working experience. To help students find proper jobs, appropriate recommendation systems are becoming a necessity. However, since most students start to find jobs in a very short time, it is difficult for a recommender system due to the lack of history information. To solve this problem, in this research we proposed a rating prediction mechanism by considering the feedback from graduates who have offers and also provided ratings to the employers. By calculating the similarity between the students, a rating prediction method is proposed to generate a list of potential employers for the students. Furthermore, we also take into account the factor of student's interest into the recommendation list's generation to further polish the overall performance. Experimental study on real recruitment dataset has shown the model's potential.

Keywords: Student profile · Job recommendation · Rating · Prediction

1 Introduction

When people started to find a new job it is normally a great challenge as the current job market has become more competitive [20]. Different from experienced job hunters, college students will find it more challenging and more fierce in the job market since as new professionals, they will find more continuously changed requirement from potential employers [29]. As a result it has become an essential problem for the students to smoothly go through the "school-to-work" transition period [24].

During the process of helping college student in finding their first jobs, a lot of efforts can be conducted. For example, students are encouraged to gather basic working experience from part-time job during their stay in the campus [14]. They can also obtain help and suggestion from their family [15]. Guidance from their mentors and/or supervisors are also particularly useful for students in having potential recruitment information [24]. Friends from the students' social network are another kind of information source [9,28].

© Springer International Publishing Switzerland 2016
O. Gervasi et al. (Eds.): ICCSA 2016, Part V, LNCS 9790, pp. 453–467, 2016.
DOI: 10.1007/978-3-319-42092-9_35

Though these different channels have proven the potential in helping students find jobs, there is another important mechanism for the student to find a job: get potential job information from college's support [5]. For example, in China nearly every college has an employment office. One of its main responsibilities is to help organise a so called "job fair" [19], which is a platform for the employers and students to find jobs in a relatively short time.

Though such "job fair" can provide an efficient mechanism for students and employers in job seeking, there are also several challenges among which information overload is a typical one [21]. It is expected in a very short time, the students will have to find a huge number of job descriptions from diverse employers. For employers, attracting huge number of resumes will also take a lot of efforts to filter potentially matching students. To solve this problem, job recommendation technique are consequently proposed and proven usefulness in the previous applications [33]. Currently the most widely used recommender technique is collaborative filtering [1]. It employs the previous rating information on different items by different users to predict if the user will be interested in a certain items. Though recommender systems have been proven capability in diverse applications, there still a lot of challenges and one of them is cold start [27], which means it will be difficult to recommend new items which has no rating scores and to new users who do have previous selection on certain items.

During the "job fair" event, to recommend students with proper job positions is also facing cold start problem. In fact, most of the students will not have any relevant activity data in the job recommendation system as they normally only use the system during the job seeking season. Therefore, how to recommend students who have few activity history has become an essential task of great significance. Though the fresh graduate students do not have history data in the job marketing, the universities do accumulated a lot of recruitment data from previous graduate students. It is argued that such information can help recommend new professionals in their job seeking process. As such in this research, we proposed a student similarity mechanism to modelling the students' similarity in terms of their personal profile. Afterwards, a rating based mechanism is proposed to predict the students' potential interest and capability in certain employers and then recommend them to those possible employers.

The rest of the paper is structured as follows. Section 2 will introduce the related work about job recommendation. Section 3 will present the proposed university student job recommendation framework in detail. Experimental study based on real employment data will be elaborated in Sects. 4 and 5 will conclude this paper and point out possible future work.

2 Related Work

2.1 Recommender System

With the development of Internet, more and more services are becoming available online, thereby making selection of a proper service a difficult task [30]. To overcome such information overload problem, advanced recommendation technique has become a necessity [25]. During the past decades, collaborative filtering has proven its potential in help implement recommender system [1].

One of the most widely used models is the neighbourhood oriented approaches [1]. Due to its simpleness and efficiency, it has proven the capability in providing accurate and personalised recommendation from large volume information. Neighbourhood-based recommendation model includes user-based and item-based model, which calculate the similarity between users and items using Pearson correlation coefficient or cosine similarity. Afterwards a similarity matrix is constructed to describe the similarity between the two users or items, and then predict if a certain user will select certain item according to the neighbour's activities.

Besides neighbourhood based approaches, model based recommender systems have also attracted much attention and one example is singular value decomposition (SVD) [26]. The singular value decomposition model can extract latent factors from the rating matrix using matrix factorisation, and then describe users and items by factor vectors. The item factor reflects the relation degree between items and factors, while the user factor reflects the degree of user preferences for these factors.

Though the collaborative filtering based recommender systems have achieved great success, there are also several challenges among which a typical one is referred as cold start [27]. For newly added users and items, there is no history activities attached to them. Without scoring information it will become difficult for the system to give a proper recommendation. Such difficulty is also great in job recommendation, particularly in assisting student's job seeking. As discussed before, particularly in "job fair" event, it will become a significant difficulty for employers and students since most students will not have job seeking history.

2.2 Job Recommendation

Job recommendation has been declared as an applicable mechanism to meet the gap between applicants and employers [16], since there is always ambiguity for the employers and job hunters to presenting their needs appropriately [23]. To better recommend job hunters, the first challenge is then becoming to match job description with the user's preference, which is usually referred to content based job recommendation [18]. Features in matching employers' job position and users include job description, job title [2], user's personal profile [12], users' skill [3]. Through these methods, folksonomy of user's expertise can be extracted for more effective matching [4]. However, sometimes user's profile will not be updated regularly, it will become inconvenient to help employers find most suitable candidates [12]. How to take into account user's activity in job seeking is now a new challenge. Furthermore, if more similar persons with similar background can be considered in the job seeking process, it might be more helpful in matching the employers and job hunters [6], thereby making collaborative filtering based job recommender system a promising direction.

Collaborative filtering technique has been proven great success in general recommender system, not except job recommendation. Based on group knowledge, it is possible to locate similar users with similar background and similar job intention, thereby recommending potential jobs to the most suitable users. In job recommendation, a user-job matrix can be constructed with the score provided by the job hunters. The score can be user's behaviour, such as revisit, read time and usage pattern [23]. User's application record can be also used to implement item-based collaborative filtering based job recommendation system [32].

Collaborative filtering job recommendation also has limitations, e.g., cold start. As such hybrid job recommendation mechanism is consequently proposed [1]. It is believed that integration of different type recommendation system can better serve user's needs. Such example is integration of logic matchmaking and also similarity based ranking model [10]. Some other applications to use the result from one model to polish another model's output [8], since it is argued that this approach can better serve the dynamic feature of user's profile and job requirement [12].

3   Rating Based Job Recommendation Framework

Inspired by previous work in the literature, in this research, we proposed a rating prediction based job recommendation service for students in seeking their first job. Its goal is to study how to recommend companies to students without the their history rating for the employers. The main idea is to use the similar student ratings on the employer and the employability attributes similarity between them to generate candidate employers. Here we present the proposed job recommendation framework Fig. 1.

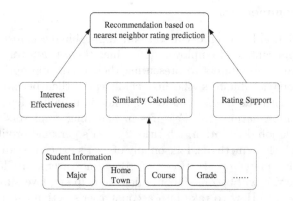

Fig. 1. Rating based job recommendation framework

In this framework, we firstly calculate the similarity between students based on the students' personal profile, and then predict the rating of the employers according to the student nearest neighbours' rating to the employers.

The predictive rating can reflect the students' interest in these employers to some extent. Finally, the framework will recommend suitable employers to the students according to the predicted ratings. Considering that the students' predicted ratings of the employers may come from different number of the neighbours' ratings, we introduce the rating support to achieve the purpose of optimising the recommended effect. Due to the attractiveness of different employers to students, as well as the size of the recruitment, the probability of students' resumes get passed is different. As such this model introduced the interest effectiveness in order to further optimise the recommended results.

3.1 Students Profiling and Similarity Matching

To better describe student, a proper profiling mechanism is necessary. In this research, we consider a student as a set of features include basic attributes and achievement attributes. The basic attributes mainly related to his/her static information of individual background, such as where he/she come from, the gender and the major and courses he/she has taken. The achievement attributes are usually referred as the student's academic achievement, e.g., scores, studentship, award and etc. Therefore, in this research, a student can be formally described as:

$$u = \{a_1; a_2; ...; a_n\} \tag{1}$$

where u is the student and a_i represents the users i-th attribute.

After getting the students' profiles, the next step is to use these profile to get their similarity. In this research, we used weighted similarity matching schema to calculate two different students and attached different weights to different features. The similarity is calculated as:

$$S(u, g) = \sum_{i=1}^{n} w_i \cdot sim(u_{a_i}, g_{a_i}) \tag{2}$$

where u and g are the student who is finding job and graduate who has offer, respectively, and w_i is the i-th attributer's weight in similarity calculation, while a_i is the i-th attribute.

A student's feature can be either bool value or discrete value. For example, whether two students come from same home town can result in the corresponding value as true or false. As as the similarity of such bool value oriented attributes can be defined as:

$$sim\left(u_{a_i}, g_{a_i}\right) = \begin{cases} 1 & u_{a_i} = g_{a_i} \\ 0 & u_{a_i} \neq g_{a_i} \end{cases} \tag{3}$$

For the discrete attributes, the similarity can be defined as:

$$sim(u_{a_i}, g_{a_i}) = 1 - \frac{|u_{a_i} - g_{a_i}|}{a_{imax} - a_{imin}} \tag{4}$$

where max and min mean the value range. An example is the student's course score, which ranges from 0 to 100. The similarity is described as the score different divided by the whole range, i.e., 100.

There is another point related to the relation between major and course. It is possible that only using major for student similarity is not enough since the granularity might be too coarse. Students having different majors may study same courses to some extent. Furthermore, students are also allowed to select different courses even under same major. As such in this research we employ course instead of major to calculate the students' similarity, which can be defined as:

$$sim_{course}(u, g) = \frac{|C(u) \cap C(g)|}{|C(u) \cup C(g)|} \tag{5}$$

where $C(x)$ indicated all courses the student x have studied.

3.2 Job Recommendation Based on Rating Prediction

To better recommend students with employers, it will be helpful if we can predict the possible rating by this student. Considering the previous students' employment history data have been accumulated, it is feasible to predict the student's rating on potential employers by utilising previous recruitment history.

Algorithm 1. Similarity based rating prediction for job recommendation

---

Input: G, u
Output: E
1: **for** each graduate g_i in G **do**
2:     Calculate similarity with the student $s_{u,g_i} = S(u, g_i)$
3: **end for**
4: Sort s_{u,g_i} in descending order
5: Obtain top k graduates to form G_k graduate set
6: Obtain all employers to employer set E related to G_k
7: **for** each employer e_j in E **do**
8:     **for** each employer g_i in G_k **do**
9:         **if** $r_{g_i,e_j} > 0$ **then**
10:             $rate_{acc} = rate_{acc} + S(u, g_i)r_{g_i,e_j}$
11:             $sim_{acc} = sim_{acc} + S(u, g_i)$
12:         **end if**
13:         $r_{u,e_j} = rate_{acc}/sim_{acc}$
14:     **end for**
15: **end for**
16: Sort E in descending order of r_{u,e_j}
17: **return** E

In this research, we will firstly calculate the similarity between u (the student who is finding job) and G (graduates who have offer) based on the student's attributes, and then sort the students set G by the similarity. Afterwards we will select top K students in G and obtain each employer e from very graduate student g in G to form a employer set E. After that we will calculate the weighted average rating for each employer e in E. The weighted average rating is seemed to reflect the student u preference to employers set E. Finally, the model sort the candidate employers by the weighted average rating and choose top N employer to recommend to the student u. The whole algorithm is listed in Algorithm 1.

3.3   Job Recommendation Based on Rating Support

The effect of neighbourhood-based recommendation model is closely related to dataset sparsity. Herlocker [11] et al. study shows that neighbourhood-based recommendation can lead very absurd results when there are only little training samples. To address this problem, extensive research have been carried out and presented control reduction method to support the small sample situation [13].

In this research, student job recommendation model also applies similar students' job-hunting attributes to calculate similarity. As such the proposed model may suffer the same problem. Predicted ratings may be derived from k nearest neighbour ratings to the employers, and it may also be derived from only one nearest neighbour rating to the employer. If the nearest neighbour hits a high rating to some employer, and this nearest neighbour is the only one who marks the employer, then user u ratings is equal to k nearest neighbour rating. It is possible that the credibility of the rating is not high enough. It is common that some employers receive a great number of resumes. However, some employers receive few resumes. In order to solve this problem, in this research we propose a rating prediction support parameter, which defined as follows:

Definition 1. *Predicting rating support parameter means the difference between the number of rating prediction samples and the minimum number of all rating predicting samples divide the difference between the maximum number of all rating predicting samples and the minimum number of rating predicting samples. The formula shows bellows, where sn represents the number of sample support.*

$$Support = \frac{sn - sn_{min}}{sn_{max} - sn_{min}} \tag{6}$$

Therefore, the job recommendation with rating support can be defined as:

$$sort = r_{u,e_j} \times Support \tag{7}$$

3.4   Job Recommendation Based on Interest Effectiveness

The purpose of rating prediction is to recommend items interested to the users. For instance, the goal of recommendation is that the user will eventually select the recommended items In practice, it is often that users show interest in a certain item in recommendation list. But after looking through the details, they do not select the item. In this case, though the recommendations accord to the students' interest, they do not fit the real demand well, which is believed inefficient.

The same problem also exists in job recommendation. We recommend employers to students based on their interests. If the employer will fits the students' interest, they would likely deliver their resumes. However, being employed is the final purpose for students to deliver their resumes, if not, the recommendation is inefficient. Due to the fact that job-hunting and recruiting are a reciprocal process, being hired depends on the match in demand of employers and students' capacity rather than his/her desire. So it is a common phenomenon that students delivered resumes but do not get hired. Job recommendation system should consider not only students' interest, but also whether the student will be finally hired. If a recommendation system recommends a great deal of employers which interest the student but will not employ him/her, it is inefficient in the purpose of job recommendation, and it will also waste students a lot of time and energy. Owing to the diversity in scale, reputation, profession and location, employers differs greatly in attractiveness. A position in some employers attracts hundreds of thousands of candidates, while some other employers only have a few candidates. Therefore, the ease of being hired depends on the intensity of competition. To address this problem, this paper presents interest effectiveness parameter, which is defined as follows:

Definition 2. *Job recommendation interest effectiveness means the probability of a student submitting a resume to an employer and getting hired, namely, the ratio of the number of employment and the number of resume. The formula shows below, where n_{offer} represents the number of recruitment, and n_{resume} represents the number of resume.*

$$IEfficiency = \frac{n_{offer}}{n_{resume}} \tag{8}$$

Therefore, the job recommendation with rating support and interest efficiency can be defined as:

$$sort' = r_{u,e_j} \times Support \times IEfficiency \tag{9}$$

4   Experimental Study

4.1   Dataset

To validate and evaluate the proposed model, we conduct experimental analysis on real job recruitment data obtained from Beihang University in 2016, which includes 658 Master graduates and their 3830 ratings to 856 employers that they had submitted resumes and whether they were employed after delivering resumes. In experiment, we randomly selected 80 % ratings as the training set and select 20 % as the test set. The distribution of training set and test set illustrated in Table 1. The training set includes 526 students, their 2858 ratings to 748 employers that they have delivered resumes and 1096 recruit data; the test set contains 132 students, their 972 ratings to 400 employers that they have delivered resumes and 292 recruit data.

Table 1. Cold start ratings experimental data partition dataset

	#Student	#Employers	#Rating	#Acceptance
Training dataset	526	748	2858	1091
Test dataset	132	400	972	292

4.2 Evaluation Metrics

As for employer recommendation, inspired by the evaluation metric of $P@n$ in the field of information retrieval [7], which emphases the top n items in the return list based on the assumption that people will probably not be interested in the items beyond top n candidates [17], in this paper, we constitute a metric called $S@n$, which is defined as below:

$$S@n = \sum_{i=1}^{n} hit(u_i) \tag{10}$$

where $hit(u_i)$ represents a real employer of the student in the test set has been correctly included in the recommended list with the k employers. As such the $hit(u)$ is defined as:

$$hit(u_i) = \begin{cases} 1, e_{u_i} \in E \\ 0, \text{ otherwise} \end{cases} \tag{11}$$

where E is the returned list defined in Algorithm 1.

Furthermore, we will also evaluate the hit of resume delivery and recruitment. To evaluate the efficiency of job recommendation, this paper proposes efficiency criterion $Efficiency$:

$$Efficiency = \frac{S_{offer}@n}{S_{resume}@n} \tag{12}$$

where $S_{offer}@n$ represents the number of recruitment hit, and $S_{resume}@n$ represents the number of resume delivery hit.

Similarly, we also proposed a metrics called $Coverage$ to represent if the recommended list can cover as many as employers:

$$Coverage = \frac{\cup_{u \in U} R(u)}{|E|} \tag{13}$$

4.3 Student Similarity Oriented Rating Prediction Based Job Recommendation

This experimental study will study how the similarity can affect the rating prediction in job recommendation. As discussed in Sect. 3, here we select student's course, home town and GPA as features to calculate similarity. Afterwards the similarity will be used for rating prediction. We firstly analyse the different weight of the features and also the different selection of k for the overall performance. The result is listed as below.

Table 2. Similarity based rating prediction for job recommendation

Major	Home town	GPA	S_{offer}@10	S_{resume}@10	Efficiency %
0.5	0.25	0.25	17	69	24.64
0.6	0.20	0.20	22	72	30.56
0.7	0.15	0.15	23	86	26.74
0.8	0.10	0.10	24	83	28.92
0.9	0.05	0.05	23	85	27.06

From Table 2, it is found that when the weights of course set, home town and average grade are respectively taken 0.8, 0.1, 01, the value of employment hits is the highest, which is 24, while the value of efficiency is 28.92 %. However, when these three properties are respectively taken 0.6, 0.2, 0.2, the value of efficiency is the highest, which is 30.56 %, and the value of employment hits is 22.

In view of the weights of course set, student source and average grade are respectively taken 0.8, 0.1, 01 or 0.6, 0.2, 0.2, and the two tests all achieve better prediction accuracy, so we use two sets of parameters to conduct a hyper-parameter experiment with the nearest neighbour, and the value are respectively taken 1, 3, 5, 10, 15, 20. When the weight of students' job-hunting ability takes 0.8, 0.1, 0.1, the test result are shown in Fig. 2. It is found that when the value of k nearest neighbour set 3, the value of employment hits is 33, significantly higher than the other values of employment hits, and the value of efficiency is 26.61 %, slightly lower than the highest 31.08 %. When the weight of students' job-hunting ability takes 0.6, 0.2, 0.2, the test result are shown in Fig. 3. When K also takes 3, the value of employment hits is up to 40, also significantly higher than the other values of employment hits, and the value of efficiency is 28.37 %, slightly lower than the highest 32.00 %. As a result, in this research, we set the weight for course, home town and GPA to 0.6, 0.2, 0.2 in the next experimental study.

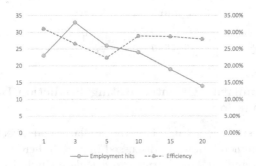

Fig. 2. Relationship between the number of neighbours and recommendation result A (Color figure online)

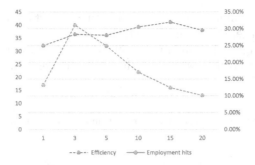

Fig. 3. Relationship between the number of neighbours and recommendation result B (Color figure online)

4.4 Job Recommendation with Rating Support

Here we will test the effectiveness of rating support introduced in Sect. 3 and also consider the recruitment scale and the popularity of the employer, the ranking basis of the recommendation enterprises has a great change. For the selection of the nearest neighbour's value, we carry out a hyper-parameter experiment to obtain the best recommendation effect. Then, we compared the effect of the recommendation based on the rating support and the recommendation effect based on the rating prediction. The experimental results of the hyper-parameter for the nearest neighbour K are shown in Fig. 4. It is found that when $k = 80$, the recommended effect is the best, the value of employment hits is up to 61, the value of efficiency is 30.35 %.

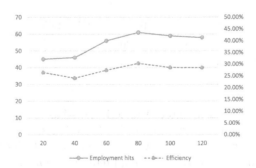

Fig. 4. Effect of neighbour number (Color figure online)

Table 3 is the comparison of rating prediction against rating support. From the value of employment hits, the value of resume hits, the value of efficiency and the value of enterprise coverage, the value of employment hits of the recommendation based on the rating support is up to 61, the value of resume hits is up to 201, the value of efficiency is 30.35 %, the effect is significantly better than the effect of recommendation based on rating prediction. However, in terms of enterprise coverage, the value of enterprise coverage in the job recommendation based on rating prediction is up to 50.93 %, and this value is much higher than the value of enterprise coverage in the job recommendation based on rating support 8.56 %.

Table 3. The comparison between rating prediction approach and rating support-based approach

Method	S_{offer}@10	S_{resume}@10	Efficiency	Coverage
Rating prediction	40	141	28.37 %	50.93 %
Rating support	61	201	30.35 %	8.56 %

4.5 Job Recommendation with Interest Efficiency

In this study, the optimal recommended effect in previous experiment was chosen as the experimental parameters, the weight of students' Job-hunting ability took 0.6, 0.2, 0.2, the value K of nearest neighbor took 80, on the basis, introducing interest-effectiveness parameter. Comparing before and after the introduction of interest-effectiveness parameter, evaluating changes in the indicators, which included the value of employment hits, the value of resume hits, the value of efficiency and the value of enterprise coverage of the four indicators.

Table 4. The comparison between interest-effectiveness-based approach and rating support-based approach

Method	S_{offer}@10	S_{resume}@10	Efficiency	Coverage
Rating support	61	201	30.35 %	8.56 %
Interest effectiveness	61	167	36.53 %	11.50 %

The comparison between interest-effectiveness-based approach and rating support-based approach was shown in the Table 4. After the introduction of interest-effectiveness parameter, the value of employment hits was similar with the recommended method based on rating prediction, both were 61, at the same time, the value of resume hits dramatically dropped, thus the value of efficiency was effectively promoted, up to 36.53 %, the value of enterprise coverage also had been effectively improved, up to 11.50 %.

5 Conclusion

Job recommendation is an essential application for people when they wish to move to a new job, participially for college student since they normally do not have much working experience. Though the recommendation system techniques have been widely studied, the student job recommendation is still a challenge task since the cold start problem can not be easily avoided. To this end, in this research a student profile mechanism is proposed and different similarity detection mechanism between students and companies are proposed. Based on this, a rating prediction mechanism is proposed to help generate employer candidate list. Furthermore, we also take into account the student's interest to further polish the generated list. Experimental study on real job market during the last few years have shown the framework's potential. It is expected that this work can provide an insight for future job recommendation for college students.

This research mainly studied the job recommendation from student's perspectively. However, job recommendation should be able to satisfy both job hunters and employers [22], since it is also tedious for employers to filter huge number of resumes in a relatively short period. This reciprocal requirement deserves further research in the future work [31], thereby saving students and employers efforts simultaneously.

Acknowledgments. This work was partially supported by the National High Technology Research and Development Program of China (No. 2013AA01A601), the National Natural Science Foundation of China (No. 61472021), and the Fundamental Research Funds for the Central Universities.

References

1. Adomavicius, G., Tuzhilin, A.: Toward the next generation of recommender systems: a survey of the state-of-the-art and possible extensions. IEEE Trans. Knowl. Data Eng. **17**(6), 734–749 (2005)
2. Al-Otaibi, S.T., Ykhlef, M.: Job recommendation systems for enhancing e-recruitment process. In: Proceedings of 2012 Information and Knowledge Engineering conference (2012)
3. Almalis, N.D., Tsihrintzis, G.A., Karagiannis, N.: A content based approach for recommending personnel for job positions. In: Proceedings of the 5th International Conference on Information, Intelligence, Systems and Applications, pp. 45–49 (2014)
4. Bastian, M., Hayes, M., Vaughan, W., Shah, S., Skomoroch, P., Kim, H., Uryasev, S., Lloyd, C.: Linkedin skills: large-scale topic extraction and inference. In: Proceedings of the 8th ACM Conference on Recommender Systems, pp. 1–8 (2014)
5. Breaugh, J.A.: Employee recruitment: current knowledge and important areas for future research. Hum. Resour. Manag. Rev. **18**(3), 103–118 (2008)
6. Cheng, Y., Xie, Y., Chen, Z., Agrawal, A., Choudhary, A., Guo, S.: Jobminer: A real-time system for mining job-related patterns from social media. In: Proceedings of the 19th ACM SIGKDD International Conference on Knowledge Discovery and Data Mining, pp. 1450–1453 (2013)
7. Cormack, G.V., Lynam, T.R.: Statistical precision of information retrieval evaluation. In: Proceedings of the 29th Annual International ACM SIGIR Conference on Research and Development in Information Retrieval, pp. 533–540 (2006)

8. Drigas, A., Kouremenos, S., Vrettos, S., Vrettaros, J., Kouremenos, D.: An expert system for job matching of the unemployed. Expert Syst. Appl. **26**(2), 217–224 (2004)
9. Ellison, N.B., Steinfield, C., Lampe, C.: The benefits of Facebook "friends:" social capital and college students' use of online social network sites. J. Comput.-Mediat. Commun. **12**(4), 1143–1168 (2007)
10. Fazel-Zarandi, M., Fox, M.S.: Semantic matchmaking for job recruitment: an ontology-based hybrid approach. In: Proceedings of the 8th International Semantic Web Conference (2009)
11. Herlocker, J., Konstan, J.A., Riedl, J.: An empirical analysis of design choices in neighborhood-based collaborative filtering algorithms. Inf. Retr. **5**(4), 287–310 (2002)
12. Hong, W., Zheng, S., Wang, H.: Dynamic user profile-based job recommender system. In: Proceedings of 8th International Conference onComputer Science and Education, pp. 1499–1503 (2013)
13. Koren, Y.: Factorization meets the neighborhood: a multifaceted collaborative filtering model. In: Proceedings of 14th ACM SIGKDD International Conference on Knowledge Discovery and Data Miningk, pp. 426–434 (2008)
14. Lin, M.C., Ching, G.S.: College student' employability: implications of part-time job during college years. In: Advances in Public, Environmental and Occupational Health, pp. 101–106 (2014)
15. Liu, D.: Parental involvement and university graduate employment in China. J. Educ. Work **29**(1), 98–113 (2016)
16. Malinowski, J., Keim, T., Wendt, O., Weitzel, T.: Matching people and jobs: a bilateral recommendation approach. In: Proceedings of the 39th Annual Hawaii International Conference on System Sciences (2006)
17. McLaughlin, M.R., Herlocker, J.L.: A collaborative filtering algorithm and evaluation metric that accurately model the user experience. In: Proceedings of the 27th Annual International ACM SIGIR Conference on Research and Development in Information Retrieval, pp. 329–336 (2004)
18. Miller, B.N., Albert, I., Lam, S.K., Konstan, J.A., Riedl, J.: Movielens unplugged: experiences with an occasionally connected recommender system. In: Proceedings of the 2003 International Conference on Intelligent User Interfaces, pp. 263–266 (2003)
19. Obukhova, E.: Motivation vs. relevance: using strong ties to find a job in urban China. Soc. Sci. Res. **41**(3), 570–580 (2012)
20. Paparrizos, I.K., Cambazoglu, B.B., Gionis, A.: Machine learned job recommendation. In: Proceedings of the 2011 ACM Conference on Recommender Systems, pp. 325–328 (2011)
21. Peterson, A.: On the prowl: how to hunt and score your first job. Educ. Horiz. **92**(3), 13–15 (2014)
22. Pizzato, L., Rej, T., Chung, T., Koprinska, I., Kay, J.: Recon: a reciprocal recommender for online dating. In: Proceedings of the 4th ACM Conference on Recommender systems, pp. 207–214 (2010)
23. Rafter, R., Bradley, K., Smyth, B.: Automated collaborative filtering applications for online recruitment services. In: Brusilovsky, P., Stock, O., Strapparava, C. (eds.) AH 2000. LNCS, vol. 1892, pp. 363–368. Springer, Heidelberg (2000)
24. Renn, R.W., Steinbauer, R., Taylor, R., Detwiler, D.: School-to-work transition: mentor career support and student career planning, job search intentions, and self-defeating job search behavior. J. Vocat. Behav. **85**(3), 422–432 (2014)

25. Resnick, P., Varian, H.R.: Recommender systems. Commun. ACM **40**(3), 56–58 (1997)
26. Sarwar, B., Karypis, G., Konstan, J., Riedl, J.: Incremental singular value decomposition algorithms for highly scalable recommender systems. In: Proceedings of 5th International Conference on Computer and Information Science, pp. 27–28 (2002)
27. Schein, A.I., Popescul, A., Ungar, L.H., Pennock, D.M.: Methods and metrics for cold-start recommendations. In: Proceedings of the 25th Annual International ACM SIGIR Conference on Research and Development in Information Retrieval, pp. 253–260 (2002)
28. Skeels, M.M., Grudin, J.: When social networks cross boundaries: a case study of workplace use of Facebook and Linkedin. In: Proceedings of the 2009 International ACM SIGGROUP Conference on Supporting Group Work, pp. 95–104 (2009)
29. Staton, M.G.: Improving student job placement and assessment through the use of digital marketing certification programs. Mark. Educ. Rev. **26**(1), 20–24 (2015)
30. Xiao, B., Benbasat, I.: E-commerce product recommendation agents: use, characteristics, and impact. MIS Q. **31**(1), 137–209 (2007)
31. Yu, H., Liu, C., Zhang, F.: Reciprocal recommendation algorithm for the field of recruitment. J. Inf. Comput. Sci. **8**(16), 4061–4068 (2011)
32. Zhang, Y., Yang, C., Niu, Z.: A research of job recommendation system based on collaborative filtering. In: Proceedings of 7th International Symposium on Computational Intelligence and Design. pp. 533–538 (2014)
33. Zheng, S., Hong, W., Zhang, N., Yang, F.: Job recommender systems: a survey. In: Proceedings of 7th International Conference on Computer Science and Education, pp. 920–924 (2012)

Development of M-Health Software for People with Disabilities

Suren Abrahamyan[✉], Serob Balyan, Avetik Muradov, Vladimir Korkhov,
Anna Moskvicheva, and Oleg Jakushkin

Saint Petersburg State University, 7/9 Universitetskaya nab, St. Petersburg 199034, Russia
suro7@live.com, {vkorkhov,o.yakushkin}@spbu.ru,
serob.balyan@gmail.com, avet.muradov@gmail.com,
annainternest@gmail.com

Abstract. There are disabilities like Autism, Down Syndrome, Cerebral Palsy which can be reason for disorders in speech and writing. Special card-pictograms are widely spread for people with such disabilities to help them to communicate in daily and educational purposes.

Today, the digital communication developed and different methods exist (applications, programs) for distance messaging, but people with such disorders may suffer a social exclusion given by their condition. For example, autists may have strong difficulties while typing on mobile device's keyboard. Thus, it would be useful to create a mobile application that allows people with speech and writing disorders to exchange their ideas and feelings in real time remotely by the help of digital analogues of card-pictograms mentioned above. This could be extremely important, particularly in Russian Federation, where solutions like this do not exist.

In this article we describe the initial version of the software working on conventional mobile devices, represent further development line of program, and also show performed tests results for target groups.

Keywords: M-Health · E-Health · Real-time messaging · Mobile communication

1 Introduction

Nowadays several types of people disorders can be treated in some level using modern information technologies and electronic communication infrastructures. This kind of solutions lay under the concept of E-Health. This concept spans across various services like electronic health records, clinical decision support, telemedicine, medical research using grids, health knowledge management [1] etc. which stay very close to both medicine/healthcare and information technology at the same time.

The use of mobile technology and its services are gaining momentum at many areas of human activity and extend benefits of remote delivery. Healthcare area is not an exception and its combination with mobile infrastructures is designated by the term M-Health, which is monitoring and communicating medical conditions using mobile technology [2]. M-Health, with mentioned above services, is also a part of E-Health. With the development of mobile technologies sphere of use of M-Health is also growing and

© Springer International Publishing Switzerland 2016
O. Gervasi et al. (Eds.): ICCSA 2016, Part V, LNCS 9790, pp. 468–479, 2016.
DOI: 10.1007/978-3-319-42092-9_36

as a result of its combination with telemedicine (remote diagnosis and treatment of patients by means of telecommunications technology) toolkits are created that solve such problems as consultations, patients monitoring, treatments on distance. There are also disabilities that cause problems which cannot be treated but can be minimized. One of them is speech and writing disorder which causes communication problems. Such problem can be minimized with the help of special mobile software described in this article later.

2   Actual State in E-Health

2.1   Telemedicine

Telemedicine defined as the use of medical data exchanged between two sites with the help of electronic communication systems to improve a patient's health status. Telemedicine includes a growing variety of services and applications using different types of telecommunication technologies such as smart phones, two-way videos, wireless tools etc. [3].

Health officials, doctors and patients are facing some problems due to the need of using telecommunication technologies in health care. In all the countries of the world, high-tech medical care is possible only in big medical centers and specialized institutions. The problems of local healthcare officials and institutions are the shortage of doctors-specialists, medical personnel, the remoteness of provincial towns and villages from medical centers and more often the inefficient organization of emergency medical aid for the population of remote regions. The achievement of telecommunication and information technologies eliminates the urgency of doctor's physical presence. Normally there are cases when for giving medical care the personnel need the consultation of skilled specialists, who will, considerably, raise the effectiveness of treatment and reduce its duration. Today the information and telecommunication technologies are able to provide connection with subscribers at any time in the most remote and hard-to-reach regions. Telemedicine (TM) appeared "at the junction" of information and communication technologies and medical spheres, gradually moving to become a separate branch in health care. Of course, there are many problems here: the information security, international standards, ethical and legal aspects, teaching issues, etc. However, it can be argued that the information revolution radically transforms the technology of medical service.

Telemedicine is a medicine practiced at a distance. R. G. Mark introduced the term "telemedicine" in 1974. Starting out more than forty years ago with demonstrations of hospitals extending care to patients in remote areas, the use of telemedicine has spread rapidly and is now becoming integrated into the ongoing operations of hospitals, home health agencies, specialty departments as well as consumer's homes or workplaces [4].

In 1995, WHO and International Telecommunications Union (ITU) signed a "Memorandum of Comprehension" which stated that the two organizations unite their efforts in the sphere of information and communication technologies for the improvement of medical care quality to people living in villages or in remote regions.

There are at least four basic aspects of using TM:

1. Administrative – TM can help in solving administrative issues concerning the policy of health care development and reformation.
2. Reinforcing national structure of healthcare – TM can help to improve the connections between regional and national hospitals and international leading clinical centers using telecommunications.
3. Education – TM can provide continuous education to doctors and medical personnel from separate regions of the country.
4. Quality and effectiveness of medical service – TM can help in reducing sickness and mortality rates among population due to the improvement of diagnosis, treatment, prevention and the management of health care system.

Separate elements of TM are very often found in everyday medical practice. However, to give a definition to this discipline from a more general point of view, it's necessary to emphasize its complex, systematic character, expecting:

- The use of special equipment with the help of which collection, conversion and transmission of medical information is practiced;
- Availability of telecommunications system providing connection between "suppliers" and "consumers" of medical information;
- Use of program security connecting all the elements of the system into a unified set;
- Availability of staff of specialists (physicians, programmers, electronics engineers, operators), providing professional and technical aid of the set, its effective use in resolving medical issues.
- Besides, in working with TM system certain modes of operation of equipment must be used, specific formats of medical data and records of information exchange should be applied, etc.

According to specialists, the basic and foremost problem of TM is the distant diagnosis. Medical diagnosis, in modern terms, always demanded visual information. In one word, modern information means were needed for the appearance of TM which would allow the doctor "to see" the patient. Though TM, first of all, remains distant diagnosis, its potential abilities are considerably wide. Network technologies provide opportunity to deliver medical history in case of transferring the patients from one clinic to another; strategic decisions of insurance and payment issues; new opportunities of increasing the qualification of doctors; wide introduction of new medical technologies and methods; distant medical consultations, teleconferences and telemanipulations (distant control of equipment and even surgical intervention at a distance). Also high performance multiprocessor supercomputer technologies can be used in telemedicine purposes [5]. For example, ADEPT-C (Advanced Data Evaluation Parallel Tools - in Cardiology) operates on client-server architecture based on multiprocessor supercomputers. These tools can be the basis for creating a new generation of RTCES (Real-Time Cardiology Expert System) operating on high performance supercomputers [6]. Quite a tendency of TM development is the formation of regional TM networks. Such networks, on the one hand, will extend deep into territories, covering more medical institutions. On the other hand, these networks will allow them unite with each other. In establishing TM networks, nearly all means of communications will be used: land and satellite, fiber-optic and wireless, wide-brimmed net for data transmission and mobile phone.

However, the two main aspects of TM are distant consultations and education. The reason for this is the recommendation of ITU about the strategy of TM development in developing countries [7]. The analysis of experiments shows that the above-mentioned directions are advisable to develop taking into consideration the economic and social conditions of the countries with transitional economies. That said the main usable directions of TM technologies are:

- Telemedical consultation/telementoring during diagnostics and treatment process, evacuation events or trainings;
- Telemedical lecture/seminar and distant testing/examinations;
- Telemedical conference/concilium/symposium during distant board meetings (conferences, councils), medical conciliums, scientific sessions;
- Telemonitoring (telemetry) – distant control of vital functional indices of body.

Today, more than 250 TM projects are in the world, which, by their nature, are divided into clinical, educational, informational and analytic. The vast majority are clinical projects. The rest are presented in a very small quantity. Clinical projects are focused mainly on solving the treatment and diagnosis issues, though they can also be used for educational and administrative purposes. Among TM programs, a special weight is given to projects connected with radiology, surgery, emergency medical care, cardiology, dermatology, neurology, psychiatry and pediatrics.

Most promising direction of TM development is telemonitoring or distant biomonitoring – systems for distant dynamic observation of patients suffering chronic diseases, and first of all, diseases of cardiovascular and pulmonary systems. Availability of mobile communications and Internet services allows developing such direction as "home telemedicine." This is a distant medical care to the patient being out of the medical institution and receiving treatment at home. The perspectives of TM are associated with the further miniaturization of control and measuring means, introduction of smart-technologies, robotics, the latest achievements of science and applied aspects of nanotechnologies.

Promising areas of use of TM may also be such specific areas as medicine of catastrophe, prevention of infectious diseases and control of vaccination among population. Here TM sessions may be used "post factum," after natural or man-made catastrophes, as for assistance and rescue of victims, and also for informing the medical staff about the possibility of disaster occurrence, planning and coordination of rescue services' actions, i.e. the main meaning of TM is to minimize the risk to human lives.

The using of TM technologies gave unique international experience in extreme situations like natural and manmade disasters: Spitak earthquake in Armenia (1988) and gas pipe explosion near Ufa (1989). Telebridges (using audio-, video- and facsimile communications) were established between disaster zones and leading US medical centers. Space segment and satellite earth stations were also included under the auspices of Soviet-American commission of space biology and medicine, which resulted in a large-scale international project called "Telemedicine Space Bridge." Specialists from leading Russian clinics and US medical centers participated in teleconsultations and videoconferences. Consultations were carried out among ambustial, psychiatric and other patients. During 12 weeks' telebridge, which included 34 4-h videoconferences, 247 specialists from USSR (Armenia and Russia) and 175 American specialists participated. 209 clinical cases in 20 medical specialties have been considered, significant changes were made in diagnostic and therapeutic process, new therapeutic methods were introduced and significant quantity of information has been transferred. Therefore, diagnosis was changed in 33 % cases, additional diagnostic measures were recommended in 46 % cases, in 21 % cases treatment strategy was changed and new treatment methods were introduced in 10 % cases.

Thus, the use of modern information and telecommunication technologies allows to provide access and high quality of organizing medical care for population. It showed its effectiveness in carrying out necessary medical services and controlling epidemic situations in rural communities and hard-to-reach regions that have underdeveloped health care structure and shortage of medical institutions and personnel.

2.2 E-Health

"Cost-effectiveness of telemedicine and Telehealth enhances extensively when they are part of an incorporated utilization of telecommunications and information technology in the health sector." [8].

This has resulted to the identification of "E-Health" as a generic term, with determinations such as "a new term needed to describe the combined use of electronic communication and information technology in the health sector... the use in the health sector of digital data - transmitted, stored and retrieved electronically - for clinical, educational and administrative purposes, both at the local site and at distance" [9].

WHO describes E-Health as a transmission of healthcare and health resources by electronic medium. It encompasses three main areas:

- Specialized health information delivery for both health professionals and consumers via Internet and telecommunication technologies.
- Using the information technology tools and electronic commerce to improve public health services, e.g. through the education and training of health workers.
- The use of electronic commerce and electronic business instruments in health systems management.

E-Health provides a new method for using health resources - such as information, money, and medicines - and in time should help to improve efficient use of these resources. The Internet also provides a new medium for information dissemination, and for interaction and collaboration among institutions, health professionals, health providers and the public [10].

E-Health, as a logical successor of telemedicine, includes various forms of services and systems, like:

- EHR (Electronic Health Records),
- Consumer health informatics,
- Health knowledge management,
- Telemedicine,
- M-Health (we'll talk about it later), etc.

Talking about main approaches between Telehealth and E-Health, Telemedicine oriented in more hardware-centric aspects, which rely on the traditional equipment sales model, while E-Health is apparently oriented to service delivery, which is more interesting from the business side.

Telemedicine stays bounded to medical professionals, whereas E-Health is controlled by non-professionals – patients with their interests drive new services even in the healthcare field - mostly for their clearance through admission to information and knowledge.

2.3 M-Health

M-Health, which is also known as mobile health, is the most advanced interpretation of E-Health and a form of telemedicine where wireless devices and cell phone technologies are playing the main role. It is useful to think of M-Health as a tool, "a medium", through which telemedicine can be practiced. M-Health is a notably potent development because it delivers clinical care through consumer-grade hardware and allows for greater patient and provider mobility. ATA (American Telemedicine Association) has a Special Interest Group dedicated to the practice and development of M-Health [3].

Here are some common application uses of M-Health:

- Education and awareness
- Treatment support and diagnostic
- Disease and epidemic tracking
- Healthcare supply chain management
- Remote data collection and remote monitoring
- Telehealth/Telemedicine/E-Health
- Chronic disease management

There's no doubt that people are going mobile – from wearables to tablets and smartphones, and this makes implementation and adoption of M-Health services faster. Even healthcare providers are turning to mobile, according to the Research2Guidance report, 80 % of doctors and physicians use smartphones and medical apps [11].

In M-Health, biggest compliance risk with mobile devices, however, is data leakage. Because the patient secret information is stored on a mobile device, there is some risk having that device lost or stolen, and therefore leaking that patient's secret data. However, with a proper healthcare IT plan, compliance can be made possible.

With the implementation and use of modern IT infrastructures in healthcare, researchers and physicians had a huge move forward in support of remote diagnosis and treatment. More and more applications and software is published to online stores and markets, and millions of people are downloading them to their portable devices. This is very helpful for people in regular movement to get consultations and medical treatments on the go. But another problem exists, which we think, extremely important to solve as soon as possible.

There are some illnesses where the main need of involving information technologies is not in the monitoring or support of stable contact between doctor and patient, but eliminating communication barrier for patient itself.

3   Our Solution

Here are the most common types of various diseases which cause disorders of speech and writing:

Autism Spectrum Disorder (ASD): studies in Europe, Asia and North America have identified individuals with ASD with an average prevalence of between 1 % and 2 % [12].

Down Syndrome: the estimated incidence of Down Syndrome is about 0,2 % worldwide [13]

Cerebral Palsy (CP): it is the most common movement disorder in children and occurs in about 0,21 % of live births [14]

In the world practice for children (people) with such disorders in everyday communication and educational purposes so-called card-pictograms are widely used (Fig. 1).

Fig. 1. Usual card-pictograms

These cards help to overcome the difficulties of communication and may represent objects, actions, characteristics of objects, as well as words that can be necessary for conversation. It is proved that such people perceive the information of these cards much better than spoken or written language [15]. A system that allows a child with speech and writing disorders to communicate with the help of card-pictograms is called PECS — Picture Exchange Communication System. The main advantages of using this system are:

- PECS is a method that allows to quickly acquire basic functional communication skills
- With PECS it possible to teach a child to show initiative and say the words spontaneously much faster than with learning names of objects or vocal imitation.
- With PECS communication for a child with others becomes more affordable and thus, it becomes possible to generalize acquired verbal skills

Nowadays, with the growth of digital technologies different programs, mobile applications are developed for distance real-time communication, but people with speech and writing disorders may suffer a social exclusion by the reason of their condition. For example, they may have strong difficulties while typing on mobile device's keyboard may not understand the meaning of read text message. Thus, it would be useful to create a mobile application that allows people with such disorders to exchange their ideas, feelings and wishes in real time remotely with the integration of PECS system in it. This will minimize their social exclusion in on-line area and let them to stay in touch with their family members and have consultation with their doctors.

We have created an application that not only allows its users to interact with each other via special card-pictograms but also has GPS tracking possibility designed for children's parents and doctors to make them aware about children's location and react fast if something unexpected happen or their help is needed. We named this application Sezam (http://sezamapp.ru/). Sezam is a comfortable, easy to use mobile software of alternative communication. Program is designed mainly for children with speech and writing disorders, but can be used also by adults with the same disorders or with temporary disorders (e.g. after a stroke). This is the first application in Russian which allows to communicate through special card-pictograms (translations to other languages will be done in upcoming versions). It is possible to create complete sentences from these card-pictograms. It is a unique opportunity for people with such disorders to engage in dialogue with someone remotely. Application can help people with limited communicative abilities and their families to have a chance to mitigate the stressful situation caused by problems of distant communication.

Existing version of the program is developed for Android OS and designated to run on any conventional mobile device (smartphone, tablet) running OS v4.1 and above. Communication takes place through the social network VKontakte (http://vk.com), which creates new opportunities for people with disabilities in communication and social adaptation. VK API allows to exchange messages between network users [16]. Sentences created via card-pictograms are encoded and sent as a text messages and receiver's application decodes them back. There are more than 500 black and white card-pictograms of international standard in the application. Most of them are translated into Russian versions of pictograms provided by the website http://www.pictogram.se/. Also it is possible to combine them in a message with letters or signs to create a united grammatical structure. All card-pictograms are located in functional groups that represent their common characteristics. For example, "Objects", "Question" "Emotions", etc. (Fig. 2). This kind of division allows users to easily locate desired card-pictograms.

Group "Emotions"

Group "Questions"

Fig. 2. Grouped card-pictograms in the application

There are several approaches to develop a mobile application. The most stable approach is traditional way of coding (or use of native SDK). In this case, developers use special IDE's and SDK's provided by platform developers which are optimized for certain platform only. By using a native approach, developers focused only on one mobile OS with its own programming language, such as Objective C/Swift for iOS, Java for Android etc. Applications which are developed in this way are not compatible with other mobile platforms [17]. So, to make application to run on other operating systems it needs to be rewritten for these systems with their native languages, which is the main problem of this approach.

There are also cross-platform, and hybrid (combining two mentioned methods) approaches. For Sezam developing native approach was chosen. One of the main reasons of this choice is application performance: dynamic loading of more than 500 images (card-pictograms) in one user interface, if cross-platform or hybrid approach was chosen, would significantly reduce the efficiency of the application. On the other hand, the first social network that we used for user login and exchanging data was VKontakte, which API is fully integrated with Android native SDK.

4   Testing

To evaluate the effectiveness and convenience of the program was decided to carry out tests for the target groups. Testing was done on the base of non-profit educational institution for children with special developmental needs. 26 children from 5 to 18 years old with different types of disabilities of speech and writing took part in the test (we renewed already done test [17] but with more number of participants to improve previous results).

Participants were asked to answer different questions with the help of the program:

– First question was the simplest, implying the answers "Yes" and "No" (e.g. "Do you want a candy?" or "Do you want some juice?"). As a result, all children answered to this question.
– Second one was more complicated than the first one and implied a choice of several actions in the application: "What do you want to do now?". Again all children gave answers (Fig. 3).

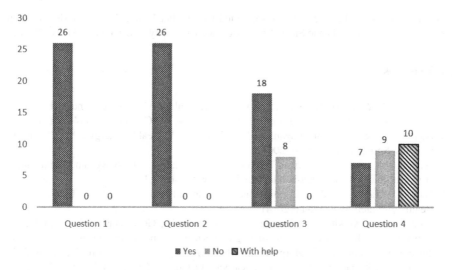

Fig. 3. Test results

– Third question implied a choice of several actions, but also gave an opportunity to make a detailed sentence: "What did you do yesterday?". Most of the participants (18 people) succeeded. With experienced users of gadgets was even established an extensive dialogue. Failed children had insufficient familiarity with PECS in general.
– The last task itself was not a question but an offer to make a choice from the groups "emotions" and/or "adjectives": "Describe your state." 7 children could describe their condition without any help, 10 - with minimal help and 9 participants failed. Again failures mostly were due to lack of understanding of emotions as such or insufficient familiarity with PECS.

Test showed that work with the program went well even for participants with severe disorders if they were familiar with pictographic signs in an appropriate level.

The gadget allows users with disabilities to communicate with someone "from the other room", expands communication possibilities and makes it "safer" from the point of view of the child. Mentioned are the main advantages of the program.

5 Conclusion

The combination of use of telemedical systems and M-Health solutions led to dramatic improvement in healthcare worldwide last few decades. Although a diversity of mobile applications that are available for distance health care, there are still problems that disabled people may encounter in their daily elementary communication process. Solution we propose eliminates the verbal "gate" between disabled persons, giving a chance to freely and easily communicate and share information with as less actions as possible. With the help of digital pictograms, application users are able to reach desired words and phrases in group-based alignment. Conducted test results were very satisfying, proving application efficiency and ease of use for targeted groups of people.

Acknowledgment. The research was supported by Russian Foundation for Basic Research (projects No. 16-07-01111) and Saint Petersburg State University (project No. 0.37.155.2014).

References

1. Zubair, M.F., Jahan, H., Rahaman, S., Raina, R.: M-Health An Emerging Trend An Empirical Study. Computer Science and Information Technology, AIRCC, pp. 167–174 (2016)
2. Istepanian, R.S.H., Laxminarayan, S., Pattichis, C.S.: M-Health Emerging Mobile Health Systems. Springer, Heidelberg (2005)
3. What is Telemedicine. American Telemedicine Association. http://www.americantelemed.org/about-telemedicine/what-is-telemedicine#.VwjSA6R96Uk
4. Fabio, C., Naimoli Andrea, E., Giuseppe, P.: Telemedicine for Children's Health, TELe-Health. Springer, Switzerland (2014)
5. Gankevich, I., Korkhov, V., Balyan, S., Gaiduchok, V., Gushchanskiy, D., Tipikin, Y., Degtyarev, A., Bogdanov, A.: Constructing virtual private supercomputer using virtualization and cloud technologies. In: Murgante, B., Misra, S., Rocha, A.M.A., Torre, C., Rocha, J.G., Falcão, M.I., Taniar, D., Apduhan, B.O., Gervasi, O. (eds.) ICCSA 2014, Part VI. LNCS, vol. 8584, pp. 341–354. Springer, Heidelberg (2014)
6. Guskov, V.P., Gushchanskiy, D.E., Kulabukhova, N.V., Abrahamyan, S., Balyan, S., Degtyarev, A.B., Bogdanov, A.V.: An interactive tool for developing distributed telemedicine systems. Comput. Res. Model. **7**(3), 521–528 (2015)
7. Авакян [и др.], М.Н.: К вопросу развития телемедицины в Республике Армения. Медицинское образование и профессиональное развитие. №. 4. С. 42–54 (2013)
8. Mitchell, J.: Increasing the cost-effectiveness of telemedicine by embracing e-health. J. Telemed. Telecare **6**(Suppl. 1), S16–S19 (2000)
9. Mitchell, J.: From telehealth to e-health: the unstoppable rise of e-health. Commonwealth Department of Communications, Information Technology and the Arts (DOCITA), Canberra, Australia (1999)
10. E-Health: World Health Organization. http://www.who.int/trade/glossary/story021/en/
11. Is Mobile Healthcare the Future? [Infographic]. Greatcall official website. http://www.greatcall.com/greatcall/lp/is-mobile-healthcare-the-future-infographic.aspx
12. Autism Spectrum Disorder (ASD) Data and Statistics. Centers for Disease Control and Prevention. http://www.cdc.gov/ncbddd/autism/data.html
13. Genes and human disease: World Health Organization. http://www.who.int/genomics/public/geneticdiseases/en/index1.html

14. Oskoui, M., Coutinho, F., Dykeman, J., Jetté, N., Pringsheim, T.: An update on the prevalence of cerebral palsy: a systematic review and meta-analysis. Dev. Med. Child Neurol. **55**(6), 509–519 (2013)
15. Frost, L., Bondy, A.: PECS: The Picture Exchange Communication System. Pyramid Educational Consultants, Inc. (2002)
16. Android SDK: VK developers. https://vk.com/dev/android_sdk
17. Balyan, S., Abrahamyan, S., Ter-Minasyan, H., Waizenauer, A., Korkhov, V.: Distributed collaboration based on mobile infrastructure. In: Gervasi, O., et al. (eds.) ICCSA 2015. LNCS, vol. 9158, pp. 354–368. Springer, Heidelberg (2015)

Using Formal Concepts Analysis Techniques in Mining Data from Criminal Databases and Profiling Events Based on Factors to Understand Criminal Environments

Quist-Aphetsi Kester[1,2,3,4(✉)]

[1] Ghana Technology University, Accra, Ghana
Kester.quist-aphetsi@univ-brest.fr, kquist@ieee.org
[2] Lab-STICC (UMR CNRS 6285), University of Brest, Brest, France
[3] Directorate of Information Assurance and Intelligence Research,
CRITAC, Accra, Ghana
[4] Department of Computer Science and Information Technology,
University of Cape Coast, Cape Coast, Ghana

Abstract. As population increases and technological developments advances, criminal activities become complex and sophisticated. Understanding relationships between criminal activities, social factors, geographical locations, crime types, communications links, etc. becomes very crucial. Physical surveillance and other conventional modes of analysis of crime data does not show relationships between direct criminal activity variables and create visualization and interactivity between them. This has yielded difficulties in controlling indirect and direct factors that are likely to provide an environment for such activities. Within this paper, we proposed and engaged Formal Concept Analysis which is based on Galois lattice theory to create relationships between criminal activities, social factors, geographical locations, crime types, communications links, etc. This created a better visualization of relationships between geographical areas and provided a better view in combating crime as well as providing intelligence on the environmental factors based on geography.

1 Introduction

In every society, one of the critical things policy makers tries to do is to optimize activities in their environment so as to help them make better and informed decisions. As data becomes more generated from disparate information systems, policy makers turn to rely more on such data in helping them to better understands trends in the society [1]. These collected data may consist of corporate, educational data, geographical and economic information, etc. Knowledge extraction through data analytics techniques of such data has helped in making informed decisions. This can be seen in "Big data" initiatives and discussions that aimed to mine data from disparate multiple databases varying from a wide range of government and non-government agencies with the objectives of transforming data into the delivery of effective social services [2].

© Springer International Publishing Switzerland 2016
O. Gervasi et al. (Eds.): ICCSA 2016, Part V, LNCS 9790, pp. 480–496, 2016.
DOI: 10.1007/978-3-319-42092-9_37

In line with that, our proposed approach, seeks to use and analyze data from different sources from a geographical area and relate such data with other geographical data in order to extract relationships between events in the context of crime. In our work, we presented an analytic and visualization approach of direct and indirect factors that create an enabling environment for crime activities. This helped to better understand criminal environments for effective policy making and crime combating. It further help project possible crime possibilities based on variations in economic and social factors and hence providing a better advanced intelligence on a geographical environment.

2   Related Works

In combating criminal activities, data collection and analysis is critical in making informed decisions so as to better understand critical patterns and relationships between factors likely to create an environment for possible criminal activity. Hence today's data management information systems must poses the capacity of analyzing, predicting events based on trends and extracting vital knowledge in a form of mined information from data. This can engage machine learning techniques that are based on predefined rules with supervised or unsupervised learning techniques. This calls for effective and advanced computational models and mathematical approaches in the domain of data science, data mining and machine learning. These approaches are as a result of advances in artificial intelligence in current information systems. For an effective decision to be made with data collected, the retrieved knowledge should be used to better interpret the criminal environment and positively influence policy and decision making as seen in the figure illustrated below [3] (Figure 1).

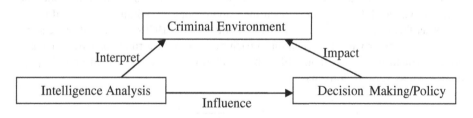

Fig. 1. *3-i model*

Formal concept analysis (FCA) is a discipline of applied mathematics which focuses on the mathematical formalization of concepts and concepts hierarchy in classification of objects and their properties. This has paved way for formal representation, analyses, and construction of conceptual structures and models [4]. This has produced a systematic presentation of the mathematical foundations that focuses on classifications of entities and their relations in data analysis and knowledge processing therefore yielding graphical methods for representing conceptual systems [5]. FCA has generally been used in analysis and applied in different fields especially in the field of software engineering to support the activities of effective software maintenance and also in object-oriented class identification tasks [6–8]. This helped in building structures and ontology [9, 10] as well as analysis of relationships between objects [11–14]. It has also been applied in mining and extraction of data for knowledge discovery [15–17].

Engagement of FCA tools and techniques in the extraction of the hierarchical relations of data values in the database has allowed the visualization, realization of patterns and relationships among multiple variables not previously detectable [18]. This has resulted in the advancement of knowledge integration tools for large data sets and also applied in terrorist-related information. [20] described some useful applications of FCA in the detection of crime, monitoring of crime activities and for finding multiple instances of an activity associated with Organized Crime as well as identifying dependencies between Organized Crime attributes, and new indicators of Organized Crime from the analysis of existing data from information systems.

Crime activities are geospatial phenomena and factors in societal environments can create conducive environment for certain criminal activities. Such actors can be low education, high unemployment, poor social policies, etc. We based our approach on such factors to extract more knowledge in order to understand the criminal environment.

3 Methodology

The introduction of Formal Concept Analysis (FCA) was initially developed as a subsection of Applied Mathematics based on concepts hierarchy, where a concept is constituted by its extension, comprising of all objects which belong to the concept, and its intension, comprising of all attributes (properties, meanings) which apply to all objects of the extension [19] and this was done in the 1990s by Rudolf Wille and Bernhard Ganter [20]. They developed their approaches by building on applied lattice and order theory developed by Birkhoff and others in the 1930s. The set of objects and attributes, together with their relation to each other, form a formal context, which can be represented by a cross table [19]. This method of data analysis then grew in popularity across various domains.

	y_1	y_2	y_3	$\cdots$
x_1	×	×	×	
x_2	×	×		$\vdots$
x_3		×	×	
$\vdots$		$\cdots$		$\ddots$

Fig. 2. Tables with logical attributes: crisp attributes

A table with logical attributes, as shown in Fig. 2, can be represented by a triplet (X, Y, I). where I is a binary relation between the fields X and Y. The elements of X are called objects and the elements of Y are called and for $x \in X$ and $y \in Y$, $(x, y) \in I$

	Latin America	Europe	Canada	Asia Pacific	Middle East	Africa	Mexico	Caribbean	United States
Air Canada	X	X	X	X	X		X	X	X
Air New Zealand		X		X					X
All Nippon Airways		X		X					X
Ansett Australia				X					
The Austrian Airlines Group	X	X	X	X	X				X
British Midland		X							
Lufthansa	X	X	X	X	X	X	X		X
Mexicana	X		X				X	X	X
Scandinavian Airlines	X	X		X	X				X
Singapore Airlines		X	X	X	X	X			X
Thai Airways International	X	X					X	X	
United Airlines	X	X	X	X			X	X	X
VARIG	X	X	X		X		X	X		X

Fig. 3. A formal context about the destinations of the Star Alliance members.

indicates that object x has attribute y while $(x, y) \notin I$ indicates that x does not have y [27]. Figure 3 shows a formal context about the destinations of the Star Alliance members.

3.1   The Context

Definition 1. A (formal) context is a triple $K := (G, M, I)$, where G is a set whose elements are called objects, M is a set whose elements are called attributes, and I is a binary relation between G and M (i.e. $I \subseteq G \times M$). (g, m) 2 I is read "object g has attribute m".

Definition 2. For $A \subseteq G$, let

$$A' := \{m \in M \mid \forall g \in A : (g, m) \in I\}$$

and, for $B \leq M$, let

$$B' := \{g \in G \mid \forall m \in B : (g, m) \in I\}$$

3.2   The Concept

A Formal Concept in a given formal context (G, M, I) is a pair (A, B) with $A \subseteq G$, $B \subseteq M$, A' = B and B' = A. The set A is the extent of the concept and the set B is the intent of the concept. A formal concept is, therefore, a closed set of object/attribute relations, in that its extension contains all objects that have the attributes in its intension, and the intension contains all attributes shared by the objects in its extension [19]. The sets A and B are called the extent and the intent of the formal concept (A; B), respectively. The subconcept-superconcept relation is formalized by [19]

$$(A1, B1) \leq (A2, B2) :\Leftrightarrow A1 \subseteq A2 (\Leftrightarrow B1 \supseteq B2)$$

And this can be seen in Fig. 3, a formal context about the destinations of the Star Alliance members. We can observe in the cross-table that Air Canada and Austrian Airlines fly to both USA and Europe. However, this does not constitute a formal concept because both airlines also fly to Asia Pacific, Canada and the Middle East. Adding these destinations completes (closes) the formal concept [19]:

({Air Canada, Austrian Airlines}, {Europe, USA, Asia Pacific, Canada, Middle East}).

3.3   The Galois Connections

A Galois connection implies that "if one makes the sets of one type larger, they correspond to smaller sets of the other type, and vice versa" [22]. Using the formal concept above as an example, if Africa is added to the list of destinations, the set of airlines reduces to {Austrian Airlines} [19].

Definition 3 (Galois connection). A Galois connection between sets X and Y is a pair (f, g) of f : $2^X \rightarrow 2^Y$ and g : $2^Y \rightarrow 2^X$ satisfying for A, A1, A2 $\subseteq$ X, B, B1, B2 $\subseteq$ Y [27] :

$$A1 \subseteq A2 \Rightarrow f(A2) \subseteq f(A1),$$
$$B1 \subseteq B2 \Rightarrow g(B2) \subseteq g(B1),$$
$$A \subseteq g(f(A)),$$
$$B \subseteq f(g(B)).$$

3.4   The Lattice

The set of all formal concepts of a context K together with the order relation $\leq$ is always a complete lattice, called the concept lattice of K and denoted by B(K). This can be visualized in a Concept Lattice (Fig. 4), which is an intuitive way of discovering thus far undiscovered information in data and portraying the natural hierarchy of concepts that exist in a formal context. A concept lattice consists of the set of concepts of a formal context and the subconcept-superconcept relation between the concepts.

The extent of a concept consists of all objects whose labels are attached to sub concepts, and, dually, the intent consists of all attributes attached to super concepts. For example, the concept labeled by `Middle East' has {Singapore Airlines, The Austrian Airlines Group, Lufthansa, Air Canada} as extent, and {Middle East, Canada, United States, Europe, Asia Pacific} as intent. Besides British Midland and Ansett Australia, all airlines are serving the United States. Those two airlines are located at the top of the diagram, as they serve the fewest destinations | they operate only in Europe and Asia Pacific, respectively.

The concept lattices provide richer information than from looking at the cross-table alone. This type of hierarchical intelligence that is gleaned from FCA is not so readily available from other forms of data analysis.

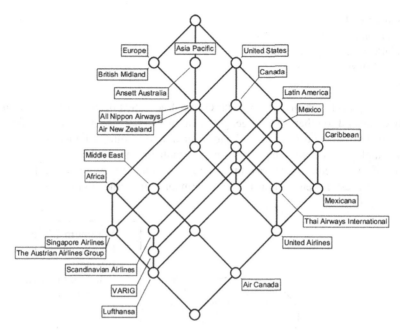

Fig. 4. A lattice corresponding to Fig. 3 [19].

3.5   The Criminal Data and Environmental Data

In processing data obtained from sources with FCA, it is necessary to consider what are suitable to be classified as the objects of study and attributes of those objects as well [19]. A formal context can be created from recorded instances of crime, geography or from other datasets. After achieving the acquisition of the needed data from appropriate data sources, formal contexts can then be created using existing software tools and techniques [22–25]. Then, using the formalisms and tools available in FCA and its related disciplines, it will be possible to carry out analyses to evaluate:

– Geographical Areas that have a specific kind of crime more rampant due to high increase in other factors such as high dropouts in education.
– Finding dependencies between crime attributes based on association rules [26].
– Finding new environmental indicators from existing and external data sources based on Machine Learning/Classification methods [19].
– Developing and using an indicator association rules to better understand crime types in geographical areas.

The following datasets were considered:

• Geographical locations
• Criminal activities in the geographical area and this include crime types.
• Minimum Wage
• Employment data
• Business Establishment Data

- Standard of Living
- Educational Index
- Social/Community Activities
- Social Lifestyles
- Starting Parenting Ages

A geographical area may have some criminal activities occurring more within its locality compared to others. These geographical areas may also have some indications of low, medium or high Minimum Wage comparative to others. Other characteristics such as employment data, types of business establishments in the geographical area, standard of living, educational rate turnout rates, community Activities such as youth clubs, social lifestyles etc. can also give information on crime environments.

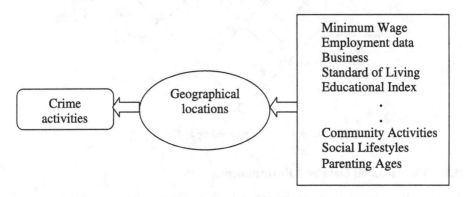

Fig. 5. Depiction of variable effects.

The lists of factors in the rectangular box in Fig. 5 can be obtained from various information systems. These indicators can be mapped to geographical areas and then aligned with crime types within geographical areas to see relationships between geographical crime variations.

The following information systems can be considered sources of such data:

- Registrar General Information Systems: Birth and Death rates, Registration of businesses
- Health Information System
- Police Departments Information System
- Ministry of Justice Information Systems
- Regional Education Information Systems
- City and Town Planning Information System
- Accountant General's Department Information System
- Statistical Service Information System etc.

3.6   Obtaining the Dependencies

A geographical location is picked from X, where X = {X1, X2, X3............Xn} and the geographical areas having factors A such that A = {A1, A2, A3,.............An}. And each member of A has a value between the ranges of 1-5 based on their ratings. Let the crime C also be C = {C1, C2, C3, ...Cn}, where each element of C is a different kind of crime occurring.

For instance, crime types from C at geographical areas from X will have a representation as follows.

The datasets reorientations of the processed extracted data based the definitions above is represent in the context form in the Table 1, which is table of X (Geographical locations) objects and A (factors) attributes, and Table 2, which is the Table X (Geographical locations) and (Crime types) attributes.

Table 1. Table of X (Geographical locations) and A (factors) attributes

	A1					A2					.					An				
	1	2	3	4	5	1	2	3	4	5						1	2	3	4	5
X1		x								x	.	.	.	.	.					x
X2	x					x					.	.	.	.	.			x		
X3		x								x	.	.	.	.	.	x				
X4			x				x				.	.	.	.	.			x		
X5		x								x	.	.	.	.	.		x			
X5	x										.	.	.	.	.					
.	.	.	.	.	.	.	.	.	.	.	.	.	.	.	.	.	.	.	.	.
.	.	.	.	.	.	.	.	.	.	.	.	.	.	.	.	.	.	.	.	.
.	.	.	.	.	.	.	.	.	.	.	.	.	.	.	.	.	.	.	.	.
.	.	.	.	.	.	.	.	.	.	.	.	.	.	.	.	.	.	.	.	.
Xn		x								x	.	.	.	.	.		x			

The table above represents geographical area of formal context (G, M, I) where (A, B) with A ⊆ G, B ⊆ M, A' = B and B' = A.
And this means (X, A) with X ⊆ G, A ⊆ M, X' = A and A' = X.

Table 2. Table Objects C(Crime types) and attributes of X (Geographical locations)

	X1	X2	X3	X4	X5	X6	.	.	Xn
C1	x		x	x	x	x	.	.	X
C2		x					.	.	
C3	x						.	.	X
C4		x	x		x	x	.	.	X
C5	x						.	.	
C5		x		x	x	x	.	.	.
.	.	.	.	.	.	.	.	.	.
.	.	.	.	.	.	.	.	.	.
.	.	.	.	.	.	.	.	.	.
.	.	.	.	.	.	.	.	.	.
Cn	x		x	x		x	.	.	X

The table above represents geographical area of formal context (G, M, I) where (A, B) with A ⊆ G, B ⊆ M, A' = B and B' = A.
And this means (C, X) with C ⊆ G, X ⊆ M, C' = X and X' = C.

4   Generating the Lattice

In generating the lattice, sets of formal concepts of context was built Table 1 and the results were as follows. Ten selected geographical areas were used as shown in the figure below. (Figure 6)

```
<?xml version="1.0" encoding="UTF-8"?>
<BIN name="CRIMINAL ENVIRONMENTContext.slf" nbObj="10" nbAtt="40"
type="BinaryRelation">
   <OBJS>
    <OBJ id="0">X1</OBJ>
    <OBJ id="1">X2</OBJ>
    ............................
    <OBJ id="8">X9</OBJ>
    <OBJ id="9">X10</OBJ>
   </OBJS>
   <ATTS>
    <ATT id="0">A1-1</ATT>
    <ATT id="1">A1-2</ATT>
    ..............................
    <ATT id="38">A8-4</ATT>
    <ATT id="39">A8-5</ATT>
   </ATTS>
   <RELS>
    <REL idObj="0" idAtt="0" />
    <REL idObj="0" idAtt="6" />
    ....................................
      <REL idObj="9" idAtt="30" />
    <REL idObj="9" idAtt="39" />
   </RELS>
</BIN>
```

The following index value ratings were used for the geographical locations: minimum wage, employment ratings, standard of living, educational index, activeness of youth and community associations, social lifestyles, starting parenting age, and acculturation. The crime considered was Burglary, Computer Crime, Drug Trafficking, Hate Crime, Kidnapping, Money Laundering, Murder, Prostitution, Robbery and Theft.

The geographical location were labeled with X, where X = {X1, X2, X3............ Xn} and the geographical areas having index value ratings as A such that A = {A1, A2, A3,.............An}. And each member of A has a value between the ranges of 1–5

Fig. 6. A map of selected areas.

Table 3. Table of X (Geographical locations) and A (factors) attributes with their ratings

	A1	A2	A3	A4	A5	A6	A7	A8
X1	1	2	1	5	1	1	5	0
X2	1	3	2	3	3	2	4	1
X3	2	2	1	2	1	2	2	2
X4	1	1	5	1	3	2	2	2
X5	3	3	5	3	3	2	2	2
X6	0	4	4	4	4	3	3	2
X7	3	3	5	3	5	4	3	3
X8	2	5	4	3	5	5	3	2
X9	5	1	5	4	4	5	2	4
X10	1	2	4	1	3	5	1	5

	A1-1	A1-2	A1-3	A1-4	A1-5	A2-1	A2-2
X1	X						X
X2	X						
X3		X					X
X4	X					X	
X5			X				
X6							
X7			X				
X8		X					
X9					X	X	
X10	X						X

Fig. 7. Sample of the binary data generated based on Tables 1 and 3.

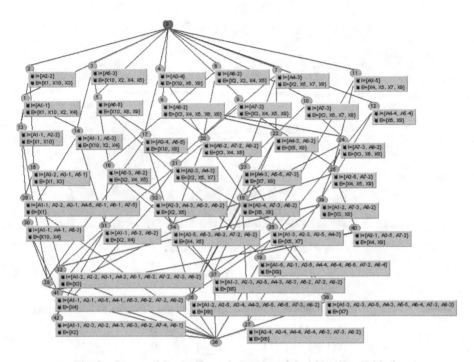

Fig. 8. Generated Lattice from the processed data based on Table 3.

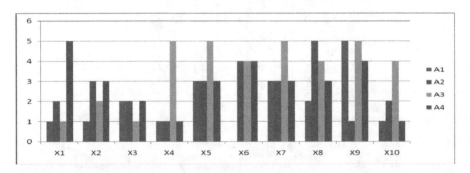

Fig. 9. A graph of geographical Areas with Attributes of A1–A4. (Color figure online)

based on their ratings as shown in Table 3 below. Let the crime C also be C = {C1, C2, C3, ...Cn}, where each element of C is a different kind of crime occurring.

For instance, crime types from C at geographical areas from X will have a representation as follows (Figures 7, 8, 9, 10, 11, 12, 13, and 14).

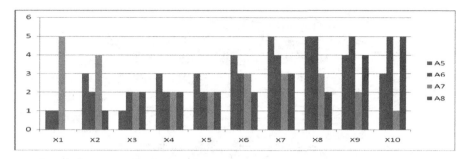

Fig. 10. A graph of geographical Areas with Attributes of A5–A8 (Color figure online)

Table 4. Table of C (crimes) and X (Geographical locations) attributes with their ratings

	X1	X2	X3	X4	X5	X6	X7	X8	X9	X10
C1	1	1	1	5	4	2	5	0	4	2
C2	5	1	4	5	5	2	2	1	4	5
C3	3	1	3	2	5	3	3	2	4	3
C4	1	1	1	2	4	2	1	1	5	5
C5	1	2	4	3	2	1	4	2	2	3
C6	5	2	1	2	4	4	2	2	1	2
C7	1	1	1	1	3	2	3	1	2	2
C8	2	5	3	2	5	2	2	1	2	2
C9	1	1	3	1	4	3	1	1	3	4
C10	2	1	1	2	3	2	4	5	1	4

	X1-1	X1-2	X1-3	X1-4	X1-5	X2-1	X2-2	X2-3
C1	X					X		
C2					X	X		
C3			X			X		
C4	X					X		
C5	X						X	
C6					X		X	
C7	X					X		
C8		X						
C9	X					X		
C10		X				X		

Fig. 11. Sample of the binary data generated based on Tables 2 and 4.

The datasets reorientations of the processed extracted data based the definitions above is represent in the context form in the Table 1, which is table of X (Geographical locations) objects and A (factors) attributes, and Table 2, which is the Table X (Geographical locations) and (Crime types) attributes.

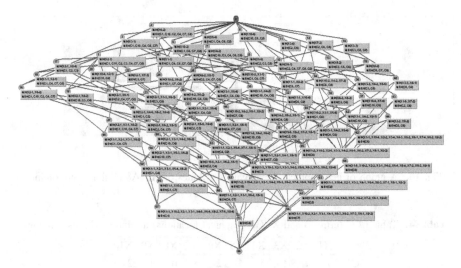

Fig. 12. Generated Lattice from the processed data based on Table 4.

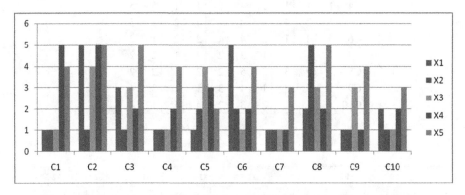

Fig. 13. A graph of crime types with Attributes of X1–X5. (Color figure online)

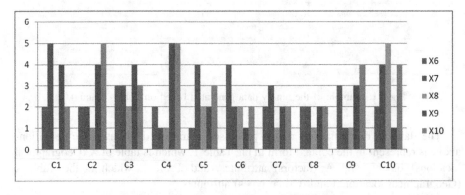

Fig. 14. A graph of crime types with Attributes of X16–X10. (Color figure online)

From the general case we picked geographical area X1–X5 and considered attributes A1–A3. The sub lattice yielded is shown in Fig. 15 below.

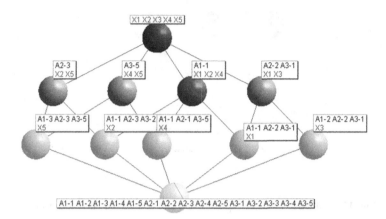

Fig. 15. A Lattice of geographical area X1–X5 and considered attributes A1-A3.

From the Fig. 15, it can clearly be seen that geographical location X4 and X5 have a high standard of living, A1 but with low and medium minimum wage respectively. A graph of the situation is shown in Fig. 16 below.

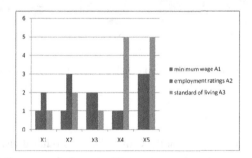

Fig. 16. A graph of geographical area X1–X5 and considered attributes A1–A3. (Color figure online)

From the lattice and the graphs, it can be observed that there were higher crime rates in geographical locations where there is low minimum wage ratings and high standard of living (Figures 17, 18).

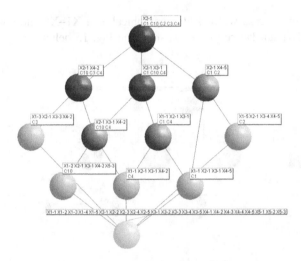

Fig. 17. A Lattice of crime types C1–C4 and considered attributes X1–X5.

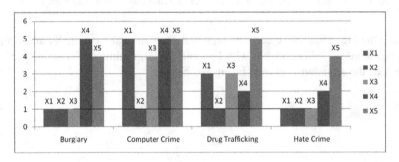

Fig. 18. A Graph of crime types C1–C4 and considered attributes X1–X5. (Color figure online)

5   Conclusions

From our work we have realized that the Gallois lattice produced visualized situation of the environments that the various crimes occurred with their associated attributes. This creates a platform for interactive data analytic approach and knowledge discovery of the environment to provide a good basis for a well informed policy formulation. Crime types are been observed to have higher ratings in environments where certain community development activities, and educational activities are low with high standard of living.

The Formal Concept Analysis Technique provided a flexible with high level insight into the data. One can observe an activity in an environment and see how other related activities associate with it. And a more indebt understanding can be deduced from the lattice. This created a better visualization of relationships between geographical areas and provided a better view in understanding and combating crime as well as providing intelligence on the environmental factors based on geography.

References

1. Stefanowski, J., Japkowicz, N.: Final remarks on big data analysis and its impact on society and science. In: Stefanowski, J., Japkowicz, N. (eds.) Big Data Analysis: New Algorithms for a New Society. Studies in Big Data, vol. 16, pp. 305–329. Springer International Publishing, Switzerland (2005)
2. Gillingham, P., Graham, T.: Big data in social welfare: the development of a critical perspective on social work's latest "electronic turn". Austr. Soc. Work 1–13 (2016)
3. Ratcliffe, J.H.: Intelligence-Led Policing. Willan Publishing, Cullompton, Devon (2008)
4. Ganter, B., Wille, R.: Formal Concept Analysis: Mathematical Foundations. Springer Science & Business Media, New York (2012)
5. Ganter, B., Stumme, G., Wille, R.: Formal Concept Analysis: Foundations and Applications, vol. 3626. Springer, Heidelberg (2005)
6. Tilley, T., Cole, B., Becker, P., Eklund, P.: A survey of formal concept analysis support for software engineering activities. In: Ganter, B., Stumme, G., Wille, R. (eds.) Formal Concept Analysis. LNCS (LNAI), vol. 3626, pp. 250–271. Springer, Heidelberg (2005)
7. Tonella, P., Ceccato, M.: Aspect mining through the formal concept analysis of execution traces. In: 2004 IEEE Proceedings of 11th Working Conference on Reverse Engineering, pp. 112–121. IEEE (2004). doi:10.1109/WCRE.2004.13
8. Snelting, G., Tip, F.: Understanding class hierarchies using concept analysis. ACM Trans. Program. Lang. Syst. 22, 540–582 (2000)
9. Cimiano, P., Hotho, A., Stumme, G., Tane, J.: Conceptual knowledge processing with formal concept analysis and ontologies. In: Eklund, P. (ed.) ICFCA 2004. LNCS (LNAI), vol. 2961, pp. 189–207. Springer, Heidelberg (2004)
10. Godin, R., Valtchev, P.: Formal concept analysis-based class hierarchy design in object-oriented software development. In: Ganter, B., Stumme, G., Wille, R. (eds.) Formal Concept Analysis. LNCS (LNAI), vol. 3626, pp. 304–323. Springer, Heidelberg (2005)
11. Hesse, W., Tilley, T.: Formal concept analysis used for software analysis and modelling. In: Ganter, B., Stumme, G., Wille, R. (eds.) Formal Concept Analysis. LNCS (LNAI), vol. 3626, pp. 288–303. Springer, Heidelberg (2005)
12. Cellier, P., Ducassé, M., Ferré, S., Ridoux, O.: Formal concept analysis enhances fault localization in software. In: Medina, R., Obiedkov, S. (eds.) ICFCA 2008. LNCS (LNAI), vol. 4933, pp. 273–288. Springer, Heidelberg (2008)
13. Priss, U.: Formal concept analysis in information science. Arist 40(1), 521–543 (2006)
14. Tilley, T.: Tool support for FCA. In: Eklund, P. (ed.) ICFCA 2004. LNCS (LNAI), vol. 2961, pp. 104–111. Springer, Heidelberg (2004)
15. Tourwe, T., Mens, K.: Mining aspectual views using formal concept analysis. In: 2004 Fourth IEEE International Workshop on Source Code Analysis and Manipulation, pp. 97–106 (2004). doi:10.1109/SCAM.2004.15
16. du Boucher-Ryan, P., Bridge, D.: Collaborative recommending using formal concept analysis. Knowledge-Based Systems, Elsevier 19(5), 309–315 (2006)
17. Beydoun, G.: Formal concept analysis for an e-learning semantic web. Expert Syst. Appl. 36(8), 10952–10961 (2009)
18. Voss, S., Joslyn, C.: Advanced knowledge integration in assessing terrorist threats. LANL Technical report, LAUR 02-7867 (2002)

19. Andrews, S., Akhgar, B., Yates, S., Stedmon, A., Hirsch, L.: Using formal concept analysis to detect and monitor organised crime. In: Larsen, H.L., Martin-Bautista, M.J., Vila, M.A., Andreasen, T., Christiansen, H. (eds.) FQAS 2013. LNCS, vol. 8132, pp. 124–133. Springer, Heidelberg (2013)
20. Ganter, B., Wille, R.: Formal Concept Analysis: Mathematical Foundations. Springer, Heidelberg (1998)
21. Priss, U.: Formal concept analysis in information science. Ann. Rev. Inf. Sci. (ASIST) **40**, 521–543 (2008)
22. Andrews, S.: In-Close2, a High Performance Formal Concept Miner. In: Andrews, S., Polovina, S., Hill, R., Akhgar, B. (eds.) ICCS-ConceptStruct 2011. LNCS, vol. 6828, pp. 50–62. Springer, Heidelberg (2011)
23. Andrews, S., Orphanides, C.: FcaBedrock, a formal context creator. In: Croitoru, M., Ferré, S., Lukose, D. (eds.) ICCS 2010. LNCS, vol. 6208, pp. 181–184. Springer, Heidelberg (2010)
24. Becker, P., Correia, J.H.: The ToscanaJ suite for implementing conceptual information systems. In: Ganter, B., Stumme, G., Wille, R. (eds.) Formal Concept Analysis. LNCS (LNAI), vol. 3626, pp. 324–348. Springer, Heidelberg (2005)
25. Yevtushenko, S.A.: System of data analysis "concept explorer". In: Proceedings of the 7th National Conference on Artificial Intelligence KII-2000, pp. 127–134 (2000). (in Russian)
26. United Nations: Global programme against transnational organized crime. Results of a pilot survey of forty selected organized criminal groups in sixteen countries. Technical report, Office on Drugs and Crime, United Nations (2002)
27. Outrata, J.: Drawing lattices with a geometric heuristic. In: 2008 4th International IEEE Conference on Intelligent Systems, IS 2008, vol. 2, pp. 15–35. IEEE, September, 2008

Network System Design for Combating Cybercrime in Nigeria

A.O. Isah[1], J.K. Alhassan[1], Sanjay Misra[2(✉)], I. Idris[1],
Broderick Crawford[3,4], and Ricardo Soto[3,5,6]

[1] Department of Cyber Security Science,
Federal University of Technology Minna, Minna, Nigeria
[2] Department of Computer Science, Covenant University, Otta, Nigeria
Sanjay.misra@covenantuniversity.edu.ng
[3] Pontificia Universidad Católica de Valparaíso, Valparaíso, Chile
[4] Universidad Central de Chile, Santiago, Chile
[5] Universidad Autónoma de Chile, Santiago, Chile
[6] Universidad Cientifica del Sur, Lima 18, Peru

Abstract. This research work is to bring to light, the danger posed by Cyber Crime in the world generally and Nigeria in particular with the hope that policy makers will work with the recommendations and practical combating framework design of this research work. In order to achieve this, the following approaches were adopted; survey of some common Cybercrime in Nigeria with the frequency of occurrence and design of a frame work system to combat the crime. The system design controls and track cyber criminals on the Nigerian cyber space. The paper proposes the establishment of National Cybercrime Control Center (NCCC) to effect this system. The Security Agents could obtain tracked information from NCCC as evidence to arrest and prosecute cybercriminals.

Keywords: Network · Cybercrime · Framework · Combating

1 Introduction

Generally, crime means "a legal wrong that can be followed by criminal proceedings which may result into punishment" whereas Cyber Crime may be "unlawful acts wherein the computer is either a tool or target or both" [13].

Nigeria is adjudged as the most populous black nation in the world and by the figures of the last population census, Nigeria's population has grown slightly above 140 million [10]. This has no doubt increase commercial activities and hence attracted more providers of services locally and internationally, one of these services is the computer and internet revolution that has rapidly increased over the years. Cyberspace is already woven in to the fabric of our society. Our security and economy can no longer do without the cyberspace and online access is the most dominant part of the cyberspace, it is already viewed by many as the 'fourth utility', a right rather than a privilege [6]. In less than 11years, the number of global web users has exploded by more than a hundred-fold, from 16 million in 2003 to more than 1.7 billion presently [9].

© Springer International Publishing Switzerland 2016
O. Gervasi et al. (Eds.): ICCSA 2016, Part V, LNCS 9790, pp. 497–512, 2016.
DOI: 10.1007/978-3-319-42092-9_38

While cyberspace provides Nigeria with massive opportunities, the dangers associated from our increasing dependence on it are huge in the last 10 years. There have
been more interconnected devices than ever; everything from mobile phones, computers, cars and surveillance systems, are networked across homes, offices and class-
rooms across the geopolitical zones of the country [9].

The advancement of man in terms of Information Technology which surpasses any
previous generation has usher in a new dimension in crime. The Cyber space has
become a fertile land where various crimes are being committed per second.

Cyber Crime is a computer-based crime which involves all criminal activities that
are carried out in the Cyber-Space [13]. The Internet revolution formed the backbone
for this crime as it merged the Universe into a global village. This crime could be
committed against individual, businesses and governments.

In the era of cyber world as the usage of computers has became more widely
deployed, with the advancement in supporting technology devices as well. The term
'Cyber' became more familiar. The rapid growth of Information Technology (IT) gave
birth to the cyber space wherein internet provides equal opportunities to all the people
to access any information, data storage, analysis, with the use of high technology. Due
to increase in the number of users, misuse of technology in the cyberspace was
clutching up which gave birth to cyber crimes at the domestic and international level as
well [10].

Cyber Crimes Actually Means: It could be hackers attacking your site and trying to
gain undue privilege to your information, with the use of internet. It can also include
'denial of services' and virus attacks preventing regular traffic from hitting your site.
These crimes are not committed by outsiders unless in case of viruses and with respect
to security related cyber crimes that is usually committed by the employees of particular company who can easily access the password and data storage of the company
for selfish gain [13]. Cyber crimes also includes criminal activities done with the use of
computers like plagiarism, online advanced fee frauds, pornography, online gambling,
piracy and other cyber crimes.

Problem statement of this work is the fact that Impact of Cybercrime on National
image and Security cannot be over-emphasized. Any Nation, be it developed or
developing like Nigeria depend on interaction with other countries of the world. The
integrity of data and communication at the private, business and official level forms the
bedrock for any meaningful development in terms of policy formation, security, education, economy, social tribes.

Therefore, any false modification of data and fraudulent activities on the nation's
cyber space may spell doom on such country. For an example, if someone gained
access into the data storage of a certification of competency to fly a plane, if cyber
criminals succeeded in defrauding intending investors in the country of all their money,
the consequences both on safety, economic downfall and the country's image can be
imagined.

This problem is increasing globally and in Nigeria; hence this designed frame work
as proposed by this paper shall reduced this problem drastically and as a deterrent to
cyber criminals.

The research is aimed at combating cybercrime using centralized control system
with the objectives to;

- Identify the various cyber crimes in Nigeria and measure the frequency of their occurrences.
- Design a system to combat cyber crimes by way of monitoring and controlling users.
- Sum all registered users in a control centre with a mathematical model
- Be able to track and apprehend criminals on the Nigerian cyber space.

This research work is significant in the sense that it serves as an automated system to control and checkmate cyber criminals' activities in Nigeria and the world at large. Therefore, the author(s) of this paper shall be willing to partner and make this framework available to partner the security agencies of government to be able to make right policy that will curb the menace of Cyber Crime and by so doing improve the security of the nation's cyber space.

The scope of the work is to cover some common Cybercrimes that poses serious threat to the financial and material Security and Nigeria image as a whole. The proposed system design recognizes any IP addressable devices on the cyber space under consideration but can only produce full detail logs of only the screen-based devices since the imagery part of the logs is the most important evidence of cyber crime.

Although so many things could be integrated on the proposed frame work design, but for the purpose of this very work, the system is limited to control only screen-based and camera embedded devices used on the Nigerian cyber space.

Since it is required that the network installs the auto imagery controller in user's device in a matter of seconds, the complex programs to synchronize the imagery capturing devices with the logs recorder at the control center can still be perfected further to be able to discriminate users with obscure face or any form of shades to prevent recognition.

2 Review of Existing Framework for Combating Cybercrime

Researchers in the IT profession has been working seriously to come up with frame work to combat cybercrime globally and locally, but various frame work has always been with its own limitations in the fight to combat cyber crime.

Legislative approaches, administrative measures and Technical measures were the solutions suggested by [7]. Technical measures which are of interest to the authors of this paper, was theoretical and advisory as the author could not provide practical implementation of such technical measures. Curriculum should include courses on cyber-management, crime and its prevention. Education is a most vital weapon, as inculcating the right culture will create a high level of awareness among all stakeholders, seminars and workshop should be organized from time to time with emphasis on cyber safety so that the individuals will learn to keep their personal information safe and flee cybercrime. Some youths are misguided and misdirected by peers and uncensored films; unless they are guided they may not realize the inherent danger in the act. In the work of [2], the authors suggested building of database of phone numbers and faxes of fraudsters relying on Criminal Act Sect. 419 of the Nigeria Criminal code

Capp 777 of 1990 that prohibits advance fee fraud. This is to allow the authority to shut down the phone numbers and cafes in case of criminal activities. Considering the argument of the author [5], maintains that "Fighting cybercrime requires not just IT knowledge but IT intelligence on the part of the security agencies. After all, in [1] the authors said a little stringent measure on individual's internet activities can assist in reducing cybercrime.

Cyber crime is information and intelligence based. Curiously, the criminals have the technological advantage which the fighters lack sometimes. To outsmart the criminal, having the necessary skills and intelligence are sine qua non. Nigeria government should do more than just enactment of laws and to start prosecuting offenders. [11], according to the author, fighting cybercrimes goes beyond bill boards at strategic places as warnings to cybercriminals in Nigeria without any serious government policy to deal with cybercrime offenders. The effort of the Nigeria Cyber Working Group (NCNG) formed by the Federal Government of Nigeria in 2004 has also not yielded any positive fruit. In another work [4], the authors came up with a frame work using packet attestation which can establish whether or not a given packet is sent by a particular subscriber. This has the capability of allowing network operator to verify the source of malicious traffic and it will also help to validate complaints. However, the limitation of this frame work bothers on the availability and credibility of the packet attestation itself. The availability and credibility depends on the ISPs of the originating messages or packets. Again, this system cannot trace a multi stage attacks by itself back to the source.

The impact of cyber crimes on Nigerian economy is the focus of [12] in which survey and statistical analysis are used as a methodology for accessing how prevalent the cyber crime menace is, and to sensitive the Nigerian masses against risks that are evident in the cyberspace. The drawback of this study is the inability to define and suggest a framework for fighting cybercrime in our cyberspace.

Accessing Cyber Crime and E-banking in Nigeria uses Social Theories as a methodology for proffering policy modulation in combating cybercrime in electronic banking services in Nigeria [15]. It is limited with the fact that banks in Nigeria have implemented bank verification numbers, a policy that adds captured biometrics details of individual bank customer means cyber security solutions have gone beyond social theories to a more practical approach and model.

In [3], the causes, effects and way out of cybercrimes in Nigeria were suggested. The methodology lies on policy definition, it lacks a specific technically model for achieving security in the cyberspace.

However, the review of [16] reveals that it is the same as [15] but with a different title with same authors and contents.

Models from technological innovation with public health, law of sea, aviation law automotive regulation and coordinated ecosystem change were part of the frame work designed to combat cybercrime [14]. The work dealt mostly on laws and regulations of various sector that concerns cyber space, but the limitation of this framework lie in the fact that practical solutions are not achievable where those laws and regulations are not implemented or where cyber criminals can cleverly by pass laws and regulations.

In [8], authors proposed the risk-based approach which work on the principle of assumption that unauthorized user can gain access to the system and compromised data. The design responses based on the data that could be compromised. This approach is meant to prioritize risks and Categorizing the most valuable data. The paper also proposes the approach of developing actionable cyber threat intelligence with a model of cyber intelligence acquisition and analysis.

This frame work has the limitation of not being able to apprehend the cyber criminals in most cases. The prioritizing of data may have some consequence because, a less important data taken from a place can be very useful to compromise more important data somewhere and thereby used to commit serious cyber crime.

In summary, all existing framework reviewed lacked little or no system designs that can could ensure practical implementation leading to monitoring, arrest or prosecution of cyber criminals. This paper seeks to bridge that gap.

3 Proposed System Design to Combat Cybercrime

This is a system design where all devices that uses internet can be monitored. Transactions, communications and information can be traced to the originator of such transactions and communication with the clear image of the perpetuator for arrest and prosecution.

3.1 Methodology of the Research

The survey and system design model was used in this research. The survey was used to identify the common cyber crimes and the frequency of their occurrence in Nigeria; then the proposed system model was designed to reduce the frequency of the cyber crime to the barest minimum and to serve as practical control and preventive mechanism against cybercrime as well as evidence to arrest cyber crime perpetuators. Also, a mathematical model for summing up all registered user is included, bearing in mind that is important to know the number of people using the network with their individual bio data saved in the database. This research is proposing the establishment of NCCC that will be the custodian of the implementation of this framework design. Several Network security technologies and configurations shall be implemented under very strict network security policy. The model was designed and simulated with a network design tool (packet tracer).

3.1.1 Survey of Cybercrimes in Nigeria

The study identified some common cyber crime in Nigeria such as: Online Advance-fee fraud, pornography, software piracy, software cracking, ATM fraud, spam e-mail, website hacking, and personal identification theft (PIT). A total of two hundred experience respondents were selected to give the general rating of cyber crimes mentioned as either L – Low, H – High or VH – Very High. The results were presented in tables below.

Table 1. Cybercrimes rating

Cybercrime	Rating	Outcomes	%
1. Online advance-fee fraud	L	20	10
2. Pornography	H	40	20
3. Software piracy,	VH	140	70
4. Software cracking,	Total	200	100
5. ATM fraud			
6. Spam e-mail,			
7. website hacking			
8. Personal identification theft (PIT).			

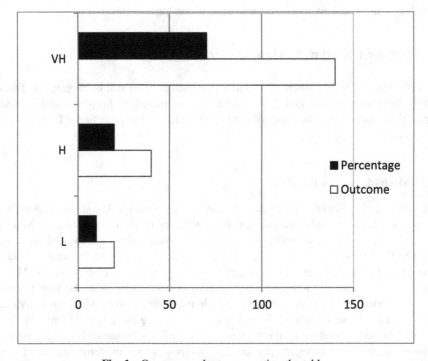

Fig. 1. Occurrence chart representing the table.

From Table 1 above, it clearly shows that all the identified cybercrimes in Nigeria has had a sharp rise going by the percentage of respondents in terms of the very high (VH) rating parameter. While the least of respondents on the table are 2, that rated cyber crime occurrences low and that are just 2 %, this shows that most enlightened Nigerians understands the negative effect of various cybercrime in the country. The high rise in online Advance-fee fraud and other cybercrimes as enumerated above could be seen through a lot of social media such as Facebook and free e-mail services such as Yahoo mail. Most Nigerians now have access to the internet through their phones via the GSM service provider. Figure 1 above shows the occurrence chat from the table.

3.1.2   Designed Model for the Proposed System

Figure 2 below shows the proposed design system model. An anonymous user can login from any location in the Nigerian cyber space and start surfing and other activities without being aware of all the complex processes that are going on within the network system via the NCCC. This shall be explained later in the paper.

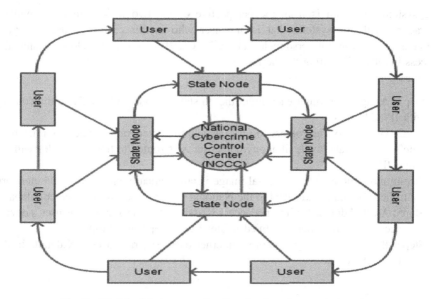

Fig. 2. Model of the proposed national cybercrime control center

The control center also needs to login and perform several actions to establish connections with the anonymous user via the State nodes or direct to the user's devices.

3.1.3   Mathematical Model for the System

Given Users (U) is a function of individually registered X users, using personal computing devices.

$$U = \left(X_1, X_2, X_3 \ldots X_{n-3}, X_{n-2}, X_{n-1}, X_n\right) \qquad (1)$$

At the State level (S), the system calculate the sum of anonymous users to access the internet. This shown mathematically as;

$$s = \sum_{ci=1}^{n} U(Xi) \qquad (2)$$

Finally, at the National Cybercrime Control Centre (N), the sum of anonymous users in the country's cyberspace is calculated by adding users in all the states as follow;

$$N = \sum_{i=1}^{37} S\left(\sum_{i=1}^{n} U(Xci)\right) \tag{3}$$

3.2 The System Algorithm

The system algorithm is from the perspective of the two main actors that will be interacting with the system; that is, the user who in this case, the targeted cyber criminal and the control personnel(s) at the NCCC. The algorithm below explains the process involved in setting up the NCCC.

Start

Step1: Assign one router each to all the 36 states (Nodes) plus FCT in the country with IP addresses

Step2: Use one of the routers to serve as admin server router linked to a database

Step3: At the state level, all anonymous user X is routed to the state nodes with its router IP.

Step4: anonymous user X's facial image is automatically captured by the network controlled web cam and saved to the National Cybercrime Control Center database.

Step5: An IP Address for internet access is assigned to the anonymous user's device else access is automatically denied if Step3 and Step4 are skipped.

Step6: the Control Center monitors internet uses and trace a user X details in the event of cybercrime.

Stop

- The system algorithm is from the perspective of the two main actors that will be interacting with the system; that is, the user who in this case, the targeted cyber criminals and the control personnel at the NCCC.

USER

- Login
- Networks controlled-auto-Imagery program activated
- Profile created
- Access network

NCCC

- Login
- Access active nodes
- View logs at the nodes
- View user's logs at the node
- Display or print user's logs

3.3   Flow Chat for the Proposed System Design

The Fig. 4 below presents the system flow chart for the proposed National Cybercrime Control System. Several anonymous users can attempt to login to internet on the Nigeria cyberspace from any of the state nodes. The auto-imagery program from the NCCC activates the camera on the users' devices via the state node (Fig. 3).

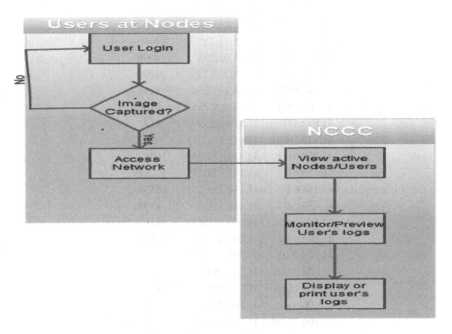

Fig. 3. Flow chart of the system

3.4   The Network Design Concept for the System

This paper execute the idea of Internet Protocol (IP) location based system steering framework where IP version4 (IPV4) or IP version 6(IPV6) can be used; in light of the fact that IPV4 and IPV6 has turned out to be a vigorous and adaptable convention for Internet directing. IPV4 utilizes 32-bit IP address, and with 32 bits, the number of IP locations can reach 4,294,967,296. This is more than four billion. But the Ipv6 shall also be considered in the implementation of this system design in view of the foreseen geometrical increase of devices on the Internet, or the requirement for more IP locations than IPV4 could supply. The system design work on the supposition that each electronic gadgets connected to the internet must have an IP for Identification. In this system design, we are considering only the screen-based devices like the desktop, Portable Computer, PDAs, and convenient telephone gadgets.

IPV6 is utilizing 128 bits which gives a hypothetical location space of $3.4 * 10^{38}$ locations. This is 3.4 trailed by 38 zeros, or 3,400,000,000,000,000,000,000,000,000,000,000,000,0

Table 2. Major IP address blocks for Nigeria

.From IP	To IP	Total IPs
41.58.0.0	41.58.255.255	65536
41.67.128.0	41.67.191.255	16384
41.71.128.0	41.71.255.255	32768
41.73.0.0	41.73.31.255	8192
41.73.128.0	41.73.159.255	8192
41.73.224.0	41.73.255.255	8192
41.75.16.0	41.75.31.255	4096
41.75.80.0	41.75.95.255	4096
41.75.192.0	41.75.207.255	4096
41.84.160.0	41.84.191.255	8192
41.86.128.0	41.86.159.255	8192
41.87.64.0	41.87.95.255	8192
41.138.160.0	41.138.191.255	8192
41.139.64.0	41.139.127.255	16384
41.155.0.0	41.155.127.255	32768
41.184.0.0	41.184.255.255	65536
41.189.0.0	41.189.31.255	8192
41.190.0.0	41.190.31.255	8192
41.203.64.0	41.203.95.255	8192
41.203.96.0	41.203.127.255	8192
41.204.224.0	41.204.255.255	8192
41.205.160.0	41.205.191.255	8192
41.206.0.0	41.206.31.255	8192
41.206.224.0	41.206.255.255	8192
41.211.192.0	41.211.255.255	16384
41.216.160.0	41.216.175.255	4096
41.217.0.0	41.217.127.255	32768
41.219.128.0	41.219.191.255	16384
41.219.192.0	41.219.255.255	16384
41.220.64.0	41.220.79.255	4096
41.221.112.0	41.221.127.255	4096
41.221.160.0	41.221.175.255	4096
62.173.32.0	62.173.63.255	8192
62.193.160.0	62.193.191.255	8192
80.248.0.0	80.248.15.255	4096
80.250.32.0	80.250.47.255	4096
82.128.0.0	82.128.127.255	32768
105.196.0.0	105.199.255.255	262144
105.235.192.0	105.235.207.255	4096
193.189.0.0	193.189.63.255	16384

(*Continued*)

Table 2. (*Continued*)

.From IP	To IP	Total IPs
195.166.224.0	195.166.255.255	8192
196.1.176.0	196.1.191.255	4096
196.27.128.0	196.27.255.255	32768
196.29.208.0	196.29.223.255	4096
196.40.192.0	196.40.255.255	16384
196.45.48.0	196.45.63.255	4096
196.200.64.0	196.200.79.255	4096
196.200.112.0	196.200.127.255	4096
196.207.0.0	196.207.15.255	4096
196.220.0.0	196.220.31.255	8192
196.220.64.0	196.220.95.255	8192
196.220.192.0	196.220.207.255	4096
196.220.224.0	196.220.239.255	4096
196.220.240.0	196.220.255.255	4096
196.222.0.0	196.222.255.255	65536
197.149.64.0	197.149.127.255	16384
197.156.192.0	197.156.255.255	16384
197.159.64.0	197.159.79.255	4096
197.210.0.0	197.210.255.255	65536
197.211.32.0	197.211.63.255	8192
197.214.96.0	197.214.111.255	4096
197.240.0.0	197.240.255.255	65536
197.242.96.0	197.242.127.255	8192
197.242.240.0	197.242.255.255	4096
197.244.0.0	197.244.255.255	65536
197.253.0.0	197.253.63.255	16384
197.255.0.0	197.255.63.255	16384
197.255.160.0	197.255.175.255	4096
197.255.208.0	197.255.223.255	4096
212.100.64.0	212.100.95.255	8192
217.14.80.0	217.14.95.255	4096
217.117.0.0	217.117.15.255	4096

By this, we are convinced that the number of possible ip addresses will be enough to be assigned to any number of the computer devices in Nigeria for a very long time to come.

With the table below showing the IP address blocks already allocated to the Nigerian cyber space, it will be very convenient for the propose system to work effectively (Table 2).

3.4.1   Network Design for the System and Working Principle

From Fig. 4 above, the design entails complex configurations from the NCCC to all the state nodes. The network and devices shown on the Federal Capital Territory node (node 1) is same for all the remaining 36 nodes. All user devices like desktops, laptops, and other screen–based PDAs connected to any node via the node switch are anonymous. The NCCC can ping any node or any device; the green LED indicators shows successful communications when tested.

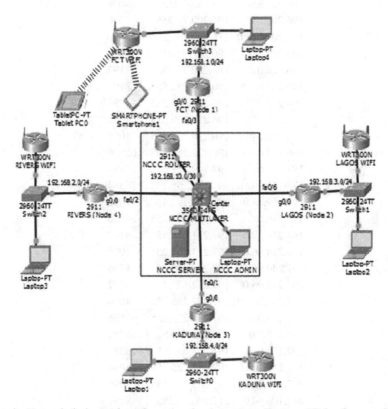

Fig. 4. Network design and configuration for the proposed system (Color figure online)

3.5   Proposed National Cybercrime Control System Interactive Interfaces

3.5.1   The Users

The graphical user interface of the proposed system is shown in the Fig. 5 below. The anonymous user login and the system sends request to control server at the NCCC via the core node of the network where the user is connected.

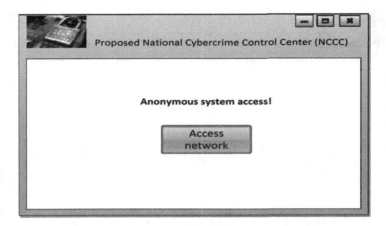

Fig. 5. Anonymous system login

Figure 6 below is the anonymous system access where the request of the user to access the network is granted upon the capture of the user's clear imagery. The system shall deny access upon obscured imagery. The clocking circuit is also activated.

Fig. 6. Anonymous system access

3.5.2 The Control Center

The control center oversees the activities of users in the national cyber space. So like earlier explained, they can track users' foot prints in the network.

Figure 7 below shows the node access interface at the NCCC. The control center login and can access all the nodes at a glance, a particular node could also be access for view, for example, the NCCC may want to view the Lagos node only. The detail log on that node could also be accessed.

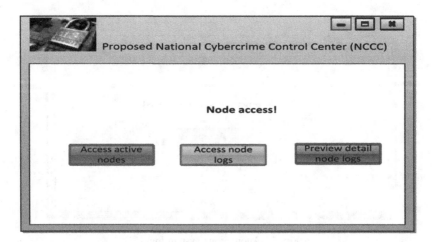

Fig. 7. Node access

Figure 8 below shows the users access interface at the control center. The NCCC could by pass any node to access any user on that particular node, access the logs and preview detail logs of the user.

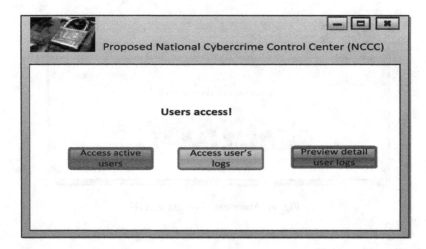

Fig. 8. Users access

Figure 9 below is the user log display interface at the control center. The NCCC could display the detail logs of any user with the user's captured image, the particulars of the screen based device that was used by the user. The anonymous with question mark in the user log display is actually the criminal whose captured image appeared along with other details, that is, user's device number, node number (state node in this case), computer's mac address, login time, logout time and the date the date that the user's device access the internet.

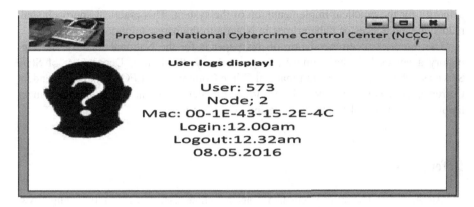

Fig. 9. User logs display

4 The Discussion of the System Design

The system design was based on the concept of IP addressing system where every device can be assigned an IP address. All devices on the network can be tracked through the IP address given to it.

Each of the 37 states of Nigeria was assigned to be a node. From the network shown in 3.2 above, node 1 was assigned to the Federal Capital Territory (FCT), node 2 for Lagos state, node 3 for Kaduna state and node 4 for Rivers State and so on to the 36th State which shall be assigned node 37. Every screen-based systems like laptop, desktop, ipads and phones that attempt to connect to the internet in any of the nodes shall prompt the NCCC server which shall automatically activate the imagery device to have the user's image captured and thus allow access to the internet. The sequence of the operations of any anonymous user and the control center is as explained and presented in Figs. 4, 5, 6, 7, 8 and 9.

The authority simply needs to implement the policy that ensure all Internet Service Providers (ISPs), GSM operators rout their network through the NCCC thereby putting all nodes on the same network of the NCCC.

The configuration and implementation of the network security Virtual Private Network (VPN) and the tunnelling technology will ensure the NCCC access to every nodes and users on the Nigerian cyber space.

5 Conclusion and Recommendations

From our investigation on cybercrimes, we observed its threat to the economy of a nation and even peace and security. This paper has been able to identify some common cyber crimes in Nigeria and their frequency of occurrences and came up with a design framework to combat them. The design concept was successfully simulated with a CISCO network tool. In this system, utmost secrecy, policy definition and its enforcement is of very high essence. Therefore, the authors of this paper shall cooperate with the authority for the practical implementation of the system. This practical approach shall reduce the incidences of cybercrimes in Nigeria to the barest minimum.

The foremost recommendation of this paper is for the National security and regulatory agencies like the Military and paramilitary, the Police, Department of State Services (DSS), Economic and Financial Crime Commission (EFCC), and Independent Corrupt Practices Commission (ICPC) to as a matter of urgent, see to the implementation of this practical frame work design system as proposed by this paper.

References

1. Abhatise, E.J.: Cybercrime definition, computer crime research center (2008). http:/www.crime-research.org/articles/joseph06/2 Accessed 4 Feb 2016
2. Aluko, M.: 17 ways of stopping financial corruption in Nigeria (2004)
3. Anah, B.H., Funmi, D.L., Makinde, J.: Cybercrime in Nigeria: causes, effects and the way out. ARPN J. Sci. Technol. (2012). www.ejournalofscience.org
4. Andreas, H.: University of Pennsylvania, Fighting Cybercrime with Packet Attestation (2011)
5. Ayantokun, O.: Fighting cybercrime in Nigeria (2016). http://archive.cert.uni-stuttgart.de/isn/2006/06/msg00034.html
6. CGI White paper: Cyber security in Modern Critical Infrastructure Environments (2014) http://www.canadianinstitute.com/files/pdf/marketing/CGI-Whitepaper.pdf
7. Chawki, M.: Nigeria tackles advance fee fraud. J. Inf. Law Technol. (2009). http://www2.warwick.ac.uk/fac/soc/law/elj/jilt/2009_1/chawki/
8. Deloitte Development LLC: USA, Combating the fastest growing cyber security threat (2010)
9. Internet Society Global Internet report 2014 (2015). https://www.internetsociety.org/
10. Kofo, A.A.: Significance of 2006 Head Counts and House Census to Nigerian sustainable development. Int. J. Basic Appl. Sci. 01(02) (2012). http://www.insikapub.com/
11. Longe, O.B., Chiemeke, S.C.: Cyber crime and criminality in Nigeria what roles are internet access points in playing? Eur. J. Soc. Sci. 6(4), 132–139 (2008)
12. Maitanmi, O., Ogunlere, S., Ayinde, S., Adekunle, Y.: Impact of cyber crimes on Nigerian economy. Int. J. Eng. Sci. (IJES) 2(4), 45–51 (2013). www.theijes.com
13. Marco, G.: Understanding cybercrime: phenomena, challenges and legal response (2012)
14. Michael, B.: PayPal. www.pdffiller.com/en/project/63758181.htm?form_id=43555194. USA, Combating Cybercrime: Principles, Policies and Programs (2011)
15. Wada, F., Odulaja, G.O.: Assessing cyber crime and its impact on E-Banking in Nigeria using social theories. Afr. J. Comput. ICT 5(1), 69–82 (2012). www.ajocict.net
16. Wada, F., Odulaja, G.O.: Electronic banking and cyber crime in Nigeria - a theoretical policy perspective on causation. Afr. J. Comput. ICT 4(2), 69–82 (2012). www.ajocict.net

Clustering of Wikipedia Texts Based on Keywords

Jalalaldin Gharibi Karyak[1], Fardin Yazdanpanah Sisakht[1],
and Sadrollah Abbasi[2(✉)]

[1] Technical and Vocational University, Yasooj, Iran
{gharibi.jalal92,fyazdanpanah14}@gmail.com
[2] Department of Computer Engineering,
Iran Health Insurance Organization, Yasouj, Iran
sadrollahsadeghi@gmail.com

Abstract. The paper presents application of spectral clustering algorithms used for grouping Wikipedia search results. The main contribution of the paper is a representation method for Wikipedia articles that has been based on combination of words and links and it has been used to categorize search result in this repository. We evaluate proposed approach with Primary Component Analysis and show, on a test data, how usage of cosine transformation to create combined representations influence a data variability. On a sample test datasets we also show how combined representation improves the data separation that increases overall results of data categorization. We gave the review of the main spectral clustering methods and we compare them using external validation criteria with standard clustering quality measures. Discussion on descriptiveness of evaluation measures and performed experiments on test datasets allows us to select the one spectral clustering algorithm that has been implemented in our system. We gave a brief description of the system architecture that groups on-line Wikipedia articles retrieved with specified keywords. Using the system we show how clustering increases information retrieval effectiveness for Wikipedia data repository.

Keywords: Documents · Text · Spectral clustering · IR · Wikipedia · HCI

1 Introduction

The documents categorization is one of the tasks of automatic knowledge organization. This approach is based on computation of similarities between objects that finds many applications in the areas where a text should be analyzed; e.g. in Information Retrieval domain [1] introducing particular similarities between text documents allows building structures that organize the documents repository and thus improves searching for relevant information.

In the paper we describe our method for text representation based on combined approach of words and hyper references between articles (links). We show on visualizations made with Principal Component Analysis projections, how introduction of combined representation improves separation of the data that helps to differentiate

© Springer International Publishing Switzerland 2016
O. Gervasi et al. (Eds.): ICCSA 2016, Part V, LNCS 9790, pp. 513–529, 2016.
DOI: 10.1007/978-3-319-42092-9_39

articles that belong to different categories. Cumulative analysis of components variability shows that combined representation allows representing the data with much smaller number of features than using single representations, which considerably improves the efficiency of data processing.

In the experiments presented in this article we evaluate spectral clustering methods applied to identification of groups of similar Wikipedia articles that allow organizing search results into hierarchy that increases effectiveness of information retrieval in this repository. We test three clustering algorithms on datasets constructed from Wikipedia articles. The categories of these articles made by Wikipedia's editors have been used 34 as relevance sets used for algorithms evaluation. The evaluation allows us to select the best clustering method for our purposes and it has been implemented in the form of web portal. The system organizes Wikipedia search results into clusters that represent groups of conceptually similar topics. Binding search results into clusters is an alternative approach for presenting large collections of data where groups of similarities are used instead of displaying ranked list of articles and it allows the user to effectively review the content of the search results.

2 Spectral Clustering

Spectral clustering algorithms are one of the most efficient groups of clustering methods, mainly because they do not suffer from the problem of local optima [2]. The analyzed data can be represented in the form of a graph that is usually constructed by the object's similarities. In that approach clustering is based on cutting the graph using methods of spectral analysis [3]. In the recent years this theory has been intensively developed [4], especially in direction of graph clustering algorithms, where the most well known are: ShiMalik (2000), KannanVempalaVetta (2000), JordanNgWeiss (2002) and MeilaShi (2000).

If we consider clustering in terms of graph theory, we can introduce a measure that describes partition of graph nodes into two subsets. If we have weighed graph $G = (V; E)$, its partition into two sets of nodes A and B ($A \cap B = \emptyset$ and $A \cup B = V$) can be specified with cutset number [5] defined as (1).

$$\text{Cut }(A,B) = \sum_{u \in A, u \in B} W(u,v) \tag{1}$$

Representing the objects that are to be clustered with graph nodes and the w weights of the graph edges with objects similarities, the partitioning problem is reduced to finding optimal cutset. Sometimes cutset gives the results where clusters are different than intuitive graph partition, so different measures are introduced:

Normalized cut (NCut) (2), that increases its value for clusters that have nodes with small sum of edge weights.

$$\text{NCut}(A, B) = \frac{\text{Cut }(A, B)}{\text{Vol A}} + \frac{\text{Cut }(A, B)}{\text{Vol B}} \tag{2}$$

where

$$\text{Vol } A = \sum_{u \in A} \sum_{u \in V} W(u,v) \tag{3}$$

Multi-way Normalized Cut (MNCut) (4) promotes stronger relationships between elements in one cluster and better describes more separated clusters.

$$\text{MNCut}(\Delta) = \sum_{i=1}^{k} \left(1 - \frac{\text{Cut }(C_i, C_1)}{\text{Vol } C_i} \right) \tag{4}$$

It is known the spectral clustering methods give high quality partitions [6] and have good, polynomial computational complexity [5, 7]. There are many variants of spectral algorithms. Mainly they differ the way of construction transformation space, where eigenvectors are calculated, and their further usage. What is common – they treat source objects as graph nodes that are mapped onto d-dimensional space. This space is constructed using spectral analysis and there essential clustering is performed. We can select three main steps of spectral algorithms [8]:

1. Data Preprocessing: At this stage the data is preprocessed into its computational representation. If we are clustering the text documents, typically Vector Space Model (VSM) [9] is used. In this approach documents using particular representation method (see Sect. 3) are mapped into feature space where further computations are performed. During the preprocessing phase the text is smoothed: so called stop words, special signs (as e.g. punctuation marks) are removed from the content of documents. Also in this stage the words can be turned into their basic form. Word's basic form is extracted with usage tools such as stemmers and lemmatizers. The first tool is based mainly on a fixed set of derivation rules, the second employs additionally the data from the attached dictionary. Usually lematizers obtain better results (e.g. the word "were" will be converted to "be", while stemmer returns "were").
2. Spectral mapping: This stage distinguished spectral approach. Using the data from the first step some normalizations and adjustments are performed (e.g. create of Laplacian matrix), then appropriate number of eigenvectors is calculated.
3. Clustering: The objects represented with spectral mapping are divided into two or more sets. Sometimes it is enough to find appropriate cut of the n-th element in the sorted collection which divides this collection into two clusters. In other methods this step is more complicated and partitioning is performed in new representation space (provided by spectral mapping) using one of standard clustering algorithms e.g. k-means [10].

In our experiments we evaluate three spectral clustering algorithms:

1. Shi–Malik algorithm [11] (SM), is realized in the following steps:
   a. Calculate eigenvectors of similarity Laplacian graph.
   b. Sort elements of the dataset according to the second smallest eigenvector value, which is denoted as $x_1, x_2, \ldots, x_n$,

   c. Calculate partition $\{\{x_1,x_2,\ldots,x_i\},\{x_{i+1},x_{i+2},\ldots,x_n\}\}(1 \leq i \leq n-1)$ having the smallest NCut.

   d. If given partition has NCut value smaller than given a priori value (that means it is better) then this method is run again in each of the divided sets, otherwise the algorithm stops.

2. Kann–Vempala–Vett algorithm (8) (KVV) is a heuristic method that finds graph cut which minimizes quality function, called conductance. Considering partition $(S,\bar{S})$ of graph $G = (V;E)$, where $\bar{S} = V \backslash S$ conductance is the function defined with Formula (5).

$$\emptyset(S) = \frac{\mathrm{Cut}(S,\bar{S})}{\min(\mathrm{Vol}(S),\mathrm{Vol}(\bar{S}))} \tag{5}$$

In comparison to the previous one, the algorithm instead of Laplacian matrix operates on normalized similarity matrix and uses the second biggest (instead the second smallest) eigenvector. The algorithm goes as follows:

   a. Normalization performed by dividing each of row elements by the sum of elements in each row.

   b. For each of clusters C_i created so far, similarity matrix W_i is created from the cluster nodes.

   c. Normalize each matrix W_i by inserting at the position on diagonal value that complete sum of the elements in this row to 1.

   d. Calculate the second biggest eigenvector of v_2^i of matrix W_i.

   e. For each C_i sort its elements according to the value of respective coordinate of v_2^i. We denote as $x_1^i, x_2^i, \ldots, x_{ni}^i$ sequence of ordered objects form cluster C_i.

   f. For each C_i find cut in the form of $\left\{\left\{x_1^i, x_2^i, \ldots, x_j^i\right\}, \left\{x_{j+1}^i, x_{j+2}^i, \ldots, x_{ni}^i\right\}\right\}$ which has the smallest conductance.

   g. Find cluster C_i with the smallest conductance for a given cut, divide it according to this cut and replace cluster C_i with two new clusters.

   h. If the given number of clusters has not been reached, go to (b).

3. Jordan–Ng–Weiss algorithm [6] (JNW) is a partitioning algorithm – in one iteration it creates flat clusters, given by a priori K parameter. This parameter denotes also the number of used eigenvectors.

The algorithm performs the following steps:

   a. Calculate Laplacian of similarity matrix

   b. Calculate K biggest eigenvectors of Laplacian matrix $v_1, v_2, \ldots, v_k$.

   c. Perform spectral mapping from set V (original nodes) to R^K:

      a. $\mathrm{map}(i) = [v_1(i), v_2(i), \ldots, v_k(i)]^T (1 \leq i \leq n)$

   d. Perform clustering of the points $\mathrm{map}(i)$ in space R^K.

   e. Return result of partitioning of entrance elements into K clusters.

Because JNW is a partitioning approach, the last step is not as simple as in previous algorithms, where binary split was performed. In this step, JNW uses the other clustering method to select the groups in spectral space. In its initial version the authors of the algorithm apply to that task one of the simplest approaches – k-means [12].

3   Text Representation

In Information Retrieval two main approaches to documents representation are used: (a) based on text content, (b) based on document relations with others.

The first one typically uses the analysis of word frequencies. The representation that employs words as features is called Bag of Words (BoW) because it does not take into account semantics of the utterances, but is based only on frequencies of words occurrences in the documents. A sever limitation of this representation is not taking into account neither word's order nor simple grammatical constructions, but its application for simple classification and clustering tasks is known to work well. Other approaches to representation based on text content use analysis of letter distributions, n-grams or successive n-words [13]. This approach deals with some issues related to the usage of single words, but they do not change processing quality very significantly and they are known to introduce additional problems, e.g. increasing dimensionality that usually leads to higher computational costs. The second representation method employs the fact that the documents are often related to each other. Using references between documents the representation of the particular document is constructed with its associations to the others. In this method the document is described by other documents it is related to. Below we describe these two methods more in details. Then we describe our approach that is the combination of methods based on words and references.

3.1   Text Content

The preprocessed words are called terms, and in the BoW text representation they are used as features. The descriptiveness strength of a term for a given document is estimated by the w value that associates the term and the text. Typically for n-th term and k-th document w value is calculated as a product of two factors: term frequency tf and inverse document frequency idf, given by $w_{k;n} = tf_{k;n} \cdot idf_n$ [14]. The term frequency is computed as the number of its occurrences in the document and is divided by the total number of terms in the document. The frequency of a term in a text determines its importance for document content description. If a term appears in the document frequently, it is considered to be more important. The inverse document frequency increases the weight of terms that occur in a small number of documents. The idf_n factor describes the importance of the term for distinguishing documents from each other and is defined as $idf_n = \log(k/k_{term(n)})$, where k is the total number of documents, and $k_{term(n)}$ denotes the number of documents that contain term n.

Features (terms) and weights w that associate them with the collection of documents allow to represent each document by a single point in the Vector Space Model (VSM) [15]. The documents similarity is then computed using different distance measures such as cosine or Euclidean measures [16].

3.2   Links

Representations of a text based on terms leads to high-dimensional feature spaces. Without preprocessing and dimensionality reduction [17] the number of features is equal to the total number of the distinct words that appear in all documents. While large datasets are considered, operation on term based representations is very computationally intensive. More compact way to create numerical representation of texts is the usage of references that appear between objects that are to be computed. For the articles and books the list of references and bibliographical notes contain useful information. If HTML documents are considered, their hyperlinks can be used as features. Feature space based on hyperlinks or citations may be constructed in several ways. In the simplest approach each reference states as a new dimension and the document representation creates a binary vector, where 1 denotes the presence of the link (reference) to another document, and 0 indicate its lack. Documents on similar topics tend to link to similar set of other documents and cite the same references. Possible extensions of this representation involve frequency of references, various forms of weighting e.g.: based on the position of a link in the document, the use of directed links ($\pm$ 1 for links from or to the document). These modifications haven't been considered here, only binary representations of references between articles have been used in the experiments.

3.3   Combined Representation

No redundant and informative data representation should prevent data distribution while it is represented with smaller number dimensions. Also it should allow distinguishing objects belonging to one particular category from others. To describe our method for combined representation we show how the variability of the data with successive raw representations based on words and links change while it is reduced. We also depict it to show how different representations change the distribution of data categories. For our experiments we find Wikipedia *dumps* very useful to evaluate our approach for documents categorization. These data are easy accessible as well as cover wide spectrum of human knowledge thus provide very varied set of documents in different domains. What is the most beneficial Wikipedia dumps provide also human made categories that can be used during evaluation process.

To demonstrate our method of text representation we selected 419 articles that belong to 4 arbitrarily selected categories from one super category Philosophy. In Table 1 we present categories used for the demonstration, and the amount of the articles they contain. We also provide information about the size of the features spaces according to the text representation which has been used. To reduce dimensionality of features spaces we remove the features that are related only to one article – it is articles that are referenced only by one other article (in representation with links), and the words that appear only in one article (in representation with terms). It allowed us to reduce considerably (approximately three times) the size of representation spaces during preprocessing stage. In Table 1 we marked this adjustment with $\downarrow$.

Table 1. The data used for demonstration of combined representation

Category name	Symbol & Color	# of articles	# of features in text representation	
			Links	terms
Philosophers	Magenta*	110	2523	18 411
Ethics	Green+	131		
Logic	blue ▫	170	↓	↓
Epistemology	black.	8	945	5819

It is comfortable to have a rough view of the data. To see how it is distributed we present in Figs. 2 and 1 2D data visualization of the data using the two highest principal components (details of the PCA are described further in Sect. 3.3). On the left hand, we show plots of the raw data – which has been prepared directly from the text. On the right hand figures we provide view of the data processed with cosine distance applied to raw data. The cosine distance (calculated using the formula 6) is especially useful in Information Retrieval where the data are sparse (18). While the data are high dimensional, introducing representation based on similarity is an easy way for reducing dimensions. Note that if we have raw data of the sizes $n \times m$ where $n \gg m$ calculating the similarity matrix, we are able to operate on data of the sizes $n \times n$.

$$d(x, y) = \frac{\sum_i x_i y_i}{\sqrt{\sum_i x_i^2} \sqrt{\sum_i y_i^2}} \qquad (6)$$

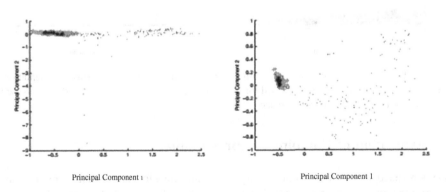

Principal Component 1   Principal Component 1

Fig. 1. View of the data represented with terms in 2D, created with two highest principal components. On the left hand the visualization of the raw data has been presented, the right hand – the data processed with cosine distance

What can be seen by comparing the left hand with the right hand visualizations in Figs. 2 and 1, mapping the data into metric space using cosine distance produces objects distribution that can be much easier separated. Also it should be noticed that the visualization is performed in 2D and some relations between objects can't be presented in this space, but further computations are performed using more dimensions (Fig. 3).

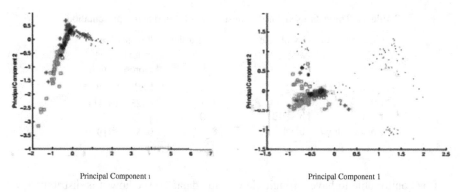

Principal Component 1   Principal Component 1

Fig. 2. View of the data represented with links in 2D, created with two highest principal components. On the left hand the visualization of the raw data has been presented, the right hand – the data processed with cosine distance

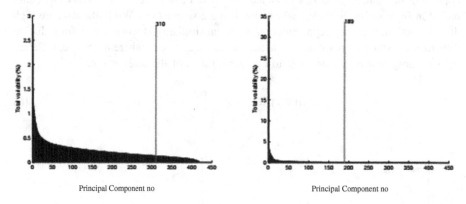

Principal Component no   Principal Component no

Fig. 3. The % of variability for succeeding principal components using terms representation. Left hand graph shows the raw data used for PCA, and right hand graph shows the data processed with similarity based on cosine distance

4   Evaluation of Combined Representation

Demonstrated in previous section, usage of cosine transformation that combines two representations has been evaluated measuring the influence of the particular representations on categorization task. An objective of the experiment is the better results are obtained by the same categorization method - the better representation is.

Very popular approach to text categorization is Naive Bayes classifier e.g.: [18]. We use this wide studied approach to perform classification of the text within ten test data packages. The evaluation dataset has been created in a similar way to the dataset presented in Table 1. The ten packages for the experiment have been created using the data taken from articles form 10 arbitrary selected Wikipedia categories. The categories for each data package have been selected randomly from the sets of categories having the same upper category. The process of building each of the data package was based on selection of the root category and then random selection of its ten sub categories. As Wikipedia category structure is very irregular, it was decided, to select for each of the category in data package around 100 [10] articles, having similar length. This guarantee uniform class distribution during categorization thus it does not bias the results (Fig. 4).

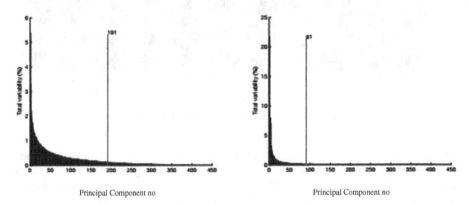

Principal Component no   Principal Component no

Fig. 4. The % of variability for succeeding principal components for representation based on references. Left hand graph shows the raw data used for PCA, right hand – the data processed with similarity based on cosine distance

In Fig. 5 we present the results of performed classifications using different methods of representation based on words, links and their combination. What can be observed almost in all data packages combined representation leads to improvement of classification. Also aggregated average results indicate that the combined representations work better than usage of single representation.

In our research we aim at evaluation of usage of unsupervised spectral methods for automatic text categorization. The data for the experiments presented in this work has been generated using MATRIX'u software [19]. The software allows preparing Wikipedia content in a form that can be processed using machines. Among many functionalities it allows selecting Wikipedia categories that narrow the set of articles and generate for them a set of representing features (according to a selected method of text representation). In the experiments presented here we use representations based on words and links but the application supports other approaches: based on references between the articles, suffix trees and common substrings [20], information content computed by compression [21]. The application is available to download on-line[2] and free for academic use.

In our experiments we evaluated the results according to the external criteria, which are known to be a harder task than to evaluate them according to internal criteria [22]. External cluster validation criteria measure the similarity of the structure of the clusters provided by the algorithm to a priori known structure that is expected to be achieved [23, 24].

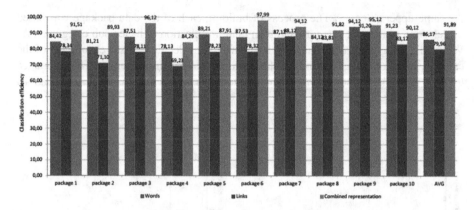

Fig. 5. Evaluation of combined representation

Table 2. Test packages. Descriptions: n – number of nodes (in brackets — unisolated), l – non zero elements in neighborhood matrix, N – number of upper categories, k – number of all categories, d – depth of category tree), P – number of categories to Wikipedia root category

Num	Name	N	l	N	k	d	P	Comment
1	Z01	575(575)	67099	2	18	1	3	2 distant categories: Distance Education i Science Experiments
2	Z02	1157 (1156)	323901	5	35	1	3	5 distant categories: Calligraphy, General Economics, Military logistics, Evolution, Analytic number theory
3	Z03	3905 (3903)	2260919	8	102	1	3	8 distant categories: Geometric Topology, Epistemology, Rights, Aztec, Navigation, Clothing companies, Protests, Biological Evolution
4	Z04	3827 (3826)	3195963	2	204	6	4	Two distant categories at the same hierarchy level: Criticism of journalism and Corporate crime
5	Z05	3647 (3644)	1682361	6	213	6	5	6 distant categories at high level of abstraction: DIY Culture, Emergence,

(Continued)

Table 2. (*Continued*)

Num	Name	N	1	N	k	d	P	Comment
								Military transportation, Formal languages, Geology of Australia, Computer-aided design
6	Z06	4750 (4747)	2568378	9	289	6	5	9 distant categories at high level of abstraction: DIY Culture, Emergence, Military transportation, Formal languages, Geology of Australia, Computer-aided design, Special functions, History of ceramics, Musical theatre companies.
7	Z07	4701 (4701)	4230139	2	298	6	4	2 neighboring categories at high abstraction level: Computer law and Prosecution
8	Z08	5717 (5716)	11288283	4	893	6	4	4 neighboring categories at high abstraction level: Impact events, Droughts, Volcanic events, Storm

To evaluate our results we use standard clustering quality measures. In Fig. 6 we show metrics values of external validation criteria formed with methods SM, KVV and JNW. Each figure corresponds to one test package (presented in Table 2), colors denote lines for quality metrics: Rand statistics (R) – blue, Jaccard index (J) – red, FowlkesMallows index (FM) – green and Hubert statistics (H) – orange. Different clustering algorithms have been denoted with different line patterns.

5   Results Discussion

What can be seen from the graphs are strong correlation of Rand (R) with Hubert (H) statistics as well as FowlkesMallows (FM) with Jaccard index (J). It is caused by high value of a parameter which denotes the number of pairs in one cluster in validated

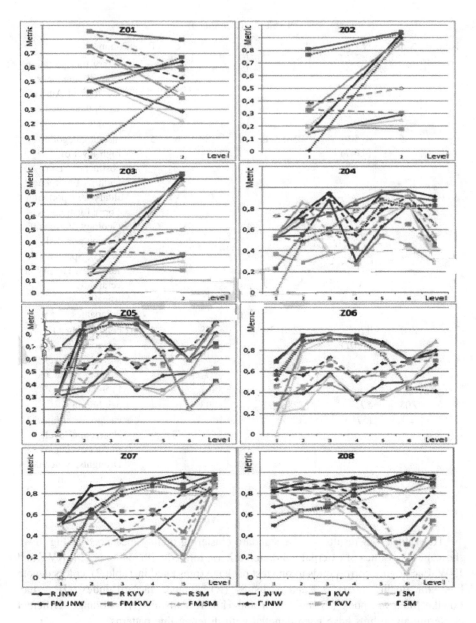

Fig. 6. Results of cluster validation in packages Z01–Z08. (Color figure online)

and referential structure. The second parameter, that has high influence on metrics is d which denotes a number of objects that were assigned to different clusters in referential and in validation set. Metrics J and FM do not use parameter d, statistics R and Γ involve it in dominator and nominator which cause similarity of these measures.

In some areas we can see differences between metric pairs J and FM as well as R and H e.g. for test package Z04 at level 4 J value for KVV algorithm is low while Γ is high. It is because the structure of the metrics J does not use d parameter, but Γ does. In comparison at 3rd level of hierarchy in this test d value grows (from 3.652.328 to 3.819.940), but a value decreases (from 939.463 to 374.220). Because of the increased d value Γ also increases, while the decrease of a caused the decrease of J value. Similarly there can be substantiated decrease Γ while FM grows (e.g. at level 6 of test package Z06). In this case we can see the increase of a and decrease of d. It suggests that on level 6 in referential structure there are fewer categories than at level 5. Indeed number of categories at level 5 is 895 and at level 6 – 503.

The highest values of R and J parameters have been obtained with JNW algorithm. Only one case, when other methods gave better results, is a case described above (level 4 in package Z04). Also FM measure pointed out JNW algorithm as the best. The highest values of in packages Z07 and Z08 we obtain using JNW algorithm, the others have been obtained using KVV.

Based on JNW algorithm we have created a system named WikiClusterSearch (WCS). As algorithm is a partitioning approach to introduce the structure of clusters we bind them using typical Hierarchical Agglomerative approach (HAC) [25]. In the approach to clusters binding we use modified linkage method: to average linkage [26] method where l denotes minimal distance between two centroids we add additional factor ε. In standard HAC only two nearest clusters are bind together. In our approach we bind together nearest clusters and additionally join new elements, if their distance is smaller than $\varepsilon = \frac{l(n + 1)}{2n}$ where n denotes hierarchy level. This simple modification binds more actively clusters together on lower hierarchy levels. As we go higher into hierarchy structure the method aggregates the clusters in a more similar way to HAC. Also it allows forming non uniform structures, as sometimes is it required to put more than two clusters on the same hierarchy level (that is not possible using standard HAC algorithm). The modification causes the hierarchy is more bushy-like than regular, deep tree (produced by HAC), that is more natural for lexical data [27, 28].

To evaluate quality of clusters in terms of their commonsense, we use standard information retrieval metrics: precision and recall [1]. Clustering precision is defined as fraction of relevant text documents (Wikipedia articles) in cluster that is described in Formula (7)

$$P = \frac{\text{Number of relevant documents in cluster}}{\text{Total number of documents in cluster}} \qquad (7)$$

Recall is defined by the Formula (8) as a percentage of all relevant documents that were aggregated into one cluster.

$$R = \frac{\text{Number of relevant documents in cluster}}{\text{Total number of relevant documents for cluster}} \qquad (8)$$

F-measure is a composition using weighted harmonic mean Precision and Recall values defined by Formula (9).

$$F_\beta = \frac{(\beta^2 + 1)PR}{\beta^2 P + R} \qquad (9)$$

where $\beta \in [0, \infty)$ is a weight coefficient. For $\beta = 1$ F-measure balances P and R. By increasing we β put emphasis on Precision. Most common values for β are 1, 3 and 5. Evaluation of the system quality for a particular query has been performed for each of the cluster separately and final measure per query is normalized averaged value with achieved for all clusters. It should be noticed that calculation of 'Total number of relevant documents for cluster' is a particularly time consuming task, as it requires to look trough all documents in the repository. Because of that, for each of test queries we extend the result set using query synonyms taken form WordNet. As articles in Wikipedia belong to the categories we can select thematic domains of retrieved information. The articles from the categories indicated by the articles from the result set have been used to extend original result set created with keywords. Additionally we extend this set by adding articles related with them by hyperlinks. This extension significantly increases the number of articles that have to be analyzed for each query but it keeps their thematics relatively related to the specified by the user keywords. Thus we do not need to look through whole articles repository to create relevance set. During the evaluation process set of 'Total number of relevant documents for cluster' we could create only according to retrieved documents but it wouldn't give us information about misretrieved items. Also because of the fact, we perform retrieval using different number of items in result set used for clustering, this set should be review again for each search that would make hard to repeat the experiments. It should be noticed the construction of relevance set ('Total number of relevant documents for cluster') in the proposed way causes obtaining lower values of recall measure. It is caused by the fact the relevance set is more detailed than the one created only by search phrase.

As mentioned we plan to develop clustering algorithms for text documents and test approaches based on densities [29] computed in spectral spaces. This approach should allow us to break through a problem of creating clusters only having convex shapes as well as it will eliminate the requirement for k parameter specification.

We also plan to run our software for English Wikipedia that requires to introduce some improvements into system architecture as the scale of the data requires more effective approach for storage and processing. The long term goal is to join the method of retrieving the information based on clusters of Wikipedia categories with classifier [30] that allows categorizing linear search results returned by search engine into these categories. Currently we are develop a new implementation that offers extension of clustering algorithm with functionality of narrowing search result by interaction with the user and extending the clusters content using text similarity measures. The new version of the system is available at http://kask.eti.pg.gda.pl/BetterSearch and it supports English Wikipedia.

We also plan to perform experiments with large scale clustering – it is on the whole Wikipedia. Here instead of performing clustering within limited dataset (with specified by the user keywords) we plan to compute the whole Wikipedia. The size of the dataset in such experiments have to be run in parallel environment and it will be performed off - line.

The experiments using well tuned clustering algorithm would allow improving category system of Wikipedia finding missing and wrong assignments articles to categories.

As text representation is crucial for obtaining good results we plan to develop methods that are able to capture elementary semantics. We plan to introduce to a system some background linguistic knowledge. Mechanism inspired by cognitive theories of a language [31] such as brain process called spreading activation [32] should allow adding concepts not explicitly present in a text, but fundamental for categorization. This process may be partially captured in algorithms provided with the usage of large ontologies or semantic networks [33]. Our approach is to map words into network of senses where we use Wordnet [36] synsets. First results of creating representations based on synsets are promising – for now we achieved 65 % of proper disambiguations [35, 36].

Evaluation dataset created during the tests formed a kind of initial golden standard dataset for retrieving information in Wikipedia. Currently it is only for Polish language, but our initiative is to develop it and create datasets similar to TREC data. Creating uniform datasets for evaluation of information retrieval methods in Wikipeda should increase the number of people whose research is involved in processing this huge repository of human knowledge.

References

1. Manning, C., Raghavan, P., Schütze, H.: Corporation, E.: Introduction to information retrieval, vol. 1. Cambridge University Press, Cambridge (2008)
2. Yang, P., Zhu, Q., Huang, B.: Spectral clustering with density sensitive similarity function. Knowl.-Based Syst. **24**, 621–628 (2011)
3. Cvetkovic, D., Doob, M., Sachs, H.: Spectra of Graphs-Theory and Applications, III revised and enlarged edn. Johan Ambrosius Barth Verlag, Heidelberg-Leipzig (1995)
4. Von Luxburg, U.: A tutorial on spectral clustering. Stat. comput. **17**, 395–416 (2007)
5. Vazirani, V.: Algorytmy aproksymacyjne. WNT Warszawa, Warszawa (2005)
6. Ng, A., Jordan, M., Weiss, Y.: On spectral clustering: analysis and an algorithm. Adv. Neural Inf. Process. Syst. **2**, 849–856 (2002)
7. Kannan, R., Vetta, A.: On clusterings: good, bad and spectral. J. ACM (JACM) **51**, 497–515 (2004)
8. Verma, D., Meila, M.: A comparison of spectral clustering algorithms. Technical report, University of Washington UW-CSE-03-05-01 (2003)
9. Salton, G., Wong, A., Yang, C.: A vector space model for automatic indexing. Commun. ACM **18**, 613–620 (1975)
10. Wagstaff, K., Cardie, C., Rogers, S., Schrödl, S.: Constrained k-means clustering with background knowledge. In: Proceedings of the Eighteenth International Conference on Machine Learning, vol. 577, p. 584. Citeseer (2001)
11. Shi, J., Malik, J.: Normalized cuts and image segmentation. IEEE Trans. Pattern Anal. Mach. Intell. **22**, 888–905 (2000)
12. Hartigan, J., Wong, M.: Algorithm as 136: A k-means clustering algorithm. J. Roy. Stat. Soc.: Ser. C (Appl. Stat.) **28**, 100–108 (1979)
13. Damashek, M.: Gauging similarity with n-grams: language-independent categorization of text. Science **267**, 843 (1995)

14. Salton, G., Buckley, C.: Term-weighting approaches in automatic text retrieval. Inf. Process. Manage. **24**, 513–523 (1988)
15. Wong, S.K.M., Ziarko, W., Wong, P.N.: Generalized vector spaces model in information retrieval. In: Proceedings of SIGIR 1985, pp. 18–25. ACM Press, New York (1985)
16. Steinbach, M., Karypis, G., Kumar, V.: A comparison of document clustering techniques. In: KDD workshop on text mining, vol. 400, pp. 525–526. Citeseer (2000)
17. Korenius, T., Laurikkala, J., Juhola, M.: On principal component analysis, cosine and Euclidean measures in information retrieval. Inf. Sci. **177**, 4893–4905 (2007)
18. Jiang, Y., Lin, H., Wang, X., Lu, D.: A technique for improving the performance of naive bayes text classification. In: Gong, Z., Luo, X., Chen, J., Lei, J., Wang, F.L. (eds.) WISM 2011, Part II. LNCS, vol. 6988, pp. 196–203. Springer, Heidelberg (2011)
19. Szymański, J.: Wikipedia articles representation with matrix'u. In: Hota, C., Srimani, P.K. (eds.) ICDCIT 2013. LNCS, vol. 7753, pp. 500–510. Springer, Heidelberg (2013)
20. Grossi, R., Vitter, J.: Compressed suffix arrays and suffix trees with applications to text indexing and string matching. In: Proceedings of the Thirty-Second Annual ACM Symposium on Theory of Computing, PP. 397–406 (2000)
21. Bennett, C., Li, M., Ma, B.: Chain letters and evolutionary histories. Sci. Am. **288**, 76–81 (2003)
22. Eldridge, S., Ashby, D., Bennett, C., Wakelin, M., Feder, G.: Internal and external validity of cluster randomised trials: systematic review of recent trials. BMJ **336**, 876 (2008)
23. Yeung, K., Haynor, D., Ruzzo, W.: Validating clustering for gene expression data. Bioinformatics **17**, 309 (2001)
24. Provost, F., Fawcett, T., Kohavi, R.: The case against accuracy estimation for comparing induction algorithms. In: Proceedings of the Fifteenth International Conference on Machine Learning, vol. 445. Citeseer (1998)
25. Zepeda-Mendoza, M.L., Resendis-Antonio, O.: Hierarchical agglomerative clustering. In: Dubitzky, W., Wolkenhaue, O., Cho, K.-H., Yokota, H. (eds.) Encyclopedia of Systems Biology, pp. 886–887. Springer, New York (2013)
26. Krebs, C.J.: Ecological Methodology, vol. 2. Benjamin/Cummings, Menlo Park (1999)
27. Wang, C., Duo, C.: An improved density-based DBSCAN clustering algorithm. J. Guangxi Norm. Univ. Nat. Sci. Edit. **25**, 104 (2007)
28. Frey, B., Dueck, D.: Clustering by passing messages between data points. Science **315**, 972 (2007)
29. Kriegel, H., Pfeifle, M.: Density-based clustering of uncertain data. In: Proceedings of the Eleventh ACM SIGKDD International Conference on Knowledge Discovery in Data Mining, p. 677. ACM (2005)
30. Szymański, J.: Towards automatic classification of wikipedia content. In: Fyfe, C., Tino, P., Charles, D., Garcia-Osorio, C., Yin, H. (eds.) IDEAL 2010. LNCS, vol. 6283, pp. 102–109. Springer, Heidelberg (2010)
31. Duch, W.: Neurocognitive informatics manifesto. In: Series of Information and Management Sciences (2009)
32. Collins, A., Loftus, E.: A spreading-activation theory of semantic processing. Psychol. Rev. **82**, 407 (1975)
33. Duch, W., Matykiewicz, P., Pestian, J.: Neurolinguistic approach to natural language processing with applications to medical text analysis. Neural Netw. **21**(10), 1500–1510 (2008)
34. Miller, G.A., Beckitch, R., Fellbaum, C., Gross, D., Miller, K.: Introduction to WordNet: An On-line Lexical Database. Cognitive Science Laboratory, Princeton University Press, Princeton (1993)

35. Szymański, J., Mizgier, A., Szopi ński, M., P., L.: Ujednoznacznianie słów przy uzyciu słownika WordNet. Wydawnictwo Naukowe PG TI 2008 18 89–195 536 (2008)
36. Szymański, J., Duch, W.: Annotating words using wordNet semantic glosses. In: Huang, T., Zeng, Z., Li, C., Leung, C.S. (eds.) ICONIP 2012, Part IV. LNCS, vol. 7666, pp. 180–187. Springer, Heidelberg (2012)

Measure-Based Repair Checking by Integrity Checking

Hendrik Decker[1(✉)] and Sanjay Misra[2]

[1] PMS, Inst. f. Informatik, LMU, München, Germany
hdecker@pms.ifi.lmu.de
[2] Covenant University, Canaanland, Ota, Nigeria

Abstract. Quality damage in databases can be measured by seizing extant inconsistency, i.e., by quantifying the amount of violation of integrity constraints. A repair is an update that reduces inconsistency and hence improves data quality. Repair checking finds out if a given update is a repair or not. Repair checking can be done by checking if the undo of the update increases the amount of integrity or not. To do so, sound measure-based integrity checking methods can be used. To do so well, the used methods should also be complete. Repair checking by integrity checking is an attractive alternative to conventional repair checking approaches. However, the completeness of measure-based integrity checking may be a problem, in general. We build on concepts, techniques and results as presented in the first author's previous work.

1 Introduction

Repair checking (abbr. *RCh*) is the problem to find out if a given update repairs violations of database integrity [1,5,6,20]. Conventional RCh evaluates the integrity of each constraint brute-force. Alternatively, we propose to compute RCh by measure-based integrity checking (abbr. *ICh*) [8,11,12].

The main conceptual difference between brute-force and measure-based RCh: the latter is inconsistency-tolerant, i.e., it accepts reductions of integrity violations that may not totally eliminate inconsistency. As opposed to that, brute-force RCh disqualifies each update that does not yield total consistency. The main technical difference: measure-based RCh can be implemented by simplified ICh [7], while brute-force RCh evaluates all constraints brute-force. The main practical difference: If implemented by simplified ICh which exploits the incrementality of updates, measure-based RCh is more efficient.

The two main pillars of our approach are inconsistency-tolerant ICh [9,15] and inconsistency measures [4,16,17,19]. Both have been combined to measure-based ICh in [8,14], which we now propose to use for RCh. Obviously, RCh is related to repairing, i.e., the problem to eliminate inconsistency from a database D such that consistent parts of D are preserved [10,22]. Repairing may be in need of RCh, as broached in [6,20]. Yet, repairs can be processed as goals to be solved, e.g., as view update requests to be satisfied, as done in [12,15].

© Springer International Publishing Switzerland 2016
O. Gervasi et al. (Eds.): ICCSA 2016, Part V, LNCS 9790, pp. 530–543, 2016.
DOI: 10.1007/978-3-319-42092-9_40

By combining integrity-preserving update methods with measure-based ICh, inconsistency-tolerant repairing can be achieved, as described in [10].

In fact, those methods usually do not check if repairs are minimal, as required by consistent query answering (CQA) [2,5], because most of the repairs computed by the mentioned methods are minimal already. On the other hand, if an arbitrarily given update needs to be checked to be or not to be a repair, then the satisfaction of some minimality criterion is indispensable.

Our main contributions: a measure-based inconsistency-tolerant generalization of repairs, of their minimality, and of RCh, as well as an implementation of RCh by measure-based ICh. Proofs of all results are available in the full paper.

2 Key Concepts

As in [7–14], we use common database research terminology and formalisms. See [10] also for definitions of cases and causes of integrity violations, inconsistency measures, measures $\iota, |\iota|, \zeta, |\zeta|, \kappa, |\kappa|$ based on cases or causes, and measure-based inconsistency-tolerant integrity checking methods. Let D, IC, U, μ always stand for a database, an integrity theory, an update, and, resp., an inconsistency measure. We may use ";" for delimiting clauses in sets, instead of ",".

In Subsect. 2.1, we define measure-based repairs. In Subsect. 2.2, we characterize repair checking of updates U as a function, which is decomposed into checking if U reduces inconsistency, and if it does, then checking if U is also minimal, in some sense.

2.1 Repairs

Measure-based repairs generalize total repairs: they do not insist that all inconsistency is eliminated. Thus, a measure-based repair is an update U of a database that is inconsistent with its constraints, such that the updated database becomes less inconsistent, and there is no subset of U that could achieve the same or a larger amount of inconsistency reduction. Below, we define total and measure-based repairs as inconsistency reductions that are minimal.

Definition 1 *(total and measure-based repair).*

(a) U is a *total inconsistency reduction* of (D, IC) if $\mu(D^U, IC) = 0$.
(b) U is a *total repair* of (D, IC) if U is a total inconsistency reduction of (D, IC), and there is no proper subset U' of U with that property.
(c) U is a *μ-based inconsistency reduction* of (D, IC) if $\mu(D^U, IC) < \mu(D, IC)$.
(d) U is a *μ-based repair* of (D, IC) if U is a μ-based inconsistency reduction, and there is no proper subset U' of U such that $\mu(D^{U'}, IC) \leq \mu(D^U, IC)$.

Clearly, total repairs do not depend on any inconsistency measure, as opposed to inconsistency reductions and measure-based repairs. In the literature, total repairs are defined without recurring explicitly on any measure, by requiring that each constraint in IC is satisfied in D^U. However, minimality needs to be measured, and indeed, there are several alternative, non-equivalent definitions of the minimality of total repairs. Yet, subset minimality as in Definition 1a is the one most often referred to.

Obviously, each total and each μ-based repair is a total and, resp., μ-based inconsistency reduction, but not vice-versa.

The following example features an update U that satisfies $\mu(D^U, IC) < \mu(D, IC)$ as in Definition 1c, but not the minimality condition of 1d. Typically, updates of that kind contain elements that do not contribute to the reduction of inconsistency. Example 1 also features an update of D that is a measure-based repair of (D, IC).

Example 1. Let $D = \{p, q, r\}$, $IC = \{\leftarrow q\}$, $U = \{delete\ q,\ insert\ s\}$. It is easy to verify that, for each measure $\mu \in \{\iota, |\iota|, \zeta, |\zeta|, \kappa, |\kappa|\}$, $\mu(D^U, IC) < \mu(D, IC)$ holds, i.e., U is a μ-based inconsistency reduction. However, U is not a μ-based repair of (D, IC), since its subset $U' = \{delete\ q\}$ clearly is a μ-based repair of (D, IC) that yields the same amount of inconsistency reduction in a minimal way, since $\mu(D^{U'}, IC) = \mu(D^U, IC)$ holds.

In fact, the minimality condition of measure-based repairs in Definition 1d should not be weakened so as to simply require that there is no proper subset U' of U such that $\mu(D^{U'}, IC) < \mu(D, IC)$, as illustrated by Example 2.

Example 2. Let $\mu \in \{\iota, |\iota|, \zeta, |\zeta|, \kappa, |\kappa|\}$. Further, let $D = \{p, q, r, s\}$, $IC = \{\leftarrow q,\ \leftarrow r,\ \leftarrow s\}$, $U = \{delete\ r,\ delete\ s\}$, and $U' = \{delete\ r\}$. Clearly, U and U' are μ-based repairs such that $D^U = \{p, q\}$ and $D^{U'} = \{p, q, s\}$. Moreover, U' is a proper subset of U and $\mu(D^{U'}, IC) < \mu(D, IC)$. However, also $\mu(D^U, IC) < \mu(D^{U'}, IC)$ holds, i.e., U reduces inconsistency more than U', i.e., U' is not preferable to U.

The following corollaries of Definition 1 interrelate measure-based repairs, total repairs and inconsistency reductions.

Corollary 1. For each measure μ and each inconsistent pair (D, IC), each total repair of (D, IC) is a μ-based repair.

Corollary 2. If U is a total inconsistency reduction of (D, IC), then some subset of U is a total repair of (D, IC).

Corollary 3. Each singleton update U such that $\mu(D^U, IC) < \mu(D, IC)$ is a μ-based repair of (D, IC).

Corollary 3 devises an iteration of updates for iterated repairs by singleton updates. For example, let $\{I_1, I_2, \ldots, I_n\}$ be the set of violated basic cases in (D, IC). Then, for $\mu \in \{\zeta, |\zeta|\}$, a total elimination of inconsistency in (D, IC) can be achieved by iteratively applying singleton μ-based repairs $U_1, U_2, \ldots, U_n$ such that, for $U_0 = \emptyset$, U_i eliminates the violation of I_i in $D^{U_{i-1}}$, $1 \leq i \leq n$, until all violations in (D, IC) are gone in D^{U_n}. If some U_i also eliminates the violation of I_j ($j > i$), then $U_j = \emptyset$.

Example 3a and b illustrate that, for two measures μ, μ', a μ-based repair U is not necessarily a μ'-based repair, since μ and μ' may measure the effect of U differently.

Example 3. Let $D = \{p, q, r\}$ and $IC = \{\leftarrow q, \leftarrow s\}$.

(a) Let $U_a = \{delete\ q,\ insert\ s\}$. Clearly, U_a is a μ-based repair of (D, IC) for each μ that assigns a higher weight of inconsistency to the violation of $\leftarrow q$ than to the violation of $\leftarrow s$. However, U_a is not a $|\iota|$-based repair of (D, IC) ($|\iota|$ counts the violated constraints in IC), since $|\iota|(D^{U_a}, IC) = |\iota|(D, IC) = 1$.
(b) Let $U_b = \{insert\ o\}$, and μ be the measure that counts the facts in D that contribute to some integrity violation and then divides that count by the cardinality of D. Since $\mu(D^{U_b}, IC) = 1/4$ and $\mu(D, IC) = 1/3$, U_b is a μ-based repair, by Corollary 3. However, for each $\mu' \in \{\iota, |\iota|, \zeta, |\zeta|, \kappa, |\kappa|\}$, U_b is clearly not a μ'-based repair of (D, IC), since $\mu'(D, IC) = \mu'(D^{U_b}, IC) = 1$.

2.2 Repair Checking

We distinguish between *total* and *measure-based* repair checking. The total repair checking problem is to find out if an update U is a total repair of (D, IC). Hence, each total repair checking method can be described as a function rc_t that maps triples (D, IC, U) to $\{yes, no\}$. Similarly, the μ-based repair checking problem is to find out if U is a μ-based repair of (D, IC). Hence, each μ-based repair checking method can be described as a function rc_m that maps triples (D, IC, U) to $\{yes, no\}$.

By Definition 1, repair checking of an update U proceeds in two phases. First, to check if U reduces inconsistency totally (1a) or if U is an inconsistency reduction (1c). We call this phase the *inconsistency reduction check*. If U has passed the inconsistency reduction check, the second phase of repair checking is to check if U is minimal, in the sense of Definition 1b and d, respectively. We call this phase the *minimality check*. The soundness and completeness of repair checking methods is defined as follows.

Definition 2 *(soundness and completeness of measure-based repair checking methods).* Let μ be an inconsistency measure, and rc a function that maps triples (D, IC, U) to $\{yes, no\}$. rc *is* called a sound, resp., complete, μ-based repair checking method if (*), resp., (**) holds, for each triple (D, IC, U).
   (*) $rc(D, IC, U) = yes \Rightarrow U$ is a μ-based repair
   (**) U is a μ-based repair $\Rightarrow rc(D, IC, U) = yes$

In words, rc is sound if its output $rc(D, IC, U) = yes$ correctly identifies U as a μ-based repair of (D, IC), and complete if each μ-based repair U of (D, IC) is checked correctly by rc. Soundness and completeness of total repair checking is defined analogously: ignore μ and replace each occurrence of "μ-based" by "total" in Definition 3.

Corollary 4, below, is going to be useful for valuating results in Sect. 3.

Corollary 4.

(a) If U is not a μ-based repair of (D, IC), then each sound μ-based repair checking method rc outputs $ir(D, IC, U) =$ no.

(b) If a complete repair checking method rc outputs $rc(D, IC, U) =$ no, then U is not a μ-based repair of (D, IC).

According to the first of the two phases of repair checking as identified above, the following definition characterizes sound and complete measure-based inconsistency reduction checking. (Analogously, total integrity reduction checking could be defined.)

Definition 3 *(measure-based inconsistency reduction checking).* Let μ be an inconsistency measure, and ir a function that maps triples (D, IC, U) to $\{yes, no\}$. ir is called a *sound,* resp., *complete,* μ-based inconsistency reduction checking method if (*), resp., (**) holds, for each triple (D, IC, U).

(*) $ir(D, IC, U) = yes \;\Rightarrow\; \mu(D^U, IC) < \mu(D, IC)$

(**) $\mu(D^U, IC) < \mu(D, IC) \;\Rightarrow\; ir(D, IC, U) = yes$

In words, ir is sound if $ir(D, IC, U) = yes$ correctly indicates that U reduces the inconsistency of (D, IC) measured by μ, and complete if each U that reduces inconsistency is checked correctly by rc.

Next, we define the soundness and completeness of the second phase of measure-based repair checking, viz. measure-based minimality checking, according to Definition 1d.

Definition 4 *(measure-based minimality checking).*

Let μ be an inconsistency measure, and mc a function that maps triples (D, IC, U) to $\{yes, no\}$. mc is called a *sound,* resp., *complete,* μ-based minimality checking method if (*), resp., (**) holds, for each triple (D, IC, U) such that U is an μ-based inconsistency reduction.

(*) $mc(D, IC, U) = yes \;\Rightarrow\;$ for each $U' \subsetneq U$, $\mu(D^{U'}, IC) \not\leq \mu(D^U, IC)$

(**) for each $U' \subsetneq U$, $\mu(D^{U'}, IC) \not\leq \mu(D^U, IC) \;\Rightarrow\; mc(D, IC, U) = yes$

In words, mc is sound if $mc(D, IC, U) = yes$ correctly indicates that U is a minimal inconsistency reduction, according to Definition 1d, and mc is complete if the minimality of each μ-based repair of (D, IC) is checked correctly by rc.

The following result is a straightforward consequence of Definitions 3, 4 and 5. It effectively says that repair checking can be realized by inconsistency reduction checking, followed by minimality checking, if needed. (Minimality checking is of course not needed if the output of the inconsistency reduction check is negative.)

Theorem 1. Let μ be an inconsistency measure, ir a sound (resp., complete) μ-based inconsistency reduction method, mc a sound (resp., complete) μ-based minimality checking method, and rc a function that maps triples (D, IC, U) to $\{yes, no\}$. rc is a sound (resp., complete) μ-based repair checking method if it is defined by the following equivalence.

$$rc(D, IC, U) = yes \;\Leftrightarrow\; ir(D, IC, U) = yes \text{ and } mc(D, IC, U) = yes$$

In Sect. 3, we are going to see how each of the two phases of measure-based repair checking can be implemented by measure-based integrity checking.

3   The Main Results

In Subsect. 3.1, we show that inconsistency reduction can be computed by integrity checking. In Subsect. 3.2, we show that also minimality can be verified by integrity checking. This leads to the main result in Subsect. 3.3, that repair checking can be computed by integrity checking.

3.1   Inconsistency Reduction Checking by Integrity Checking

We are going to show that inconsistency reduction checking, i.e. $ir(D, IC, U)$, can be computed by $ic(D^U, IC, \overline{U})$, where $\overline{U}$ is the update that undoes U, as formalized below. We denote consecutive updates U, U' of D and then D^U by $D^{UU'}$. Hence, $D^{U\overline{U}} = D$.

Definition 5. Let $\overline{U}$ denote the *undo* of U: for each element of the form *insert X* or *delete Y* in U, $\overline{U}$ contains *delete X* or, resp., *insert Y*, and nothing else.

The following lemmata serve to identify a way to check inconsistency reduction by a complete integrity checking method. Lemma 1 says that complete measure-based inconsistency reduction checking of an update can be realized by measure-based integrity checking of its undo. Lemma 2 says that soundness of inconsistency reduction checking can be obtained by an additional integrity check of the update itself. Lemma 3 says that sound and complete inconsistency reduction checking is realizable by checking the undo for integrity preservation alone, in case the used integrity checking method is complete and the range of the measure on which it is based is totally ordered.

Checking inconsistency reduction by integrity checking is more efficient than by evaluating each constraint in IC brute-force against D^U, inasmuch as integrity checking simplifies the constraints to be checked by number and complexity [7].

Lemma 1. Let ic be a sound μ-based integrity checking method, and ir a function that maps triples (D, IC, U) to $\{yes, no\}$. ir is a complete μ-based inconsistency reduction checking method if ir is defined by the following equivalence.

$$ir(D, IC, U) = yes \iff ic(D^U, IC, \overline{U}) = no$$

An immediate consequence of Lemma 1 is that each update that does not reduce inconsistency is detected by checking the undo for integrity preservation with a complete integrity checking method.

The following lemma states that U reduces inconsistency if and only if the output of checking the integrity of $\overline{U}$ is negative and the output of checking the integrity of U is positive; if the outcome for $\overline{U}$ was positive, U is not a repair, by Lemma 1.

Lemma 2. Let ic be a sound and complete μ-based integrity checking method, and ir a function that maps triples (D, IC, U) to $\{yes, no\}$. ir is a sound and complete μ-based inconsistency reduction checking method if ir is defined by the following equivalence.

$$ir(D, IC, U) = yes \iff ic(Di^U, IC, \overline{U}) = no \text{ and } ic(D, IC, U) = yes$$

Lemma 3, below, sharpens the equivalence of Lemma 2: it states that, if the range of the measure μ is totally ordered, then $ir(D, IC, U)$ can be computed by applying any sound and complete μ-based integrity checking to $(D^U, IC, \overline{U})$.

Lemma 3. Let μ be an inconsistency measure with a totally ordered range, ic a sound and complete μ-based integrity checking method, and ir a function that maps triples (D, IC, U) to $\{yes, no\}$. Then, ir is a sound and complete μ-based inconsistency reduction checking method if ir is defined by the following equivalence.

$$ir(D, IC, U) = yes \iff ic(D^U, IC, \overline{U}) = no$$

The following example shows that incomplete integrity checking is deficient for inconsistency reduction checking, i.e., Lemmata 2 and 3 cannot be weakened by dropping the completeness requirement of integrity checking.

Example 4. Let $D = \{p \leftarrow q;\ p \leftarrow r;\ q\}$, $IC = \{\leftarrow p\}$, $U = \{insert\ r\}$, ic be the well-known integrity checking method in [18], and inconsistency be measured by ι. It is shown in [15] that ic is a sound but incomplete ι-based integrity checking method. It is easy to see that $ic(D^U, IC, \overline{U}) = no$, although we have that $\iota(D^{U\overline{U}}, IC) = \iota(D, IC) = \iota(D^U, IC)$, i.e., U is not a repair of (D, IC).

The next example illustrates that Lemma 3 cannot be weakened by waiving the total order requirement of the range of μ.

Example 5. The range of κ (the measure that maps (D, IC) to the set of causes of violations of IC in D) is partially but not totally ordered by the subset-or-equal relationship. Now, let ic be Decker's extension of Nicolas' well-known method to deductive databases. ic is a sound and complete κ-based method for definite propositional databases and constraints. For $D = \{p \leftarrow q;\ r \leftarrow s;\ q\}$, $IC = \{\leftarrow p,\ \leftarrow r\}$ and $U = \{delete\ q,\ insert\ s\}$, it is easy to see that $\overline{U} = \{insert\ q,\ delete\ s\}$, $\kappa(D, IC) = \{q\}$, $D^U = \{p \leftarrow q; r \leftarrow s; s\}$, $\kappa(D^U, IC) = \{s\}$, and $ic(D, IC, U) = ic(D^U, IC, \overline{U}) = no$.

Hence, by Lemma 2, $ic(D, IC, U) = no$ correctly indicates that U is not a κ-based repair of (D, IC). Yet, Lemma 3 cannot be used to obtain that result, since the range of κ is not totally ordered. Indeed, waiving that requirement in Lemma 3 would lead to the wrong conclusion that U was an κ-based repair. If, instead of κ, $|\kappa|$ would be used, for which ic is sound but incomplete, we'd have $|\kappa|(D, IC) = |\kappa|(D^U, IC) = 1$. By Lemma 1, we could define two complete inconsistency reduction methods, based on κ or, resp., $|\kappa|$, via the output of ic, both of which however would not be sound, since they both would wrongly indicate that U was a repair of (D, IC).

3.2   Minimality Checking by Integrity Checking

By Lemmata 4 and 5, we show that also the minimality check of measure-based repairs can be accomplished by integrity checking.

Lemma 4. For each inconsistency measure μ, each sound μ-based integrity checking method ic, each database D, each integrity theory IC, and each μ-based inconsistency reduction U of (D, IC), the following holds.

(a) U is not a μ-based repair of (D, IC) if there is $U' \subsetneqq U$ such that $ic(D^U, IC, \overline{U}'') = yes$, where $U'' = U \setminus U'$. If, additionally, ic is complete, then

(b) U is not a μ-based repair of (D, IC) iff there is $U' \subsetneqq U$ such that $ic(D^U, IC, \overline{U}'') = yes$, where $U'' = U \setminus U'$.

The contrapositive of Lemma 4 is the following corollary, which provides a way to check the minimality of a measure-based inconsistency reduction by measure-based integrity checking, without the necessity of computing measures.

Corollary 5. For each inconsistency measure μ, each sound μ-based integrity checking method ic, each database D, each integrity theory IC, and each μ-based inconsistency reduction U of (D, IC), the following holds.

(a) U is a μ-based repair of (D, IC) only if, for each $U' \subsetneqq U$, $ic(D^U, IC, \overline{U}'') = no$, where $U'' = U \setminus U'$. If, additionally, ic is complete, then

(b) U is a μ-based repair of (D, IC) if and only if, for each $U' \subsetneqq U$, $ic(D^U, IC, \overline{U}'') = no$, where $U'' = U \setminus U'$.

Corollary 5 enables us to define sound and complete minimality checking methods on the basis of complete and, resp., sound integrity checking methods, as follows.

Lemma 5. Let μ be an inconsistency measure, ic a sound μ-based integrity checking method, and mc a function that maps triples (D, IC, U) to $\{yes, no\}$. mc is a complete μ-based minimality checking method if mc is defined by the following equivalence.

$$mc(D, IC, U) = yes \iff \text{for each } U' \subsetneqq U, \, ic(D^U, IC, \overline{U}'') = no,$$
$$\text{where } U'' = U \setminus U'.$$

If, additionally, ic is a complete μ-based integrity checking method, then mc also is a sound minimality checking method.

3.3   The Main Theorems

Lemmata 1, 2, 3 and 5 provide handles for unfolding Theorem 1 into the main results of this paper, as stated in Theorems 2, 3 and 4, below. Essentially, each of these results states that measure-based repair checking can be done by integrity checking, without having to compute the measures. Theorem 2 is the weakest of the three, in that it only devises a way to see that a given update U is not a repair (cf. Corollary 3). By requiring that the used integrity checking method is not just sound but also complete, each of Theorems 2 and 3 devises a sound and complete way to check if U is or is not a repair. The difference between the two results is that Theorem 3 additionally requires that the range of the used inconsistency measure is totally ordered, but in turn requires one run of integrity checking less than Theorem 2.

Theorem 2. Let ic be a sound μ-based integrity checking method, and rc a function that maps triples (D, IC, U) to $\{yes, no\}$. rc is a complete μ-based repair checking method if rc is defined by the following quivalence.

$$rc(D, IC, U) = yes \iff ic(D^U, IC, \overline{U}) = no, \text{ and for each } U' \subsetneq U,$$
$$ic(D^U, IC, \overline{U}'') = no, \text{ where } U'' = U \setminus U'.$$

Theorem 3. Let μ be an inconsistency measure, ic a sound and complete μ-based integrity checking method, and rc a function that maps triples (D, IC, U) to $\{yes, no\}$. rc is a sound and complete μ-based repair checking method if rc is defined by the following equivalence.

$$ir(D, IC, U) = yes \iff ic(D^U, IC, \overline{U}) = no, \ ic(D, IC, U) = yes, \text{ and for each}$$
$$U' \subsetneq U, \ ic(D^U, IC, \overline{U}'') = no, \text{ where } U'' = U \setminus U'.$$

Theorem 4. Let μ be an inconsistency measure with a totally ordered range, ic a sound and complete μ-based integrity checking method, and rc a function that maps triples (D, IC, U) to $\{yes, no\}$. rc is a sound and complete μ-based repair checking method if rc is defined by the following equivalence.

$$rc(D, IC, U) = yes \iff ic(D^U, IC, \overline{U}) = no, \text{ and for each } U' \subsetneq U,$$
$$ic(D^U, IC, \overline{U}'') = no, \text{ where } U'' = U \setminus U'.$$

4   Computing Repair Checking

In Subsect. 4.1, we outline how to compute total repair checking. In Subsects. 4.1 and 4.3, we do the same for measure-based repair checking. The approach in Subsect. 4.1 is "naive"; it computes inconsistency measures. The one in Subsect. 4.3 uses measure-based integrity checking; we call it "repair checking by (measure-based) integrity checking". For comparing the three, note that the approaches in Subsects. 4.1 and 4.3 are inconsistency-tolerant, and therefore much more realistic than total repair checking. Moreover, we are going to see that measure-based repair checking tends to be far less costly than both naive and total repair checking.

4.1   Computing Total Repair Checking

After describing the computation of total repair checking, we assess its cost. Then, we describe how total repairs can be computed from a given inconsistency reduction.

4.1.1   Two Phases for Computing Total Repair Checking

By Definitions 1a and b, total repair checking may verify or falsify, for triples (D, IC, U), that (D^U, IC) is consistent, and that U is minimal, in two phases.

Phase 1 (*inconsistency reduction check*):
Check if U is a total inconsistency reduction of (D, IC), by querying each $I \in IC$ against D^U. If some I is not satisfied in D^U, then U is not a total inconsistency reduction and hence not a total repair of (D, IC). If each $I \in IC$ is satisfied in D^U, then U is a total inconsistency reduction, hence proceed to Phase 2.

Phase 2 (*minimality check*):
For each $U' \subsetneq U$, check if U' is a total inconsistency reduction of (D, IC), by querying each $I \in IC$ against $D^{U'}$. If each I is satisfied in $D^{U'}$, then U' is a total inconsistency reduction of (D, IC). Hence, U is not a total repair of (D, IC). If some I is not satisfied in $D^{U'}$, then U' is not a total inconsistency reduction of (D, IC). If no proper subset U' of U is an inconsistency reduction of (D, IC), then U is a total repair of (D, IC).

4.1.2   The Cost of Total Repair Checking

We now assess the cost of computing Phases 1 and 2. Let n be the cardinality of U and $m = 2^n$. Thus, there are m subsets $U_1, \ldots, U_m$ of U; one of them empty. Hence, easily up to $m - 1$ brute-force integrity checks of (D^{U_i}, IC) $(i = 1, \ldots, m\text{-}1)$ are needed for deciding if U (or any of its proper non-empty subsets) is a total repair of (D, IC). If k is the cardinality of IC, then that amounts to $k \times (m - 1)$ evaluations of arbitrarily complex constraints in IC.

4.1.3   A Third Phase for Computing Total Repairs

If, in Phase 1, U turns out to be an inconsistency reduction of (D, IC), and in Phase 2, U turns out to not be a total repair of (D, IC), then, by Corollary 2, at least one total repair of (D, IC) can be found in a third phase, as follows.

Phase 3 (*identify total repairs*):
Let $\mathcal{U} = \{U' \mid U$ is inconsistency reduction of (D, IC) detected in Phase 2, $U' \subsetneq U\}$. The subset-minimal elements in $\mathcal{U}$ are total repairs of (D, IC).

Phases 1–3 can also be presented uniformly, as follows: For each $U' \subseteq U$, check if U' is a total inconsistency reduction of (D, IC), by querying each I in IC against $D^{U'}$. U' is a total inconsistency reduction of (D, IC) if and only if each I is satisfied in $D^{U'}$. Let $\mathcal{U}$ be the set of subsets of U that are a total inconsistency reduction of (D, IC). If $\mathcal{U}$ is empty, neither U nor any of its subsets is a total repair of (D, IC). Otherwise, each subset-minimal set in $\mathcal{U}$ is a total repair of (D, IC).

4.2   Computing Naive Repair Checking

In this subsection, we first describe a naive way of computing total repair checking. Then, we assess the cost of that computation. Last, we describe how measure-based repairs can be computed from a given inconsistency reduction.

4.2.1   Two Phases for Computing Naive Repair Checking

According to Definition 1d, measure-based repair checking can be implemented naively in two phases: inconsistency reduction checking and minimality checking.

Phase 1: To check if U is a μ-based inconsistency reduction of (D, IC), i.e., to check $\mu(D^U, IC) < \mu(D, IC)$, the values of $\mu(D, IC)$ and $\mu(D^U, IC)$ can be computed and then compared. By that comparison, U is or is not an inconsistency reduction.

Phase 2: To check if a μ-based inconsistency reduction U of (D, IC) is minimal, the measure $\mu(D^{U'}, IC)$ of each proper non-empty subset U' of U has to be computed and compared to $\mu(D, IC)$, as already computed in Phase 1. If $\mu(D^{U'}, IC) \leq \mu(D^U, IC)$ does not hold for any such U', then U is a μ-based repair of (D, IC). If, for some such U', $\mu(D^{U'}, IC) \leq \mu(D^U, IC)$ holds, then (U' is an μ-based inconsistency reduction of (D, IC), and) U is not a μ-based repair of (D, IC).

The part of the description of Phase 2 between $(\dots)$ does not contribute to the decision if U is a total repair of (D, IC) or not, but is of interest in the context of an additional third phase, as addressed in Subsubsect. 4.2.3.

4.2.2   The Cost of Naive Measure-Based Repair Checking

Clearly, $\mu(D^{U'}, IC)$ has to be computed for each subset U' of U: for subsets $\{\}$ and U in Phase 1, and for proper non-empty subsets in Phase 2. For $|U| = n$, that amounts to the computation of 2^n measures, one for each subset U' of U. The cost of the computation of measures obviously depends on the definition of μ. To compute $\mu(D, IC)$ for any of the measures $\iota, |\iota|, \zeta, |\zeta|, \kappa, |\kappa|$ involves the evaluation of each I in IC against D, for $\zeta, |\zeta|, \kappa, |\kappa|$ also an analysis of the proof tree, for determining the violated cases or the causes of integrity violation. Thus, the order of magnitude of evaluating constraints for naive measure-based repair checking is about the same as for total repair checking. However, depending on the specific measure to be computed, the total cost of computing naive repair checking may easily be higher than that of total repair checking.

4.2.3   A Third Phase for Computing Measure-Based Repairs

If U turns out in Phase 1 to be an inconsistency reduction and in Phase 2 to be not an μ-based repair of (D, IC), then, analogously to Subsubsect. 4.2.3, at least one such repair can be identified, as in the following third phase.

Phase 3 *(identify total repairs)*: Let $\mathcal{U} = \{U' \mid U$ is inconsistency reduction of (D, IC) detected in Phase 2, $U' \subsetneq U\}$. The subset-minimal elements in $\mathcal{U}$ are total repairs of (D, IC).

4.3   Computing Repair Checking by Integrity Checking

A computation of measure-based repair checking is suggested by Theorems 2, 3 and 4. According to Theorem 2, the output $rc(D, IC, U) = no$, obtained by a measure-based integrity checking method that is only sound but not complete, indicates that U is not a μ-based repair. For U to be a μ-based repair, the output $rc(D, IC, U) = yes$, computed according to Theorem 2, is only necessary, but not sufficient. Necessary and sufficient conditions to identify updates as repairs are given by Theorems 3 and 4, which require that a sound and complete μ-based integrity checking method is used.

We are going to assess measure-based repair checking according to Theorem 3, again by the two phases of inconsistency reduction and minimality checking. Note, however, that the cost of measure-based repair checking by Theorem 4 is lower than as by Theorem 3, since the totally ordered range of μ, as required in Theorem 4, enables a less costly inconsistency reduction checking phase.

Phase 1: To check if U is a μ-based inconsistency reduction of (D, IC), compute $ic(D^U, IC, \overline{U})$ and $ic(D, IC, U)$ according to the equivalence in Lemma 2. Thus, at most 2 runs of ic are needed.

Phase 2: To check if U is minimal according to Definition 1d, it suffices, by Lemma 5, to check if $ic(D^U, IC, \overline{U}'') = no$, for each $U' \subsetneq U$ such that $U' \neq \emptyset$, where $U'' = U \setminus U'$. Thus, if n is the size of U, at most $2^n - 2$ runs of ic are needed.

So, for Phases 1 and 2, maximally 2^n runs of ic are needed. Thus, the actual cost of repair checking by integrity checking depends on ic. If a sound and complete method for simplified integrity checking is available, then running such a method tends to be much less costly than brute-force integrity checking, as employed for total repair checking and for naive measure-based repair checking.

Recall from Subsect. 4.1 that the cost of total repair checking was $k \times (m-1)$ unsimplified constraint evaluations, where $m = 2^n$ and k is the cardinality of IC. For ease of comparison, suppose that all constraints in IC are expressed by a single constraint formula I (the conjunction of all constraints). Then, for total repair checking, we'd have in the order of m-1 evaluations of I against D^{U_i} ($1 \leq i \leq m - 1$), where U_i is one of the non-empty subsets of U. Compared to that, $m - 2$ runs of a simplified form of I for simplified repair checking obviously tends to be significantly more advantageous.

5   Related Work

Related work on integrity checking [7,8,11,12,14,15,18], inconsistency measuring [4,8,11,12,14,16,17,19] and repairing [6,9–12,20,22] has been duly referenced already. Maybe more relevant is the work on brute-force repair checking [1,3,5,6,20,21], as already addressed in previous sections, particularly in Sect. 4.

Work on conventional repairing and repair checking has always focused on complexity issues in relation with CQA. As opposed to that, our alternative to CQA in [13], is not hung up in the complexity nexus between CQA, repairing and repair checking.

The discussion of complexity issues related to CQA focuses grosso modo on certain classes of constraints, e.g., those expressed as conjunctive queries over flat relational databases. The related complexity classes are amorphously treated, e.g., conjunctive queries have polynomial-time complexity. However, such an undifferentiated treatment misses the point of where cost issues of integrity checking may hurt most.

For example, in a relational database with very large extensions of p, q, r, the evaluation of the conjunctive constraint $I = \leftarrow p(x, y), q(y, z), r(x, y, z)$ may be considered prohibitively costly, although I falls into a relatively "harmless" complexity class. However, for repair candidate U that consists of the insertion of $r(a, b, c)$, our approach to repair checking does not have to evaluate I brute-force. Rather, for most integrity checking methods that are amenable to our way of repair checking, it suffices to evaluate the simplified instance $\leftarrow p(a, b), q(b, c), r(a, b, c)$, by simple look-ups. This shows that our proposal is significantly different from related work on repair checking.

6 Conclusion

Deploying inconsistency-measure-based ICh, we have elaborated a non-standard approach to RCh, that compares favorably to conventional brute-force RCh.

Measure-based RCh does not need to compute measures, nor does it have to know which constraints are violated, nor why or how they are violated. Nor does brute-force RCh. However, the latter does check each constraint brute-force, and gives up upon the least bit of remaining consistency impairment. As opposed to that, measure-based RCh is inconsistency-tolerant, permits measured comparisons of the quality increase of repair alternatives, and is less costly since its integrity checks are simplified. In general, however, sound repair checking by integrity checking requires a complete measure-based integrity checking method. Although most known measure-based integrity checking methods are incomplete, there are significant classes of databases, integrity theories, updates and inconsistency measures for which the completeness of integrity checking can be guaranteed, or relaxed while preserving the soundness of measure-based repair checking. Ongoing work is concerned with identifying such classes.

Acknowledgement. John Grant had provided valuable comment on early drafts of this work.

References

1. Afrati, F., Kolaitis, P.: Repair checking in inconsistent databases: algorithms and complexity. In: 12th ICDT, pp. 31–41. ACM Press (2009)
2. Arenas, M., Bertossi, L., Chomicki, J.: Consistent query answers in inconsistent databases. In: Proceedings of 18th PoDS, pp. 68–79. ACM (1999)
3. Arming, S., Pichler, B., Sallinger, E.: Combined complexity of repair checking and consistent query answering. In: 19th International Conference on Database Theory (ICDT), vol. 48, pp. 21:1–21:18. LIPIcs (2016)

4. Besnard, P.: Revisiting postulates for inconsistency measures. In: Fermé, E., Leite, J. (eds.) JELIA 2014. LNCS, vol. 8761, pp. 383–396. Springer, Heidelberg (2014)
5. Chomicki, J.: Consistent query answering: five easy pieces. In: Schwentick, T., Suciu, D. (eds.) ICDT 2007. LNCS, vol. 4353, pp. 1–17. Springer, Heidelberg (2006)
6. Chomicki, J., Marcinkowski, J.: Minimal-change integrity maintenance using tuple deletions. Inf. Comput. **197**(1/2), 90–121 (2005)
7. Christiansen, H., Martinenghi, D.: On simplification of database integrity constraints. Fundam. Inf. **71**(4), 371–417 (2006). IOS Press
8. Decker, H.: Quantifying the quality of stored data by measuring their integrity. In: Proceedings of DIWT 2009, Workshop SMM, pp. 823–828. IEEE (2009)
9. Decker, H.: Causes of the violation of integrity constraints for supporting the quality of databases. In: Murgante, B., Gervasi, O., Iglesias, A., Taniar, D., Apduhan, B.O. (eds.) ICCSA 2011, Part V. LNCS, vol. 6786, pp. 283–292. Springer, Heidelberg (2011)
10. Decker, H.: Partial repairs that tolerate inconsistency. In: Eder, J., Bielikova, M., Tjoa, A.M. (eds.) ADBIS 2011. LNCS, vol. 6909, pp. 389–400. Springer, Heidelberg (2011)
11. Decker, H.: New measures for maintaining the quality of databases. In: Murgante, B., Gervasi, O., Misra, S., Nedjah, N., Rocha, A.M.A.C., Taniar, D., Apduhan, B.O. (eds.) ICCSA 2012, Part IV. LNCS, vol. 7336, pp. 170–185. Springer, Heidelberg (2012)
12. Decker, H.: Measure-based inconsistency-tolerant maintenance of database integrity. In: Schewe, K.-D., Thalheim, B. (eds.) SDKB 2013. LNCS, vol. 7693, pp. 149–173. Springer, Heidelberg (2013)
13. Decker, H.: Answers that have quality. In: Murgante, B., Misra, S., Carlini, M., Torre, C.M., Nguyen, H.-Q., Taniar, D., Apduhan, B.O., Gervasi, O. (eds.) ICCSA 2013, Part II. LNCS, vol. 7972, pp. 543–558. Springer, Heidelberg (2013)
14. Decker, H., Martinenghi, D.: Modeling, measuring and monitoring the quality of information. In: Heuser, C.A., Pernul, G. (eds.) ER 2009. LNCS, vol. 5833, pp. 212–221. Springer, Heidelberg (2009)
15. Decker, H., Martinenghi, D.: Inconsistency-tolerant Integrity Checking. IEEE Trans. Knowl. Data Eng. **23**(2), 218–234 (2011)
16. Grant, J.: Classifications for inconsistent theories. Notre Dame J. Form. Log. **19**(3), 435–444 (1978)
17. Hunter, A., Konieczny, S.: Approaches to measuring inconsistent information. In: Bertossi, L., Hunter, A., Schaub, T. (eds.) Inconsistency Tolerance. LNCS, vol. 3300, pp. 191–236. Springer, Heidelberg (2005)
18. Lloyd, J., Sonenberg, E., Topor, R.: Integrity constraint checking in stratified databases. J. Log. Program. **4**(4), 331–343 (1987)
19. Mu, K., Liu, W., Jin, Z., Bell, D.: A syntax-based approach to measuring the degree of inconsistency for belief bases. IJAR **52**, 978–999 (2011). Elsevier
20. Staworko, S.: Declarative inconsistency handling in relational and semi-structured databases. Ph.D. thesis, May 2007
21. ten Cate, B., Fontaine, G., Kolaitis, P.: On the data complexity of consistent query answering. Theory Comput. Syst. **57**(4), 843–891 (2015)
22. Wijsen, J.: Database repairing using updates. ACM Trans. Database Syst. **30**(3), 722–768 (2005)

Towards Temporal Analysis of Integrated Scenarios for Sustainable Innovation

Alfredo Cuzzocrea[1]([⊠]), Ilaria D'Elia[2], Antonio De Nicola[2], Hugo Maldini[2], and Maria Luisa Villani[2]

[1] DIA Department, University of Trieste and ICAR-CNR, Trieste, Italy
alfredo.cuzzocrea@dia.units.it
[2] ENEA, CR Casaccia, Rome, Italy
{ilaria.delia,antonio.denicola,marialuisa.villani}@enea.it,
hugo.maldini@gmail.com

Abstract. We propose the TEMPORANA framework for analysis of integrated scenarios aiming at studying the possible effects on the society of sustainable innovation policies. The framework consists of a set of advanced temporal queries for scenario analysis and a software application. Integrated scenarios are possible future states of different aspects of the world (e.g., energy, environment, technology, economy, societal system) representing plausible conditions under different assumptions. Temporal scenario queries aim at detecting a specified behavior for the system over time and, hence, at verifying that a temporal property holds.

Keywords: Environmental modelling · Temporal analysis · Scenarios management

1 Introduction

Sustainable innovation polices concern different interlinked aspects of the society related to the environment, energy, technologies, economy, and the societal system. Before defining new policies, policy makers need to define different scenarios in order to take aware decisions. Scenarios are possible future states of the world representing plausible conditions under different assumptions [8]. They represent future projections of a system obtained by means of some computational model. Policy makers involved in the definition of sustainable innovation policies need tools to envision the consequences of their decisions and the potential impact of the adopted measures.

To this purpose various tools exist that allow definition of possible scenarios, each from a specific perspective. For instance a scenario from the energy and technology perspective can be obtained through the MARKAL-TIMES [9] model. Scenario for the definition of atmospheric pollution policies could be defined with the GAINS [6] model, used to describe possible paths of evolution for greenhouse gases (GHGs) emissions. Finally a scenario for macroeconomic analysis can be generated through methods like Social Accounting Matrix (SAM)

© Springer International Publishing Switzerland 2016
O. Gervasi et al. (Eds.): ICCSA 2016, Part V, LNCS 9790, pp. 544–551, 2016.
DOI: 10.1007/978-3-319-42092-9_41

[10] and GTAP model [7]. However these systems have been conceived from a sectoral perspective and, mainly, aim at generating new scenarios.

According to [8], the scenario development process consists of five phases: scenario definition, scenario construction, scenario analysis, scenario assessment, risk management [8]. Here we aim at supporting scenario analysis as the above-mentioned tools provide support mainly for scenario definition and construction and, only partially, for scenario analysis. In particular, they do not provide support for the analysis and comparison of multiple integrated scenarios and of the temporal evolutions of their variables. Furthermore experts need tools allowing also to support integrated scenarios analysis.

For these reasons we present the TEMPORal ANAlysis (TEMPORANA) framework consisting of a set of advanced temporal queries for scenario analysis and a software application. It allows analyzing the temporal evolution of the energy, technological, economic and environmental variables of multiple scenarios by means of temporal queries inspired from temporal logic. It supports the analysis of scenarios generated by complex systems (as MARKAL-TIMES and GAINS). To avoid the analysis of obsolete scenarios, as new scenarios are defined by experts they should be entered in TEMPORANA.

TEMPORANA was conceived to support the work of different stakeholders: policy makers, who need automatic support for impact analysis of measures and for decision making; experts, who need to define energy/environmental/socio-economic scenarios; and, finally, industrial stakeholders, who have extended their information systems with scenario data to quantify the impact of regulations and political measures over their business, and to regard the sustainable policies as a business opportunity [2].

2 TEMPORANA

In this Section, we provide principles, functionalities and architecture of TEM-PORANEA.

2.1 Integrated Scenarios

We define an *integrated system* a collection of n variables from the energetic, economic and environmental fields $Sys = \{v_1, \ldots, v_n\}$. Given a temporal range $T_{base,target} = < y_1 = y_{base}, \ldots, y_r = y_{target} >$, with $y_i < y_{i+1}, i = 1, \ldots r - 1$ of r years from a starting year y_{base}, called *base year*, to a *target year* y_{target}, for each variable we define $v_j(T) = < v_j(y_1), \ldots, v_j(y_r) >$, where $v_j(y_i) \in R$, for $i = 1, \ldots r$. A *scenario* $Scen(Sys, T = T_{base,target}, C, Obj) = \{v_1(T), \ldots v_n(T)\}$ represents one future projection of the integrated system over the specified time period. Figure 1 shows an example of scenario defining projections of *CO2 cost, employment, energy consumption, CO2 emissions, industrial production,* and *GDP* variables.

An integrated scenario is computed by combining the results of domain-specific computational models under a set of input assumptions $C = \{c_i\}_i$,

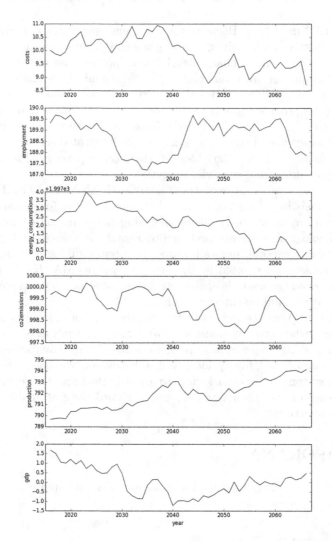

Fig. 1. Example of scenario defining projections of *CO2 cost, employment, energy consumption, CO2 emissions, industrial production,* and *GDP* variables.

which, for example, might be related to policies for sustainability or to economic trends, and towards satisfaction of some specified objectives $Obj = \{o_j\}_j$ to be reached on the target year, such as pollution reduction or economic growth. Therefore, multiple scenarios for an integrated system Sys over a temporal range T may result from different assumptions and objectives.

A *scenario repository* $Rep \subseteq \cup_k Scen_k$, with $Scen_k = Scen(Sys, T^k, C^k, Obj^k)$, collects all the scenarios of the system over which a temporal analysis takes place. Thus, a scenario repository is a subset of all scenarios of the system. In this work, we focus on the collective analysis of scenarios defined on the same temporal range,

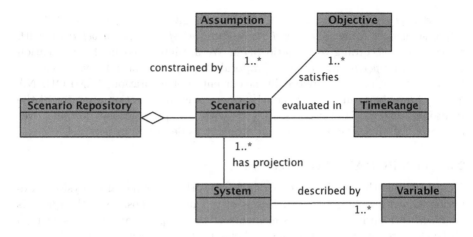

Fig. 2. System model class diagram.

i.e., $T^k = T$, for a given T, for all scenarios of the repository, which we denote by $Rep(T)$.

Figure 2 shows a conceptual model of the setting just described.

2.2   Temporal Queries

TEMPORANA is a tool devoted to solving temporal queries over a scenario repository $Rep(T) = \{Scen_1, \ldots, Scen_m\}$, for a given T. A *temporal query* is a query $Q_{Prop}(\{Scen_i\})$ aiming at detecting scenarios from the repository which exhibit a specified behavior for the system over time, as described in *Prop*. This behavior is also specified by a *temporal property* that must hold on the scenarios as result from the query. Temporal properties can be represented by temporal logic formulas, for example LTL [11], and verified on a formal representation of the scenario behavior. Although a formal verification by existing LTL model checkers has been foreseen in our framework, in this paper we focus on simple properties, that we present in Sect. 2.4, which are verified by a built-in module of TEMPORANA.

To any time-range $T' = < y_1', \ldots, y_m' > \subseteq T$ we associate a sequence of time intervals $I = < t_1, \ldots, t_{m-1} >$, where $t_i = [y_i', y_{i+1}']$, for $i = 1, \ldots, m - 1$. Let $Scen \in Rep(T)$ be a scenario of the repository, and ϕ be a propositional formula over $v(T) \in Scen$. We define a mapping $F_\phi(v(T), I) = < s_1, \ldots, s_{m-1} > \in \{true, false\}^{m-1}$, with $s_i = true$ if $v(t_i)$ satisfies ϕ and $s_i = false$ otherwise, for $i = 1, \ldots, m - 1$. For example, set: $T = < 2010, 2015, 2020, 2025, 2030, 2035, 2040 >$, $v = CO2cost$, $T' = < 2020, 2030, 2040 >$. We get $I = < t_1 = [2020, 2030], t_2 = [2030, 2040] >$. Suppose $\phi(v) = $ "$CO2cost > 0.1$", and $CO2cost(T') = < 0.1, 0.1, 0.20 >$. We have $F_\phi(v(T), I) = < false, true >$.

We say that F_ϕ associates a *temporal behavior* (sequence) within I to each $v(T)$ with respect to ϕ. In this work, we focus on analysis of set of temporal sequences $\{F_\phi(v_{i_1}(T), I), \ldots, F_\phi(v_{i_k}(T), I)\}$, for given ϕ, I and $v_{i_j}(T) \in Scen$. These sequences form an *abstract representation of the system behavior*, with respect to ϕ, according to scenario *Scen*. We are interested in the verification of *temporal properties* like Linear Temporal Logic (LTL) formula over such system representation(s). Therefore, in the current implementation, TEMPORANA queries have the form $\sigma_{Prop}(\{Scen_i\})$, where σ is the selection operator of the relational algebra [1], $Prop = <\phi, \chi>$ with ϕ a propositional formula over $v_j \in Sys$, for all j, and χ a temporal property as described in Sect. 2.4.

2.3   TEMPORANA Architecture

Figure 3 shows the architecture of our framework for temporal analysis of scenarios. Given a temporal query, this is solved in two steps. First the query is analyzed to identify the various elements of the property and to construct an abstract representation of the system behavior with respect to those elements. This consists of a set of temporal sequences, each associated with a variable of the system and the property, as defined in Sect. 2.2. Then the temporal property is checked on of the system behavior representation.

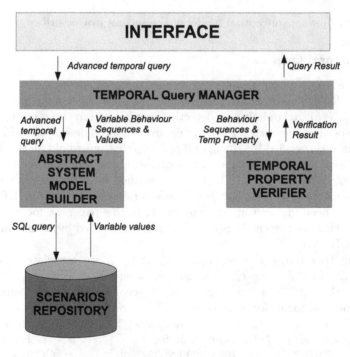

Fig. 3. TEMPORANA architecture.

Thus, beside the user interface, the architecture highlights the components responsible for these two activities, and a coordination component named *Temporal Query Manager*. The decoupling of system representation functions from properties verification provides flexibility for deciding from various formalisms and verification methods implementations.

The role of the *Temporal Query Manager* is: to delegate the construction of the query-specific system model behavior to the *(Abstract) System Model Builder*; to require from the *Temporal Property Verifier* the analysis of temporal property(ies) derived from the query on that model; finally, to build up the user query solution.

The *(Abstract) System Model Builder* achieves the objective by means of the following functions:

1. extracting the variables of interest from the query and the conditions to be held on their values (the ϕ formula(e));
2. constructing SQL queries to select variables values from the scenarios repository and run them;
3. create an abstract behavior representation for each variable and scenario with respect to the ϕ property.

The result of this activity is to provide a formal description of the system behavior in each scenario to be supplied as input to the *Temporal Property Verifier*, together with the temporal property χ, which is responsible to model check that property.

2.4   An Example of Advanced Temporal Scenario Query

In this section we present implementation of a temporal scenario query defined by field experts. Starting from their requirements, we recognized the need for defining a number of temporal properties, each involving one or more temporal operators from the Linear Temporal Logic (LTL) [11]. In the following, we present the temporal properties types we have currently identified, which might be considered as building blocks for more complex queries. In particular, for all of the following queries we take, as examples, $\phi(v) = increase(v, t = [y_i, y_{i+1}]) := (v(y_i) < v(y_{i+1}))$, and/or $\phi = increaseBy(x, v, t = [y_i, y_{i+1}]) := (v(y_{i+1}) = v(y_i) + x)$, $x \in R$. If $I =< t_1, \ldots, t_m >$ is a sequence of time intervals, we set $\phi(v, I) := F_\phi(v(T), I) =< \phi(v, t_1), \ldots, \phi(v, t_m) >$.

The *Always Increase / Decrease Verification* method allows to verify that a variable value always increases/decreases over a given temporal range. This means that the condition should apply for every available shortest time interval in a given temporal range. E.g., The $CO2Emissions$ value always increases from 2020 to 2026. For this example, with the notation of Sects. 2.1 and 2.2, we have: $T' =< 2020, 2021, 2022, 2023, 2024, 2025, 2026 >$, $I =< t_1 \ldots, t_6 >$, the sequence of intervals of length 1, e.g., $t_1 = [2020, 2021]$, $t_2 = [2021, 2022], \ldots$, and $v = CO2Emissions$.

In the example of Table 1 the increase relation is verified for scenario sc#a and not-verified for scenario sc#b due to the decreasing values between the year 2022 and 2023.

Table 1. Increase verification of the `CO2Emissions` variable in the temporal range from 2020 to 2026 for the scenarios `sc#a` and `sc#b`.

	2020	2021	2022	2023	2024	2025	2026
CO2_Em (sc#a)	843.53	853.22	854.35	887.34	901.23	905.64	910.13
CO_2Em (sc#b)	843.53	853.22	854.35	807.34	901.23	905.64	910.13

For $Scen$ =sc#a, we get $\phi(v, I)$ =$< true, true, true, true, true, true >$, whereas for $Scen$ =sc#b, $\phi(v, I)$ =$< true, true, false, true, true, true >$. The final result of the query $\sigma_{Prop}(Rep(T))$ over $Rep(T) = \{sc\#a, sc\#b\}$, and $Prop$ =$< increase(CO2Emissions, I)$, globally $increase(CO2Emissions, I) >$ is sc#a, as we have:

```
Increase [CO2Emissions, 2020-2026, sc#a] = True
Increase [CO2Emissions, 2020-2026, sc#b] = False
```

3   Conclusions and Future Work

Analysis of integrated environmental scenarios for studying the effects on the society of sustainable innovation policies is a relevant issue for policy makers and scenario experts. To this aim, in this paper, we proposed the TEMPORANA framework consisting of a set of advanced temporal queries for scenario analysis and a software application implementing them.

As temporal scenario queries require the verification of a temporal property, as future work, we intend to provide a new implementation of TEMPORANA based on the Linear Temporal Logic and on model checking techniques. On another side of research, we are addressing how to further developing our framework in order to cope with emerging *big data management* (e.g., [5,12]) and *analytics* (e.g., [3,4]) topics.

Acknowledgments. This work has been inspired by an ongoing project activity in favour of the Italian Ministry of Economic Development (MiSE). Many enlightening discussions on these topics with Antonio Bartoloni (MiSE) are kindly acknowledged. Then we wish to acknowledge several ENEA colleagues who contributed to improving our work with their constructive comments: Natale Massimo Caminiti, Umberto Ciorba, Enrico Cosimi, Flavio Fontana, Maria Gaeta, Stefano La Malfa, Sergio La Motta, Tiziano Pignatelli, Giangiacomo Ponzo, Marco Rao, Maria Velardi, and Maria Rosa Virdis.

References

1. Abiteboul, S., Hull, R., Vianu, V.: Foundations of Databases, vol. 8. Addison-Wesley, Reading (1995)
2. Camporeale, C., De Nicola, A., Villani, M.L.: Semantics-based services for a low carbon society: an application on emissions trading system data and scenarios management. Environ. Softw. **64**, 124–142 (2015)

3. Cuzzocrea, A.: Analytics over big data: Exploring the convergence of dataware-housing, OLAP and data-intensive cloud infrastructures. In: 37th Annual IEEE Computer Software and Applications Conference, COMPSAC 2013, Kyoto, Japan, July 22–26, 2013, pp. 481–483 (2013)
4. Cuzzocrea, A., Bellatreche, L., Song, I.-Y.: Data warehousing and OLAP over big data: current challenges and future research directions. In: Proceedings of the Sixteenth International Workshop on Data Warehousing and OLAP, DOLAP 2013, San Francisco, CA, USA, 28 October 2013, pp. 67–70 (2013)
5. Cuzzocrea, A., Saccà, D., Ullman, J.D.: Big data: a research agenda. In: 17th International Database Engineering and Applications Symposium, IDEAS 2013, Barcelona, Spain, 09–11 October 2013, pp. 198–203 (2013)
6. GAINS. Gains (2016). Accessed 24 Mar 2016
7. Hertel, T.W., Hertel, T.W.: Global Trade Analysis: Modeling And Applications. Cambridge University Press, Cambridge (1997)
8. Mahmoud, M., Liu, Y., Hartmann, H., Stewart, S., Wagener, T., Semmens, D., Stewart, R., Gupta, H., Dominguez, D., Dominguez, F., Hulse, D., Letcher, R., Rashleigh, B., Smith, C., Street, R., Ticehurst, J., Twery, M., van Delden, H., Waldick, R., White, D., Winter, L.: A formal framework for scenario development in support of environmental decision-making. Environ. Model. Softw. **24**(7), 798–808 (2009)
9. MARKAL-TIMES. Markal-times (2016). Accessed 24 Mar 2016
10. Pyatt, G., Round, J.I.: Social accounting matrices for development planning. Rev. Income Wealth **23**(4), 339–364 (1977)
11. Prasad Sistla, A., Clarke, E.M.: The complexity of propositional linear temporal logics. J. ACM (JACM) **32**(3), 733–749 (1985)
12. Byunggu, Y., Cuzzocrea, A., Jeong, D.H., Maydebura, S.: On managing very large sensor-network data using bigtable. In: 12th IEEE/ACM International Symposium on Cluster, Cloud and Grid Computing, CCGrid 2012, Ottawa, Canada, May 13–16, 2012, pp. 918–922 (2012)

An Innovative Similarity Measure for Sentence Plagiarism Detection

Agnese Augello[1], Alfredo Cuzzocrea[1,2]($\boxtimes$), Giovanni Pilato[1], Carmelo Spiccia[1],
and Giorgio Vassallo[3]

[1] Istituto di Calcolo E Reti Ad Alte Prestazioni (ICAR),
Italian National Research Council (CNR), Palermo, Italy
{augello,pilato,spiccia}@pa.icar.cnr.it
[2] University of Trieste, Trieste, Italy
alfredo.cuzzocrea@dia.units.it
[3] Universitá degli Studi di Palermo, Palermo, Italy
giorgio.vassallo@unipa.it

Abstract. We propose and experimentally assess *Semantic Word Error Rate* (SWER), an innovative similarity measure for sentence plagiarism detection. SWER introduces a complex approach based on latent semantic analysis, which is capable of outperforming the accuracy of competitor methods in plagiarism detection. We provide principles and functionalities of SWER, and we complement our analytical contribution by means of a significant preliminary experimental analysis. Derived results are promising, and confirm to use the goodness of our proposal.

Keywords: Sentence similarity measure · Plagiarism detection

1 Introduction

Nowadays, a great attention is devoted to *sentence plagiarism detection*, specially in the context of scientific texts (e.g., [34]). The problem of effectively detecting plagiarism has a relevant tradition (e.g., [35–37]), also due to important consequences it introduces in the scientific and industrial communities (e.g., as related to scientific articles, industrial patents, and so forth). More generally, the relevance of the plagiarism problem has been further highlighted in the broader context of education and science, and, symmetrically, a wide family of commercial plagiarism detection tools has been proposed recently. Some are: *PlugScan*, *PlagTracker*, *The Plagiarism Checker*.

A solution for supporting plagiarism detection is based on *similarity detection among sentences*. Automatic evaluation of the similarity among two different sentences is generally accomplished by means of similarity measures. Most measures take into account the number and the size of common sequence of words (ngrams). Sentences similarity measures have applications in several tasks, including: Machine Translation [9], Paraphrase Identification [10], Speech Recognition [5], Question-answering [6] and Text Summarization [7,8].

© Springer International Publishing Switzerland 2016
O. Gervasi et al. (Eds.): ICCSA 2016, Part V, LNCS 9790, pp. 552–566, 2016.
DOI: 10.1007/978-3-319-42092-9_42

In all the aforementioned tasks, the definition of "sentence similarity" differs subtly from human cognition. In particular, let us consider the following three sentences:

1. I like cats.
2. I like dogs.
3. I like astrophysics.

Three sentence pairs are possible: $(1, 2)$, $(1, 3)$ and $(2, 3)$. To a human reader it should be obvious that the first two sentences are more similar to each other than those in the pairs $(1, 3)$ or $(2, 3)$. However, no sentence similarity measure to date is able to elicit this difference automatically: they would all attribute the same similarity score to each pair. This happens because in the aforementioned tasks "sentence similarity" is defined as a measure of "equivalence" rather than "resemblance". In fact, these tasks requires to assert whether or not two sentence have the same meaning. This has two important implications:

(a) The *form of the sentence* is not deemed important: as long as two sentences have the same meaning they can be reworded/paraphrased to the extent of having a very different appearance; however, they will be considered semantically equivalent and therefore be awarded the maximum similarity score. While no state-of-the-art measure is totally independent by the form of the sentence, this implication have affected the design of datasets for testing the performance of sentence similarity measures. As a consequence, the design of the measures has been *indirectly* affected.

(b) *Word relatedness* is not deemed important: a few similarity measures handle synonyms and word inflections when comparing the words of two sentences, e.g. "dog", "dogs" and "puppies" may be considered equivalent. However, no sentence similarity measure will consider "cats" more similar to "dogs" than to "astrophysics", since they all convey different meanings, regardless of their similarity or relatedness. This assumption *directly* affected the design of sentence similarity measures.

While this definition of sentence similarity and their corresponding implications are reasonable for the aforementioned tasks, they nevertheless limit the applicability of sentence similarity measures to other tasks where resemblance can be considered more important than equivalence. Affected applications include Plagiarism Detection and Computational Creativity tasks. We therefore propose a new sentence similarity measure specifically designed to evaluate resemblance, so to handle situations similar to the previous example better.

In Sect. 2 we describe the current state of the art on sentence similarity. Section 2.1 illustrates the most notable measures. In Sect. 2.2 current available datasets for sentence similarity are compared. Section 2.3 shows how sentence similarity measures are employed in several applications. In Sect. 3 the rationale for a new similarity measure is discussed, alongside with some possible applications. In Sect. 4 the proposed sentence similarity measure is described. In Sect. 5 experimental results are shown and discussed. Finally, Sect. 6 reports on conclusions and future work of our research.

2   Related Work

2.1   Sentence Similarity Measures

Given a sentence pair (s_1, s_2), the Jaccard similarity coefficient [11] is a popular metric in Information Retrieval that can be used to compare s_1 and s_2:

$$J = \frac{|s_1 \cap s_2|}{|s_1 \cup s_2|} \tag{1}$$

where the notation $s_1 \cap s_2$ and $s_1 \cup s_2$ refers respectively to the intersection and to the union of the set of the words of s_1 and s_2.

In 2002 IBM developed BLEU [12], a family of metrics designed for the evaluation of machine translation algorithms. Given two sentences, i.e. a candidate translation s_1 and a reference translation s_2, the precision p_n of the translation is obtained by counting how many ngrams of size n of s_1 are present in s_2: if a given ngram is present more than once in s_1, it is counted multiple times, up to the number of occurrences of the ngram in s_2. This total is divided by the number of words of s_1. When multiple reference translations exists, the one which maximizes the precision is considered. The precision p_n represents the similarity among the two sentences s_1 and s_2. On this basic similarity measure, the BLEU score is obtained by computing, for different values of n, a weighted geometric average of p_n over the entire corpus and then multiplying by a brevity penalty (BP) factor.

An improved variant of BLEU is the NIST measure [13]. Major changes includes: the replacement of the geometric mean with the arithmetic mean; an increased weight for rare ngrams; a revised brevity penalty factor.

The Levenshtein distance [4] is a common edit distance to measure the difference among two words. Instead of using it to compare two words, it can be modified to compare two sentences: character-level operations are replaced by word-level operations. This leads to a measure known as Word Error Rate (WER) [5]. The measure works as follows. Given two sentences s_1 and s_2, three types of operations are allowed to transform s_1 into s_2: words can be inserted, deleted or substituted. The WER is computed by minimizing the following function:

$$\text{WER}(s_1, s_2) = \frac{I + D + S}{|s_2|} \tag{2}$$

where I are the number of insertions, D the number of deletions and S the number of substitutions. The similarity will be therefore:

$$\text{Similarity}(s_1, s_2) = 1 - \text{WER}(s_1, s_2) \tag{3}$$

Translation Error/Edit Rate (TER) [14] enhances WER by adding a new operation: blocks of words shifting. Since adding move operations renders optimal calculation of the edit distance NP-Complete [31], heuristic restrictions are employed to calculate a close approximation.

TER-Plus (TERp) [15] enhances TER by relaxing some shifting constraints and allowing less-than-unitary word substitution cost if at least one of three criteria is satisfied: the two words share the same stem according to the Porter algorithm [16]; the two words are synonymous according to Wordnet; the two words (or blocks of) may be paraphrases according to a precomputed table.

The METEOR measure [17] compares two sentences s_1 and s_2 by attempting to match each word of s_1 to a word of s_2. Each word can be matched only once. Three steps are employed: first of all, exact matches are found; secondly, unmatched words are compared by their stem according to the Porter algorithm; thirdly, remaining words are compared for synonymy according to Wordnet. Given the number of matches m, a similarity score F is computed as follows:

$$\text{METEOR}(s_1, s_2) = \frac{P \cdot R}{\alpha P + (1 - \alpha)R}(1 - \gamma F^{\beta}) \tag{4}$$

where P is the precision $\frac{m}{|s_1|}$, R is the recall $\frac{m}{|s_2|}$, F is the fragmentation $\frac{c}{m}$, c is the number of chunks (i.e. the minimum number of groups of adjacent matched words) and α, β and γ are three free parameters. It represents a parametrized F-measure, i.e. a weighted harmonic average of precision and recall, discounted by a penalty factor to minimize fragmentation.

Normalized Phrasal Overlap is a similarity measure developed by Banerjee et al. [29] and modified by Pozzetto et al. [30] who added normalization:

$$\text{NPO}(s_1, s_2) = \tanh \frac{\sum_{i=1}^{m} \sum_{m} i^2}{|s_1| + |s_2|} \tag{5}$$

As a preparatory step for finding novel sentences from a collection of documents about a topic, in 2003 Allan et al. [18] developed a similarity function A for finding relevant sentences from the collection:

$$
\begin{aligned}
A(s_1, s_2) = \sum_{w \in s_1 \cap s_2} &\log(\text{tf}(w, s_1) + 1) \\
&\cdot \log(\text{tf}(w, s_2) + 1) \log \frac{|C| + 1}{0.5 + \text{df}(w)}
\end{aligned}
\tag{6}
$$

where $\text{tf}(w, X)$ returns the frequency of the word w in the sentence X, $\text{df}(w)$ returns the number of documents containing the word w and $|C|$ is the total number of documents in the corpus.

The Hoad and Zobel Identity Measure [19] was developed to compare documents for the purpose of eliciting plagiarism. Metzer et al. [20] modified the similarity measure to compare small documents, i.e. sentences, better:

$$
\begin{aligned}
\text{IM}(s_1, s_2) = &\frac{1}{1 + \frac{\max(|s_1|, |s_2|)}{\min(|s_1|, |s_2|)}} \\
&\cdot \sum_{w \in s_1 \cap s_2} \frac{\frac{\log N}{\text{df}(w)}}{1 + |\text{tf}(w, s_1) - \text{tf}(w, s_2)|}
\end{aligned}
\tag{7}
$$

Table 1. Sentence similarity measures

Measure	Symmetry	Synonyms/Stems	Word order
Jaccard	Yes	No	No
BLEU	No	No	ngrams
NIST	No	No	ngrams
WER	No	No	Subs, Del, Ins
TER	No	No	Subs, Del, Ins, Shift
TERp	No	Yes	Subs, Del, Ins, Shift
METEOR	$\alpha = 0.5$	Yes	Crossings, Chunks
NPO	Yes	No	ngrams
A	Yes	No	No
IM	Yes	No	No

Table 1 compares the ten aforementioned similarity measures. Five of them are not symmetric: BLEU, NIST, WER, TER, TERp. Furthermore, METEOR is symmetric only if $\alpha = 0.5$. It can be noted that these six measures have been applied mainly in Machine Translation tasks. Only two measures consider synonymy or stemming during word matching: TERp and METOR use both, while all the other measures handle exact word match only. No measure takes into account similar or related words with different meanings, e.g. "dogs" and "cats". Finally, word order is ignored by three measures (Jaccard, A, IM); three measures takes into account only ngrams (BLEU, NIST, NPO); four measures are more robust in considering word order since they can handle more complex alterations of the sentence structure (WER, TER, TERp, METEOR).

2.2 Sentence Similarity Datasets

In 2003 Barzillay and Lee [21] released a dataset, thereafter referred as BL-PC, composed by 118 sentence pairs: each pair was obtained by extracting a sentence from a text corpus about Middle East violence and by generating a paraphrase of the sentence automatically via an algorithm. For each pair, two judges were asked whether the two sentences had the same meaning.

In 2005 the Microsoft Research Paraphrase Corpus (MSR-PC) [22] was published: it contains 5801 sentence pairs extracted from online news sources. The pairs were generated by selecting sentence random pairs having unnormalized WER (Word Error Rate) less or equal than eight. Each pair is tagged for being a paraphrase or not: two human judges were employed for the task and a third judge resolved the disagreements. The dataset is composed by 4076 training sentence pairs (67.5 % positive for paraphrasing) and 1725 test sentence pairs (66.5 % positive).

In 2006 Li et al. [23] developed a new dataset, STSS-65, composed by 65 sentence pairs: each pair was evaluated by 32 humans for assigning a similarity rating running from 0 to 4. For each pair the average rating and the standard deviation is provided. The sentence pairs were obtained by an existing list of word pairs: each word has been replaced by its first definition retrieved from a dictionary. The dataset was expanded in 2013 with the addition of 64 new sentence pairs [24], leading to a new dataset, STSS-131. The sentence pairs were proposed by 32 participants in the study, on the basis of some provided stimulus words. Sentence pairs were than selected by three judges and provided to 64 participants in order to be rated for similarity.

In 2012 a dataset, thereafter referred as SEV-12, was developed as part of the task 6 of SemEval-2012 [25, 26]. It includes sentence pairs extracted from three different corpora: the Microsoft Research Paraphrase Corpus (750 pairs), the Microsoft Resarch Video Description Corpus (750 pairs) [27] and WMT-2008 Europarl (734 pairs) [28]. The first corpus has already been illustrated. The second corpus consists in a collection of 120,000 sentences describing 2,000 short videos; each video is described several times by different viewers; hence, for a given video a set of similar sentences aimed at describing it is available; sentence pairs have been formed by pairing both same-video and different-video sentences. The third corpus consists in a list of sentences, extracted from European Union Parliament transcriptions; for each sentence a set of different translations is provided; since these translations are similar to each other, sentence pairs were generated from them. The resulting corpus, obtained from these three corpora, is composed by 2234 sentence pairs. Each pair has been evaluated for paraphrasing by five humans via Mechanical Turk, a crowdsourcing platform.

Table 2 compares the main characteristics of each dataset in terms of: number of sentence pairs; rating scale; number of raters; source of the sentences. Even though STSS-131 is a superset of STSS-65, its table entry refers only to the newly added 64 sentences. Regarding the source of the sentences, STSS-131 is the most generic dataset: in fact, each new sentence has been generated by participants in the study with a very high degree of freedom regarding to the actual content. However, its size and the size of BL-PC are very limited, compared to MSR-PC and SEV-12. It must be stressed that each dataset has been designed to test equivalence, rather than resemblance. In fact, human raters were provided with a specific definition of similarity before starting the evaluation:

- BL-PC: "determine whether the two units are matching in that they are two valid options of describing the same event" and "identify the pair as a match if one of the phrases can generally be substituted for the other".
- MSR-PC: "determine if two sentences express the same content" (from "'Equivalent' vs. 'not equivalent' content" in "Detailed Tagging Guidelines").
- SEV-12: "score how similar two sentences are to each other according to the following scale. The sentences are: (5) Completely equivalent, as they mean the same thing. (4) Mostly equivalent [...]".
- STSS-131: "rate the similarity of meaning of these sentence pairs" and "look at the two sentences and ask yourself 'How close do these two sentences come to meaning the same thing?'".

Table 2. Sentence similarity datasets

Dataset	Pairs	Ratings	Raters	Source
BL-PC	118	Yes/No	2	News
MSR-PC	5801	Yes/No	3	News
STSS-65	65	0–4	32	Dictionaries
STSS-131	64	0–4	64	Users
SEV-12	2234	0–5	5	News, Users, EU Parl

2.3 Sentence Similarity Measures Applications

In this Section we describe the main applications of sentence similarity measures.

In Machine Translation [9] these measures are employed to quantify the quality of translation algorithms. This can be achieved by employing a dataset composed by sentences in a given language and, for each sentence, by one or more translations in a target language, produced by human professional translators. The automatic translation is compared to the translations produced by human translators: a sentence similarity measure is used. High similarity values means are correlated with high translation quality.

In Paraphrase Identification [10] two documents are compared for conveying the same information content. If each document is a sentence, a sentence similarity measure can be employed: when a similarity threshold is exceeded, the two sentences are considered paraphrases. If any document is composed by more than one sentence, it is split into sentences first: sentence-level paraphrases can then be identified; the order of the sentences may or may not be taken into account.

In Speech Recognition [5] the quality of the recognition algorithms can be evaluated through sentence similarity measures. The hypothesis generated by an algorithm are compared to the sentences transcribed by human professionals. The higher the similarity among the sentences, the higher the quality of the algorithm is assumed.

In Question-answering [6] a question asked by a user can be compared by a software to each element of a finite set of already-answered questions. A sentence similarity measure is used for the comparison, in order to find the best match. The corresponding answer is then provided to the user.

In Text Summarization [7,8] two sentences of different length can be compared via a similarity measure. If they can be considered paraphrases according to a similarity threshold, the shorter sentence is preferred to the longer one.

Table 3 compares the aforementioned applications by their usage of sentence similarity measures. In particular, it must be noted that in three cases (Paraphrase Identification, Question Answering, Text Summarization) similarity measures are employed in production to actually execute a task; however in the two remaining cases (Machine Translation, Speech Recognition) sentence similarity measures are used for the testing phase only.

Table 3. Sentence similarity applications

Application	Phase
Machine translation	Test
Paraphrase identification	Production
Speech recognition	Test
Question answering	Production
Text summarization	Production

3   Rationale

Previously, we have shown how current state of the art measures for sentence similarity takes into account only exact word matches, with the notable exception of METEOR and TERp. These two measures relies on Wordnet for handling synonymy and on the Porter algorithm for removing inflections from the words to match. However they are incapable to assess and exploit the degree of similarity among two words with different meanings. We have also shown the major fields of application of sentence similarity measures. Since all the corresponding tasks need to assess sentence equivalence rather than resemblance, current available datasets have been developed to test for semantic equivalence, leaving little or no space for sentence structure similarity and for words with similar but different meanings.

There are however applications where taking in consideration the form of the sentence and the degree of similarity among words can be advantageous. Let us consider the following example:

1. Prince Joseph kissed his wife and left the castle to hunt the red dragon.
2. King Arthur hugged the queen and abandoned his room to fight the black knight.

The two sentences are very similar, but current similarity measures are not able to elicit this similarity, since there are almost no words in common, nor synonyms or stems. However, the structure of the sentence is the same and words have been replaced by related terms. This obvious plagiarism will not fool a human observer: yet it would be undetected by a computer. In fact, current state of the art algorithms for Plagiarism Detection use similarity measures that do not take in account word relatedness, e.g. IM (Eq. 7). Plagiarism Detection is not the only viable application for a resemblance-senstitive sentence similarity measure. The task can be reversed: the similarity measure can be employed to assess the quality of texts generated by Compuational Creativity algorithms. This is an important, current problem to address, since the evaluation of algorithm-generated textual artifacts can be highly subjective. Such a similarity measure can be, for example, a valuable tool for assessing the originality of new artifacts produced by a given algorithm with respect to its previously generated texts.

4   SWER Similarity Measure: Principles and Functionalities

To devise a novel sentence similarity measure capable of assessing resemblance more than equivalence, we have chosen to satisfy the following requirements:

1. the degree of similarity among words with different meanings must be taken into account;
2. the sentence structure in general and word order in particular must affect the measure in a significant way;
3. the measure must be symmetric;
4. the similarity value must be bound in the interval [0,1].

The reasons behinds the first two requirements have been discussed previously: word similarity and sentence structure are key factors for measuring resemblance.

The third requirement arises from past studies on human cognition of similarity: given a sentence pair, the impact of the order of the two sentences on the perceived similarity is not statistically significant [32].

The aim of the fourth requirement is two-fold. First of all, the meaning of the similarity value should be self-explanatory, i.e. being interpretable as a percentage. Some measures, like WER or TER, are not even bound to consistent interval: the limits depends on the length of the sentences involved. Secondly, given four sentences, s_1, s_2, s_3 and s_4, it should be straightforward to answer the question: are the two sentences in the pair (s_1, s_2) more similar to each other than the sentences in the pair (s_3, s_4)? Obviously, this cannot be done with similarity measures whose interval is not consistent, since the similarity values will not be comparable.

4.1   Semantic Word Error Rate

In order to satisfy the aforementioned requirements, we chosed to modify one of the state of the art sentence similarity measures shown in Table 1. First of all, with regard to the second requirement, we restricted the choice to the measures that takes into account word order most: WER, TER, TERp, METEOR. We discarded METEOR for one main reason: a fuzzy word match based on word similarity, as the first requirement implies, makes ill-defined the term "chunk" (i.e. group of adjacent matched words). While TER and TERp are based on WER, they introduce an extra "shifting" operation which adds considerable complexity to the measure: even though heuristics have been developed to reduce this complexity, we favoured WER for its superior simplicity.

As described in Sect. 2.1, Word Error Rate (WER) derives from the Levenshtein distance, but compares sentences instead of words. In particular, given two sentences s_1 and s_2 the idea is to count the minimum number of edit operations needed to make s_1 equal to s_2. Three types of operations are allowed:

insertion, deletion and substitution of words. With regard to the three remaining requirements, we improve on WER by defining our proposed Semantic Word Error Rate (SWER) as follows:

$$\text{SWER}(s_1, s_2) = \frac{I + D + \sum S(w_i, w_j)}{\max(|s_1|, |s_2|)} \tag{8}$$

where I are the number of insertions, D the number of deletions and $S(w_i, w_j)$ is the similarity among the words w_i and w_j. With respect to Eq. 2 the following modifications have been applied.

First of all the previous term S, i.e. the number of substitutions, has been replaced by a sum of all the similarity values $S(w_i, w_j)$ involved in each substitution (w_i, w_j). This allows to satisfy the first requirement: words will be matched for substitution according to their degree of similarity.

Secondly, the normalization factor has been replaced. Instead of dividing by $|s_2|$, the total cost of the operations is divided by $max(|s_1|, |s_2|$. This has two implications:

1. Since the word similarity function S will be symmetric, the SWER measure will also be symmetric. The third requirement will therefore be satisfied.
2. Since the word similarity function S will be bound in the interval [0,1], the SWER measure will also be bound in the interval [0,1]. The fourth requirement will therefore be satisfied.

Similarly to Eq. 3, the similarity among two sentences will be:

$$\text{Similarity}(s_1, s_2) = 1 - \text{SWER}(s_1, s_2) \tag{9}$$

It should be noted that this value will also be bound in the interval [0,1], while WER-based similarity can be negative. In fact, since the normalization factor of WER depends only on the length of s_2, the WER measure can be greater than one.

4.2 Latent Semantic Analysis

So far, we have not considered how the word similarity function $S(w_i, w_j)$ can be implemented. However, we have assumed it to be symmetric and bound in the interval $[0, 1]$. To address the problem we have employed Latent Semantic Analysis (LSA) [33].

Given a lexicon L constituted by a set of m words:

$$L = \{w_1, w_2, ..., w_m\} \tag{10}$$

a document D is defined as a finite sequence of words:

$$D = <t_1, t_2, ..., t_{|d|}> \tag{11}$$

where $t_i \in W$ and $|D|$ is the length of the document.

Let C be a corpus of n documents:

$$C = \{D_1, D_2, ..., D_n\} \tag{12}$$

A word-document frequency matrix $\mathbf{A}$ is built: each cell a_{ij} represents the number of occurences of the word w_i in the document D_j. By employing Truncated Singular Value Decomposition the best r-rank approximation is achieved:

$$\mathbf{A}^{(r)} = \mathbf{U}^{(r)}\mathbf{\Sigma}^{(r)}\mathbf{V}^{(r)^T} \tag{13}$$

Each row $\mathbf{u}_i^{(r)}$ of the matrix $\mathbf{U}^{(r)}$ contains the coordinates of a point representing the word w_i in a semantic space of size r. Thereafter, we will refer to $\mathbf{u}_i^{(r)}$ as just $\mathbf{u}_i$.

Given two words w_i, w_j, their similarity can therefore be computed by evaluating the distance among their corresponding points in the semantic space. While several distance measures exists, two measures have particular significance: euclidean distance and cosine similarity. In Information Retrieval, cosine similarity has been largely established as a the most valid way to measure semantic similarity. Furthermore, the euclidean distance is not bound in the interval [0, 1], while the absolute value of the cosine similarity is. Since the cosine similarity among two points can be computed as the scalar product of their vectors, we define the similarity function $\mathrm{S}(w_i, w_j)$ as follows:

$$\mathrm{S}(w_i, w_j) = \frac{|\mathbf{u}_i \cdot \mathbf{u}_j|}{|\mathbf{u}_i||\mathbf{u}_j|}. \tag{14}$$

where w_i, w_j are two words and $\mathbf{u}_i$, $\mathbf{u}_j$ are their corresponding vectors in the semantic space.

5   Experimental Analysis

This Section shows some experimental results that prove the effectiveness of the proposed approach. A semantic space has been built according to the Latent Semantic Analysis methodology, using as texts corpus the training set of the Microsoft Sentence Completion Challenge RIF. The training set is composed of about N. 19th century novels from Project Gutenberg RIF (**Zweig and Burges, 2012**). Therefore, to validate the proposed SWER methodology, a corpus composed by 113 sentence triples has been created. Each triple of the dataset is composed of two resemblant sentences and a dissimilar one.

The results, summarized in Table 4 shows a remarkable improvement in the accuracy with respect to the WER methodology.

Let us consider some examples in order to show also the typologies of sentences created to test the approach. As shown in the results, in most cases WER and SWER are the best strategies to recognize the meaningful differences with respect to a simple LSA similarity measure, that lacks of the sentence structure information. An example is given by the following triple:

Table 4. A comparison among LSA, WER and the proposed SWER approach

Method	Accuracy
SWER	81 %
WER	65 %
LSA	35 %

1. From the helicopter my house seemed very small.
2. From the airplane my house seemed very small.
3. From my house the airplane seemed very small.

where, according to the LSA, the most different sentence is the first one, and not the third, as correctly suggested by both WER and SWER.

However, even if WER gives best results with respect to LSA, in many cases it does not succeed to detect meaningful differences among the sentences. Instead, these differences are well detected by a semantic approach. As an example, in the following triple, while WER does not discern among the sentences, LSA and SWER recognizes correctly that the intruder is the third phrase.

1. Everybody needs somebody to love.
2. Everybody needs somebody to hate.
3. Everybody needs somebody to call.

Another meaningful example is represented by the triple specified in the introduction:

1. I like cats.
2. I like dogs.
3. I like astrophysics.

According to a LSA-based similarity measure the sentence detected as dissimilar is the first one, according to SWER methodology the intruder sentences is the third one, while the WER does not consider meaningful differences among the sentences.

In this other example, all the three methodologies recognizes correctly the most different sentence in the third one.

1. A big cat drank the milk in the cup.
2. A small mouse ate the cheese over the plate.
3. A young boy crossed the road on a skateboard.

6   Conclusions and Future Work

In this paper, we have proposed and experimentally assessed SWER, an innovative similarity measure for sentence plagiarism detection. As we demonstrated through the paper, SWER introduces a complex approach based on latent semantic analysis, which is capable of outperforming the accuracy of competitor methods in plagiarism detection. We described principles and functionalities of SWER, by also providings several examples that contribute to better show how the metric works. We have also provided a preliminary experimental analysis, whose results have confirmed the goodness of our proposal, and its superiority over comparison approaches. Future work is mainly oriented towards extending the proposed metric as to deal with novel requirements dictated by emerging (*i*) *big data management and processing* (e.g., [38]) and (*ii*) *big data analytics* (e.g., [39,40]).

References

1. Dolan, B., Quirk, C., Brockett, C.: Unsupervised construction of large paraphrase corpora: exploiting massively parallel news sources, In: Proceedings of the 20th International Conference on Computational Linguistics (COLING), Geneva, Switzerland, pp. 350–356 (2004)
2. Hassan, S.: Measuring semantic relatedness using salient encyclopedic concepts. Ph.D. thesis, University of Texas (2011)
3. Ji, Y., Eisenstein, J.: Discriminative improvements to distributional sentence similarity. In: Proceedings of Empirical Methods in Natural Language Processing (EMNLP), Seattle, Washington, USA, pp. 891–896 (2013)
4. Levenshtein, V.I.: Binary codes capable of correcting deletions, insertions, and reversals. Sov. Phys. Dokl. **10**(8), 707–710 (1966)
5. Morris, A.C., Maier, V., Green, P.: From WER and RIL to MER and WIL: improved evaluation measures for connected speech recognition. In: INTERSPEECH (2004)
6. Burke, R.D., Hammond, K.J., Kulyukin, V., Lytinen, S.L., Tomuro, N., Schoenberg, S.: Question answering from frequently asked question files: experiences with the FAQ finder system. AI Mag. **18**(2), 57–66 (1997)
7. Aliguliyev, R.M.: A new sentence similarity measure and sentence based extractive technique for automatic text summarization. Expert Syst. Appl. **36**(4), 7764–7772 (2009)
8. Erkan, G., Radev, D.R.: LexRank: graph-based lexical centrality as salience in text summarization. J. Artif. Intell. Res. **22**, 457–479 (2004)
9. Liu, D., Gildea, D.: Syntactic features for evaluation of machine translation. In: Proceedings of the ACL Workshop on Intrinsic and Extrinsic Evaluation Measures for Machine Translation and/or Summarization, pp. 25–32 (2005)
10. Madnani, N., Tetreault, J., Chodorow, M.: Re-examining machine transla-tion metrics for paraphrase identification. In: Proceedings of the 2012 Conference of the North American Chapter of the Association for Computational Linguistics: Human Language Technologies (NAACL HLT 2012), pp. 182-190. Association for Computa-tional Linguistics (2012)
11. Niwattanakul, S., Singthongchai, J., Naenudorn, E., Wanapu, S.: Using of Jaccard coefficient for keywords similarity. In: Proceedings of the International MultiConference of Engineers and Computer Scientists (IMECS), vol. 1, pp. 380–384 (2013)

12. Papineni, K., Roukos, S., Ward, T., Zhu, W.J.: BLEU: a method for automatic evaluation of machine translation. In: Proceedings of the 40th Annual Meeting on Association for Computational Linguistics (ACL), pp. 311–318. Association for Computational Linguistics (2002)
13. Doddington, G.: Automatic evaluation of machine translation quality using n-gram co-occurrence statistics. In: Proceedings of the Second International Conference on Human Language Technology Research (HLT), pp. 138–145 (2002)
14. Snover, M., Dorr, B., Schwartz, R., Micciulla, L., Makhoul, J.: A study of translation edit rate with targeted human annotation. In: Proceedings of the Seventh Conference of the Association for Machine Translation in the Americas (AMTA), pp. 223–231 (2006)
15. Snover, M., Madnani, N., Dorr, B., Schwartz, R.: TERp system description. In: Proceedings of Metrics MATR Workshop at the Eighth Conference of the Association for Machine Translation in the Americas (AMTA), vol. 555 (2008)
16. Porter, M.F.: An algorithm for suffix stripping. Program **14**(3), 130–137 (1980)
17. Lavie, A., Denkowski, M.J.: The METEOR metric for automatic evaluation of machine translation. Mach. Transl. **23**(2–3), 105–115 (2009)
18. Allan, J., Wade, C., Bolivar, A.: Retrieval and novelty detection at the sentence level. In: Proceedings of the 26th Annual International ACM SIGIR Conference on Research and Development in Informaion Retrieval (SIGIR 2003), pp. 314–321 (2003)
19. Hoad, T.C., Zobel, J.: Methods for identifying versioned and plagiarised documents. J. Am. Soc. Inform. Sci. Technol. **54**, 203–215 (2003)
20. Metzler, D., Bernstein, Y., Croft, W.B., Moffat, A., Zobel, J.: Similarity measures for tracking information flow. In: Proceedings of the 14th ACM International Conference on Information And Knowledge Management (CIKM), pp. 517–524 (2005)
21. Barzilay, R., Lee, L.: Learning to paraphrase: an unsupervised approach using multiple-sequence alignment. In: Proceedings of the Conference of the North American Chapter of the Association for Computational Linguistics on Human Language Technology (NAACL), vol. 1, pp. 16–23 (2003)
22. Dolan, W.B., Brockett, C.: Automatically constructing a corpus of sentential paraphrases. In: Proceedings of the Third International Workshop on Paraphrasing (IWP) (2005)
23. Li, Y., McLean, D., Bandar, Z.A., O'Shea, J.D., Crockett, K.: Sentence similarity based on semantic nets and corpus statistics. IEEE Trans. Knowl. Data Eng. **18**(8), 1138–1150 (2006)
24. O'Shea, J., Bandar, Z., Crockett, K.: A new benchmark dataset with production methodology for short text semantic similarity algorithms. ACM Trans. Speech Lang. Process. (TSLP) **10**(4) (2013). Article no. 19
25. Agirre, E., Diab, M., Cer, D., Gonzalez-Agirre, A.: Semeval- task 6: A pilot on semantic textual similarity. In: Proceedings of the First Joint Conference on Lexical and Computational Semantics, vol. 1, Proceedings of the Main Conference and the Shared Task. Proceedings of the Sixth International Workshop on Semantic Evaluation, vol. 2, pp. 385–393 (2012)
26. Agirre, E., Cer, D., Diab, M., Dolan, B.: SemEval-2012 task 6 corpus. University of York (distributor) (2012). https://www.cs.york.ac.uk/semeval-2012/task6/
27. Microsoft Research, Microsoft Resarch Video Description Corpus. Microsoft Corporation (2010). http://research.microsoft.com/en-us/downloads/38cf15fd-b8df-477e-a4e4-a4680caa75af/

28. Callison-Burch, C.: Workshop on statistical machine translation at ACL 2007 - development data. Johns Hopkins University (2008). http://www.statmt.org/wmt08/shared-evaluation-task.html

29. Banerjee, S., Pedersen, T.: Extended gloss overlaps as a measure of semantic relatedness. In: Proceedings of the 18th International Joint Conference on Artificial Intelligence (IJCAI), vol. 3, pp. 805–810 (2003)

30. Ponzetto, S.P., Strube, M.: Knowledge derived from wikipedia for computing semantic relatedness. J. Artif. Intell. Res. **30**, 181–212 (2007)

31. Shapira, D., Storer, J.A.: Edit distance with move operations. In: Apostolico, A., Takeda, M. (eds.) CPM 2002. LNCS, vol. 2373, pp. 85–98. Springer, Heidelberg (2002)

32. Lee, M.D., Pincombe, B.M., Welsh, M.B.: An empirical evaluation of models of text document similarity. In: Proceedings of the 27th Annual Conference of the Cognitive Science Society, pp. 1254–1259 (2005)

33. Deerwester, S., Dumais, S.T., Furnas, G.W., Landauer, T.K., Harshman, R.: Indexing by latent semantic analysis. J. Am. Soc. Inf. Sci. **41**(6), 391–407 (1990)

34. Sochenkov, I., Zubarev, D., Tikhomirov, I., Smirnov, I., Shelmanov, A., Suvorov, R., Osipov, G.: Exactus like: plagiarism detection in scientific texts. In: Ferro, N., et al. (eds.) ECIR 2016. LNCS, vol. 9626, pp. 837–840. Springer, Heidelberg (2016)

35. Velásquez, J.D., Covacevich, Y., Molina, F., Marrese-Taylor, E., Rodríguez, C., Bravo-Marquez, F.: DOCODE 3.0 (DOcument COpy DEtector): a system for plagiarism detection by applying an information fusion process from multiple documental data sources. Inf. Fusion **27**, 64–75 (2016)

36. Chae, D.-K., Ha, J., Kim, S.-W., Kang, B., Im, E.G., Park, S.: Credible, resilient, and scalable detection of software plagiarism using authority histograms. Knowl.-Based Syst. **95**, 114–124 (2016)

37. Jaric, I.: High time for a common plagiarism detection system. Scientometrics **106**(1), 457–459 (2016)

38. Cuzzocrea, A., Saccà, D., Ullman, J.D.: Big data: a research agenda. In: Proceedings of the 17th International Database Engineering and Applications Symposium (IDEAS), pp. 198–203 (2013)

39. Cuzzocrea, A., Bellatreche, L., Song, I.-Y.: Data warehousing, OLAP over big data: current challenges and future research directions. In: Proceedings of the 16th International Workshop on Data Warehousing and OLAP (DOLAP), pp. 67–70 (2013)

40. Cuzzocrea, A.: Analytics over big data: exploring the convergence of data warehousing, OLAP and data-intensive cloud infrastructures.In: Proceedings of the 37th Annual IEEE Computer Software and Applications Conference (COMPSAC), pp. 481–483 (2013)

Critical Success Factors for Implementing Business Intelligence System: Empirical Study in Vietnam

Quoc Trung Pham[1], Tu Khanh Mai[1], Sanjay Misra[2(✉)],
Broderick Crawford[3,4], and Ricardo Soto[3,5,6]

[1] School of Industrial Management, HCMC University of Technology,
Ho Chi Minh City, Vietnam
[2] Department of Computer and Information Sciences,
Covenant University, Ota, Nigeria
sanjay.misra@covenantuniversity.edu.ng
[3] Pontificia Universidad Católica de Valparaíso, Valparaíso, Chile
[4] Universidad Central de Chile, Santiago, Chile
[5] Universidad Autónoma de Chile, Santiago, Chile
[6] Universidad Científica del Sur, Lima, Peru

Abstract. Although the application of business intelligence (BI) system has increased recently, the critical success factors (CSFs) of BI system implementation remain poorly understood. Therefore, understanding CSFs for BI system is necessary for a company to be successful in implementing a BI system. Currently, the number of business successfully implemented BI in Vietnam is still limited. In order to understand the current situation of BI implementation in Vietnam, case study method has been used. Based on the research framework developed by Yeoh & Koronios, in-depth interviews have been conducted with 4 Vietnamese companies, who already implemented BI system. The main result of this study helps to extract the lessons learnt from 4 cases to rank the importance of CSFs for BI implementation in Vietnam context. Based on this result, recommendations were made for increasing the probability of successfully implementing BI system in Vietnam and other countries with similar conditions.

Keywords: Critical success factor · Business intelligence · Information system implementation · Empirical study · Vietnam

1 Introduction

Since 1990s, the world economy has transitioned toward the knowledge-based economy. The important role of information/knowledge had been realized by many businesses. In this scene, Business Intelligence (BI) has emerged as a new method to support companies in extracting knowledge for improving their performance. According to Gartner [20], there would be a significant shift from application Business Intelligence solutions from information analyst to the business executive and management. Management of enterprises would use about 10 % of the time to use BI.

O. Gervasi et al. (Eds.): ICCSA 2016, Part V, LNCS 9790, pp. 567–584, 2016.
DOI: 10.1007/978-3-319-42092-9_43

Indeed, more and more firms apply BI into their business and BI becomes one of the most crucial parts of the solution to provide businesses the necessary information for decision-making to be able to ensure competitive advantages. According to Gartner's surveys five years ago (2011), BI applications have been dominating the technology priority list of many CIOs of many companies all over the world.

Vietnam also catches up with this trend of the world. According to a Director of Business Intelligence Solution of Oracle Vietnam [20], business should invest to BI because BI solution can provide many detail reports, access reports anytime, anywhere, help departments in company communicate better, control quality processes, and prevent risk instead of solving issues. There are several enterprises in Vietnam, which already implemented BI system. Most of them are multinational corporations, which aware the significant roles of Business Intelligence for a long time. The rest is some Vietnamese companies in retailer, food and beverage industry, which require accurately and timely response of information.

Although many companies have already implemented Business Intelligence Systems (BIS), the rate of failures is still high. Besides, BI is more complexity than Enterprise Resource Planning (ERP) in the meaning. To be successful in BI solution requires more information as well as more knowledge about how to use and manage information effectively.

In general, many companies aware the vital role of BI but they are still afraid of the failure. They need some lesson learns, experience sharing, and/or what else to secure the success of BI system implementation. Therefore, understanding the critical success factors (CSFs) of BI system is crucial and meaningful in the current situation to Vietnamese enterprises.

According to the report of Gartner [20], the rate of failure in implementing BI is about 70–80 %. When estimating to the year 2014, although many businesses consider the importance role of BI, the rate of failure still does not decrease a lot, approximately 65–70 %. There is no report or document about the situation of BI implementation in Vietnam; however, the rate surely higher than the number two-third as mentioned. According to a Vice Director of a gold partner of Oracle in Vietnam [20], the rate of the company indeed implementing BI in Vietnam is only about 10 %. Most of them just use a part of BI, such as Management Report or Reporting System although they invest in the whole infrastructure for BI. That can be considered as failure cases because they do not use the strength of BI. To understand why the rate of failure is so high and how the rate success for BI implementation in Vietnam is increased, it is important to raise these following questions: (1) Why are some of the enterprises in Vietnam successful while others fail in implementing BI system? (2) What must be improved in management and process for implementing BI system?

The particular objectives of this research include: (1) To understand the situation of BI system implementation in Vietnam; (2) To get an understanding of the CSFs affecting the implementation of BI system in companies in Vietnam; (3) To provide suggestions for successfully implementation of BI system in Vietnam. The structure of this paper is organized as follows: (2) Theoretical background; (3) Research method; (4) Data collection & analysis; (5) Findings; (6) Discussion & Conclusion.

2   Theoretical Background

2.1   Business Intelligence Systems

BI is understood as the set of techniques and tools using in the transformation of the raw data into the meaningful and useful information for business purposes. These technologies are capable of handling large amounts of unstructured data to help identify, develop and otherwise create new strategic business opportunities. BI is aimed to allow for the natural interpretation of these large volumes of data. BI also provide businesses with a competitive market advantage and long-term stability by its ability in helping corporate to identify new opportunities and implementing an effective strategy based on insights [1].

BI is processes and technologies that businesses use to manage their enormous raw data in the history and current, transform to meaningful data. The result of these meaningful data can help business to predict the future. With the outcome from what predicts, corporate can build a set of actions (strategy) which useful to be enabled to make the decision effectively.

Additionally, BI can be used to support a broad range of business decisions from operational to strategic [2]. Basic operating decisions include daily execution actions such as pricing or material planning. Strategic business decisions include long-term impact activities such as whether to remain a business (priorities), goals and directions at the high level. The business needs information from both external (data derived from the market) and internal (such as financial and operations data). When combined, these data can provide a complete picture that, in effect, creates an "intelligence" that any singular set of data cannot provide.

To summary, BIS is a system used for finding patterns from existing data from operations but not simple as a report set of an IT system. Software's reports just help to show raw data or raw-processed from one or two systems cannot be considered as BIS. BIS is more complexity than any other IT systems: a source of data combines both internal and external; data purpose is both operational and strategic, and data must be processed with complication techniques.

2.2   Business Intelligence Systems in Vietnam

In recent years, the requirement for human resources skilled in data analyst and using tool for BIS in Vietnam increases dramatically. For example, a new graduate student who is still need to be trained more, after one year training to use tool for ETL (extract, transform and load data), creating BI report/analyze/dashboard easily to find a job with salary triple his 1 year ago. Besides, the headcount of Oracle BI salesforces also rises from 2–3 members 5 years ago to 6–7 members now.

All of these facts can be assumed that BI market in Vietnam has been growing intensely and many companies in Vietnam aware the important role of BI. Even so, most of those do not actually apply complete BI solution, they just use a part of it, such as report functions, not use either analytics or forecast. One of the reasons is that those businesses do not know how to forecast, or their companies' characteristics are unpredictable. Vietnamese culture also impacts for the reason for not using advanced parts of BI. Most of the Vietnamese managers do not have the habit of making decisions by viewing the numbers or reports on the computer. They like to hold meeting, discuss and make decisions. They like to read reports onpaper, not by selecting parameters or difference style on the computer.

There are some enterprises invest for BIS with solutions from consultants, high performance IT infrastructure, a sufficient support team, including Vinamilk (Food and Beverage), Tan Hiep Phat (Beverage), Pepsi (Beverage), Masan (Food), Bibica (Food), Kinh Do (Food), Big C Supermarket (Retailer), Coop-mart (Retailer), Vietsov Petro (Petrol), Phu My Fertilizer (Fertilizer), Vietnammobile (Telecom), Banks, etc. In which, Masan is an outstanding firm for success. Other companies choose to implement by internal technical and business team, such as Annam Group (Beverage and Cosmetics), VNG Corporation (Online Entertainment), Vietnam work, etc.

Most of these companies apply commercial solutions from vendors SAP, Oracle, Microsoft, and IBM. Nowadays, the trend for small and medium enterprises applying BI solutions from open sources such as Pentaho, Odoo, Python also goes up due to the low license cost. However, the cost of developing and maintaining the stability of the system is not small in comparison with commercial solutions.

Some consultant companies for BI solution in Vietnam can be listed as follows: FPT, CSC, Accenture Vietnam, IBM Vietnam, HP Vietnam, Cybersoft Vietnam, Elite Technology, Gimasys, SSG, Global Cybersoft (Viettel), Vietsoft, Pythis, etc.

A BI project in Vietnam is typically implemented in six months to one year. After that, they review the needs and decide to continue to upgrade the system or not. Some of the companies have invested infrastructure, framework, etc. for a full solution of BI although they just use a small part of BI. Not many businesses in Vietnam admit that they fail in implementing BI. However, based on the meaning of BI, providing meaningful information for users to make decisions, there are only about 10 % of firms implementing BI in Vietnam meet these indicators.

2.3   Critical Success Factors

Critical success factor is defined as the term for an element that an organization or project need to satisfy to achieve its mission. It is required for ensuring the success of a company as a key factor or activity. This concept was first represented by [26]. There is some limited number of areas once being satisfactory; they will ensure successful competitive performance for the organization.

CSFs are not the same between industries, even in the same industry; CSFs are also diverse between the various enterprises. They vary from strategic, managerial to operational and are divided into 3 aspects: organizational, industry and environmental. They can exist at all levels of the company: corporate, division, plant, and department. Sometimes CSFs are even necessary consider of individual employees [19].

2.4   Critical Success Factors for Business Intelligence

There are many CSFs studies for conventional application-based on an IT project (such as an operational or transactional system). Implementing a BI system is a set of complex activities demanding suitable infrastructure and resources over periods [4, 5]. BI system implementation is regarded as an "organic cycle" that develops continuously. CSFs in the context of BIS can be perceived as a set of tasks and procedures, which are either adopted (if they had already occurred) or worked out (if they were nonexistent), that should be taken to make sure BI systems achievement.

There are several studies on BI success factors. Some authors classify CSFs for BI in 3 dimensions: environment, organization, and planning of the project. Hwang et al. [16] found there is substantial support for organizational factors. Lately, Ariyachandra and Watson [17] analyzed CSFs for BIS implementation's measurement into two key dimensions: process performance, and infrastructure performance. Process performance can be assessed in the time schedule and budgetary considerations. Infrastructure performance is evaluated by the quality of system and information as well as this system use. Yeoh and Koronios [5] classified CSFs for BIS implementation into three dimensions: organization, process, and technology. An organizational dimension contains committed management support and sponsorship, clear vision, and well-established business case. The process dimension includes business-centric championship and balanced team composition, business-driven and an iterative development approach as well as user-oriented change management. The last dimension, technology, focuses on these elements: business-driven, scalable and flexible technical framework, sustainable data quality and integrity.

Lately, some other researches, e.g. [6–8], generally agree with Yeoh and Koronios [5] and build upon those CSFs for implementing enterprise-level BI. Up to now, the number of academic research on the CSFs of implementing BI systems is still rare and limited in scope of analysis.

3   Research Method

3.1   Research Purpose

This research is based on exploratory research as the author wish to verify and validate the results obtained from analysis of CSFs in the literature review. In the process of achieving the overall objective of the project, the author have experienced practical study as interviews with relevant participants in companies implementing BI and also gone through various literatures and have done analysis of the CSFs of the BI implementation process. The purpose of this research is to give more understanding of the quality of data gained and analyzed, the context of it and relevance to the current project.

3.2   Research Framework

This research follows [5], who classify seven CSFs for BIS to three key dimensions: Organization, Process, and Technology. This framework pointed out how a set of CSFs contributes to success of BIS implementation: there is a set of multi-dimensions CSFs that influences the success of implementing BI systems that are assessed through infrastructure performance and process performance. The infrastructure performance can be measured with the three major Information System success variables: system quality, information quality, and system use [27]. The process performance can be assessed in terms of time-schedule and budgetary considerations.

Each factor contains several contextual elements. While some elements defining in [9] were still used in [5], some were removed. In this research, all elements both in [9, 5] are used. Besides, during the interviews, there are some elements were emerged and need to add into research framework for further analysis including Involvement of top management (Committed management support and sponsorship), Change management (User-oriented change management), Performance considerations (Business-driven, scalable and flexible technical framework). Finally, seven critical success factors were described in the table below (Table 1).

Table 1. CSFs and their contextual elements and descriptions

	Critical success factor	Contextual elements
Organization	Committed management support and sponsorship	- Committed top management support - Adequate resources are provided - Involvement of top management
	A clear vision and a well-established business case	- Aligning the BI project with org. business vision - Well-established business case
Process	Business-centric championship and a balanced team composition	- Existent of a business-centric champion - Use of external consultant at early phase - Committed expertise from business domain - The team is cross-functional
	Business-driven and iterative development approach	- Adoption of iterative development approach - Project scope is clearly defined - Project scheduled to deliver quick wins
	User-oriented change management	- Formal user involvement throughout the lifecycle - Foundation education, training,& support are in place - Change management
Technology	Business-driven, scalable and flexible technical framework	-Stable source systems are in place - Establishment of strategic scalable and flexible technical framework - Performance considerations
	Sustainable data quality and integrity	- High quality of data at source system - Business-led establishment of common measures and classifications - Sustainable dimensional and metadata model - Business-led data governance

Success criteria for implementing a BIS was defined as below [5]:

- Each criterion was measured on 3 levels: Good, Acceptable, Poor.
- A case will be evaluated "Successful" only if all criteria of Infrastructure performance are "Good" and criteria of Process performance are "Good" or "Acceptable".
- If there is any "Acceptable" and no "Poor", the case will be assessed Partially Successful.
- A case will be appraised "Unsuccessful" if there is any "Poor" criterion.

3.3 Research Process

This research use qualitative (case study) method to achieve a deep understanding of the problems. The case study methodology provides better explanations of the examined topic which would be lost in quantitative designs [10]. This study has been carried out through three steps as follows:

Step 1: Through the literature review of the BIS, a list of the success factors identified or found in the previous study was summarized. The summarized list of key success factor, then it was then being used as the check-list for testing against the findings from the interviews in the selected cases.

Step 2: Then, based on interviews with project members of 4 cases, this research analyzed, evaluated, and benchmarked to confirm the existence/importance of CSFs.

Step 3: Finally, the conclusion is withdrawn from the study, and the recommendations for the management and the project team in implementing BI solution are suggested.

4 Data Collection & Analysis

4.1 Sample Design

This research takes place in enterprises implementing BI system in Vietnam. For this study, each case study is an empirical analysis that allows replicating logic leading to analytic generalization. Case studies in this research should be considered as experiments and not merely respondents in a survey [10]. Therefore, selecting cases must focus on relevance rather than representativeness. Several studies recommend the number to attain "theoretical generalizability" [10, 11] should be at least 4 and should not over 15 cases to enable comfortable understanding of "local dynamics" [11]. Four cases is chosen for this study falling within the recommended range. Table below lists the background of four cases (see the appendix for more details). Their descriptions have been masked slightly to preserve the anonymity of the participants (Table 2).

Table 2. Case background

Case	Code	Type of organization	Annual revenue	No. of staff	BI system owner	Implementation success level
S	A	Importation, wholesale, retail and distribution	M	700	MIS Department	Successful
P1	B	Retail, distribution, and E-commerce	M	600	MIS department	Partially successful
P2	C	Digital content	M	1500	Customer service department	Partially successful
F	C	Digital content	M	1500	Technical support department	Unsuccessful

Note. Vietnam Govt: Annual revenue is Medium (M) if between 0.5 and 2.5 mil. USD.

4.2   Data Sources

Primary sources information and data has been collected through the semi-structure interviews. There are total 18 interviewees, and the author herself was the interviewer for all interviews. The author tries to contact participants by calling, emailing to the relationship before for asking they join the interview and using snowballing for filling in some missing informants of research design.

There are five face-to-face interviews and recorded in the audio tape. One interview is a short communication happening when the author accidently met the interviewee in the company lounge, so there is no audiotape was recorded. The rest 12 interviewees accept to take part in the interview via some internet message tools including Skype, Facebook Messenger and Google Hangout due to they cannot arrange the time for the face-to-face interview.

The interviews were taken place any time the participants are available and not continuity. Sometimes, the author reviews the data and need to make clear some information and suggest for a short interview to confirm or check data. During the interview, the author was introduced the systems such as the way to use systems, the information systems provide, the schedule of the projects.

Overall, 18 interviews was taken place for all four cases. Interviewees are people who join in the process of implementing BI system including project sponsors, project managers, technical people and key users (Table 3).

4.3   Data Analysis

This study uses a cross-case analysis approach to understand about the findings more accurately [11]. Examining similarities and differences in relationships within the data helps to search the pattern and varying the order in which case data are arranged enables patterns to become more evident [12].

Four cases were categorized to successful (S), partially successful (P), or unsuccessful (U) by applying the research framework with a set of success criteria of. The extent of implementation success was examined against two key dimensions:

Table 3. Summary of interviewees

Case	Position of participants	Project role	Function	Method of interview
S(4)	Managing director	Project sponsor	Business	Email
	MIS manager	Project manager	Business/IT	Face-to-face
	System leader	Project member	IT	Internet message
	Planning manager	Project member	Business	Internet message
P1(5)	Finance controller	Project sponsor	Business	Internet message
	MIS manager	Project manager	IT	Internet message
	MIS specialist	Project member	IT	Internet message
	Budget owner	Project member	Key user	Internet message
	Budget controller	Project member	Key user	Internet message
P2(5)	CEO	Project sponsor	Business	Short conversation
	Head of customer service	Project member	Business/IT	Internet message
	Process & operation leader	Project member	Key user	Face-to-face
	Technical leader	Project member	IT	Face-to-face
	Product manager	Project member	Key user	Internet message
F(4)	COO	Project sponsor	Business/IT	Face-to-face
	Tech manager	PM assistant	IT	Face-to-face
	Management accountant	Project member	Key user	Internet message
	PQA	Project member	Key user	Internet message

(1) Infrastructure performance, which was viewed through the lens of system quality, information quality, and system use; (2) Process performance, which budgetary considerations and time-schedule measures were taken into account.

After the interview results for all four cases have been analyzed, one success cases emerged, two partially success cases, and one failure case. The success case was described as stable, dynamic, extendable, and the time of responsive within expecting. Likewise, the information provided by the BI system was considered timely, accurate, complete, and relevant to most participants. Additionally, the implementation processes were completed with an acceptable delay and within budget. Table below shows in more detail how the four cases measured up against the previously introduced list of success measures (Table 4).

Table 4. Implementation success level of four cases.

Case	Infrastructure performance			Process performance		Overall
	System quality	Information quality	System use	Budget	Time schedule	
S	G	G	G	A	G	S
P1	G	A	A	A	G	P
P2	G	G	A	A	A	P
F	A	X	X	G	A	U

Note: G = Good, A = Acceptable, X = Poor, S = Successful, P = Partial Successful,
U = Unsuccessful

5   Findings

5.1   Background to Implementing the BIS

For better understanding the relevance of the CSFs, we need to think through the background and motivations for implementation the BI system across all cases (Table 5).

Table 5. Background to and motivations for implementing the BI

Case	Background to and motivation for implementing the BI systems
S	• BI system was used for advance analysis of exists data from ERP system and to leverage the business such as estimating sale requirement, planning for resources, business performance appraisal, finding focus sale items/sale markets, cost determine • BI help to improve business performance and shortage period of planning due to information was updated every day: Planning for sale requirement and importation weekly, analyzing for sale markets monthly, planning for budget quarterly... • BI system was implemented to meet the requirement of top management: having more meaningful data to get deeper understanding of the business situation and improving the processes of the enterprise • The accuracy of reports is also an important thing: to get data directly from its container, not be processed or cooked by anyone
P1	• BI system was implemented to use information from ERP in another way: saving time to get data and make reports, over viewing the entire organization, managing data in an easy-to-use interface: all information show on the dashboard, analyze with what if forecast business situation and make decisions • BI system to plan finance indicators with a small variation, therefore, helping to control finance performance effectively, supporting managers easier • Solving problems: user used to take much time to collect data from ERP and Budgeting System then cooking it to make reports and send to Budget Owners. Sometimes they said that data is not what-they-want such as the template is hard to sense the situation or more suitable for financing view not for strategy aspect, data is not exactly as they remember • Saving cost for human resources: • Technical people: time to develop reports, e.g., one layout need one report takes about 1-2 MD. Each type of user needs the different layout of the reports. That needs to take time for development. Users can build reports analytic on BI dynamically by the way of pulling data sources and creating analytics by some indicators they care • Business users: some of summary report were daily sent to users; therefore, they can aware their situation without login to the BI system or waiting report from finance department • Accuracy and in-time report: Users need real-time and updated data for accurate reports. If users export data from the transactional system (ERP) and create reports, it takes them much time, lead to reports not accurate anymore at the time the top managers receive them

(*Continued*)

Table 5. (*Continued*)

Case	Background to and motivation for implementing the BI systems
	• Providing more analytics for different types of user, for example, one just needs some key information, while his subordinators need information in detail. A BI system must stipulate this requirement
P2	• Solving problems: • There are many data but still diffuse from many systems and not be standardization Therefore, those data are not exploited and used effectively. When users need information, they do not know whether data exist, where the data are from or what specific data they need to request the data owner provide. Hence, it is critical to implementing a BI system to integrate all of the related systems and give an overview picture of the business, from which people can make an important decision • BI Tool was created to use existing data effectively, to analyze and exploit data in many views. From which, they can make the questions and find the answers from the BI system. For example, the system does not only provide information about revenue increase or decrease but also give the reason. BI system has meaning if it help users base on it to make decisions • Saving time to make reports or aware and answer for a phenomenon: data was consolidated, therefore making reports or analyzing business situation do not make users take the time to find data
F	• Solving problems: • Lack of necessary information • Lack of close link between departments in corporate • No history for data due to reports are kept in Excel • Waiting much time for reports • Request from Managers for a BI system for new industry Game online • Request from finance controller to control the cash flow in products

5.2 Case Analysis Based on Seven CSFs of Research Framework

Based on the scoring of each CSF's element, the overall appraisement for each CSF is summarized in table below. By multiplying scores of elements with the case's coefficient (successful: 2, partially successful: 1, unsuccessful: 0), the ranking of CSFs' elements was calculated in Table 7. By calculating the average score for each CSF via its elements, the ranking of CSFs also is listed in Table 8.

Table 6. Summary of overall appraisement for CSFs in four cases

	Critical success factors	S	P1	P2	F
Org.	Committed management support and sponsorship	Y	Y	P	N
	A clear vision and a well-established business case	Y	P	P	N
Proc.	Business-centric championship and a balanced team composition	Y	P	Y	N
	Business-driven and iterative development approach	Y	P	Y	N
	User-oriented change management	Y	P	P	N
Tech.	Business-driven, scalable and flexible technical framework	Y	Y	P	N
	Sustainable data quality and integrity	Y	Y	Y	N

Table 7. Ranking of CSF elements in four cases

CSF elements	Ranking
Adoption of iterative development approach	1
Project scope is clearly	1
The team is cross-functional	2
Change management	2
Establishment of strategic scalable and flexible technical framework	2
Committed top management support	3
Existent of a business-centric champion	3
Committed expertise from business domain	3
Formal user involvement throughout the lifecycle	3
Aligning the BI project with organizational business vision	4
Well-established business case	4
Stable source systems are in place	4
Adequate resources is provided	5
Involvement of top management	5
Foundation education, training and support are in place	5
High quality of data at source system	5
Sustainable dimensional and metadata model	5
Performance considerations	6
Business-led data governance	6
Business-led establishment of common measures and classifications	7
Use of external consultant at early phase	8
Project scheduled to deliver quick-wins	8

Table 8. Ranking of CSFs in four cases

CSF	Ranking
User-oriented change management (P3)	1
A clear vision and a well-established business case (O2)	2
Business-driven, scalable and flexible technical framework (T1)	2
Committed management support and sponsorship (O1)	3
Business-driven and iterative development approach (P2)	4
Business-centric championship and a balanced team composition (P1)	5
Sustainable Data Quality and Integrity (T2)	6

6 Discussion and Conclusion

6.1 Summary of Main Findings

In case successful (case S), all CSFs' elements were satisfied. The managing director, who is directly involved in the project, expressed her determination to implement the project at the company-wide level. The project was widely noticed and taken on high priority. This requires all participants to dedicate their resources to the project. That policy would impact the annual performance of forces people to find ways to coordinate effectively with each other to ensure the success of the project. This case has a huge advantage that there is a long-term business strategy. Clear and long-term vision helps a lot for the BIS system. With a long term strategy, the report sets that users reporting for superior are still updated and can be applied into the BIS. Users have been familiar and skillful to analyze data despite there is BIS software system or not. Therefore, they know what they want and what they must report to the higher levels. This lead to an advantage for the technical team when they get requirements and advise users, they just benchmark the requirements to the ability of the system. The budget for the project is not abundant although when necessary it will be supported. The decision to choose Microsoft solutions shows that there was the consideration of the project budget. This also indicates that the selection of appropriate technical platform and capable of mastering the technology are more important than the budget more or less. On technical factors, this project also has an advantage that the data sources from the ERP system has been standardized, and the entire data sources are all controlled the MIS department. This helps to eliminate many issues (Table 6).

In the first case of partial success (case P1), there are some similarities with the case of success (case S) such as the project sponsor is also fierce in putting the project align with the project objectives, and the system uses data from the standardized application before such as ERP, EPM, and BPM. However, the differences could be the cause for the project half success. First, the project sponsor as finance controller could not push all the business stakeholders, including the management. Although this project also made policy of putting project results into the annual individual performance appraisement, this policy was just only applied to the finance and MIS department. This leads to the second difference compared with case S, the difficulty in handling the requirements. Although the stakeholders also got the benefits from the project but they did not really take the time for the project. To be able to involve, these people in the process of taking the requirements is really the most difficult work for this project such as arranging appointment, lack of knowing about the changing of business within the next few months. The reason given is that the nature of this business is always volatility and large-scale growth of the company. In fact, the enterprise did not have a stable business strategy in a year. A bright spot is that the system is designed according to the directions easy to open and flexible, so the changes can be implemented quickly.

Next, the unsuccessful case (case F) should be analyzed to see more visible the similarities and differences between case failures and two previous cases. In case F, most of the CSFs are ignored. Although the budget for the project is quite big to invest in technical platform with the purpose of building a more integrated system of data sources and contains a large amount of data, the skip CSFs made the project to a heavy defeat and cannot be improved (case P1), which forced company either continue to use the system with the current status or construct a new system. The fact that the project manager did not work hard on the project is an imperative factor leading to the failure of the project. Top management can only support the project if any raised up need help. In case S and case P1, along with the project managers, project sponsors always actively involved in most of the activities of the project, entirely different from case F. Besides, although the vision of this project was defined, it was not satisfied since the project manager did not stands out balance between technical and business, resulting in the system went towards technical direction and could not solve the business problems. Technical-oriented also leads to obtaining requirements only at the high-level and ignoring the level below. At the testing phase, the management said that they just care the overview numbers from their subordinates, who really need more detailed information to assess the situation of the product. Until then, the system had been built, and some changes could not be implemented. In addition, the inclusion of too much redundant data pushes the project to some other issues, such as: poor performance, costs for managing the data, making the data outdated or unnecessary for business.

In the second case of partial success (case P2), the CEO is the person who launched the project and appreciated the meaning of the project. Nevertheless, the CEO and top management had not yet really confident in the success of the project, so the project was launched in two phases: the exploring phase and the upgrading phase. Because the CEO thought that the first phase was just for exploration purpose, he involved into the project at the extent of demanding the relevant departments providing data sources. The concept of exploration had limited the vision and the business case in an unclear definition. The highlight of this project is the project manager, a person capable of in-depth technical but also have the mindset toward business. This helps the effective balance between business and technology solutions in the context of the limited budget. Different from the failed project (case F), this project focused on the requirements of both high-level and low-level. The project team consists of both technical and business side. The business analyst presented the requirements collecting from the business side to the technical side. Then, which data necessary to satisfy the requirements was analyzed. After that, the redundant data were removed to prevent the increase of data size. If the business requirements conflicted with technical capacity, the project manager would stand off balance by prioritizing business or technical depending on each stage. After the completed prototype, the business analyst guided and convinced users to use the prototype to give suggestions, from which the project team would make the necessary changes. The management of the data quality must be especially focused because the systems were integrated with each other without any standard.

6.2   Discussion and Implications

From above analysis, compared with the assessment in the study of [13], there are differences in the ranking of CSFs for BIS implementation as follows (Table 9):

Based on this finding, some differences in the cases studied in Vietnam are realized and some managerial implications are suggested as follows (Table 10):

Table 9. Comparison of CSFs ranking of this research and previous one

CSF	Ranking of this research	Ranking of [13]
Business-centric championship & balanced team composition	**1**	**3**
User-oriented change management	**2**	4
A clear vision and a well-established business case	**3**	**2**
Business-driven, scalable and flexible technical framework	3	6
Committed management support and sponsorship	4	**1**
Business-driven and iterative development approach	5	5
Sustainable Data Quality and Integrity	6	7

Note. In this table, the ranking of [13] by deducting from the content of the research.

Table 10. CSFs for BIS and implications for Vietnamese enterprises

CSFs	Rank	Implications
Business-centric championship and a balanced team composition	1	In Vietnam, this is the most significant weakness for BI system deployment success. Meanwhile, Vietnam is a huge shortage of people with the ability to deeply knowledgeable about the technology and mastery of the business. In fact, it is hard to rely on a consultant because no other one understood the business of enterprise by a person in that enterprise. One of solutions is that one with technical background should take part in the professional course as well as join the business activities of the company to gradually infuse the company's business
User-oriented change management	2	While the study of [13] considers this factor rating is lower than the elements of the perspective Organization, in this study, this factor was considered the second important factors affecting success. The reason is that if the process of defining the stakeholders for project do not do well in order to involve them at the first stage, the system will potentially in a risk causing the failure such as the users refuse to use, the design is not flexible enough to able adapt to changes or new requirements, costs for maintaining the outdated data are inflated since the demand cannot be maintained
A clear vision and a well-established business case	3	This study and the study by [13] are both appreciated the level of influence to the success of BIS implementation of this factor. If the organization does not obtain clear vision, making the business case will encounter many difficulties and lead to consequences such as business side does not support the project (case F), request to change too quickly leads to the system cannot satisfy (case P2, F). The results will be worse if the design of the system is not flexible and scalable enough (case F)

(Continued)

Table 10. (*Continued*)

CSFs	Rank	Implications
Business-driven, scalable and flexible technical framework	3	While research by [13] confirms the factors related to Technology is affecting the less to the success of BIS implementation, in the cases in Vietnam, this factors are ranked equivalent to factor "A clear vision and a well-established business case". The reason is that the technology must follow the right way from the beginning to adapt to meet business demands
Committed management support and sponsorship	4	While this factor is assessed in the study of [13], at level four, it is affecting the most to success BIS implementation in Vietnam. Deeper analysis in case P2, although the degree of support only to the extent and budget for the project was not plentiful, the project team can still rotate and find appropriate solutions. In case of success we noticed the budget for the project is also only in moderation, but the stipulation is that management is the determination and appreciates the importance of the project
Business-driven & iterative dev. approach	5	The approach is not highly appreciated in cases in Vietnam due to iterative deployment model has been developed for a long time, and many companies apply. No company would use the waterfall model for deploying BI
Sustainable Data Quality and Integrity	6	The project will certainly fail if it does not control well this factor. In other countries, the majority of systems are developed stable and data quality good while in Vietnam the majority of the applications are in-house developed with many issues related to the quality of the data. Furthermore, it is necessary to guarantee for the integrated system to be smoothly to prevent the problem such as the system does not retrieve data from the source systems, do not know whether the problem involved interrupted data, etc. To resolve this issue, it is necessary to aware of the importance of data management and use the appropriate tools for data quality assurance

6.3   Recommendations

In Vietnam, all the CSFs' elements were confirmed. Besides, there are 4 new elements emerged, including: involvement of top management, change management, performance considerations, business-led data governance. Some recommendations for ensuring the success of BI system in Vietnamese enterprises could be as follows:

- They should focus on high ranking factors for ensuring the success of BI project, such as: business-centric championship & a balanced team composition, user-oriented change management, and clear vision and well-established business case.
- Besides, some management aspects to be improved including:
  - Solving the culture problem that managers do not like to use the system or prefer to read the reports on paper => Top management makes policies to force people

in the company must show reports in the meeting from the BIS. This will create the habit to use the BIS every time.

- Solving the human resources problem: low analysis capabilities of the middle managers. => Middle managers must spend more time to get a really understand about their business; Top management frequently shares the business vision, strategies to the middle managers.
- Solving the problem of the quick changes in the business environment. => Top management frequently shares the business vision, strategies to the middle managers; Project manager should be the person who strong in technology and having time to work on the business side.
- Solving the problem of getting data from sources of other departments. => Top management should make policies to force departments; Project Manager should define the right stakeholders at the beginning of the project to involve them in the meeting having the top management.

6.4 Limitations and Implications for Future Research

The number of cases in this study is still a small number (4). Future studies should examine more cases for the more generalized results; especially focus on studying cases of failure would also add valuable insights. In addition, selection of cases in this study does not regard industry and size of the organization. The industry sector and the firm size of the organizations could have influenced results on the success of BIS implementation. Thus, future research should control for firm size and industry sector. For differences attributable to organizational resources, future research should extend and examine how other elements, such as: culture, organizational structure, people, and their skills, and routines interaction to ensure the success of BIS implementation. Besides, quantitative method could also be used for evaluating the impact of these factors on the success of BI projects in practice.

References

1. Rud, O.P.: Business Intelligence Success Factors: Tools for Aligning Your Business in the Global Economy. Wiley, Hoboken (2009)
2. Coker, F.: Pulse: Understanding the Vital Signs of Your Business, pp. 41–42. Ambient Light Publishing, Bellevue (2014)
3. Turban, E., McLean, E., Wetherbe, J.C.: Information Technology for Management: Making Connections for Strategic Advantage. Wiley, New York (2001)
4. Moss, L.T., Atre, S.: Business Intelligence Roadmap The Complete Project Lifecycle for Decision Support. Addison Wesley, Reading (2003)
5. Yeoh, W., Koronios, A.: Critical success factors for business intelligence system. J. Comput. Inf. Syst. **50**, 23–32 (2010)
6. Dawson, L., Belle, J.P.V.: Critical success factors for business intelligence in the South African financial services sector. J. Inf. Manag. **15**, 12 (2013)

7. Olszak, C.M., Ziemba, E.: Critical success factors for implementing business intelligence systems in small and medium enterprises on the example of Upper Silesia, Poland. Interdisc. J. Inf. Knowl. Manag. **7**, 129–150 (2012)
8. Sangar, A.B., AIahad, N.B.: Critical factors that affect the success of business intelligence systems (BIS) implementation in an organization. Int. J. Sci. Technol. Res. **12**, 176–180 (2013)
9. Yeoh, W., Koronios, A., Gao, J.: Managing the Implementation of Business Intelligence Systems: a Critical Success Factors Framework. IGI Global, Hershey (2008)
10. Yin, R.K.: Case study research: Design and methods. Sage, CA (2003)
11. Miles, M., Huberman, A.: Qualitative Data Analysis: An Expanded. Sage, London (1994)
12. Stuart, I., McCutcheon, D., Handfield, R., McLachlin, R., Samson, D.: Effective case research in operations management: a process. J. Oper. Manag. **20**, 419–433 (2002)
13. Yeoh, W., Popovic, A.: Extending the understanding of critical success factors for implementing business intelligence systems. J. Assoc. Inf. Sci. Technol. **67**, 134–147 (2015)
14. Inanga, E., Adelegan, O.: Principles of Accounting. HEBN, Ibadan, Nigeria (2013)
15. Turban, E., Sharda, R., Aronson, J., King, D.: Business Intelligence. Prentice Hall, New Jersey (2007)
16. Hwang, H.G., Ku, C.Y., Yen, D.C., Cheng, C.C.: Critical factors influencing the adoption of data warehouse technology: a study of the banking industry in Taiwan. Decis. Support Syst. **37**, 1–21 (2004)
17. Ariyachandra, T., Watson, H.: Which data warehouse architecture is most successful. Bus. Intell. J. **11**, 4 (2006)
18. Kwak, Y.H.: Critical Success Factors in International Development Project Management. Ohio, Cincinnati (2002)
19. Turban, E., McLean, E., Wetherbe, J.: Information Technology for Management: Improving Quality and Productivity. Wiley, New Jersey (2001)
20. Gartner: Predicts 2012: business intelligence still subject to non-technical challenges. Business report (2011)
21. Fortune, J., White, D.: Framing of project critical success factors by a systems mode. Int. J. Project Manag. **24**, 53–65 (2006)
22. Reddy, G.S., Srinivasu, R., Rikkula, S.R., Sreenivasarao, V.: Management information system to help managers for providing decision making in an organization. Int. J. Rev. Comput. (2009)
23. Ranjan, R.: Business intelligence: concepts, components, techniques and benefits. J. Theoret. Appl. Inf. Technol. **9**, 60–70 (2009)
24. Thomsen, E.: OLAP Solutions: Building Multidimensional Information Systems. Wiley, New York (2002)
25. Saunders, M., Lewis, P., Thornhill, A.: Research Methods for Business Students, 2nd edn. Pearson Education Limited, Essex (2000)
26. Daniel, D.R.: Management information crises. HBR **39**, 111–116 (1961)
27. Delone, W., McLean, E.: Information systems success: the quest for the dependent variable. J. Inf. Syst. Res. **3**, 60–95 (1992)

Particle Swarm Based Evolution and Generation of Test Data Using Mutation Testing

Nishtha Jatana[1], Bharti Suri[2], Sanjay Misra[3(✉)], Prateek Kumar[1], and Amit Roy Choudhury[4]

[1] Department of Computer Science and Engineering, MSIT, New Delhi, India
[2] USICT, GGS Indraprastha University, New Delhi, India
[3] Department of Computer and Information Sciences,
Covenant University, Ota, Nigeria
sanjay.misra@covenantuniversity.edu.ng
[4] Department of Computer Engineering, Delhi Technological University,
New Delhi, India

Abstract. Adequate test data generation is a vital task involved in the process software testing. Process of mutation testing, a fault-based testing technique, generates mutants of the program under test (PUT) by applying mutation operators. These mutants can assist in finding test cases that have the potential to detect faults in the PUT. Particle Swarm Optimisation (PSO) share similar working characteristics with Genetic Algorithm (GA) which has already been applied to test data generation using mutation testing. In this paper, applicability of PSO for the generation of test data with mutation testing is explored. The results obtained by empirical evaluation of the proposed approach on benchmark C programs are presented. The evaluated results show that the test cases generated from the technique proposed kills substantial number of mutants and therefore, has a scope of exploring its performance in the area of search based test case generation.

Keywords: Particle Swarm Optimization · Mutation testing · Genetic Algorithm · Software-artifact Infrastructure Repository

1 Introduction

Mutation testing is a fault-based testing technique that was originally proposed by DeMillo and Hamlet [1,2]. The underlying principle of mutation testing is to emulate the faults in software that a proficient programmer may make during the software development phase. Jia and Harman presented a detailed survey [3] on the analysis and developments in the field of mutation testing up to the year 2009. Mutation testing was initially proposed to measure the quality of the test cases that uses 'mutation score' for evaluating the effectiveness of the test suite. The researchers later began to use it as a technique for generation of test data [3]. A recent study [4] identified various approaches used by researchers for test data generation using mutation testing.

© Springer International Publishing Switzerland 2016
O. Gervasi et al. (Eds.): ICCSA 2016, Part V, LNCS 9790, pp. 585–594, 2016.
DOI: 10.1007/978-3-319-42092-9_44

Particle Swarm Optimization (PSO) is an evolutionary computation technique developed by Kennedy and Eberhart [5]. It was modeled on the herding, flocking or swarming behavior of animals. The candidate solutions in the population are represented as a swarm of particles in a multidimensional metric and real valued spaces. PSO tries to solve an optimization problem by iteratively working on the candidate solutions. It begins with a population of candidate solution (known as particles) and moves these particles in the search space by using mathematical formulae based on the particle's current position and velocity. The movement of each particle is influenced by the particular particle's best position and the globally best particle position found by movement in the search space. The original PSO was proposed for continuous search spaces. In order to apply PSO to problems that have a discrete search space, a new version of PSO, termed as Discrete PSO (DPSO) came into existence [6]. Since then, many methods of DPSO have been proposed and applied to a number of discrete problems [7].

Genetic Algorithm (GA) has been extensively applied to the generation of test cases using mutation testing and it exhibits similar solution exploration capabilities as shown by PSO [8]. However, PSO requires less computational efforts in comparison to GA [9]. This work therefore investigates the potential of PSO in search based test data generation using mutation testing.

This paper is divided into 6 sections. Section 2 summarizes the work done in the area of search based test data generation using mutation testing. Section 3 compares the underlying principle behind GA and PSO. Section 4 explains the proposed technique. Section 5 presents the results of the experimental evaluation of the technique. Section 6 concludes the paper with plans of future work.

2   Related Work

Genetic Algorithms have been applied to Mutation Testing for test data generation and evolution by various researchers. GA was used for generation of test data for branch testing and evaluated their test data using mutation testing to reveal its effectiveness [10]. Bottaci introduced a new fitness function [11] for GA which was based on three conditions (reachability, sufficiency and necessity) that are mandatory for a test case to kill the mutant. Baudry et al. used GA for improving the quality of test suite for object oriented software components using mutation testing [12].

May et al. [13] presented a comparative study between the Immune Inspired Algorithm(IIA) for evolution of test data and the elitist Genetic Algorithm for evolution of test data. The measure of effectiveness in both these methods is the number of mutants executed to achieve a given mutation score.

Masud et al. [14] proposed a model for generation of test data using mutation testing. This model divides the program under test and its mutants into program units and it selectively tries to find faults in these smaller program units using the principal of mutation testing. If any program unit survives, then the next population of test cases is evolved with the application of genetic algorithm iteratively.

Molinero et al. [15] presented a method to produce test suite for the testing of Finite State Machines (FSMs) using an approach that uses both genetic algorithm and mutation testing. This work proposed a methodology and named it 'GAMuT' that consisted of three phases. The first two phases dealt with the learning process through evolution and specialization and the last phase dealt with the selection and reduction of test cases.

Mishra et al. [16] used elitist GA for test data generation. A comparative evaluation of Genetic Algorithm and Bacteriological algorithm for improving test data using mutation testing was done by Rad et al. [32]. GA with mutation testing has also been applied to SQL [17]. Their work aims to find a reduced data set that is capable of detecting faults in SQL instructions. Other works that applied GA with mutation testing include [18–21].

Metaheuristic techniques (GA, IIA, BA and more) have been applied for search based test data generation using mutation testing. Applicability of PSO has not yet been explored in this area. Therefore, this work intends to apply PSO for search based test data generation using mutation testing.

3   Comparing and Contrasting GA and PSO

Genetic Algorithm is one of the well known and most explored meta-heuristic algorithm [9]. The history of GA for solving computational problems has been mentioned in "An Introduction to Genetic Algorithms" [22]. The particle swarm optimisation is swarm intelligence based stochastic optimization technique inspired from the movement of organisms in animal societies. PSO as proposed by Kennedy and Eberhart requires less memory and speed and hence is computationally inexpensive [5].

GA and PSO are paradigms based on evolutionary computation. These are nature inspired metaheuristic algorithms that have been used to solve optimization problems [23]. An insight into the working of GA and PSO shows that both the paradigms iteratively attempt to improve the solution space. The initial population (solution space) is evolved by application of operators to it until some termination criteria is met. Eberhart and Shi [8] compared these two evolutionary techniques, providing insights into the effect of operators used in each algorithm and their behavioural characteristics in the search space. They have shown the two techniques to be analogous in following ways:

- A particle in PSO is analogous to chromosome in GA
- Acceleration velocity in PSO is similar in behaviour to crossover operation in GA
- Elitism property is inherent in PSO whereas GA implements elitism by picking the best candidate solution for the next population
- Both PSO and GA conclude when a termination criteria is met

Eberhart and Shi [8] differentiated GA and PSO in the following aspects:

- PSO is more ergodic in nature than GA

– PSO has no operator that is analogous to selection operation in GA
– PSO does not use one of the core concepts of evolution that is survival of the fittest (which is used in GA)

GA and its variants hold high popularity both in industry and academia due to ease of implementation and its efficiency in solving complex engineering problems. PSO is claimed to have the same effectiveness as GA [9]. Hassan et al. [9] statistically analyze the efficiency of PSO over GA and state that PSO shows better computational efficiency on a set of benchmark programs used. The authors also state that GA is inherently discrete whereas PSO is inherently continuous and thereby needs modification for handling discrete problems.

Li et al. [24] compared PSO and GA by evaluating it on Schaffer function. The simulation results of their experiment depict that GA possess high global search ability with low computational efficiency, low speed and difficult/slow convergence. In case of PSO, the convergence speed is comparatively high and entails few parameters and possesses good global search ability owing to single directional information of particles. In case of GA, the sharing of the information of the chromosomes leads to an even movement of swarm to optimal region. In PSO, the particles memorize their past best position and also move towards the global best position. There is single directional flow of information due to this particle, which leads to the convergence of the swarm more quickly in most cases.

4   Particle Swarm Optimization Based Test Data Generation Using Mutation Testing

Section 3 depicts how PSO is similar and comparable to GA in its working and also exhibits better computational efficiency. GA has been widely used for optimized generation of test cases using mutation testing [11,13,14,25]. This work proposes a discrete form of PSO for optimized generation of test cases using mutation testing. Pseudo-code of PSO used in this work is as presented by Clerc and Kennedy in [26]. The pseudo-code of PSO is shown in Algorithm 1. Algorithm 1 has been used in this work for optimized generation of test data using mutation testing. The components of PSO are defined and its corresponding mapping to the proposed method (PSO-MT) is shown. The components of the original procedure implemented [27] in PSO are:

– Particles in hyperspace
– Number of dimensions in the hyperspace
– Definition of hyperspace (specification of lower and upper bounds for each dimension)
– The function f that evaluates the fitness of each particle in hyperspace
– The position and velocity vector for the particles in each dimension
– The population size (number of particles)
– Termination condition

Algorithm 1. Particle Swarm Optimization

1: **while** Termination Criterion not met **do**
2:      Initialize Population
3:      $i \leftarrow 1$
4:      **for** $i <= PopulationSize$ **do**
5:          **if** $f(\vec{x_i}) < f(\vec{p_i})$ **then**
6:              $\vec{p_i} \leftarrow \vec{x_i}$
7:          **end if**
8:          $\vec{p_g} \leftarrow min(\overrightarrow{p_{neighbors}})$
9:          $d \leftarrow 1$
10:         **for** $d <= Dimensions$ **do**
11:             $\vec{v_{id}} \leftarrow \vec{v_{id}} + \varphi_1 * (\vec{p_{id}} - \vec{x_{id}}) + \varphi_2 * (\vec{p_{gd}} - \vec{x_{id}})$
12:             $\vec{x_{id}} \leftarrow \vec{x_{id}} + \vec{v_{id}}$
13:         **end for**
14:     **end for**
15: **end while**

The above defined system has been mapped to the procedure for test data generation proposed and implemented(PSO-MT) in this work. Table 1 shows the mapping of the components of PSO to PSO-MT.

Figure 1 depicts the procedure of the proposed technique for optimised generation of test data. Initially, the mutants are generated. After this the initial population of the test cases is generated. These test cases are executed on the program under test (PUT) and the mutants. This execution is used for evaluation of the fitness of the test cases. The velocity and the position of the parti-

Table 1. Mapping components of PSO to PSO-MT

Components of PSO	Corresponding components of PSO-MT
Particles in hyperspace	Initial test cases that are randomly selected points in the search space
Number of dimensions in the hyperspace	Number of inputs of the program under test or the number of values in a test case
Definition of hyperspace	Defining upper bound and lower bound of each dimension, thus confining the search space
Fitness function	It returns the number of mutants killed by a particular test case
Position and velocity of particles	Position of the test case in the hyperspace. Velocity vector in each dimension is the rate of change of position in that dimension with respect to time
Population size	Number of test cases in initial test suite
Termination condition	Termination condition is attained when no new mutant is killed for a fixed number of iterations. This number is obtained empirically

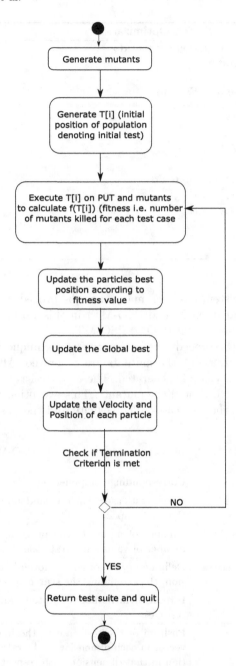

Fig. 1. Procedure of PSO-MT

cles are updated according to the particle's best position and the globally best particle position. The test cases are further evaluated for fitness. This sequence of operations continue till the termination criteria is met.

5   Results

Evaluation of the proposed PSO-MT approach for optimized test case generation, on C language programs which have been popularly used in the field of
mutation testing are presented in this section. The first two programs are the
Triangle program and Quadratic program which are one of the popular programs used in empirical study in the field of mutation testing [3]. The third
program that is used is Schedule program from the Software-artifact Infrastructure Repository (SIR) [28]. The triangle program takes the three sides of the
triangle as input and generates an output that indicates the triangle type. The
quadratic program tells us the nature of the roots of a quadratic equation. The
schedule program is a priority scheduler. Schedule is a Siemens program that
was assembled by Tom Ostrand and colleagues at Siemens Corporate Research
[29]. First order mutants of the benchmark programs are generated using Milu
[30]. The mutation operators used for mutant generation are the same as used
by MILU. MILU uses the 77 C mutation operators by as stated in the technical
report by Agrawal et al. [31]. The PSO-MT algorithm has been executed 50
times on benchmark programs. The average of the mutants killed corresponding
to the multiple executions is as presented in the Table 2.

Table 2. Benchmarks obtained on programs used

Program	LOC	Number of mutants	Number of mutants killed	Number of parti-cles generated
Triangle	50	121	102	350
Quadratic	30	71	67	849
Schedule	412	79	64	508

To further analyze the performance of PSO-MT, other performance parameters have been defined and calculated using the iterative model used in PSO-MT.
These parameters are: the iteration period (T), iteration size (P) and the particle
period (K). The definition of the terms is as follows:

Iteration Period (T): The iteration period is defined as the time it takes for
the complete execution of one iteration that is proportional to the iteration
size.

Iteration size (P): Iteration size is the number of particles that are generated
after evolution at each iteration. This is the initial population size.

Particle Period (K): The particle period can be defined as the ratio of the
iteration period to the iteration size. It is the time it takes to generate a
particle and compute its mutation score from the fitness function. The value
of K is dependent on the system processor speed. The value of K can be
improved by the use of an optimized model for test case evaluation and
also with the help of optimized computation with the help of techniques like

parallel processing. The value of K is also an indicator of the complexity of the software or program under test.

$$K = T/P \tag{1}$$

The results shown in Tables 2 and 3 have been calculated on an Intel Core i7-3610QM 2.3 GHz processor. The value of K for the benchmark programs on the above said processor is shown in Table 3.

Table 3. Other performance parameters

Program	Iteration period (T)	Iteration size (P)	$K = T/P$
Triangle	31640 ms	15	2109.33
Quadratic	16268 ms	15	1084.33
Schedule	30683 ms	15	2045.53

6 Conclusion and Future Work

In this work, we have proposed an approach for optimized generation of test data based on PSO coupled with concepts of mutation testing. Other evolutionary approaches (like GA, ACO, IIA and so on) have also been applied for generation of test data using mutation testing. PSO-MT is inspired by Kennedy and Clerc's work on particle swarm optimisation and thus, the proposed algorithm imitates the behavior of social interaction for optimized evolutionary generation of test data. Results of the empirical evaluation of the proposed approach on Triangle program, Quadratic program and Schedule program show encouraging results.

As a part of our future work, we will apply PSO-MT to larger programs in order to further evaluate its effectiveness and compare its performance with other evolutionary algorithms combined with mutation testing.

References

1. DeMillo, R.A., Lipton, R.J., Sayward, F.G.: Hints on test data selection: help for the practicing programmer. Computer **11**, 34–41 (1978)
2. Hamlet, R.G.: Testing programs with the aid of a compiler. IEEE Trans. Softw. Eng. **3**(4), 279–290 (1977)
3. Jia, Y., Harman, M.: An analysis and survey of the development of mutation testing. IEEE Trans. Softw. Eng. **37**(5), 649–678 (2010)
4. Souza, F.C., Papadakis, M., Durelli, V.H., Delamaro, M.E.: Test data generation techniques for mutation testing: a systematic mapping. In: Proceedings of the 11th IESELAW, pp. 1–14 (2014)

5. Kennedy, J., Eberhart, R.: Particle swarm optimization. In: IEEE International Conference on Neural Network, vol. 4, pp. 1942–1948. IEEE (1995)
6. Kennedy, J., Eberhart, R.C.: A discrete binary version of the particle swarm algorithm. In: IEEE International Conference on Systems, Man, and Cybernetics, vol. 5, pp. 4104–4108. IEEE (1997)
7. Jordehi, A.R., Jasni, J.: Particle swarm optimisation for discrete optimisation problems: a review. Artif. Intell. Rev. **43**(2), 243–258 (2015)
8. Eberhart, R.C., Shi, Y.: Comparison between genetic algorithms and particle swarm optimization. In: Porto, V.W., Waagen, D. (eds.) EP 1998. LNCS, vol. 1447, pp. 611–616. Springer, Heidelberg (1998)
9. Hassan, R., Cohanim, B., de Weck, O., Venter, G.: A comparison of particle swarm optimization and the genetic algorithm. In: 46th AIAA/ASME/ASCE/AHS/ASC Structures, Structural Dynamics and Materials Conference (2004)
10. Jones, B.F., Eyres, D.E., Sthamer, H.H.: Strategy for using genetic algorithms to automate branch and fault-based testing. Comput. J. **41**, 98–107 (1998)
11. Bottaci, L.: A genetic algorithm fitness function for mutation testing. In: 8th Wrokshop on Software Engineering Using Metaheuristic Innovative Algorithm (SEMINAL 2001), pp. 3–7 (2001)
12. Baudry, B., Hanh, V.L., Jzquel, J.-M., Traon, Y.L.: Trustable components: yet another mutation-based approach. In: Wong, W.E. (ed.) Mutation Testing for the New Century, vol. 24, pp. 47–54. Springer, New York (2001)
13. May, P., Timmis, J., Mander, K.: Immune and evolutionary approaches to software mutation testing. In: de Castro, L.N., Von Zuben, F.J., Knidel, H. (eds.) ICARIS 2007. LNCS, vol. 4628, pp. 336–347. Springer, Heidelberg (2007)
14. Masud, M.M., Nayak, A., Zaman, M., Bansal, N.: A strategy for mutation testing using genetic algorithms. In: Canadian Conference on Electrical and Computer Engineering, pp. 1049–1052. IEEE (2005)
15. Molinero, C., Núñez, M., Andrés, C.: Combining genetic algorithms and mutation testing to generate test sequences. In: Cabestany, J., Sandoval, F., Prieto, A., Corchado, J.M. (eds.) IWANN 2009, Part I. LNCS, vol. 5517, pp. 343–350. Springer, Heidelberg (2009)
16. Mishra, K.K., Tiwari, S., Kumar, A., Misra, A.K.: An approach for mutation testing using elitist genetic algorithm. In: 3rd IEEE International Conference on Computer Science and Information Technology (ICCSIT), vol. 5, pp. 426–429. IEEE (2010)
17. Moncao, A.C., Camilo-Junior, C.G., Queiroz, L.T., Rodrigues, C.L., Leitao-Junior, P.d., Vincenzi, A.: Shrinking a database to perform SQL mutation tests using an evolutionary algorithm. In: IEEE Congress on Evolutionary Computation (CEC), pp. 2533–2539. IEEE (2013)
18. Fraser, G., Zeller, A.: Mutation-driven generation of unit tests and oracles. IEEE Trans. Softw. Eng. **38**(2), 278–292 (2012)
19. Haga, H., Suehiro, A.: Automatic test case generation based on genetic algorithm and mutation analysis. In: IEEE International Conference on Control System, Computing and Engineering (ICCSCE), pp. 119–123. IEEE (2012)
20. Bashir, M.B., Nadeem, A.: A fitness function for evolutionary mutation testing of object-oriented programs. In: IEEE 9th International Conference on Emerging Technologies (ICET), pp. 1–6. IEEE (2013)
21. Rad, M.F., Bahrekazemi, S.: Applying genetic evolutionary, bacteriological and quantum evolutionary algorithm for improving performance optimization segment of test data sets in mutation testing method. Int. J. Soft Comput. Softw. Eng. (JSCSE) 167–186 (2014)

22. Mitchell, M.: An Introduction to Genetic Algorithms. MIT Press, Cambridge (1996)
23. Andalib Sahnehsaraei, M., Mahmoodabadi, M.J., Taherkhorsandi, M., Castillo-Villar, K.K., Mortazavi Yazdi, S.M.: A hybrid global optimization algorithm: particle swarm optimization in association with a genetic algorithm. In: Zhu, Q., Azar, A.T. (eds.) Complex System Modelling and Control Through Intelligent Soft Computations. Studies in Fuzziness and Soft Computing, vol. 319, pp. 45–86. Springer, Switzerland (2014)
24. Li, Z., Liu, X., Duan, X.: Comparative research on particle swarm optimization and genetic. Comput. Inf. Sci. $3(1)$, 120–127 (2010)
25. Baudry, B., Fleurey, F., Jzquel, J.-M., Traon, Y.L.: From genetic to bacteriological algorithms for mutation-based testing. In: International Symposium on Software Reliability, pp. 73–96. Wiley (2005)
26. Clerc, M., Kennedy, J.: The particle swarm explosion, stability, and convergence in a multidimensional complex space. IEEE Trans. Evol. Comput. $6(1)$, 58–73 (2002)
27. Shi, Y.: Feature article on particle swarm optimization. In: IEEE Neural Network Society, p. 813 (2004)
28. Do, H., Elbaum, S.G., Rothermel, G.: Supporting controlled experimentation with testing techniques: an infrastructure and its potential impact. Empir. Softw. Eng. Int. J. $10(4)$, 405–435 (2005)
29. Hutchins, M., Foster, H., Goradia, T., Ostrand, T.: Experiments on the effectiveness of dataflow- and control flow-based test adequacy criteria. In: Proceedings of the 16th International Conference on Software Engineering, pp. 191–200 (1994)
30. Jia, Y., Harman, M.: MILU: a customizable, runtime-optimized higher order mutation testing tool for the full C language. In: Testing: Academic and Industrial Conference Practice and Research Techniques (TAIC PART 2008), Windsor, pp. 94–98. IEEE (2008)
31. Agarwal, H., Demillo, R., Hathaway, R., Hsu, W., Krauser, E., Martin, R.: Design of mutant operators for the C programming language. Technical report, March 1989
32. Rad, M.F., Akbari, F., Bakht, A.J.: Implementation of common genetic and bacteriological algorithms in optimizing testing data in mutation testing. In: International Conference on Computational Intelligence and Software Engineering (CiSE). IEEE (2010)

Business Modeling and Requirements in RUP: A Dependency Analysis of Activities, Tasks and Work Products

Carina Campos[1], José Eduardo Fernandes[2(✉)],
and Ricardo J. Machado[3]

[1] Dept. de Sistemas de Informação, Universidade do Minho,
Guimarães, Portugal
carina.campos13@gmail.com
[2] Polytechnic Institute of Bragança, Bragança, Portugal
jef@ipb.pt
[3] Centro ALGORITMI, Escola de Engenharia,
Universidade do Minho, Guimarães, Portugal
rmac@dsi.uminho.pt

Abstract. Most artifacts developed during the requirements engineering process relate themselves in different ways. In order to understand in detail how they affect each other during the software development process, it is relevant to identify their interdependencies. This paper presents a systematization of the existing interdependencies between the different elements of the Rational Unified Process (RUP) in the Business Modeling and Requirements disciplines. This work, which highlights knowledge about the different interdependencies and traceability of RUP elements, is useful to avoid unconscious decisions during software the development process and also, to detect potential problems due to the violation of the existing interdependencies.

Keywords: Requirements engineering · Interdependencies · Traceability · Artifacts · RUP

1 Introduction

Software systems, one of the most complicated things developed by humankind, became an essential part of our everyday lives and from which we are completely dependent on. Their existence is so pervasive that, nowadays, society expects them to assist us either in critical activities or in our everyday activities, to have high quality, to provide exciting functionalities, to be reliable, and to be produced at low cost.

Given the increasing demand and complexity of software systems, the software engineering discipline has accumulated, over the last years, an extensive scientific body of knowledge related to theories, methods, approaches, and tools needed to construct software systems [1]. To cope with the growing complexity and diversity of engineering problems, the adoption of systematic and disciplined approaches to deal with requirements and their problems has become crucial [1]. The study and research about requirements is extensive in literature, covering diverse areas, such as the design of

© Springer International Publishing Switzerland 2016
O. Gervasi et al. (Eds.): ICCSA 2016, Part V, LNCS 9790, pp. 595–607, 2016.
DOI: 10.1007/978-3-319-42092-9_45

requirements modeling languages [2] or model transformation techniques and code generation [3]. Recently, Fernandes and Machado [1] addressed the essence, issues, and techniques for requirements in engineering projects. Requirements engineering, in the context of the development of a system through an engineering project, embodies a set of activities that permits eliciting, negotiating, and documenting the functionalities and the restrictions of that system [1]. As an engineering discipline, is closely related to the concept of project; it is throughout the project that the engineer applies his technical and scientific knowledge, to solve the problems and to achieve the objectives which he is confronted with [1]. In turn, the concept of project is closely related to the concept of process.

Most individual requirements developed during the requirements engineering process relate to and affect each other in different ways and thus cannot be treated in isolation [4, 5]. The fact that the requirements relate to and affect each other makes it necessary to identify and manage the requirements interdependencies in order to avoid potentially costly mistakes during the system development.

Requirements interdependencies are not a problem by themselves, but they influence the number of development activities and decisions made during the software engineering process [6]. Traceability is the basis for studying the requirements interdependencies during the development process [7] since it allows identifying and justifying the artifacts that implement the requirements initially formalized.

Software development produces various kinds of artifacts. The artifacts, such as requirements, do not exist in isolation; instead they relate to and affect each other [8]. During the development of solutions and also during the exploration phase for maintenance issues, frequently arises the need to introduce several changes to the project decisions previously established. These changes should be clearly identified to ensure the complete identification of the artifacts involved in the changes. To this end, it is necessary to have knowledge about how the different artifacts relate among them since it facilitates the identification of the artifacts affected.

RUP is a process that provides the best practices and guidelines for successful software development [9]. This work, in the context of Business Modeling and Requirements disciplines of RUP, analyzes and systematizes the traceability and the interdependencies that may occur between the various elements during software development projects.

This paper has the following structure: Sect. 2 presents the importance of dealing with the interdependencies and the traceability during the software development; Sect. 3 describes the interdependencies and the traceability between the different elements of the RUP; Sect. 4 presents the conclusions.

2 Interdependencies and Traceability

Requirements traceability is an issue that for long time is investigated and discussed, such as in [10]. Dahlstedt and Persson [7] refer to traceability as "a basis for addressing the requirements interdependencies". According to Genvigir [11] and Zou et al. [12], traceability is intimately associated to the software production process, specifically to the requirements and to the ability to establish links between these requirements and other artifacts that satisfy them.

Sánchez et al. [13] mention that the requirements traceability aims to help determine the impact of changes in the conception phase of software, to support their integration, preserve the knowledge and assure the quality and correction of the global system.

Requirements traceability is as a quality factor [6, 14–16]. Actively supporting traceability in a software development project can help ensuring other qualities of software, such as adequacy and understandability [15].

On the other hand, neglecting the traceability can lead to less maintainable software and to failures due to inconsistencies and omissions [15]. Dömges and Pohl [17] refer to neglecting the traceability or capture insufficient and/or unstructured traces leads to decrease in system quality, causes revisions, and hence, increases project costs and time.

Aizenbud-Reshef et al. [18] refer that, from the perspective of requirements management, traceability facilitates the interconnection of requirements to their origins and reasons. Additionally, it allows capturing the information needed for understanding the evolution of requirements and for verification of requirements fulfillment. Complete traceability allows calculate more accurately the costs, as well as to determine lists of changes, without depending on the programmer knowledge of all the areas that these changes affect [18]. All these reasons make crucial to implement traceability practices throughout the software development.

It is essential to identify and manage the interdependencies that occur throughout the system development in order to, if needed, in any context, to properly consider related artifacts and as such, to avoid potentially costly mistakes by neglecting either those relations or eventually relevant artifacts As mentioned earlier, through traceability, it is possible to manage these interdependencies; hence, traceability is fundamental to the development process.

Several works further develop traceability practices and theories [19, 20]. Marques et al. propose a traceability representation language [21] that provides "abstractions to requirements, artifacts and trace links as well as queries, through which trace links can be searched, retrieved and filtered". In addition, they also propose a requirement traceability process [22] specifying its workflow, actors, responsibilities and inputs/outputs. Rempel and Mäder [23] review the elements involved in establishing traceability in a development project and derive a quality model for the systematic assessment of requirements traceability. To facilitate traceability in the model centric paradigm, Badreddin and Sturm [24] call for representing requirements as first class entities aiming to enable software developers, modelers, and business analysts to manipulate requirements entities as textual model and code elements. In this context, they propose a Requirement-Oriented Modeling and Programming Language (ROMPL). Soonsongtanee and Limpiyakorn [25] present an approach to enhancing the requirements traceability matrix with UML state diagrams to describe the traceability states of associated requirements or life-cycle work products. Regarding requirements interdependencies, Berrocal et al. [26] present a set of profiles to allow designers to explicitly model interdependencies between elements in BPMN 2 and UML 2 Use Case diagrams. They also define ATL transformations to automatically derive these relationships from the business specification to the requirements models.

The purpose of dealing, systematically, with requirements interdependencies improves the decisions made during software development as well as to detect the potential problems that may arise because of the requirements interdependencies [6]. Managing requirements interdependencies consists in identify, store and maintain information about how the requirements relate to and affect each other [6].

Maintain traceability of the requirements interdependencies is essential in order to support various situations and activities in the system development process [7]. Traceability should be included and treated along the development projects, thus representing, an asset to their success. Knowing the whole story of the artifacts, as well as their interdependencies, will enable easier identification and management of existing interdependencies from the early stages of development. Therefore, this knowledge minimizes problems that may arise during the software development process.

3   Interdependencies and Traceability in RUP

RUP aims to ensure the production of quality software that meets the needs and expectations of its users in a predictable schedule and cost [9]. RUP guidelines entail several elements such as activities, tasks, roles and work products. Throughout the development process, at several moments, RUP elements become interconnected; by this way, a simple change in an element causes various subsequent adjustments in others. Therefore, the knowledge of existing interdependencies between the various elements is particularly useful since it allows easier identification of elements affected during a change.

As Dahlstedt and Persson [7] refer, is essential to maintain the traceability of interdependencies since it allows to know, in detail, how the elements relate, as well as to support various situations and activities in the software development. Through the traceability of various elements of RUP, it is possible to easier identify and manage the interdependencies that may occur between elements. For those practitioners that adopt RUP guidelines, it is useful to understand the interdependencies that may exist between the various RUP elements.

3.1   Dependency Analysis of Activities and Tasks

RUP is organized in various disciplines and phases. However, the study mentioned in this paper focuses in two transitions (see Fig. 1): (1) from the Business Modeling discipline to the Requirements discipline, at Inception phase; (2) from the Inception to the Elaboration phase, within the Requirements discipline.

To facilitate an overview and analysis of all tasks of RUP (for Business Modeling and Requirements disciplines), the conduction of an initial RUP review allowed the construction of information presented in Tables 1 and 2.

These tables show the different tasks of the disciplines of Business Modeling and Requirements, the activities associated with these tasks, the phase where they are performed, and the roles responsible for them. Table 1 details the activities, tasks,

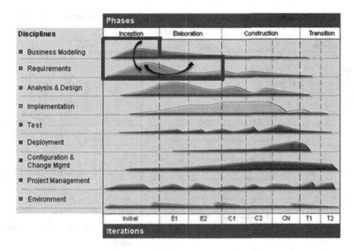

Fig. 1. Positioning of the study in the RUP (Based on [27]) (Color figure online)

Table 1. Activities, tasks, phases and roles of the business modeling discipline.

	Activities					Tasks	Phases				Role main
	B1 Assess Business Status αβ	**B2** Describe Current Business β	**B3** Define Business β	**B4** Explore Process Automation β	**B5** Develop Domain Model α		Inception	Elaboration	Construction	Transition	
	x					Assess Target Organization - ATO	B1				Business-Process Analyst
	x	x	x		x	Business Architectural Analysis - BAA	B1, B2, B5	B2, B3	B3	B3	Business Architect
			x			Business Operation Analysis		B3			Business Designer
			x			Business Operation Design		B3	B3	B3	Business Designer
			x			Business Use-Case Analysis		B3	B3	B3	Business Designer
	x	x	x		x	Capture a Common Business Vocabulary - CCBV	B1, B2, B5	B2, B3	B3		Business-Process Analyst
				x		Construct Business Architectural Proof-of-Concept		B4			Business Architect
				x		Define Automation Requirements		B4	B4	B4	Business Designer
			x			Define Business System Context		B3			Business-Process Analyst
			x		x	Detail a Business Entity - DBE	B5		B3	B3	Business Designer
			x			Detail a Business Use Case			B3		Business Designer
			x			Detail a Business Worker			B3	B3	Business Designer
		x	x			Find Business Actors and Use Cases - FBAUC	B2	B2, B3	B3		Business-Process Analyst
	x	x	x			Identify Business Goals - IBG	B1, B2	B3			Business-Process Analyst
	x	x	x		x	Maintain Business Rules - MBR	B1, B2, B5	B2, B3	B3	B3	Business-Process Analyst
			x			Prioritize Business Use Cases		B3	B3		Business Architect
			x		x	Review the Business Analysis Model - RBAM	B5	B3	B3	B3	Technical Reviewer
			x			Review the Business Use-Case Model		B3	B3		Technical Reviewer
	x	x	x			Set and Adjust Objectives - SAO	B1, B2	B3			Business-Process Analyst
			x			Structure the Business Use-Case Model		B3	B3		Business-Process Analyst

(left vertical label: Business Modeling)

	Activities
Classic RUP Lifecycle	α
Business Modeling Lifecycle	β
Classic RUP Lifecycle + Business Modeling Lifecycle	αβ

phases, and roles of the Business Modeling discipline considering both processes of Classic RUP Lifecycle and Business Modeling Lifecycle.

The column Activities presents the five activities performed in this discipline. The activities performed in the Classic RUP Lifecycle process are signaled in the table by an α, the activities performed in the Business Modeling Lifecycle process are signaled by an β and the activities performed in both processes are signaled in the table by αβ.

Table 2. Activities, tasks, phases and roles of the Requirements discipline.

Activities						Tasks	Phases				Role main
R1 Analyze the Problem	R2 Understand Stakeholder	R3 Define the System	R4 Manage the Scope of the System	R5 Refine the System	R6 Manage Changing Requirements	Requirements	Inception	Elaboration	Construction	Transition	
αβ	αβ	α	α	α	α						
x	x	x				Capture a Common Vocabulary - CCV	R1α, R2α, R3			R1β, R2β	System Analyst
				x		Detail a Use Case - DUC		R5			Requirements Specifier
				x		Detail the Software Requirements - DSR		R5			Requirements Specifier
x						Develop Requirements Management Plan - DRMP	R1α				System Analyst
	x	x		x		Develop Supplementary Specifications - DSS	R2α, R3	R5		R2β	System Analyst
x	x	x	x			Develop Vision - DV	R1α, R2α, R3, R4			R1β, R2β	System Analyst
	x					Elicit Stakeholder Requests - ESR	R2α			R2β	System Analyst
x	x	x				Find Actors and Use Cases - FAUC	R1α, R2α, R3			R1β, R2β	System Analyst
	x	x	x		x	Manage Dependencies - MDep	R2α, R3, R4	R6	R6	R6	System Analyst
			x			Prioritize Use Cases - PUC	R4				Software Architect
					x	Review Requirements - RReq		R6	R6	R6	Technical Reviewer
					x	Structure the Use-Case Model - SUCM		R6	R6	R6	System Analyst

	Activities/ Phases
Classic RUP Lifecycle	α
Business Modeling Lifecycle	β
Classic RUP Lifecycle + Business Modeling Lifecycle	αβ

Column Tasks exposes all tasks practiced in the Business Modeling discipline. Only the tasks with a gray background were studied, since the other stand in phases that are outside the scope of our study. The intersection of column Activities with the lines of column Tasks indicates (through an 'x') the tasks included in the activities.

Column Phases presents which phases include the different tasks and activities. The abbreviations B1, B2, B3, B4 and B5 (for the various activities) associate tasks and their activities to the several phases. Column Role main refers which are the roles responsible for performing the different tasks. The activities with blue background (Assess Business Status, Describe Current Business and Develop Domain Model) and the phase (Inception) refer to the activities and the phase studied in Business Modeling discipline.

Table 2 presents the activities, tasks, phases, and roles of the Requirements discipline. Column Activities presents the six activities practiced in this discipline. As before, in the table, α signals activities performed in the Classic RUP Lifecycle process, β signals activities executed in the Business Modeling Lifecycle process, and αβ signals the activities performed in both processes.

Column Tasks exposes all tasks practiced in the Requirements discipline. In this discipline, all tasks have a gray background since they are in the phases of the scope of this study and as such, covered by this study. An 'x' at the intersection of column Activities with the lines of column Tasks indicates the tasks practiced in the activities.

Column Phases presents tasks and activities performed in the different phases. Abbreviations R1, R2, R3, R4, R5 and R6 (for the various activities) associate the tasks and their activities to phases where they are performed. In the intersections, we use α for Classic RUP Lifecycle, β for Business Modeling Lifecycle and αβ for both processes. The intersections show the process where the tasks and their activities are performed.

As in the previous table, the column Role main refers to the roles responsible for performing the different tasks.

The activities with blue background (Analyze the Problem, Understand Stakeholder Needs, Define the System, Manage the Scope of the System, Refine the System Definition and Manage Changing Requirements) and the phases (Inception and Elaboration) refer to the activities and the phases studied in Requirements discipline.

The information provided in these tables is useful throughout the software development because it allows to perceive how activities, tasks, and roles relate in a particular discipline and phase.

3.2   Dependency Analysis of Work Products and Tasks

The elaboration of the previous two tables allowed to perceive the tasks and activities covered in the disciplines and phases considered in this study. Tables 3 and 4 have the purpose of clarifying the interconnection of all work products of both disciplines to their respective tasks.

The first column of Table 3 shows all the work products of the Business Modeling discipline and the second column presents all the tasks. Only the tasks with a gray background were analyzed because the other tasks are in phases that are not within the scope our study.

The intersection of these two columns depicts the work products consumed and produced in the various tasks. These intersections use the terms IN, OUT and I/O: the term IN is used to refer work products consumed by the associated task; the term OUT represents the work products produced by the task in question; the term I/O represents that the work products are both consumed and produced by the task in question. Besides these terms, the term IN* refers to work products that are an optional entry of the associated task; these work products are not necessarily consumed in the task. In the tables, the use of colors facilitate the identification of terms IN, OUT, and I/O: the term IN is represented by the green color, the term OUT by the red color and the term I/O by the yellow color.

The first column, Table 4 shows all the work products of the Requirements discipline and in the second column presents all the tasks. All the tasks of this discipline were analyzed because all the tasks are in phases that are within the scope our study. The intersection of these two columns depicts which work products are consumed and produced in the various associated tasks. These intersections use the terms IN, OUT and I/O, which were previously defined.

Tables 3 and 4 show the work products produced and consumed by the different tasks. The information available in these tables allows the identification of existing interdependencies between tasks and work products that are produced and consumed.

Figures 2 and 3 present two graphical representations that were developed to enhance the perception of the information contained in the previous tables; i.e., all the existing interdependencies between the activities and the tasks and work products of a given phase and discipline. This visualization facilitates the analysis of the existing interdependencies along the development process, thus allowing for a better understanding and management.

Table 3. Work products of the tasks of the Business Modeling discipline.

Discipline / Work Products \ Tasks	Assess Target Organization - ATO	Business Architectural Analysis - BAA	Business Operation Analysis	Business Operation Design	Business Use-Case Analysis	Capture a Common Business Vocabulary - CCBV	Construct Business Architectural Proof-of-Concept	Define Automation Requirements	Define Business System Context	Detail a Business Entity - DBE	Detail a Business Use Case	Detail a Business Worker	Find Business Actors and Use Cases - FBAUC	Identify Business Goals - IBG	Maintain Business Rules - MBR	Prioritize Business Use Cases	Review the Business Analysis Model - RBAM	Review the Business Use-Case Model	Set and Adjust Objectives - SAO	Structure the Business Use-Case Model
Business Analysis Model - BAM		OUT	I/O	IN	OUT				I/O								IN			
Business Architectural Proof-of-Concept - BAP-C		IN *			IN *		OUT													
Business Architecture Document - BAD		OUT			IN		IN									I/O				
Business Deployment Model - BDM		OUT		OUT			IN													
Business Design Model - BDesM		OUT		OUT			IN													
Business Glossary - BGl						OUT		IN												
Business Goal - BG													IN	OUT						
Business Rule - BR															OUT					
Business Use Case Model - BUCM			I/O	IN	IN				I/O				OUT			I/O		IN		I/O
Business Vision - BV		IN					IN						IN	IN	IN	IN			OUT	
Supplementary Business Specification - SBS							IN		IN				OUT	OUT						
Target-Organization Assessment - TOA	OUT								IN								IN			

OUT	Output
IN	Input
I/O	Input/Output
IN *	The work product is an optional input of the associated task

Table 4. Work products of the tasks of the Requirements discipline.

Disciplines / Work Products \ Tasks	Capture a Common Vocabulary - CCV	Detail a Use Case - DUC	Detail the Software Requirements - DSR	Develop Requirements Management Plan - DRMP	Develop Supplementary Specifications - DSS	Develop Vision - DV	Elicit Stakeholder Requests - ESR	Find Actors and Use Cases - FAUC	Manage Dependencies - MDep	Prioritize Use Cases - PUC	Review Requirements - RReq	Structure the Use-Case Model - SUCM
Glossary - Gl	OUT											OUT
Requirements Attributes - ReqA		OUT	OUT			OUT	OUT	OUT	OUT	OUT		OUT
Requirements Management Plan - RMP				OUT					I/O			
Software Requirement - SofR			OUT							OUT	IN	
Software Requirements Specification - SRS			OUT									
Stakeholder Requests - SR					IN	IN	OUT	IN				
Storyboard - St						OUT						
Supplementary Specifications - SS					OUT							OUT
Use-Case Model - UCM								OUT			IN	I/O
Vision - VI			IN			OUT				OUT		

OUT	Output
IN	Input
I/O	Input/Output
IN *	The work product is an optional input of the associated task

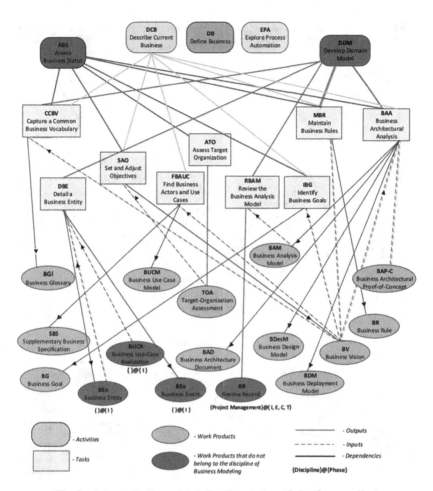

Fig. 2. Scheme Business Modeling@Inception (Color figure online)

The representation of the Tables 1 and 3. This representation refers to the Business Modeling discipline in the Inception phase. It presents the five activities belonging to this discipline. These activities interconnect to their tasks; two of these activities have no associated tasks because they are outside the scope of this study.

Each task has its associated work products. These work products may be consumed in the associated task (inputs, graphically represented by arrow green) or may be produced by that task (outputs, graphically represented by arrow red). The work products represented in yellow refers to work products belonging to the Business Modeling discipline.

The work products, represented in orange, despite being work products produced and consumed in this discipline|phase, do not belong directly to work products defined by RUP for this discipline. For these work products (in orange), a description below

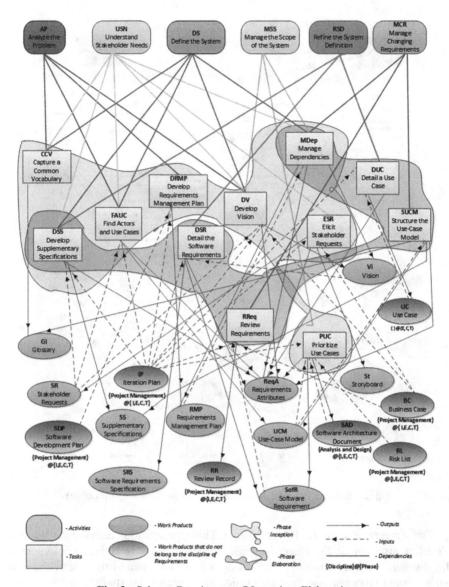

Fig. 3. Scheme Requirements@Inception, Elaboration

them indicates the discipline and the phase to which they belong; some of those do not have associated discipline because, in concrete, they do not belong to any.

The representation of the Fig. 2 is based on information gathered in the Tables 2 and 4. This representation refers to the Requirements discipline in the Inception and Elaboration phases. It presents the six activities belonging to this discipline, as well as its interconnected tasks. All these activities have associated tasks because all of them are within the scope of this study. Each task has its associated work products. These work products may be consumed in the associated task (inputs, graphically represented by arrow green) or may be produced by that same task (outputs, graphically represented by arrow red). The work products represented in yellow refers to work products belonging to the Requirements discipline.

The colored background areas allow perceiving that two tasks are handled in both phases, thus verifying that there are interdependencies between the phases. The tables built facilitate the identification of interdependencies, not only among activities, tasks, phases and the roles, but also among tasks and work products, of the disciplines under consideration.

These tables, as well as the graphical representations allow analyzing the traceability of various elements of RUP, as well as easily identifying all the existing interdependencies between those elements. This becomes particularly useful since it allows knowing in detail how the various elements of the RUP process are related.

The information provided in these tables and representations, improve the practitioner's capacity in dealing with the impact of changes and in supporting better development decisions.

This systematization of the interdependencies is also useful to compare a particular method/process model with the RUP since it allows knowing in detail how the RUP is organized. The study of traceability and of the interdependencies between the various elements of the RUP may be extended to all disciplines and phases that compose this process. The expansion of the study will allow detailing how the various elements are related throughout the whole RUP process.

4   Conclusions

RUP is a process that provides best practices and guidelines for successful software development. However, this does not provide any information that enables for easy identification of traceability and existing interdependencies between the various elements that constitute it. Throughout the software development, this can become a problem since there is no explicit native information on RUP documentation on the inter-relation of RUP elements.

Our work produced several tables and graphical representations in order to highlight how the various RUP elements are related. These tables and graphical representations allow, from the initial phases of development, an easier identification of the various interdependencies and the traceability among elements, as well as to provide a deeper knowledge about the organization of RUP. This is quite advantageous since it is possible to avoid unconscious decisions during the development process as well as to detect early potential problems due to the existing interdependencies.

Acknowledgments. This work has been supported by COMPETE: POCI-01-0145-FEDER-007043 and FCT – Fundação para a Ciência e Tecnologia within the Project Scope: UID/CEC/00319/2013.

References

1. Fernandes, J.M., Machado, R.J.: Requirements in Engineering Projects. Springer, Cham (2016)
2. Ivan, J.: The Design of Requirements Modelling Languages. Springer, Cham (2015)
3. Smialek, M., Nowakowski, W.: From Requirements to Java in a Snap. Springer, Cham (2015)
4. Carlshamre, P., Sandahl, K., Lindvall, M., Regnell, B., Nattoch Dag, J.: An industrial survey of requirements interdependencies in software product release planning. In: Fifth IEEE International Symposium on Requirements Engineering, pp. 84–91. IEEE Press (2001)
5. Regnell, B., Paech, B., Aurum, A., Wohlin, C., Dutoit, A., Nattoch Dag, J.: Requirements mean decisions! – research issues for understanding and supporting decision-making in requirements engineering. In: First Swedish Conference on Software Engineering Research and Practice (SERP 2001), pp. 49–52 (2001)
6. Dahlstedt, Å.G., Persson, A.: Requirements Interdependencies – State of the Art and Future Challenges. In: Aurum, A., Wohlin, C. (eds.) Engineering and Managing Software Requirements. LNCS, pp. 95–116. Springer, Heidelberg (2005)
7. Dahlstedt, Å.G., Persson, A.: Requirements interdependencies - molding the state of research into a research agenda. In: Ninth International Workshop on Requirements Engineering: Foundation for Software Quality, pp. 55–64 (2003)
8. Heindl, M., Biffl, S.: A case study on value-based requirements tracing. In: 10th European Software Engineering Conference, pp. 60–69. ACM, New York (2005)
9. Kruchten, P.: Tutorial: introduction to the rational unified process. In: 24th International Conference on Software Engineering (ICSE 2002), pp. 703–703. ACM, New York (2002)
10. Gotel, O.C.Z.: An analysis of the requirements traceability problem. In: 1st International Conference on Requirements Engineering, pp. 94–101. IEEE Press (1994)
11. Genvigir, E.C.: Um Modelo para Rastreabilidade de Requisitos de Software Baseado em Generalização de Elos e Atributos. Instituto Nacional de Pesquisas Espaciais (2009)
12. Zou, X., Settimi, R., Cleland-Huang, J.: Improving automated requirements trace retrieval: a study of term-based enhancement methods. Empirical Softw. Eng. 15(2), 119–146 (2010)
13. Sánchez, P., Alonso, D., Rosique, F., Álvarez, B., Pastor, J.A.: Introducing safety requirements traceability support in model-driven development of robotic applications. IEEE Trans. Comput. 60(8), 1059–1071 (2011)
14. Ramesh, B., Jarke, M.: Toward reference models for requirements traceability. IEEE Trans. Softw. Eng. 27(1), 58–93 (2001)
15. Winkler, S., Pilgrim, J.V.: A survey of traceability in requirements engineering and model-driven development. Softw. Syst. Model 9(4), 529–565 (2010)
16. Spanoudakis, G., Zisman, A.: Software traceability: a roadmap. In: Chang, S.K. (ed.) Handbook of Software Engineering and Knowledge Engineering, vol. 3, pp. 395–428. World Scientific Publishing, Singapore (2005)
17. Dömges, R., Pohl, K.: Adapting traceability environments to project-specific needs. Commun. ACM 41(12), 54–62 (1998)
18. Aizenbud-Reshef, N., Nolan, B.T., Rubin, J., Shaham-Gafni, Y.: Model traceability. IBM Syst. J. 45(3), 515–526 (2006)
19. Huang, J., Gotel, O., Zisman, A. (eds.): Software and Systems Traceability. Springer, London (2012)
20. Turban, B.: Tool-Based Requirement Traceability Between Requirement and Design Artifacts. Springer, Wiesbaden (2013)

21. Marques, A., Ramalho, F., Andrade, W.L.: TRL: a traceability representation language. In: 30th Annual ACM Symposium on Applied Computing, pp. 1358–1363. ACM, New York (2015)
22. Marques, A., Ramalho, F., Andrade, W.L.: Towards a requirements traceability process centered on the traceability model. In: 30th Annual ACM Symposium on Applied Computing, pp. 1364–1369. ACM, New York (2015)
23. Rempel, P., Mäder, P.: A quality model for the systematic assessment of requirements traceability. In: 23rd IEEE International Requirements Engineering Conference (RE), pp. 176–185. IEEE Press (2015)
24. Badreddin, O., Sturm, A.: Requirement traceability: a model-based approach. In: 4th IEEE International Model-Driven Requirements Engineering Workshop (MoDRE), pp. 87–91. IEEE Press (2014)
25. Soonsongtanee, S., Limpiyakorn, Y.: Enhancement of requirements traceability with state diagrams. In: 2nd International Conference on Computer Engineering and Technology (ICCET), pp. V2-248–V2-252. IEEE Press (2010)
26. Berrocal, J., Alonso, J.G., Chicote, C.V., Murillo, J.M.: A model-driven approach for documenting business and requirements interdependencies for architectural decision making. IEEE Lat. Am. Trans. 12(2), 227–235 (2014)
27. IBM, Rational Method Composer (version 7.1). http://www-03.ibm.com/software/products/en/rmc

Conversation About the City: Urban Commons and Connected Citizenship

Maria Célia Furtado Rocha[1], Gilberto Corso Pereira[1], Elizabeth Loiola[1],
and Beniamino Murgante[2(✉)]

[1] Federal University of Bahia, Salvador, Brazil
rochamcelia@gmail.com, {corso,beloi}@ufba.br
[2] University of Basilicata, Potenza, Italy
beniamino.murgante@unibas.it

Abstract. The analysis of conversations between Italian and Brazilian groups allows to understand meanings that inspire and motivate political demonstrations and different performances in favor of a desired city. The research takes a descriptive perspective, based on categories of analysis informed by theories, but made by the exploration of terms used in conversations caught by members of four Facebook groups. The expectation is to highlight concepts, revealed by the set of statements made by people interested in trends that city growth takes or just in the given uses of urban spaces. To provide clues for the interpretation of the talks, we use centrality indices of Social Network Analysis (SNA) in a semantic network of concepts. Thus, it is possible to establish similarities and differences between current forms of civic participation and politics on urban commons largely supported by online social networks.

Keywords: Public participation · Urban commons · Social network analysis

1 Introduction

"Public sphere" refers to the scope, area or space, socially recognized, but not institutionalized, formed by spontaneous discussion, the free movement of questions, contributions, information, views and arguments from everyday experiences of a subject. It is a network for communication content and views. Communication flows are filtered and synthesized through it, and then condensed in the public opinion [1]. Very often, the articulation of socially relevant issues is processed by sharing experiences and opinions through social media. The increase of connected people in various parts of the world, coupled with the reduction of the cost of dealing with the public allows to citizens to produce small valuable contributions [2].

These technologies make practices of social production, that alter the relationship between people and the events that surround them, possible [3]. They can give input to

---

CAPES (proc. n. 11527/13-7).

O. Gervasi et al. (Eds.): ICCSA 2016, Part V, LNCS 9790, pp. 608–623, 2016.
DOI: 10.1007/978-3-319-42092-9_46

a public debate. Networks linking human and non-human, physically present or not [4], bring together people of different cultures, diminish territories and go beyond old boundaries and scales.

Ito and Okabe [5] suggested the emergence of a visual sharing mode via portable devices characterized by a personal, penetrating and intimate nature. This movement may indicate a change in perception of context beyond rational argument, capable of bringing together different points of view. Tursi [6] recognized that current practices in social media play some functions performed by the literary public sphere [7] that has built self-awareness, critical thinkers, reputations and synergies. According to this author, the social dynamic based on sharing experiences in everyday life manifests a kind of social bond based on sharing emotions. Therefore, we can assume that digital technologies give opportunities for the development of new practices and the creation of new social actors, that come to occupy the public sphere. Through the visibility they get, social actors can expose demands accepted as a common good and of interest for everybody.

For this study, we take the public sphere as the exercise of social practices connected to various contexts and multiple scales of social action. As a sensor of public opinion [7], the public sphere, as it is here understood, does not include principles or practices necessarily anchored in rational exchange of arguments to reach consensus. Communicative relations can contribute to increase the participation and disseminate visions of democracy, and thus to support the exercise of sovereignty by a variety of local political actors, even those "geographically immobile". Therefore, individuals and communities could acquire visibility in international fora or in global politics previously exclusive space of national organizations [8].

Digital spaces where current problems in cities are discussed may open new possibilities for the development of new social practices and give opportunity for participation, although one can not ignore aspects that simultaneously curtail and augment that potential [9]. They allow the articulation of ideas but also of collective experiences that can enrich/challenge previously given meanings to the urban space. In this way, they put in evidence perspectives of various social actors that desire "to make the city".

Online social networks seem to act in the public sphere. As they spread values, habits and ideals, they provide support for culture and act on implicit substratum of shared opinions that underlie the action of groups [10]. Thus, we can assume that there is an ongoing renewal of the public sphere, either through the struggle for political participation in the city, either by proper civic actions in defense of the quality of urban space.

This article is part of a broader research project, which aims to reveal visions of urban commons of groups present in digital social networks. They run some of the public participation trajectories outlined by [11]: civic and political participation. It analyzes the conversations of four Facebook groups – two from an Italian city (Potenza, southern Italy) and two from a Brazilian city (Salvador of Bahia, northeastern Brazil) to answer the following research questions: what visions of participation, citizenship and the urban commons emerge? And how do they face visions of municipal representatives?

2   The Culture of the Commons

Appealing to fundamental rights becomes very strong [12] in the twenty-first century, with the growing of inequality all over the world. The struggle for rights extends throughout the globalized world, builds new modes of action. According to this author, the fundamental principle of equality should, first, be rethought and placed at the center of attention.

In this perspective, commons are defined as an immediate and concrete guarantee of fundamental human rights, beyond the commodification logic. They are those goods that can not be subjected to pure economic logic, such as water, clean air, healthy environment, knowledge, food, health. These assets are fundamental to human existence. The Internet, for example, would be a key resource to know what happens near or far, to dialogue with others, to participate. The common good, in this case, is a good generated by the participation of everybody and, consequently, should continue to be common [13].

Commons would have a dimension in the future: they speak of social connections, because when a good is common and we use it together with others, we must defend it together. They are characterized by a shared use; every person can enjoy them that way without needing redistributive policies or implementation of a participatory model. The non-discrimination of access to the common good, leads to the issue of equality, and therefore democracy [12].

Rodotà [12, 13] proposes the resumption of the struggle for old and new commons, such as the Internet, unifying it under the banner of struggle for the rights. However, it is certainly necessary to confront issues such as those related to the concentration of resources that promote new geographies as it is demonstrated by concentration of network infrastructures that can be measured by access to the signal of WiFi networks [14]. And so all those who think of public spaces as common goods are challenged to answer questions posed by inequalities expressed in new discontinuities established in the territory. Such inequalities are reinforced by urban policies, at least in large Brazilian and Latin American cities, which tend to favor real estate investments aimed at groups of high and middle income and increase abandon and deterioration of old central areas [15].

In the context of collective movements using social network platforms, groups seek to get public attention for the reappropriation of city areas with a focus on the enjoyment of all stakeholders; further, they oppose governance conducted by public authorities and seek to take part in choices about the allocation to be given to such areas. To a greater or lesser extent, this is the case of groups here studied.

Movements shaking the Brazilian political scene in 2013, on the eve of the World Cup, have raised social justice issues, because they asked for investments in what can be considered to be a common good to all – the right of transport, health, education. The right to urban transport is clearly connected with the access to the city and therefore with the access to diverse environments where one can exercise public personas and develop skills involved in civilization [16].

In Brazil, the right to participate in urban policy decisions is legally provided since 2001. But the inability of stakeholders acting as decision-makers ultimately nullifies the use of full conditions of that right at the time that we see a degradation of the common urban space. In Italy, the involvement of citizens in local urban policy through district contracts was strengthened in 2008 [17]. In the country, the issue of common goods (particularly water and the use of nuclear energy) mobilized 26 million people in the referendums of June 2011. At the time, Italian citizens have shown their willingness and ability to largely mobilize, independently of the media activation [18]. Today, forms of collaboration between citizens and local authorities for care, management and regeneration of urban commons have been tried in Bologna [19] and Turin [20]. Such forms promote active citizenship [21] which may turn out to be the subject of policy [22].

Our reflection leads us to assume that the informal participation practices are fundamental in urban policies, while formal instruments of participation do not collect all social instances. We must take the point of view of the expanded participation to contemplate groups who use the network communication as a mean to share visions and values to generate a knowledge of the contemporary city – the real and the desired one. And so they begin to pursue goals and to think how to achieve them. We take here the broad concept of culture that includes attitudes, mentalities and values, to study groups that constitute initiatives related to what could be called the culture of the commons. The conceptual scheme that guided the research is illustrated in Fig. 1.

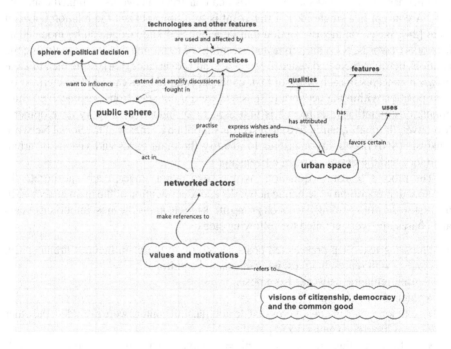

Fig. 1. The research conceptual map

3 Networked Concepts: Exploring Meanings

Conversations on Facebook pages of two Italian groups and two Brazilian groups constitute the source of the data analyzed in this paper. These groups moved ahead movements for regeneration of degraded urban areas of the city of Potenza (Italy) and Salvador (Brazil), mainly in the years 2012 and 2013.

Though the cities have very different sizes – Potenza has about 70,000 inhabitants and Salvador has 2, 7 million inhabitants – there is some mirroring between Brazilian and Italian groups. Two of them directly confront the government: the Parco del Basento (Basento's Park) [54] group in Italy and the Movimento Desocupa (Vacate Movement) in Brazil. The others two groups never directly address the power. The Italian group Il Giardino in Movimento [53] (Garden in Motion) and the Brazilian group Canteiros Coletivos (Collective Yards) take care of some urban areas while trying to develop specific kinds of sociality. We verified peculiarities in the way those four groups acted in unconventional arenas of politics in their desire to reintegrate areas to collective urban life.

To analyze their performances, we used the Network Text Analysis (NTA), a method that encodes semantic links among words and constructs networks of linked concepts [23]. A concept is a single idea; a statement is two concepts and the relation between them [24]. Based on the assumption that language and knowledge can be modeled as networks of words, NTA aims to reveal meanings of terms and themes by analyzing the relations between them (existence, frequencies, and covariance) [24]. More then linkage between concepts in each block of text, concepts and linkages can be characterized by their position within the network [23]. Since we did not start from preconceived categories, this quantitative approach is first used to an inductive category development. Moreover, the quali-quantitative analysis was supported by means of the Social Network Analysis (SNA) centrality measures to identify the main terms and visions of urban commons and the desired type of citizenship.

The process of data acquisition involves two major stages, each one iteratively developed: preparation of semantic networks and identification of the main concepts. In the first stage, after reading Facebook group messages to know the most important events and phases, the work entailed the following steps:

1. Preparing texts: tag preselected posts according to the main themes: urban space, conflict with representatives, what means to be political?
2. Getting semantic networks from posts:
   - identifying concepts;
   - extracting statements (words that do not transmit content were excluded, the other ones have been generalized [24]);
   - defining coding choices: text unit = sentence (string of words limited by full stops – period, colon, question mark, exclamation point, ellipses); window size = 2 (i.e., how distant concepts can be from each other and still have a relationship) and directionality = bi-directional;
   - representing statements as a network of shared concepts.

The semantic networks were generated with AutoMap v3.0.10.36 and first visualized using ORA NetScenes version 3.0.9.9j and later using Gephi 0.8.2. Thus comments to group posts (case of Parco del Basento and Movimento Desocupa groups) and messages (case of Il Giardino in Movimento e Canteiros Coletivos groups) were modeled as networks of connected concepts that coded in its structure a semantic field represented by the proximity of nodes [25] and communities that bring together concepts strongly connected in clusters [26].

The detection of communities was carried out equally towards all networks: randomly, without knowing the structure of communities, considering edges weight, which represents the number of occurrences of a particular relationship, and fixing in 1 the resolution, a measure of stability that considers the network partition time [27].

The second stage encompassed three steps:

3. Characterizing networks: nodes were organized according to their local connections and are contextually related to their surroundings and connection paths. The following basic statistics were used [28, 29]:
   - diameter, to know the greatest distance between any two nodes of the network;
   - density for a network cohesion index, to identify how close it is to contain all possible links;
   - average degree, to know the average number of existing links to each node;
   - average path length, to know the average distance between nodes;
   - number of communities, to know how many agglomerations are there.

4. Identifying main concepts: the prominent terms (here called concepts) were identified according to the value obtained by the concepts for each of the following centrality indices [30, 31]:
   - Degree shows the number of concepts directly connected to a given concept. It focuses on the importance of a concept (node) through the connections that it establishes with its neighbors [32].
   - Closeness centrality describes how close a concept is to all other concepts. It shows, therefore, its proximity to all other concepts of the network.
   - Eigenvector centrality tells how closely a concept is to other concepts that are important in respect to Degree. The importance of a vertex (node) increases as its neighbors are important themselves. A score is assigned for each vertex, that is proportional to the sum of the scores of its neighbors [33].
   - Betweenness centrality shows how often a concept is positioned on the shortest path (geodesic) between any other pairs of concepts. Nodes located in many geodesic lines are central in a network: they allow the flow of information through the network [33].

5. Ranking of concepts: it was obtained for each index, an ordered list of the most prominent concepts for each of the networks, particularly those that occupy the top five positions in at least three of the four indexes of centrality.

Then neighborhoods and communities of the most prominent terms have been explored, since communities represent semantic contexts. The exploration of context started from a node of interest towards its neighbors [26].

The categorization of concepts occurred throughout the process of creation of networks. It helped to appoint actions, resources and qualities mentioned for the initiatives and activities of the groups, for the city and the urban elements, the existing and desired space. Categories were: "Actor", "Action", "Event, Institution, Initiative or Resource" and "Value, Quality and Motivation" [34]. The latter was used particularly to grasp ideas about how to exercise citizenship and the ideals of democracy and commons that support the actions and activities of the groups. These ideals were found not only in the network specifically obtained for the Policy theme, but it is also found in the following thematic networks: Urban Space, Conflict with public officials. Actors category – each individual, group, professional category or role played by individuals, entity or institution – helped to identify the agents most evidenced by the groups for their political activity.

4 City and Citizenship

Each group was involved in conversations about urban areas as a commons, therefore accessible to all. The Italian group Parco del Basento and the Brazilian group Movimento Desocupa addressed to society, representatives and public officials to express their dissatisfaction with the treatment given to urban spaces and how their occupation and use were being managed. They promoted political demonstrations and directly directed issues to the political decision sphere.

Members of the Parco del Basento group aimed to exercise their civil and political citizenship. The group advocated active, participatory citizenship ("cittadinanza attiva") in particular with regard to decisions on the allocation of areas of the city of Potenza. Their political vision of citizenship had an abstract character, formal and centered on the state – the "received" citizenship [35] (Fig. 2).

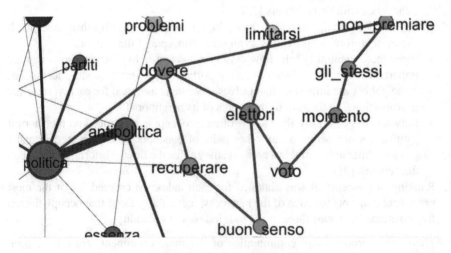

Fig. 2. Parco del Basento group – Satate-centered citizenship

However, when their members discussed the desired urban space, they assumed other perspectives. In this case, they were affiliated to an active citizenship perspective, where people propose, design and evaluate the urban space (Fig. 3). This perspective is associated to what [35] calls "acquired" citizenship, therefore it results from political struggle.

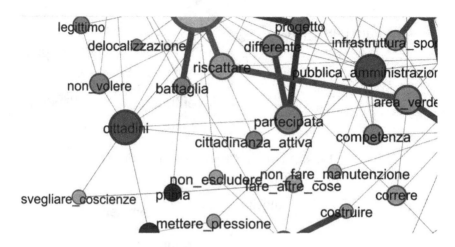

Fig. 3. Parco del Basento group – Active citizenship

In Brazil, after a period of intense protests in 2012, the growth of Movimento Desocupa faltered. Even when membership was grown, comments on the group's Facebook page mainly focused on organizing public demonstrations. Complaints brought by group members almost never gave rise to the discussion and the search for alternative solutions to the problems of Salvador City.

In a gentrification process of a traditional central area, the group called for investments and proposed the development of a plan for the neighborhood by the residents themselves.

In general the debate was poor, an atmosphere of polarization prevailed, without any opinion becoming hegemonic. The common thread was that the movement should remain nonpartisan ("política" "não partidária") (Fig. 4).

On the other hand, the Italian group Il Giardino in Movimento and the Brazilian group Canteiros Coletivos did not address the public authorities but other groups and civil society organizations to invite them to recover residual spaces forgotten by the public administration. They promoted civic participation aimed at upgrading these areas and used social media to share views and meanings about the city as commons.

Il Giardino in Movimento group valued a civic culture based on the ideal of the shared commons and the "do good" for others [35, 36]. At each meeting, it reiterated the pact about the value to be given to public spaces [37], made by Parco del Basento group. Figure 5 shows how participation ("partecipazione") is directly related to the management of the space ("gestione", "spazio").

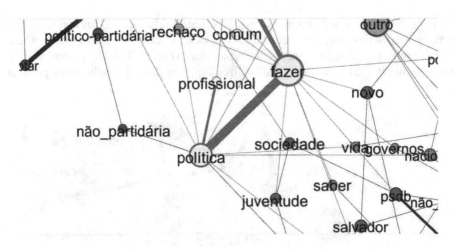

Fig. 4. Movimento Desocupa – nonpartisan

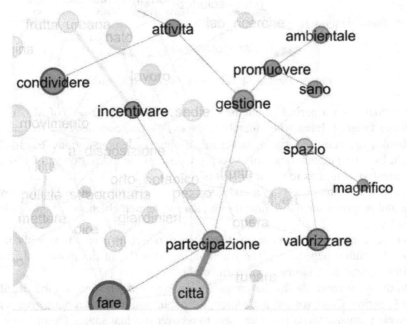

Fig. 5. Il Giardino in Movimento – participation

Regarding Canteiros Coletivos group, its participation in social projects in partnership with neighborhood communities and NGOs seems to have ensured its continuity. It aims to improve community bonds and care of urban areas by residents. While not taking a professedly political vision, the group stood against "pro-violence policy" in solidarity with the residents of a neighborhood in which they operated, after the death of some young residents. It expanded at that time its practice beyond the idea of participation and management of urban space, without directly addressing some kind of government interference.

Figure 6 shows how participation ("participação") means attending meetings, festivals ("festg") and workshops ("oficinas") for the Canteiros Coletivos group.

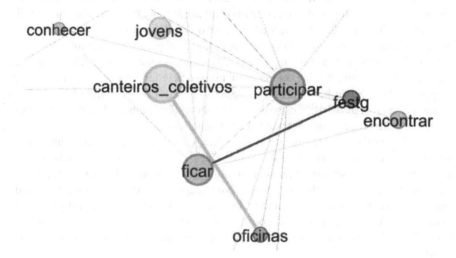

Fig. 6. Canteiros Coletivos – participation

The views of all groups on citizenship, politics and participation are summarized in Table 1.

Table 1. Citizenship, politics and participation

Group	Vision
Parco del Basento	Policy is the polis (the assembled citizens) Participation is active citizenship
Movimento Desocupa	The space for citizen political action is not the same of professional politicians Nonpartisan participation
Il Giardino in Movimento	Participation is care and management of urban space
Canteiros Coletivos	Participation is community actions to improve the quality of public space by planting and cultural interventions

5 City as Commons

Groups did actions of participation, as reported by [11], and forced the expansion of the scope of such participation, as mentioned by [38]. The defense of citizenship as a shared activity, oriented the Parco del Basento group to the realization of the common good referred to a republican tradition [39]. The Parco del Basento movement allowed a rich political experience, not only through confrontation with the public administration. Within the group itself, there was exchange of ideas about the project itself and political action strategies. This allowed its members to experience a participatory process that persists in current initiatives aiming to rescue urban areas abandoned by the public authorities of Potenza.

On the other hand, the Movimento Desocupa group since not advocated any idea of democracy, direct or representative, remained calling for legalizing each issue of conflict in a judicial dispute. This attitude leads to a deadlock. It has origins in the weakening of representation of social conflicts currently staged by parties in traditional arenas of politics, far removed from real citizenship [22]. However the movement has played an important role in strengthening the claims of civil society for holding of public hearings to approve changes in urban land use. However audiences per se certainly fail to address issues in a city that in 2010 had more then 2.5 million inhabitants [40], with a complexity of problems that need to be addressed at a metropolitan scale.

The Il Giardino in Movimento e Canteiros Coletivos groups performances are strongly linked to the recovery of symbolic values, the strengthening of social liaisons and therefore the possibility to ensure continuity of commons [12]. However, achievements for the Canteiros Coletivos group seem to depend on self-consciousness of individuals possessing a community identity. In a world without "them" the group's identity remains grounded in action, so to speak, "therapeutic". Its social practice resembles those of the Il Giardino in Movimento group, but unlike this, the meaning of participation is different.

The affiliations of the studied groups to ideals of democracy can be summarized as follows:

- The Parco del Basento group stands against real estate speculation, assumes a Republican vision: links politics to a common *praxis*, preserves the quality of urban space and its use for citizens. It mobilizes citizens of Potenza City and directs its proposal to representatives, without seeking for the mediation of politicians and parties. Still, it reinforces representative democracy and, through participatory design practices of dialogic and procedural nature, it takes part in a participatory democracy framework [41]. Its keyword is "active citizenship".

- The Movimento Desocupa comes into open conflict with the government when it repudiates political parties and politicians. It demonstrates the great dissatisfaction with the city government, but it can not draw up alternative proposals for the development of Salvador. Proposals, if any, do not become hegemonic in the movement. Without a channel of dialogue with the Executive and considering the Legislative strongly distrustful, the movement calls for the intermediation of their demands by the judiciary. Its keyword is "non-partisanship".

- Il Giardino in Movimento group strengthens the movement for the Basento Park, and assumes participatory design practices. Their idealizations for the area are collectively selected and executed. Although not exactly inserted in politics, it strengthens and educates towards active citizenship vaunted by the Parco del Basento group. Through practices of civic culture, it constitutes resources of collective life important to the exercise of citizenship in its political dimension [42]. Its keyword is "laboratory".

- The Canteiros Coletivos group has a vision of citizenship that approaches the communitarian one: it advocates the rescue of the cultural dimension of citizenship [43]. It acts against the degradation of areas of the city, while educating and propagating harmonized coexistence ties, believing that those lead people to attribute a common value to living in public space. Unlike the political clarity of prescriptions advocated by [44], it does not make clear what are the interests and values involved in the relationship "we/they". When faced with issues such as urban violence, it proposes peace without bringing out the name of the actors of urban warfare. Its keyword is "community".

In the conversations analysis of the four groups, the urban space is taken as a common good. Almost everyone joins desire and action: the meaning of what is common is produced by collective action, while space, itself, is being designed and/or produced. But this is done in different ways: while the Parco del Basento and Movimento Desocupa groups address public authorities and demonstrate clear desire to access urban areas and participate in choices about their destination, the Il Giardino in Movimento e Canteiros Coletivos groups dialogue only with their peers and sympathizers and propose to establish links between people and space by acting on it. The emphasis they place on each type of action to defend the urban commons is shown in the chart below Fig. 7.

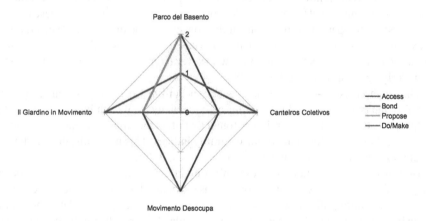

Fig. 7. Urban commons: types of action (Color figure online)

Although these groups aren't "union of people who build a project" for the whole of society, even without a comprehensive idea of the future, yet they demand things to it, and draw attention to the shared management of common space. Even if incipiently, they introduce issues as governance of commons, an institution through which the individual can be an active agent and exercises his rights of citizenship [45].

6 Conclusions

This work aimed to understand ideals and values associated to urban commons observing conversations of two Italian and two Brazilians groups Facebook pages. The central concepts were represented in semantic networks and were highlighted by centrality indices of Social Network Analysis.

The so-called "symbolic" networks [46] enabled a contextualized view of the use of terms and expressions by group members to describe their actions and thus gain insight into their views of citizenship, democracy and urban commons. If they suggest meanings attributed by the groups, certainly they can not express and communicate all meanings of speech and behavior of agents, but still they allow generalizations that reveal what motivates them.

Obtaining statements through the relationship between concepts according to a fixed distance (windows method), taking as limits the sentence, is a simplified feature. It is known that sentences combine themselves beyond these limits, in different ways, and thus guide the discourse [47]. However, we believe that the various paths of interpretation presented by semantic network, although simplistic, amplify capabilities enabled by traditional content analysis, which focuses on the co-occurrence of words [23].

If the different centrality indices contributed to highlight the most prominent concepts by the amount of times they held more central positions in each of these indices, the analysis of the particular functions of the concepts in the discourse, revealed by the indexes was left in the background. This is the case of concepts with high betweenness centrality, that can connect communities with several topics of interest and can assist in the underlying information retrieval [48].

The method used – ranking of concepts according to their position in various centrality indices – must be deepened through the analysis of each index and, then, the comparison of results. However, it was found that often the same concepts occupied the most central locations for many of the centrality indices. This may indicate that the size of networks (the largest of which had 110 nodes and 230 links) matters.

Both text encoding choices – text unit, directionality and window size – for the purpose of generation of semantic networks, as the method of selection of core concepts based on centrality indices, were adequate in relation to the level of detail needed to obtain insights to answer the research questions. Indeed, the results for the Italian groups were validated by their members. Therefore the method was effective to explore the emerging meanings of speeches, many of which remain hidden or undervalued in a traditional content analysis that "*tell us about text's fundamental building blocks, but not the structure in which those blocks are arranged*" [49]. Our focus on networks of linked concepts rather than on counted concepts [50] proved its importance to give answers and close the conceptual schema that guided the research illustrated in Fig. 1.

Through analysis of Facebook conversations we have reached a issue almost never pointed by the groups, except in the case of Movimento Desocupa during the struggle against the gentrification of a neighborhood in Salvador: after all, "who does the city?". The not raised question would have the power to relate the urban space to other actors certainly present in the public sphere, made in this case for those who treat the city as a business and are more interested in the discourse in favor of "city as a growth machine" [51].

Surely if they had unveiled the role that real estate capital currently plays in urban dynamics, groups could perform more effective ways of opposing to the transfer of control of land use and occupation and urban policy-making from the public to the private sphere [52].

The question that the groups have not formulated would lead to an answer that could finally confront ideas of possible futures for the cities of Potenza and Salvador – the future that some groups point to and the one that other city actors prepare in other forums and other ways.

References

1. Gomes, W.: Esfera pública política e comunicação em *Direito e Democracia* de Jürgen Habermas. In: Gomes, W., Maia, R.C.M. (eds.) Comunicação e Democracia. Problemas & Perspectivas, pp. 69–115. Paulus, São Paulo (2008)
2. Shirky. A cultura da participação: criatividade e generosidade no mundo conectado. Zahar, Rio de Janeiro (2011)
3. Benkler, Y.: The Wealth of Networks: How Social Production Transforms Markets and Freedom. Yale University Press, New Haven (2006)
4. Palacios, M.: A internet como mídia e ambiente. Reflexões a partir de um experimento de rede local de participação. In: Maia, R., Castro, M.C.P.S. (orgs.) Mídia, esfera pública e identidades coletivas, pp. 229–244. Editora UFMG, Belo Horizonte (2006)
5. Ito, M., Okabe, D.: Intimate visual co-presence. In: 7th International Conference on Ubiquitous Computing, UbiComp, Tokyo (2005)
6. Tursi, A.: Politica 2.0 – Blog, facebook, wikileaks: ripensare la sfera pubblica. Mimesis Edizione, Milano (2011)
7. Habermas, J.: Direito e Democracia: entre facticidade e validade, 2nd edn., vol. II. Tempo Brasileiro, Rio de Janeiro (2003)
8. Sassen, S.: Sociologia da Globalização. Artmed, Porto Alegre (2010)
9. Papacharissi, Z.: The virtual sphere: the internet as a public sphere. New Med. Soc. 4(1), 9–27 (2002)
10. Landi, S.: Opinioni silenziose. Per una storia della dimensione non discorsiva della sfera pubblica. In: Rospocher, M. (ed.) Oltre la sfera pubblica. Lo spacio della politica nell'Europa moderna, pp. 55–84. il Mulino, Bologna (2013)
11. Dahlgren, P.: Reinventare la partecipazione. *Civic agency* e mondo della rete. In: Bartoletti, R., Faccioli, F. (cura) Comunicazione e civic engagement. Media, spazi pubblici e nuovi processi di partecipazione, pp. 17–37. Franco Angeli, Milano (2013)
12. Rodotà, S.: Il diritto di avere diritti. Ed. Laterza, Roma-Bari (2012)
13. Rodotà, S.: La democrazia e il bene comune. In: Gallina, E. (cura) Vivere la democrazia, pp. 83–96. Edizioni Gruppo Abele ONLUS, Torino (2013)
14. Varnelis, K.: The Centripetal City: Telecommunications, the Internet, and the Shaping of the Modern Urban Environment, vol. 17. Cabinet Magazine (2005)
15. Carvalho, I., Pereira, G.C.: Estrutura social e organização social do território na Região Metropolitana de Salvador. In: Carvalho, I., Pereira, G.C. (org.) Salvador: transformações na ordem urbana: metrópoles: território, coesão social e governança democrática. pp. 138–177. Letra Capital/Observatório das Metrópoles, Rio de Janeiro (2014)
16. Bauman, Z.: Modernidade Líquida. Zahar, Rio de Janeiro (2001)

17. Bifulco, L.: *Governance* e territorialização: o *welfare* local na Itália entre fragmentação e inovação. Cadernos Metrópole **14**(27), 41–57 (2012)
18. Morcellini, M.: Le cose della comunicazione che non abbiamo capito. In: Bartoletti, R., Faccioli, F. (cura) Comunicazione e civic engagement. Media, spazi pubblici e nuovi processi di partecipazione, pp. 86–96. Franco Angeli, Milano (2013)
19. Comune di Bologna. Iperbole rete civica. http://comunita.comune.bologna.it/beni-comuni
20. Torino Social Inovation. http://www.torinosocialinnovation.it/a-torino-cittadini-e-comune-insieme-per-i-beni-comuni/
21. Rodotà, S.: Introduzione. In: MBS Consulting (cura). Beni Comuni, pp. 9–20. Feltrinelli, Milano (2015)
22. Sodré, M.: A ciência do comum: notas para o método comunicacional. Vozes, Petrópolis/RJ (2014)
23. Popping, R.: Computer-Assisted Text Analysis. SAGE, London (2000)
24. Diesner, J., Carley, K:M.: AutoMap1.2 – extract, analyze, represent, and compare mental models from texts, Pittsburgh, Carnegie Mellon University, School of Computer Science (2004)
25. Danowski, J.A.: Computer-mediated communication: a network-based content analysis using a CBBS conference. Commun. Yearb. **6**, 905–924 (1982)
26. Drieger, P.: Semantic network analysis as a method for visual text analytics. Procedia – Soc. Behav. Sci. **79**, 4–17 (2013)
27. Fortunato, S.: Community detection in graphs. Phys. Rep. **486**, 75–174 (2010)
28. Barabási, A-L.: Network Science Book (2012). http://barabasi.com/networksciencebook/
29. Clauset, A., Newman, M.E.J., Moore, C.: Finding community structure in very large networks. Phys. Rev. E **70**(6), 066111 (2004)
30. Freeman, L.C.: A set of measures of centrality based on betweenness. Sociometry **40**(1), 35–41 (1977)
31. Bonacich, P., Lloyd, P.: Eigenvector-like measures of centrality for asymmetric relations. Soc. Netw. **23**(3), 191–201 (2001)
32. Fadigas, I.S., Trazibulo, H., Senna, V., Moret, M.A., Pereira, H.B.B.: Análise de redes semânticas baseada em títulos de artigos de periódicos científicos: o caso dos periódicos de divulgação em educação matemática. Educação Matemática Pesquisa **11**(1), 167–193 (2009)
33. Fadigas, I.S.: Difusão do Conhecimento em Educação Matemática sob a perspectiva das Redes Sociais e Complexas. Tese de Doutoramento, Salvador (2011)
34. Rocha, M.C.F., Pereira, G.C., Murgante, B.: City visions: concepts, conflicts and participation analysed from digital network interactions. In: Gervasi, O., et al. (eds.) ICCSA 2015. LNCS, vol. 9156, pp. 714–730. Springer, Heidelberg (2015)
35. Dahlgren, P.: Media and Political Engagement: Citizens, Communication and Democracy. Cambridge University Press, Cambridge (2009)
36. Amin, A.: Collective culture and urban public space. City **12**(1), 5–24 (2008)
37. Resweber, J.-P.: A Filosofia dos Valores. Livraria Almedina, Coimbra (2002)
38. Carpentier, N.: The concept of participation. If they have access and interact, do they really participate. Revista Fronteiras - estudos midiáticos, **14**(2), 164–177 (2012)
39. Held, D.: Models of Democracy. Stanford University Press, California (2006)
40. BGE: Censo 2010. Cidades. http://cod.ibge.gov.br/GBB
41. Deplano, C.: Antropologia Urbana. Società Complesse e Democracia Partecipativa. Edicomi Edizioni, Monfalcone (2009)

42. Artieri, G.B.: Connessi in pubblico: sfera pubblica e civic engagement tra mainstream media, blog e siti di social network. In: Bartoletti, R., Faccioli, F. (cura) Comunicazione e civic engagement. Media, spazi pubblici e nuovi processi di partecipazione, pp. 97–116. Franco Angeli, Milano (2013)

43. Bakardjieva, M.: Subactivism: lifeworld and politics in the age of the internet. Inf. Soc. **25**(2), 91–104 (2009)

44. Mouffe, C.: The Democratic Paradox. Verso, London, New York (2009)

45. Sacconi, L.: Introduzione. Visione Nuova, Ragionevoli Proposte. In: Sacconi, L., Ottone, S. (cura). Beni comuni e cooperazione, pp. 7–31. Bologna, il Mulino (2015)

46. Watts, D.J.: The "new" science of networks. Ann. Rev. Sociol. **30**, 243–270 (2004)

47. Neves, M.H.: Gramática de usos de português. Ed. Unesp, São Paulo (2011)

48. Paranyushkin, D.: Visualization of Text's Polysingularity Using Nerwork Analysis. Nodus Lab, Berlin (2012)

49. Carley, K., Palmquist, M.: Extracting, representing, and analyzing mental models. Soc. Forces **70**(3), 601–636 (1992)

50. Carley, K.: Coding choices for textual analysis: a comparison of content analysis and map analysis. Sociol. Methodol. **23**, 75–126 (1993)

51. Molotch, H.: Growth machine links: up, down, and across. In: Jonas, A.E.G., Wilson, D. (eds.) The Urban Growth Machine. Critical perspectives, Two Decades Later, pp. 247–265. State University of New York, New York (1999)

52. Carvalho, I., Pereira, G.C.: A Cidade como Negócio. EURE **39**(118), 5–26 (2013)

53. Lorusso, S., et al.: Involving citizens in public space regeneration: the experience of "garden in motion". In: Murgante, B., et al. (eds.) ICCSA 2014, Part II. LNCS, vol. 8580, pp. 723–737. Springer, Heidelberg (2014). doi:10.1007/978-3-319-09129-7_52

54. Murgante, B.: Wiki-planning: the experience of Basento Park in Potenza (Italy). In: Borruso, G., Bertazzon, S., Favretto, A., Murgante, B., Torre, C. (eds.) Geographic Information Analysis for Sustainable Development and Economic Planning: New Technologies, pp. 345–359. Information Science Reference IGI Global, Hershey (2012). doi:10.4018/978-1-4666-1924-1.ch023

Remote Sensing Fire Danger Prediction Models Applied to Northern China

Xiaolian Li[1], Wiegu Song[1], Antonio Lanorte[2],
and Rosa Lasaponara[2(✉)]

[1] State Key Laboratory of Fire Science,
University of Science and Technology of China, Jinzhai 96, Hefei 230027, China
[2] CNR-IMAA, C.Da S. Loja, 85050 Tito Scalo, (PZ), Italy
rosa.lasaponara@imaa.cnr.it

Abstract. Remote sensing fire danger prediction model is applied to Northern China. This study was carried out in the Daxing'anling region, which is located in Heilongjiang Province and Inner Mongolia (50.5°–52.25° N, 122°–125.5° E), the northern China. The method integrated by dead fuel moisture content and relative greenness index, which is based on the fire potential index (FPI), was used to predict the fire danger level of the study area. The case that fire happened on the late June 2010 was used to validate the modified method. The results pointed out that the fire affected areas were located in high fire danger level on 26th, 27th, 28th June, 2010 respectively. The ROC analyses of the predicted accuracy on these days were 90.98 %, 73.79 % and 69.07 % respectively. Results from our investigation pointed out the reliability of the adopted method.

Keywords: Fire danger · Satellite · Daxing'anling region

1 Introduction

Fire represents one of the main disturbances of vegetation covers and ecosystems, bringing profound transformations at different temporal and spatial scales which affect landscapes and environments. Forest fire is a primary process that influences the vegetation causing alteration in its structure and composition, soil erosion, changes in nutrient levels, micro-climate, hydrology, vegetation succession, etc.

Climate change is expected to bring increased temperatures, prolonged droughts and heat waves that will further aggravate the risks of forest fires in the Mediterranean regions, as well as in alpine ecosystems and boreal forests, with severe environmental and economic consequences. Moreover, other climate change impacts that could add damaged or dead wood to the forest fuel load (for example, as a result of insect outbreaks, ice storms or high winds) may increase the risk of fire activity.

Satellite remote sensing can provide a useful support for natural and cultural heritage and in the case of fire monitoring the ARGON Laboratory of the IMAA-CNR developed operative tools [1–7] systematically adopted by the Protezione Civile of the Basilicata Region.

© Springer International Publishing Switzerland 2016
O. Gervasi et al. (Eds.): ICCSA 2016, Part V, LNCS 9790, pp. 624–633, 2016.
DOI: 10.1007/978-3-319-42092-9_47

Focusing on fire research in China, it is important to remind that over the years many studies [see 8] and references therein quoted] have been conducted for the characterization of the fire regime in northeastern China, as in the Heilongjiang Province and the Daxing'anling region, investigations have been also conducted on fire frequency in Mongolian pine (Pinussylvestris), in Huzhong Forest China (Northestern China) in Great Xing'an Montains and in Great Hing Mountains as well as on forest fire management policy and its effects on fuel and fire danger. These studies have brought a basic understanding of climate, forest fire danger, and fire regime in Daxing'anling in the past 30–50 years. Efforts have been also addressed to the spatial characteristics of the burned areas in 1987–2010 using on satellite remote sensing data.

China has a vast land located in the Northern Hemisphere and strongly characterized by significant differences in the climate between South China and North China. The forest coverage in China is around 16.552 % of the total land, even if the forest sources are quite limited and the forested area are located in diverse geographic location and characterized by different climatic conditions. Therefore also fire occurrence is influenced by the geographic location and climatic conditions. Three different areas and seasonal fire occurrence can be identified: (i) Northeast China and Inner-Mongolia Forest Areas, (ii) South and Southwest forest areas and (iii) Northwest Forest Area, mainly Xinjiang Autonomous Region. In the Northeast China and Inner-Mongolia fires mainly occur during spring (March to mid-June) and fall (mid-September to mid-November) and the highest fire frequency is usually during May and October. Vegetation is less prone to fire during summer and winter due to high precipitation and humid air and due to low temperature and snow accumulation, respectively. In South and Southwest forest areas, there are two seasons in a year the dry and wet season, therefore as expected fires generally occur during dry season Winter-Spring fire prevention time is suggested from Mid-November to the end of May in the following year. The most fire-hit time is February, March and April in a year. In Northwest Forest Area, mainly Xinjiang Autonomous Region, the period with the highest occurrence is summer to fall generally August and September characterized by low annual precipitation, drastic warm and cold changes typically during summer and fall when also strong monsoon strikes and dry weather occur.

2   Study Area and Data Set

2.1   Study Area

This study was carried out in the Daxing'anling region (Fig. 1), which is located in Heilongjiang Province and Inner Mongolia (50.5°–52.25° N, 122°–125.5° E), the northern China. This region is selected because the forest fires happened more and more frequently and also vast areas are seriously affected. The total area of Daxing'anling region is about 250,000 km^2. The forest coverage of Daxing'anling region is about 62 %. There is about 81 300 km^2 of the forest that cover the Inner Mongolia. The dominated land covers of this area are: coniferous forest (Larix sp., Pinus Sylvestris, Picea sp.), broadleaved deciduous forest (Quercus sp., Betula sp.) and mixed forest.

This study area experiences a cool continental monsoon climate, which is very cold in window but warm in summer. The weather is relative dry and the precipitation is low (65 mm and 80 mm) in spring and autumn. Thus, fires happen more and more frequently in these seasons as well as in summer due to the high temperature.

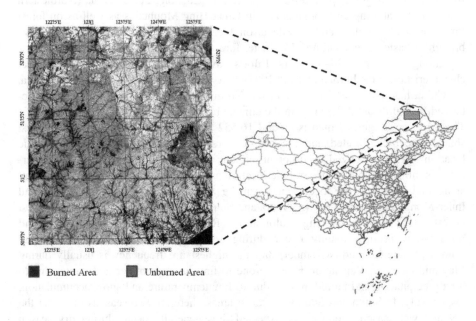

Burned Area   Unburned Area

Fig. 1. The study area

2.2 Dataset

In order to compute the fire potential index, we used land surface temperature (LST) data and relative greenness index (RGI) maps.

2.2.1 Land Surface Temperature

Land surface temperature have been demonstrated that is negatively related to fuel moisture content (FMC) because the cooling effect of evapotranspiration is reduced when plants get dry and introduce mechanisms to reduce water loss (Chuvieco et al. 2004). Thus the LST increase means that the vegetation water stress is becoming worse or the land surface is becoming dry. As a result, the forest that experiences high LST may lead to a high fire risk. Therefore, the LST was used for calculating the fire danger dynamic index (FDDI) in this study. The daily land surface temperature and emissivity product (MYD11A1 1 km spatial resolution) that was used in this study can be downloaded from the Land Processes Distributed Active Archive Center (LP DAAC http://e4ftl01.cr.usgs.gov/MOLA/).

2.2.2 Relative Greenness Index

Relative greenness index can be used as an indicator for monitoring the vegetation coverage. Several previous studies [1–3]; have demonstrated that the relative greenness has the feasibility to estimate fire susceptibility. High relative greenness indicate that the vegetation is healthy whereas low greenness values indicate that the vegetation is under stress, dry (possibly from drought), is behind in annual development, or dead. Thus, RGI was obtained from (1) the 16-days composite NDVI products (MYD13Q1 250 m spatial resolution) that were used for calculating the maximum and minimum NDVI from 2002 to 2014 and (2) MODIS level 1B data (250 m spatial resolution) for generating the FDDI. These data can be downloaded from the LP DAAC and the MODIS website or the Satellite Remote Sensing Facilities for Fire Monitoring in our laboratory (located in Hefei, China; post 2009).

3 Methodology

3.1 The Fire Potential Index

In this study, we adopted the method, which was proposed by [9–11], to assess the fire danger in northern China. This method has been demonstrate that it can be successfully used for fire danger prediction in northern China It is defined by the following equation:

$$FPI = 100 \times (1 - FMC10HR) \times (1 - VC) \tag{1}$$

where FMC10HR is the ratio of the relative humidity of dead fuel and the disappeared amount of water in 10 h. and VC is the coverage of vegetation.

However, we improve the fire potential index to assess the fire danger level in northern China. The modified fire potential index is defined by the following equation:

$$FPI_m = 100 \times (1 - DFMC) \times (1 - RG) \tag{2}$$

where DFMC is the dead fuel moisture content and RG is the relative greenness.

The results of FPIm is divided into four fire danger level (Zhou 2008), which is shown in Table 1.

Table 1. FPIm fire danger levels

FPI_m	Level	Description
<30	Very low (1)	No fire happen
$30 \leq FPI_m < 40$	Low (2)	The condition against fire occurrence
$40 \leq FPI_m < 55$	Moderate (3)	The condition benefits fire occurrence
≥55	High (4)	It is of high possibility for fire occurrence

3.1.1   The Calculation the DFMC

The DFMC is inversely related to the probability of ignition, because of the fact that part of the energy necessary to start a fire is used up in the process of evaporation before the fire starts to burn (8; Dimitrakopoulos and Papaioannou 2001). On the other hand, water content also affects fire propagation since the source of the flames is reduced with humid materials, therefore, flammability is reduced (8; Viegas et al. 1992). The DFMC varies with the change of atmospheric condition (8; Chuvieco et al. 2010; Simard 1968). The index is calculated based on the LST and relative humidity (RH) in this study by using the following equation that was developed by Sharples et al. (2009) (Sharples et al. 2009):

$$DFMC = 10 - 0.25 \times (LST - RH) \tag{3}$$

where RH is the ratio of vapor pressure (e) and saturation vapor pressure (es). Vapor pressure depends on air pressure (Pa) and specific humidity (Q), while saturation vapor pressure (es) depends on air temperature (Ta). Their relationship is shown in the following equations (Murray 1967; Peng et al. 2006):

$$RH = \frac{e}{e_s} \tag{4}$$

$$e_s = 61.1 \times \exp\left(\frac{17.2 \times T_a}{237.3 + T_a}\right) \times RH \tag{5}$$

$$e = \frac{Q \times P_a}{0.622} \tag{6}$$

where 0.622 is the ratio of the mole weight of water vapor and dry air. Also, the air temperature (Ta) is replaced by LST in this study, even though the LST is a little bit higher than Ta. So the results may have some bias. The following equation is used for calculating the specific humidity (Q) (Liu 1984; Peng et al. 2006):

$$Q = 0.001 \times \left(-0.0762 \times PW^2 + 1.753 \times PW + 12.405\right) \tag{7}$$

where PW is the precipitable water vapor, which can be estimated from the MODIS image based on the following equations (Kaufman and Bo-Cai 1992; Peng et al. 2006):

$$\tau_{(19/2)} = \rho_{19}/\rho_2 = \exp\left(\alpha - \beta\sqrt{PW}\right) \tag{8}$$

$$PW = \left(\frac{\alpha - \ln \tau_{(19/2)}}{\beta}\right)^2 \tag{9}$$

where $\tau(19/2)$ is observed transmittance, $\rho2$ and $\rho19$ are the reflectance of band 2 and band19. For mixed terrain, $\alpha = 0.020$, $\beta = 0.651$; for vegetation cover type, $\alpha = 0.012$, $\beta = 0.651$; for bare soil, $\alpha = -0.040$, $\beta = 0.651$.

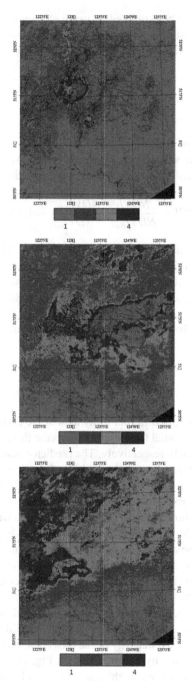

Fig. 2. The predicted results by applying the modified fire potential index. (a) the result of 26th June, 2010; (b) the result of 27th June, 2010; (c) the result of 28th June, 2010 (Color figure online)

The air pressure (Pa) can be estimated by using elevation that is derived from DEM (Peng et al. 2006). Air pressure will decrease with the increment of elevation. It can be calculated from elevation (H) according to the following experiential equation (Peng et al. 2006):

$$P_a = 1013.3 - 0.1038H \tag{10}$$

3.1.2   The Calculation of RG

The RGI indicates how green each pixel is in relation to the historical range of NDVI observations. It is calculated based on the following equations:

$$RGI = \frac{NDVI_i - NDVI_{min}}{NDVI_{max} - NDVI_{min}} \times 100\% \tag{11}$$

where $NDVI_i$ is the observation NDVI of the given pixels, $NDVI_{min}$ and $NDVI_{max}$ are the minimum and maximum NDVI values that calculated from a period of historical 16-days composite NDVI data (MYD13Q1) from 2002 to 2014. High RGI values indicate that vegetation is green and healthy whereas low RGI values demonstrate that the vegetation is under stress, dry, at the beginning/end of the vegetation process or dead. The vegetation may be in the high fire danger level under these circumstances in the study area.

4   Results

There were several fires occurred in this study area at the end of June, 2010. Thus, we applied the modified fire potential fire index to assess the fire danger level in this area on 26th, 27th and 28th June, 2010 respectively. The predicted results were shown in Fig. 2. It can be seen that the fire danger was divided into four levels. The "very low" events were indicated by green color, the "low" events were depicted in blue color, the "moderate" events were marked by yellow color while the "high" events were indicated by red color. Figure 2(a) showed that a majority of this area were under the "very low" fire danger lever on 26th June, 2010, which were depicted in green color. However, a relative small part of this area experienced the "high" fire danger level, which is located in upper-left of the image and were depicted in red color. Figure 2(b) showed that a large part of the area changed to "low" and "moderate" fire danger levels on 27th June, 2010 by comparing with Fig. 2(a), which were indicated by blue and yellow color respectively. And also the central part of the study area experienced the "moderate" and "high" fire danger level, which means that the fire occurrence in this area was of high possibility. Figure 2(c) showed that a majority of the area changed to "moderate" and "high" fire danger level on 28th June by comparing with Fig. 2(b). The area of central and upper-left of the image experienced "high" fire danger level (red color). According to the news report, there were several fires on 28th June, 2010. The locations were under "high" fire danger level in these predicted results. Thus, the modified fire potential index has the feasibility and applicability to assess the fire danger level in this study area.

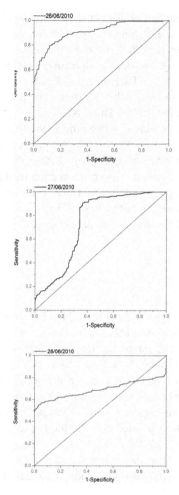

Fig. 3. Estimation accuracy of the modified fire potential index for the study area. (a) ROC curve on 26th June, 2010; (b) ROC curve on 27th June, 2010; (c) ROC curve on 28th June, 2010.

Table 2. The AUC of 26th, 27th and 28th June, 2010 by applying the proposed model

Date	26/06/2010	27/06/2010	28/06/2010
AUC of FPIm	0.9098	0.7379	0.6907

4.1   Model Validation

Model validation is an essential part in danger assessment. It refers to comparing the model predictions with a real world dataset for assessing its accuracy and predictive power (Begueria 2006). In this study, the hotspot data (MYD14A1) acquired from

MODIS is adopted to validate the fire danger results that were obtained by applying the modified fire potential index. The hotspots of 26th June, 27th June and 28th June in 2010 were extracted to compare with their predictive results for analyzing the accuracy of fire danger indices by using the receiver operating characteristics (ROC) technique. The area under the ROC curve (AUC) presents the quality of a prediction model by describing its performance to assess the occurrence or non-occurrence of predefined 'events' (Yesilnacar and Topal 2005). Thus, AUC measures the accuracy of the predictive model. If the AUC is close to one, the predictive results of the model are excellent. Inversely, if the AUC is close to 0.5, the predictive results of the model are fairer. The ROC curves of the fire danger index of these three days which several fires happened are shown in Fig. 3. All the curves were computed at 95 % confidence level. The AUC of these three days are shown in Table 2. It can be seen that the AUC is about 0.9098, 0.7379 and 0.6907 for 26th June, 27th June and 28th June respectively. It means that the accuracy of the forest fire danger is 90.98 %, 73.79 % and 69.07 % respectively.

5 Final Remarks

Fire satellite based technologies offer invaluable data ready to be applied for fire monitoring that is particularly desirable for vast and remote areas as in the case of China. In this paper we focus on in the Daxing'anling region, which is located in Heilongjiang Province and Inner Mongolia (50.5°–52.25° N, 122°–125.5° E), the northern China. Fire danger was assessed for the study area using satellite based method which integrated dead fuel moisture content and relative greenness index. The fire potential index (FPI) was estimated and validated for fire event happened on the late June 2010. The validation pointed out that the fire affected areas were located in high fire danger level on 26th, 27th, 28th June, 2010 respectively. The ROC analyses of the predicted accuracy on these days were 90.98 %, 73.79 % and 69.07 % respectively, this pointed out the reliability of the adopted method.

References

1. Lanorte, A.C., Belviso, R., Lasaponara, F., Cavalcante, F., De Santis: Satellite time series and in situ data analysis for assessing landslide susceptibility after forest fire: preliminary results focusing the case study of Pisticci (Matera, Italy). In: Computational Science and Its Applications–ICCSA 2013, 652anorte A, R Lasaponara 2012 FIRE -SAT un sistema satellitare per il monitoraggio sistematico, dinamico ed integrato degli incendi boschivi: la sperimentazione operativa nella regione Basilicata GEOmedia 16 (2013)
2. Lasapolara, R.: Geospatial analysis from space: Advanced approaches for data processing, information extraction and interpretation. Int. J. Appl. Earth Obs. Geoinf. **20**, 1–3 (2013). Lasaponara, R, Lanorte, A.: Satellite time-series analysis. Int. J. Remote Sens. 33 (15), 4649-4652 (2011)

3. Lanorte, A., Danese, M., Lasaponara, R., Murgante, B.: Multiscale mapping of burn area and severity using multisensor satellite data and spatial autocorrelation analysis. Int. J. Appl. Earth Obs. Geoinf. **20**, 42–51 (2013)

4. Lanorte, A., Lasaponara, R., Lovallo, M., Telesca, L.: Fisher-shannon information plane analysis of SPOT/VEGETATION Normalized Difference Vegetation Index (NDVI) time series to characterize vegetation recovery after fire disturbance. Int. J. Appl. Earth Obs. Geoinf. **26**, 441–446 (2014)

5. Lasaponara, R.: Inter-comparison of AVHRR-based fire susceptibility indicators for the Mediterranean ecosystems of southern Italy. Int. J. Remote Sens. **26**, 853–870 (2005)

6. Telesca, L., Lasaponara, R.: Pre- and post-fire behavioral trends revealed in satellite NDVI time series. Geophys. Res. Lett. **33**(14), 1–4 (2006)

7. Cuomo, V., Lanfredi, M., Lasaponara, R., Macchiato, M.F., Simoniello, T.: Detection of interannual variation of vegetation in middle and southern Italy during 1985–1999 with 1 km NOAA AVHRR NDVI data. J. Geophys. Res.-Atmos. **106**, 17863–17876 (2001)

8. Tian, X., Shu, L., Wang, M., Zhao, F., Chen, L.: The fire danger and fire regime for the Daxing'anling region for 1987–2010. Procedia Engineering **62**, 1023–1031 (2013)

9. Aguado, I., Chuvieco, E., Boren, R., Nieto, H.: Estimation of dead fuel moisture content from meteorological data in Mediterranean areas. applications in fire danger assessment. Int. J. Wildland Fire **16**, 390–397 (2007)

10. Burgan, R., Klaver, R., Klaver, J.: Fuel models and fire potential from satellite and surface observations. Int. J. Wildland Fire **8**, 159–170 (1998)

11. Burgan R.E., Andrews, P.L., Bradshaw, L.S., Chase, C.H., Hartford, R.A., Latham, D.J.: Current status of the wildland fire assessment system (WFAS). Fire Management Notes, vol. 27, pp. 14–17, (1997). Chuvieco E., Cocero, D., Riano, D., Martin, P., Martinez-Vega, J., de la Riva, J., et al.: Combining NDVI and surface temperature for the estimation of live fuel moisture content in forest fire danger rating. Remote Sensing of Environment, vol. 92, pp. 322–331, August 30 2004

Author Index

Printed in the United States
By Bookmasters